SYSTEMS IN MECHANICAL ENGINEERING

[FUNDAMENTALS AND APPLICATIONS]

Anup Goel

B.E. Mechanical
Post Graduation in Tool Design with CAD/CAM
Managing Director of AG Engineering Study Centre,
Akurdi, Pune.
13 Years Teaching Experience

SYSTEMS IN MECHANICAL ENGINEERING

First Edition : January 2021

Published by :

Amit Residency, Office No.1, 412, Shaniwar Peth, Pune - 411030, M.S. INDIA
Ph.: +91-020-24495496/97, Email : sales@technicalpublications.org
Website : www.technicalpublications.org

ISBN 978-93-332-2183-2

PREFACE

The importance of **Systems in Mechanical Engineering** is well known in various engineering fields. Overwhelming response to my books on various subjects inspired me to write this book. The book is structured to cover the key aspects of the subject **Systems in Mechanical Engineering**.

The book uses plain, lucid language to explain fundamentals of this subject. The book provides logical method of explaining various complicated concepts and stepwise methods to explain the important topics. Each chapter is well supported with necessary illustrations, practical examples and solved problems. All chapters in this book are arranged in a proper sequence that permits each topic to build upon earlier studies. All care has been taken to make students comfortable in understanding the basic concepts of this subject.

Representative questions have been added at the end of each chapter to help the students in picking important points from that chapter.

The book not only covers the entire scope of the subject but explains the philosophy of the subject. This makes the understanding of this subject more clear and makes it more interesting. The book will be very useful not only to the students but also to the subject teachers. The students have to omit nothing and possibly have to cover nothing more.

I wish to express my profound thanks to all those who helped in making this book a reality. Much needed moral support and encouragement is provided on numerous occasions by my whole family. I wish to thank the **Publisher** and the entire team of **Technical Publications** who have taken immense pain to get this book in time with quality printing.

Any suggestion for the improvement of the book will be acknowledged and well appreciated.

Author

Anup Goel

Dedicated to family, friends and my dear students

SYLLABUS

Unit I   Introduction of energy sources & its conversion

Energy sources : Thermal energy, Hydropower energy, Nuclear energy, Solar energy, Geothermal energy, Wind energy, Hydrogen energy, Biomass energy and Tidal energy. Grades of energy. (Numerical on efficiency calculation of thermal power plant)

Energy conversion devices : Introduction of pump, compressor, turbines, wind mills etc (Simple numerical on power and efficiency calculations) **(Chapters - 1, 2)**

Unit II   Introduction to Thermal Engineering

Laws of thermodynamics, heat engine, heat pump, refrigerator (simple numerical) Modes of heat transfer : conduction, convection and radiation, Fourier's law, Newton's law of cooling, Stefan Boltzmann's law. (Simple numerical) Two stroke and Four stroke engines (Petrol, Diesel and CNG engines). Steam generators. **(Chapters - 3, 4)**

Unit III   Vehicles and their Specifications

Classification of automobile. Vehicle specifications of two/three wheeler, light motor vehicles, trucks, buses and multi-axle vehicles. Engine components (Introduction). Study of engine specifications, comparison of specifications of vehicles. Introduction of electric and hybrid vehicles. cost analysis of the vehicle. **(Chapter - 5)**

Unit IV   Vehicle Systems

Introduction of chassis layouts, steering system, suspension system, braking system, cooling system and fuel injection system and fuel supply system. Study of electric and hybrid vehicle systems. Study of power transmission system, clutch, gear box (Simple Numerical), propeller shaft, universal joint, differential gearbox and axles. Vehicle active and passive safety arrangements: seat, seat belts, airbags and antilock brake system. **(Chapter - 6)**

Unit V   Introduction to Manufacturing

Conventional Manufacturing Processes : Casting, Forging, Metal forming (Drawing, Extrusion, etc.), Sheet metal working, Metal joining, etc. Metal cutting processes and machining operations- Turning, Milling and Drilling, etc. Micromachining. Additive manufacturing and 3D Printing. Reconfigurable manufacturing system and IOT, Basic CNC programming: Concept of Computer Numerical Controlled machines. **(Chapter - 7)**

Unit VI   Engineering Mechanisms and their Application in Domestic Appliances

Introduction to Basic mechanisms and equipment : Pumps, blowers, compressors, springs, gears, Belt-Pulley, Chain-Sprocket, valves, levers, etc. Introduction to terms : Specifications, Input, output, efficiency, etc.

Applications of : Compressors - Refrigerator, Water cooler, Split AC unit; Pumps - Water pump for overhead tanks, Water filter/Purifier units; Blower - Vacuum cleaner, Kitchen Chimney; Motor - Fans, Exhaust fans, Washing machines; Springs - Door closure, door locks, etc.; Gears - Wall clocks, watches, Printers, etc.; Application of Belt-Pulley/Chain-Sprocket - Photocopier, bicycle, etc.; Valves - Water tap, etc.; Application of levers - Door latch, Brake pedals, etc.; Electric/Solar energy - Geyser, Water heater, Electric iron, etc. (Simple numerical on efficiency calculation) **(Chapter - 8)**

TABLE OF CONTENTS

Unit - II

Chapter - 3   Thermodynamics (3 - 1) to (3 - 32)

Chapter - 4   Heat Transfer   (4 - 1) to (4 - 36)

Unit - III

Chapter - 5   Vehicles and their Specifications
(5 - 1) to (5 - 34)

Unit - IV

Chapter - 6   Vehicle Systems   (6 - 1) to (6 - 52)

Unit - V

Chapter - 7   Conventional Manufacturing Processes   (7 - 1) to (7 - 74)

Unit - VI

**Chapter - 8   Engineering Mechanisms
& their Applications in
Domestic Appliances**

(8 - 1) to (8 - 36)

Unit - I
Introduction of Energy
Sources and Its Conversion

UNIT - I

1   Energy Sources

Syllabus

Thermal energy, Hydropower energy, Nuclear energy, Solar energy, Geothermal energy, Wind energy, Hydrogen energy, Biomass energy and Tidal energy. Grades of Energy. (Numerical on efficiency calculation of thermal power plant)

Contents

Mind Map - Energy Sources

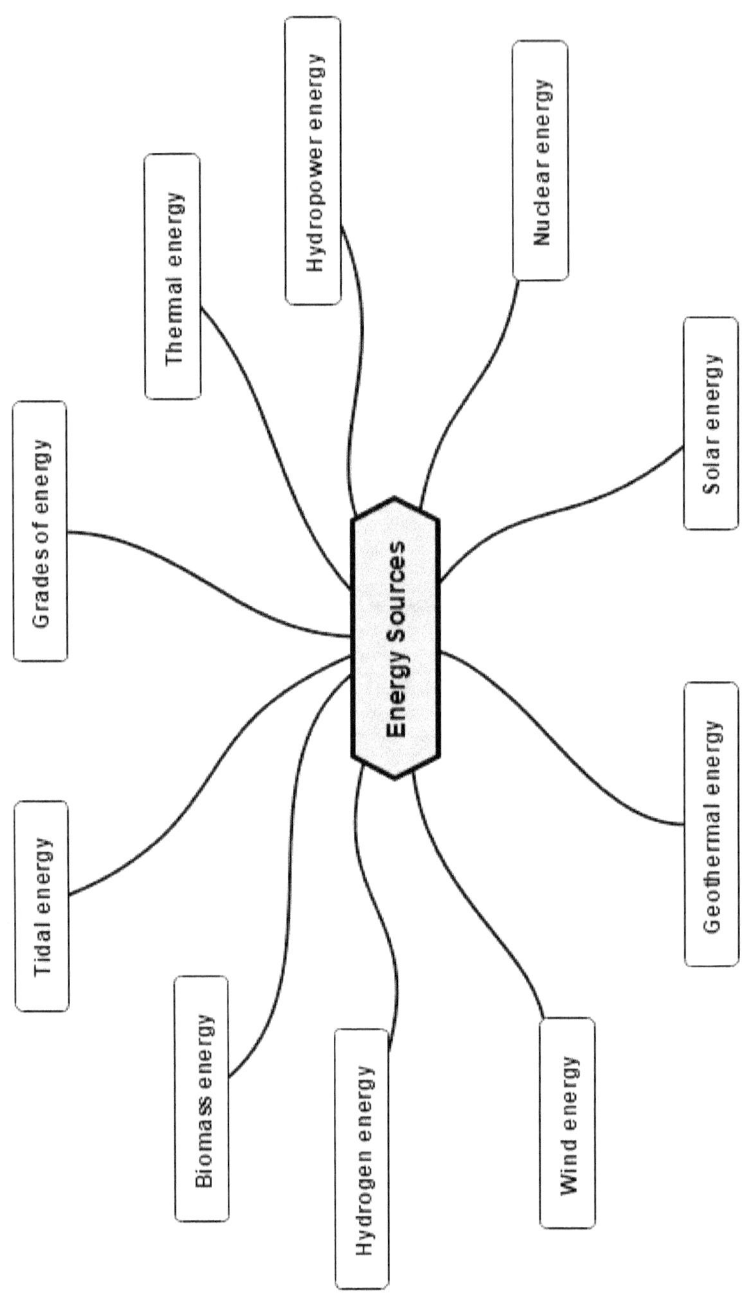

1.1 : Introduction

- Electricity or power is the only form of energy which is easy to produce, easy to use, easy to transport and easy to control.

- It plays an important role in the each and every field.

- Electricity or power consumption per capita indicates the standard of living of people of that area or country.

- Electricity or power in bulk quantities is produced or generated in power plants. The power plants can be thermal, nuclear, hydraulic, etc.

1.1.1 Introduction to Energy Resources

SPPU : Dec.-17, May-18

- Energy is the primary measure that causes changes and movement in all kinds of work by human beings and nature.

- All living things use energy constantly and hence energy is always being changed and flows from one form to another.

- The sources of energy available can be classified as follows :

> i)   Renewable energy resources
> ii) Non-renewable energy resources

i) Renewable Energy Resources

- Renewable energy is the energy which comes from resources that are naturally replenished.

- These energy resources includes solar energy, wind energy, geothermal energy, hydropower, ocean thermal energy and biomass.

- These resources are sustainable resources and never ending type.

- Renewable energy sources occurs over wide geographical areas abundantly.

ii) Non-renewable Energy Resources

- Non-renewable energy is the energy which comes from the resources that are not sustainable resources.

- Non-renewable resources does not renew itself at sufficient rate.

- They includes fossil fuels such as coal, petroleum and natural gas, earth minerals and ores, nuclear fuels.

- These resources of energy required thousands of years to form naturally and can not be replaced as fast as they consumed for use.

- Now-a-days there is increasing levels of investment in the field of sustainable energy.

- As the non-renewable sources are decline, the renewable energy has become mainstream and the future of energy production.

Types of power plants

The different types of power plants are as follows :

> 1. Thermal or steam power plants.
> 2. Hydroelectric power plants.
> 3. Nuclear power plants.
> 4. Solar-wind (Hybrid) power plants.
> 5. Geothermal power plants
> 6. Gas turbine power plants, etc.

1.2 : Thermal or Steam Power Plant

SPPU : May-07, 14, Dec.-07, 10, 11, 13

Thermal energy

- Thermal or heat energy is produced when a rise in temperature cause atoms and molecules to move faster and collide with each other.

- The energy that comes from the temperature of heated substance is called as thermal energy.

- The hotter the substance, the more its particles move and the higher its thermal energy.

- Examples of thermal energy are :

  - The warmth from the sun

  - Baking in an oven

  - Heat from a heater

  - Boiling water on a stove, etc.

- Thermal or steam power plant uses steam to produce electrical power in large amounts.

- Generally this power plant consists of boiler, steam turbine, generator, feed pump and condenser.

- In this type of power plant, chemical energy of fuel is used to convert water into steam in the high pressure boilers.

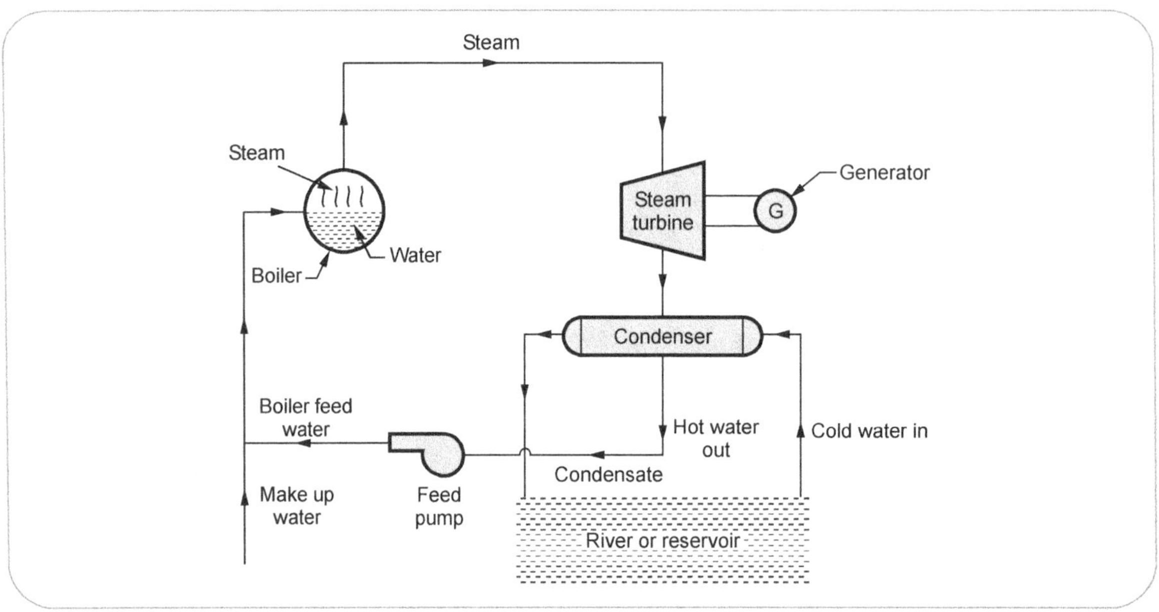

Fig. 1.2.1 Thermal or steam power plant

- This high pressure and high temperature steam is used in a turbine to produce mechanical energy. This energy is finally converted into electrical energy with the help of coupled generator. Refer Fig. 1.2.1.

- The used steam is collected in the condenser and converted into water. This water is again fed to the boiler as an input with the help of feed pump.

- The function of various elements of steam power plant is as follows :

1. **Boiler :** Boiler or **steam generator** is used to generate high pressure and high temperature steam. The input water to the boiler is supplied by feed pump (water pumped from condenser) and if required make up water or extra water is supplied. For better efficiency of power plant, boiler consists of superheater, economiser, preheater, etc.

2. **Steam Turbine :** It converts high pressure and high temperature steam supplied by the boiler into shaft work (rotation of shaft) and low temperature steam is exhausted to a condenser.

3. **Generator :** The shaft of steam turbine is coupled to the shaft of generator and the mechanical energy of steam turbine is converted into electrical energy.

4. **Condenser :** Steam exhausted from steam turbine is collected in the condenser and condensed by using recirculating cooling water. The condensed amount of water is called as **condensate.**

5. **Feed Pump :** The collected amount of condensate is again fed back to the boiler with the help of feed pump. This condensate is already a bit hot, hence there is less amount of heat required in the boiler while converting it into the steam.

Advantages of steam power plant

- Fuels used in the boiler are coal, coke, natural gas, etc. which are cheaper.

- As compared to hydraulic power plant, this power plant requires less amount of space.

- This power plant can be located near the load centre which reduces transmission losses and transmission cost.

- Initial cost of set-up of these type of plants is less as compared to other plants.

Disadvantages of steam power plant

- The time required for the plant set-up is more.

- Large amount of water is required for the operation.

- The handling of coal and ash is a serious problem.
- The maintenance cost of plant is high.

1.3 : Hydroelectric Power Plant

SPPU : May-09, 11, 13, 17, 19, Dec.10, 12

- Hydro-power (water power) is developed by allowing water to fall under the force of gravity. It is commonly used for electric power generation.
- In this case, potential energy of water is converted into mechanical energy by using a prime mover called as hydraulic turbines.
- In hydroelectric power plants the potential energy of water is utilised to drive the turbine which in turn runs the generator to produce electricity.
- The basic requirement of hydroelectric power plant is a reservoir where large amount of water is stored during rainy season and used during other seasons.
- Therefore these types of power plants are located where the water resources are available in large quantity with high head.
- During the operation, the water is supplied to the turbine (prime mover) through the penstock which is located at much lower level than the height of water in the reservoir.
- Then, the water is impinged on the turbine blades which drives the turbine shaft to generate power.

- Fig. 1.3.1 shows the general layout of hydroelectric power plant.
- The function of various elements of hydroelectric power plant is as follows :

1. **Reservoir :** Reservoir is used to store large amount of water during rainy season and supply water during other seasons. Water stored is not only used for power generation, but also for fload control, irrigation, water supply, navigation, etc. A reservoir may be natural or artificial.

2. **Dam :** Dam is a structure of considerable height built across the reservoir. It develops a reservoir to store water and it builds up a head for power generation.

3. **Trash rack :** It is provided for preventing the entry of debris (dust, dirt, etc.) from the dam, because any debris into the intake water pipe may damage the turbine blades or choke up the flow. It is generally made of steel.

4. **Gate :** It is provided to control the flow of water from reservoir to the turbine. According to the requirement, gate opening can be changed.

5. **Penstock :** It is a pipeline which carries water from reservoir to the turbine. Generally, to the

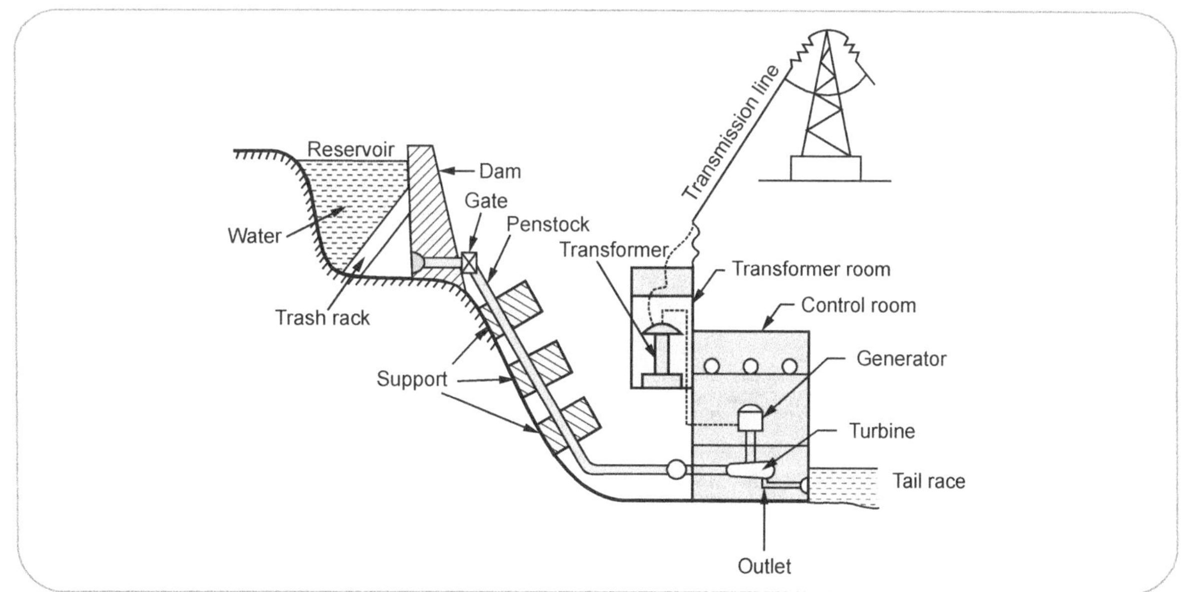

Fig. 1.3.1 Hydroelectric power plant

withstand high pressures, it is made of concrete and of 1 to 2 m in diameter.

6. **Control room :** Control room or power house consists of turbine, generator and other electric equipments to convert mechanical energy of turbine into electrical energy.

7. **Turbine :** Turbine is used to convert kinetic energy of water into mechanical energy. Generally used hydraulic turbine are Pelton turbine, Kaplan turbine, Francis turbine, etc.

8. **Tail race :** It is a path to lead the water discharged from the turbine to the river.

Advantages of hydroelectric power plant

- Water source is easily and readily available.

- No fuel is to be burnt to generate the power.

- The running cost of these power plants is very low as compared to thermal power plants.

- These power plants are more reliable.

- There is no problem of disposal of ash, emission of polluting gases and particulates.

- Life expectancy of these power plants is higher.

- Due to simple design and operation, these power plants do not require highly skilled operators. Manpower requirement is also low.

Disadvantages of hydroelectric power plant

- Power generation is dependent on the quantity of water available, which may vary from season to season.

- These power plants are located far away from the load centre which requires long transmission lines. The cost of these transmission lines and losses are more.

- The time required for the set-up of plant is more.

- Initial cost of set-up is very high.

1.4 : Nuclear Power Plant

SPPU : Dec.-04,09,10,11,12,13,18, May-05,06,07,08,09,12

- Nuclear power plant is similar to thermal or steam power plant; the only difference is that, the boiler of steam power plant is replaced by nuclear reactor.

- Nuclear power plant utilises nuclear energy produced by nuclear reactions for generation of steam.

- The heat is produced by nuclear fission of the atoms of fissionable material and it is utilised in special heat exchangers for the production of steam.

- The commonly used nuclear fuels are unstable atoms like Uranium (U^{235}), Thorium (Th^{232}) and artificial element plutonium (Pu^{239}).

- These unstable atoms liberate large amount of heat energy. The energy released by the complete fission of 1 kg of U^{235} is equal to the heat energy obtained by burning 4500 tonnes of coal or 2200 tonnes of oil.

- But there are some limitations in the use of nuclear energy like high cost of plant set-up, limited availability of raw materials, problem with disposal of radioactive waste and highly skilled operators.

- Due to these reasons, nuclear power plants are still developing in India. Only 1% to 3% energy requirement is fulfilled by nuclear power plant in countries like India, whereas in France about 30% of its energy requirement is generated by nuclear method.

- The general layout of nuclear power plant is shown in Fig. 1.4.1.

- The function of various elements of nuclear power plant is as follows :

1. **Nuclear reactor :** In the nuclear reactor, nuclear energy is produced by the nuclear fission of unstable atoms like uranium. Uranium is in the form of thin rods or plates. This energy is transferred to the circulating coolant.

2. **Heat exchanger :** The heat exchanger will act as a boiler for the power plant. The heat absorbed by the coolant in the reactor is transferred to the water and steam is generated. Generally, U-tube type heat exchanger is used. The commonly used coolants are carbon dioxide, helium, liquid metals like sodium, potessium, etc.

3. **Steam turbine :** It converts high pressure and high temperature steam supplied by the heat exchanger into shaft work and low temperature steam is exhausted to a condenser.

4. **Generator :** The mechanical energy of steam turbine is converted into electrical energy with the help of generator.

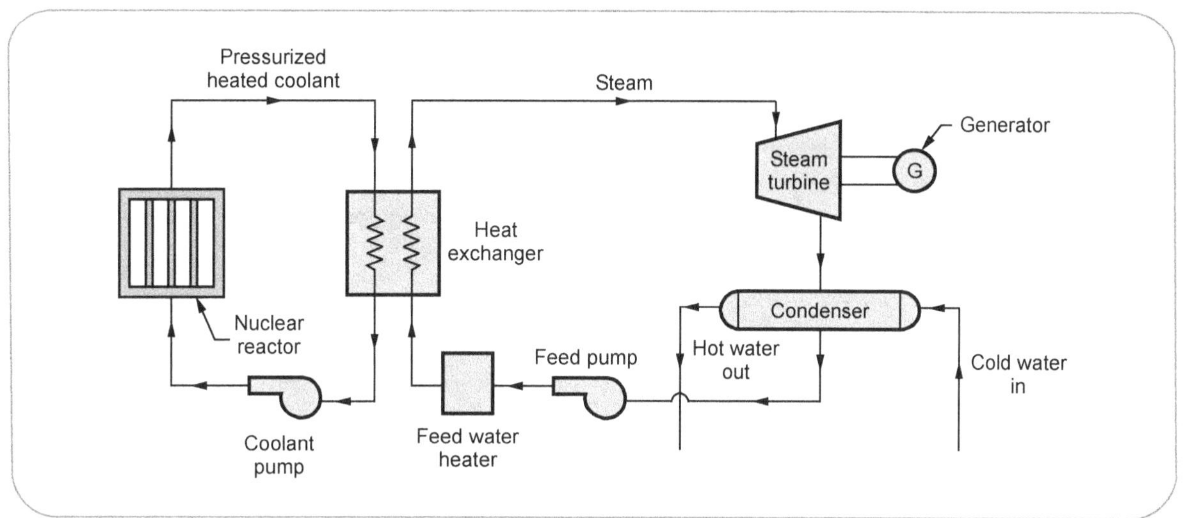

Fig. 1.4.1 Nuclear power plant

5. **Condenser** : Steam exhausted from steam turbine is collected in the condenser and condensed by using recirculated cooling water.

6. **Feed pump** : The collected amount of condensate is again fed back to the heat exchanger with the help of feed pump.

7. **Feed water heater** : The water pumped from feed pump is initially heated in the feed water heater and then supplied to the heat exchanger.

8. **Coolant pump** : The coolant after transferring the heat in the heat exchanger is again fed back to the nuclear reactor with the help of coolant pump.

Advantages of Nuclear Power Plant

• The cost of generated power is less than the thermal power plant.

• There is no problem of environmental pollution, fuel transportation, mine safety, etc.

• Nuclear power plant does not require fossil fuels like coal, coke, natural gas which can be used in other areas.

• Less space is required as compared to steam power plant.

• Large amount of energy is generated by burning small amount of fuel.

Disadvantages of Nuclear Power Plant

• Initial cost of plant set-up is high.

• Nuclear reactor fuels are not easily available.

• There is problem of disposal of radioactive waste.

• For the operation of power plant highly skilled operator is required.

• Cost of nuclear reactor fuels is also high.

• During the operation, nuclear radiations are produced hence high degree of safety is required for the operators.

1.4.1 Comparison of Thermal and Nuclear Power Plants   `SPPU : Dec.-09`

Parameter	Thermal power plants	Nuclear power plants
Initial Cost of Plant	Initial cost of plant set-up is low.	Initial cost of plant set-up is high.
Fuel	Coal is used.	Nuclear reactor fuels are used.
Cost of Generated Power	The cost of generated power is comparatively more.	The cost of generated power is less than the thermal power plant.
Fuel Cost	Cost of coal is quite low.	Cost of nuclear reactor fuels is high.

1.5 : Solar and Wind Energy Power System

1.5.1 Solar Energy Power System

SPPU : May-10, Dec.-10,16,17

- The energy produced and radiated by the sun or sun's energy reaches to the earth.

- This energy is received in the form of radiations and it can be converted directly or indirectly into other forms of energy like heat, electricity, etc.

- As the sun is expected to radiate at constant rate for a billion of years, it may be considered as an inexhaustible source of energy.

- Solar energy which falls on the earth is at considerably low density and hence it cannot be used directly for power generation.

- The energy has to be collected and converted so that it can be used to run prime mover (turbine). The equipment used to collect and concentrate the solar energy is called as **concentrating collector**.

- Concentrating collector may be flat plate type or parabolic type.

Power generation from solar system

- The basic elements of solar power plant are exactly similar to thermal power plant except the boiler is replaced by a flat plate solar collector and heat exchanger.

- Fig. 1.5.1 shows the general layout of solar power plant.

Advantages of solar power plant

- Fuel is not required for the operation.

- There are no problems for ash handling, coal transportation, environment pollution, etc.

- Solar energy is available freely and easily.

- For the operation of plant skilled operator is not required.

Disadvantages of solar power plant

- Solar energy is dilute and spread out.

- For collection of solar energy, large collection area is required.

- Intensity of solar energy is weather dependent.

- Solar energy is intermittent source because it is not available at night.

1.5.1.1 Solar Thermal Energy Harvesting

- The energy conversion in all the processes is inefficient due to the loss of energy in the form of heat.

- For examples, in case of motors, automobile engines, electric bulbs the energy is lost in the form of heat.

- There are some devices known as **harvesting devices** which recover some energy from the wasted energy and convert it into electricity.

- The large solar panels is one of the harvesting collectors which provides major alternate energy source to power grid.

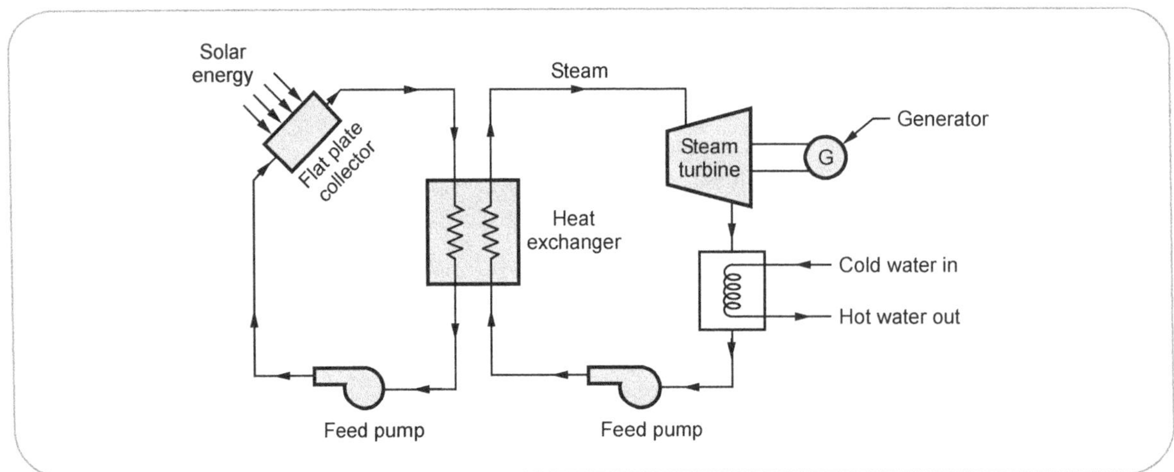

Fig. 1.5.1 Solar power plant

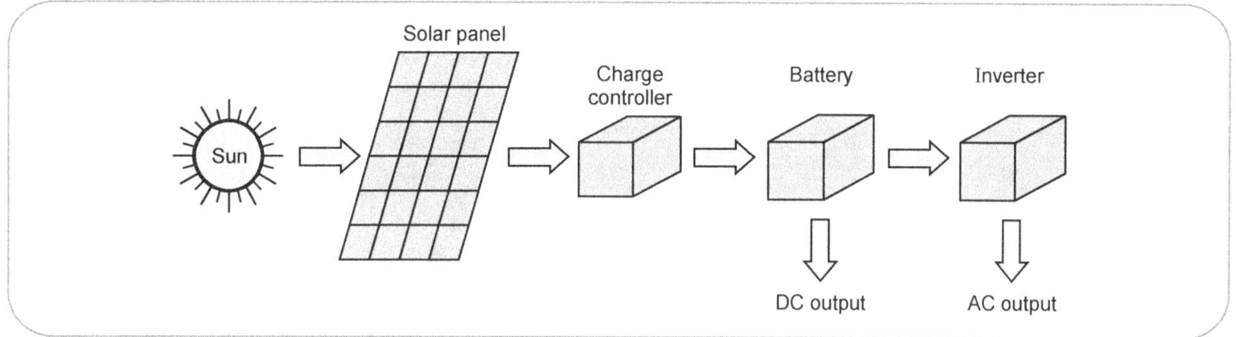

Fig. 1.5.2 : Solar thermal energy harvesting

- Fig. 1.5.2 shows the schematic diagram of solar thermal energy harvesting.
- Following are the some types of solar thermal energy harvesting devices.
  1. Liquid flat plate collector
  2. Solar air heater
  3. Cylindrical solar collector
  4. Solar pond, etc.

1.5.1.2 Liquid Flat Plate Collector

- It is a most important type of solar energy harvesting devices.
- Flat plate collector is based on the principle that, the dark surface which is exposed to solar radiations absorb the solar radiations.
- The part of this absorbed radiation is transferred by means of a media i.e. air, water, etc.

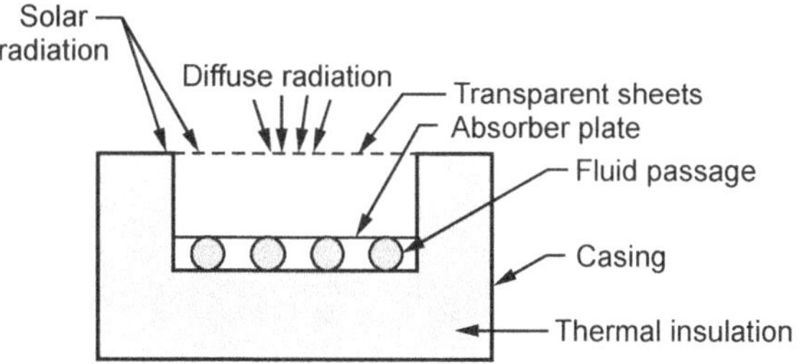

Fig. 1.5.3 : Liquid flat plate collector

The function of various elements of liquid flat plate collector is as follows :

1. **Casing :** It provides housing for the liquid passage and other parts of the collector. Thermal insulation is provided to the casing in order to avoid the loss of solar radiation.

2. **Absorber plate :** It is the most important part of the collector. The main function of absorber plate is to absorb maximum possible amount of solar radiations and allow very less amount of heat loss to the atmosphere.

3. **Transparent sheets :** These sheets are made up of glass. The transparent sheets covers the absorber plate in order to reduce the convective heat loss. The closed air passage is formed between the transparent sheets and absorber plate which helps to reduce the convective heat loss. There can be one or more transparent sheets on a liquid flat plate collector.

4. **Fluid passage :** The heat absorbed by absorber plate is transferred to the liquid present in tubes in fluid passage. In case of liquid flat plate collector, the liquid used is mostly water.

- A liquid flat plate collector is usually held tilted in a fixed position.

Advantages

- Simple in design.
- It has no moving parts such as motors, etc.
- It requires less maintenance.
- It makes use of diffuse solar radiation.
- Less expensive than concentrating collectors.

Disadvantages

- Loss of heat to the environment is more than that of other type of collectors.
- Inappropriate for high temperature applications such as steam production.

Applications

- Heating of water for domestic purpose and space heating.
- Process water heating.
- Sanitation, etc.

1.5.1.3 Solar Ponds

- Solar ponds are one of the important solar energy harvesting technique.
- A solar pond is a pool of salt water which collects and stores the solar thermal energy.
- A solar pond is based on the principles of energy transfer by convection to heat the water.
- Fig. 1.5.4 shows the schematic diagram of a solar pond.

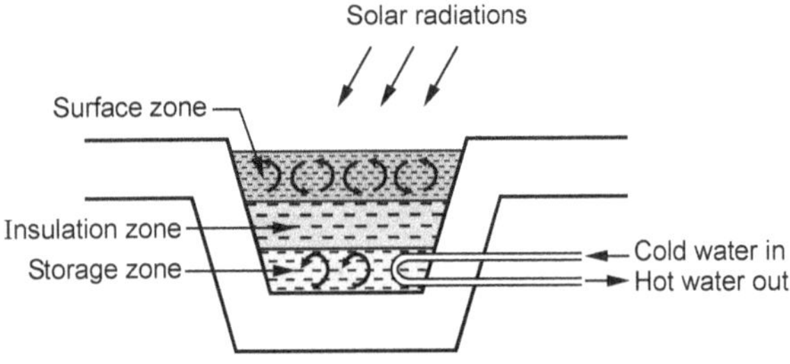

Fig. 1.5.4 : Solar pond

- The salt water in the pool naturally forms a gradient because of difference in density. The different levels in the solar ponds are as follows :

> i) Surface zone
> ii) Insulation zone
> iii) Storage zone

i) Surface zone

- The surface zone lies at the top of a the solar pond due to low density.
- It is convective due to daily heating and cooling and mixing due to wind movement.

ii) Insulated zone

- Below the surface zone, there is a non convective zone known as insulation zone.
- The density gradient also facilitates the temperature gradient and forms as an insulation.

- There is no convection in this layer because the high salt concentration at the lower layer does not allow the water level to rise.

iii) Storage zone

- It is the bottom most and dense layer of the solar pond.
- It is convective and can store a working temperature upto 80° to 90° C.
- This zone of the pond is dark in colour to absorb sun radiations.
- The various salts used to make saline water are NaCl, $MgCl_2$, sodium carbonate, sodium sulphate etc.
- The hot salt water at the bottom can be extracted and used for direct heating purpose and low temperature applications.

Advantages of solar ponds

- Simple in construction and can be located anywhere.
- Very large area collectors can be set up just on the expense of clay or plastic liner.
- There is no need of a separate collector for thermal energy storage.
- Eco-friendly system.
- Cost free operation.

Disadvantages of solar ponds

- Energy efficiency is low.
- There is need of adding new salt water continusouly as the bottom layer is continuously pumped for use.
- The accumulation of salt crystals must be avoided to prevent built up.
- The surface area required is very large.
- Location of solar pond requires solar energy access to operate.

Applications of solar ponds

- Process heating for textile and dairy industries.
- Refrigeration.
- Solar power generation.
- Desalination of drinking water.
- Heating and cooling of buildings.

- Heating animal housing and drying crops on farms.
- Heat required for biomass conversion into alchohol or methane.

1.5.1.4 Solar Photo-Voltaics

- Solar photo-voltaics refers to the conversion of solar energy into electrical energy by means of photovoltaic effect.
- The main principle of solar photo-voltaic is to generate an electromotive force by absorption of ionizing radiation.
- A photo-voltaic (PV) system consist of the following elements :

     i) Solar cell array   ii) Load leveller

     iii) Storage system   iv) Tracking system

- A photo voltaic cells or solar cells are made from semiconductors that generate electricity when they absorb light.
- When the photons from the sun are absorbed in the semiconductor, they creates free electrons.
- These free electrons have higher energy than the bonded electrons.
- There must be electric field to induce the higher energy free electrons to flow out of the semiconductor.
- Hence the solar energy is converted to the electrical energy.

Advantages of solar photovoltaic cell

- No need to work at higher temperature.
- No moving parts.
- Solar energy conversion is without pollution.
- Higher power handling capacity.
- Rapid response to changes in input radiation.
- No requirement of fuel as solar energy is free.
- They have high power to weight ratio.
- They have highly effective long life.

Disadvantages of solar photovoltaic cell

- Higher cost.
- Energy storage is required as the solar energy is unavailable at night.

Applications of solar photovoltaic cell

- Street lighting and railway signalling.
- Electrical power generation in space craft.
- In small or large power plants.
- Automatic weather monitoring.
- It is used to operate irrigation pump and drinking water supply, etc.

1.5.2 Wind Energy Power System

- Wind energy is an indirect form of solar energy, because wind is mainly induced by the uneven heating of the earth's surface by the sun.
- Winds can be classified as **planetary wind** and **local wind.**
- Planetary winds are caused by greater solar heating of the earth's surface near the equator; whereas local winds are caused by differential heating of land and water, and also by hills and mountain sides.
- During the operation, the wind approaching the blades moves the wind mill shaft thereby rotating the rotor of generator.
- In this way, kinetic energy of the wind is converted into mechanical energy by the wind mill and then converted into electrical energy by using a generator.
- Fig. 1.5.5 shows the general layout of wind mill.

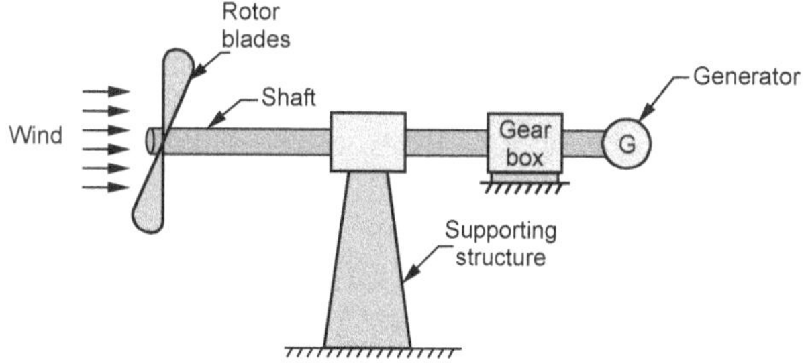

Fig. 1.5.5 Wind mill

Advantages of wind energy system

- Wind is available at no cost.
- It does not cause any kind of pollution while generating power.
- It is available in many off-shore, on-shore and remote areas which is helpful in supplying the electric power in those areas.

- It has low maintenance cost and low power generation cost.

Disadvantages of wind energy system

- It has low energy density.
- It is favourable in locations which are away from cities.
- The wind supply is variable, unsteady, intermittent and sometimes dangerous also.
- The initial cost of plant set-up is high.

1.5.3 Solar-Wind Hybrid Power System

SPPU : Dec.-09, May-10, 11, 12

- We have already discussed that, solar energy is not available at night whereas wind energy is not so strong when sun is shining.
- To overcome the drawbacks of both these plants, hybrid power system came into picture.
- Hybrid power system is a combination of wind power and solar power system, hence called as **solar-wind hybrid system**.
- These systems are still under development.
- This system can be located in flat open areas (away from forests and tall buildings) where both solar radiation and wind speeds are easily available in large quantity.
- Hence, these type of power plants ensure continuous supply of power at remote locations where grid is not available.
- Fig. 1.5.6 shows the general layout of solar-wind hybrid power system.

- During the period of high wind speed (evening time), the wind speed generator generates A.C. power which may be used directly by connecting it to A.C. applications (loads).
- The extra amount of A.C. power is converted into D.C. with the help of rectifier and stored in the battery.
- During the period of high solar energy (day time), the solar photovoltaic (PV) system converts the solar radiations directly into D.C. power which is stored in the battery.
- With the help of inverter D.C. power is converted into A.C. and supplied to various applications like street lighting, water pumping, etc.
- Battery acts as a storage device which stores the excess amount of power when energy demand by consumer is less than the generated power or supplies excess power when energy demand by the consumer is more than the generated power.

Advantages of solar-wind hybrid power system

- This type of power plant ensures continuous power supply.
- It can be used in the remote areas where transmission lines are not readily available.
- Both the sources i.e. wind and solar are available at no cost.
- This power system does not create any kind of pollution while power generation.
- Cost of maintenance and power generation is low.

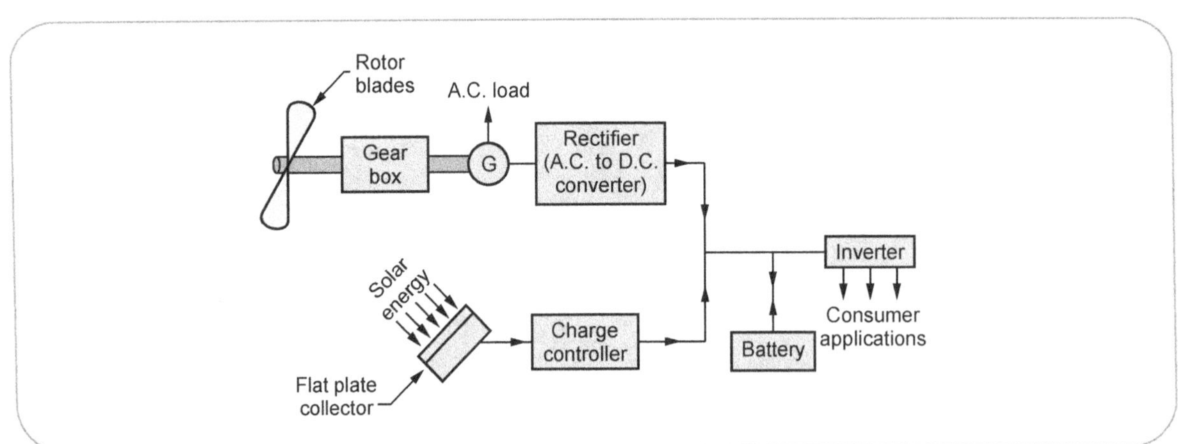

Fig. 1.5.6 Solar-wind hybrid system

Disadvantages of solar-wind hybrid power system

- This type of power plant can be located away from load centre which increases transmission loss and cost.
- Both the sources are nature dependent.
- Initial cost and time of plant set-up is very high.

1.6 : Geothermal Energy

- It is the energy which lies embedded within the earth.
- As per different theories, the earth has a molten core.
- The fact that volcanic action takes place in many places on the surface of earth, this support the different theories.
- The steam and hot water comes naturally to the surface of the earth in some locations of the earth.
- For large scale use bore holes are normally sunk with depth upto 1000 m, releasing steam and water at temperatures upto 200-300 °C and pressure upto 3000 kPa.
- There are two methods of electric power generation from geothermal energy.
- In first method, heat energy is transferred to a working fluid which operates the power cycle. This is mainly useful at places of fresh volcanic activity, where the molten interior mass of earth vents to the surface through fissures and high temperatures, (450 to 550 °C) can be found. By using coil of pipes and sending water through them can be raised.
- In second method, hot geothermal water or steam is directly used to operate the turbines.
- From the well-head the steam is transmitted by pipe line upto 3 km to the power station.
- Now-a-days only steam coming out of the ground is used to generate electriciy and hot water is discarded because it contains 30 % dissolved salts and minerals which may cause corrosion to the turbine blades.
- Research is going on to build turbines which can withstand the corrosive effects of hot water coming out of the wells.
- The sources of this form of energy are exist in USA, Japan, New Zealand, Italy and Maxico. In India, Himachal Pradesh is reported to have geothermal energy in exploitable amount.
- Fig. 1.6.1 shows a basic geothermal power plant.

Advantages:

- Geothermal is reliable source of renewable energy.
- Geothermal energy sourcing is good for the environment (No emissions).
- High efficiency of geothermal systems
- A geothermal system have low maintenance.

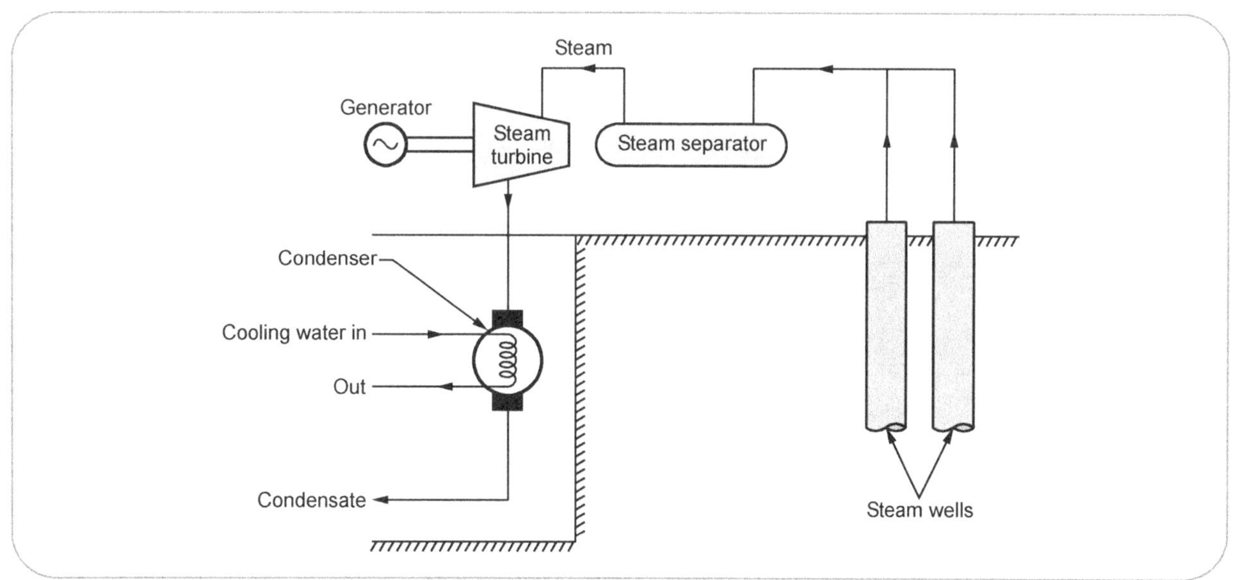

Fig. 1.6.1 Basic setup of geothermal power plant

- No outside source is required to keep the plant running.

- Power stations are smaller then the other plants.

Disadvantages :

- There is a possibility of depletion of geothermal sources.

- High investment cost of the system.

- Source is close to volcanic activity.

- This type of energy is difficult to transport.

- It can suddenly stop working (depends on source).

1.7 : Tidal Energy

- The tides in the sea are the result of the gravitational effect of heavenly bodies like sun and moon on the earth.

- Due to fluidity of water mass, the effect of this force becomes apparent in the motion of water which shows a periodic rise or fall in levels with the daily cycle of rising and setting of sun and moon.

- This periodic rise and fall of the water level of sea is called as **tide**.

- These tides are used in producing electrical power which is called as **tidal power**.

- If the water is above sea level, it is known as **flood tide** and when it is below sea level, it is known as **ebb tide**.

- The use of tides for electrical power generation is situated where the geography of an inlet or bay favors construction of a large scale hydroelectric plant.

- To harness the tides, a dam would be built across the mouth of the bay.

- It will have large gates in it and also low head hydraulic turbines are installed. A tidal basin is formed which gets separated from the sea, by dam.

- The difference in water level is obtained between the basin and sea. The basin is filled with high tide and emptied with low tide. Refer Fig. 1.7.1

- The potential energy of water stored in the basin as well as energy during high tide is used to drive the turbine which is connected to the generator.

- The arrangement shown in Fig. 1.7.1 is known as single basin plant. The plant continuous to generate power till the tide reaches its lowest level.

- Single basin plant cannot generate power continuously, though it might do so by using a pumped storage plant.

- To overcome this drawback, two basin plants are used to generate power continuously without interruption.

- In India the possible sites for tidal power plants are suggested at Gulf of Kutchh in Gujarat and Sunderban area in West Bengal.

Advantages :

- It is pollution free.

- Renewable source of energy and fuel is not required.

- It is more efficient than the wind or solar energy.

- Cost of construction is high but maintenance cost is low.

- The life of tidal energy plant is long.

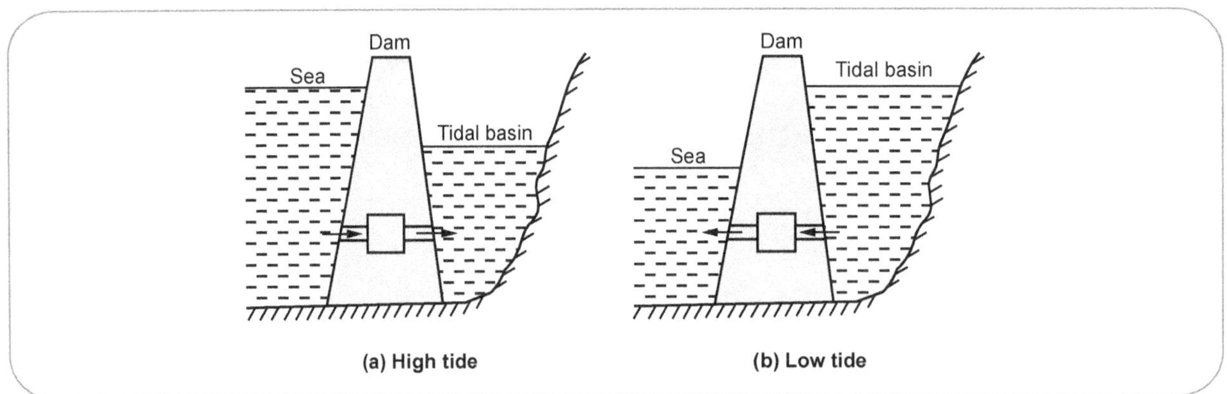

Fig. 1.7.1 Principal of tidal power plant

Disadvantages :

- It is expensive.
- Sea water is corrosive which can damage turbine.
- Power developed is limited to small scale only.
- It has uneven operation.
- Power will be developed only when high tide.
- The transmission of power is expensive and difficult.

1.8 : Hydrogen Energy

- Hydrogen as an energy carrier can play important role as an alternative to conventional fuels.
- One of the most attractive feature of hydrogen as an energy carrier is that it can be produced from water which is easily available in nature.
- Hydrogen has the highest energy content per unit mass of any chemical fuel and it can be substituted for hydrocarbons in a broad range of applications.
- Its burning process is non-polluting and it can be used in the fuel cells to produce both electricity and useful heat.
- But, the use of hydrogen as an energy source involves the following issues :

  - Production
  - Transportation and storage
  - Utilization
  - Economy
  - Safety and management

- The hydrogen can be used as fuel directly or it might be used as a raw material to produce methanol, ammonia or hydrocarbons by using either carbon dioxide or nitrogen from the atmosphere.
- The combination of hydrogen with oxygen results in the liberation of energy with water as the sole material product. Hence,

$$H_2 + \frac{1}{2}O_2 \rightarrow H_2O + Energy$$

- This reaction can be carried out and the energy made available in different ways, so that hydrogen is a versatile fuel material.
- Basically, hydrogen is chemically very reactive and hence it is not found in its free state on the earth.

- However, combined chemically with other elements it is present in water, hydrocarbons, biological materials (cellulose, starch, etc.), minerals, etc.
- Energy is supplied to those compounds to break the chemical bonds to release hydrogen.
- Actually, hydrogen is a secondary fuel that can be produced by utilizing energy from a primary source.
- The current and projected needs of hydrogen are growing rapidly. The applications like nitrogenous fertilizers, coal liquefaction, etc.
- Long term applications are for electric power producing fuel cells. It can be used as a fuel for a gas turbine and for a petrol engine.
- As the heating value of hydrogen is 28000 kcal/kg which three times that of hydrocarbon fuels, it is used in air-crafts so that reduces the take-off load.
- Fig. 1.8.1 shows the basic set-up for hydrogen fuel cell.

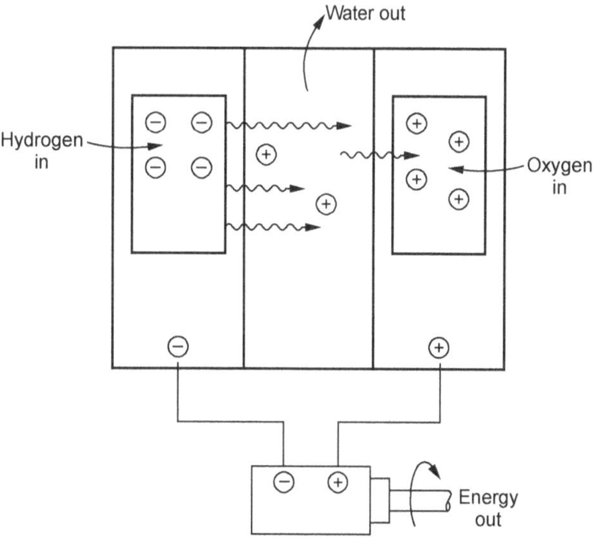

Fig. 1.8.1 Hydrogen fuel cell

Advantages

- It is a renewable energy source and bountiful (large in quantity) in supply.
- It is a clean energy source.
- It is non-toxic.
- It is more efficient than the other sources of energy.
- It is used for powering space-ships.

Disadvantages

- It is expensive.

- Due to low density of hydrogen, more storage complications.

- It's not the safest source of energy (highly flammable)

- It's transportation is difficult task.

- To separate hydrogen, from oxygen, other forms of non-renewable energy sources (natural gas, coal, oil, etc.) are required.

1.9 : Biomass Energy

- **Biomass** is a plant or animal material used for production of energy or heat.

- It is used in various industrial processes as raw material for a range of products.

- It is produced by the growth of micro organisms, plants or animals.

- This biomass can be converted into modern energy forms such as liquid and gaseous fuels, electricity and process heat to provide energy services in rural and urban areas as well as industry.

- Biomass energy includes energy from all plant matter (tree, shrub and crop) and animal dung.

1.9.1 Biomass Resources

Biomass comes from a variety of sources which include :

- Wood from natural forests and woodlands

- Forestry plantations

- Forestry residues

- Agricultural residues such as straw, stover, cane trash and green agricultural wastes

- Agro-industrial wastes, such as sugarcane bagasse and rice husk

- Animal wastes (cow manure, poultry litter etc.)

- Industrial wastes, such as black liquor from paper manufacturing.

- Sewage

- Municipal Solid Wastes (MSW)

- Food processing wastes

1.9.1.1 Agricultural Residues

- These are the non-edible stalk type materials that remain after the harvest of the edible portions of the crops, such as corn, wheat, grain and sugar cane.

- Agricultural residues also includes plant leaves, husks, some roots and stems.

- The advantage of agricultural residues is that they do not require the use of additional land space because they are grown together with the food crops.

- Agricultural residues are characterized by seasonal availability and have characteristics that differ from other solid fuels such as wood, charcoal.

1.9.1.2 Animal Waste

- There are a wide range of animal wastes that can be uses as sources of biomass energy.

- The most common sources are animal and poultry manure.

- The treatment of animal waste produces combustible methane and biogas which can then be used for heating and transportation

- The most attractive method of converting these organic waste materials to useful form is anaerobic digestion which gives biogas.

1.9.1.3 Forestry Residues

- Forestry residues are generated by operations such as thinning of plantations, clearing for logging roads, extracting stem-wood for pulp and timber and natural attrition.

- Forest residues normally have low density and fuel values that keep transport costs high, and so it is economical to reduce the biomass density in the forest itself.

1.9.1.4 Wood Wastes

- Wood wastes generally are concentrated at the processing factories, e.g. plywood mills and sawmills

- For example, the processing of wood for products or pulp produces unused sawdust, bark, branches and leaves/needles.

- These residues can then be converted into biofuels or bioproducts.

- Because these residues are already collected at the point of processing, they can be convenient and relatively inexpensive sources of biomass for energy.

1.9.1.5 Industrial Wastes

- The food industry produces a large number of residues and by-products that can be used as biomass energy sources.

- Liquid wastes are generated by washing meat, fruit and vegetables, blanching fruit and vegetables, pre-cooking meats, poultry and fish, cleaning and processing operations as well as wine making.

- Pulp and paper industry is considered to be one of the highly polluting industries and consumes large amount of energy and water in various unit operations.

1.9.2 Biomass Conversion

- Each biomass resource - wood, dung, vegetable waste, etc. can be treated in many different ways to provide a wide spectrum of useful products.

- There are various conversion technologies that can convert biomass resources into power, heat and fuels for potential use.

- Biomass conversion or bio conversion can take many forms such as

1) Direct combustion, such as wood waste and bagasse (sugarcane refuge)

2) Thermochemical conversion and

3) Biochemical conversion

- **Thermochemical conversion** takes two forms gasification and liquefaction.

   - Gasification takes place by heating the biomass with limited oxygen to produce low heating value gas.

   - It can also take place by reacting biomass with steam and oxygen at high pressure and temperature to produce medium heating value gas.

   - The fuel gas so produced can be used in **liquefaction** by converting it to methanol or ethanol or may be converted to high heating value gas.

- **Biochemical conversion** takes two forms.

   - **Anaerobic digestion :** It involves the microbial digestion of biomass under anaerobic conditions (without air or oxygen). This process takes place at low temperature (65 °C) and requires a moisture content of 80 %. This generates a gas consisting of CO_2 and methane (CH_4) which can be used as synthetic natural gas.

   - **Fermentation** is the breakdown of complex molecules in organic compound with the help of a ferment such as yeast, bacteria, enzymes, etc. suitable feedstocks include sugarcane bagasse, beet, fruit and fermentable sugars obtained from maize, wheat grain, potatoes, etc.

- After anaerobic digestion and fermentation the residue and other feed stuffs contains high protein content which can be a useful cattle feed supplement.

- A wide range of energy-rich fuels can be produced by roasting dry woody matter like straw and wood-chips.

- The material is heated in the absence of air to produce a charred residue called as charcoal. This process performed in absence of air is called as **pyrolysis**. (Refer Fig. 1.10.1 on next page)

- The end products produced are organic liquid at lower temperatures and a combustible mixture of gases at higher temperatures.

1.10 : Biomass Conversion Technologies

- There are various conversion technologies that can convert biomass resources into power, heat and fuels for potential use.

1.10.1 Biomass for Power and Heat

- Biomass power technologies convert renewable biomass fuels to heat and electricity using processes similar to those employed with fossil fuels.

- The biomass fuel is burned in a boiler to produce high-pressure steam

- This steam is introduced into a steam turbine, where it flows over a series of turbine blades, causing the turbine to rotate.

- The turbine is connected to an electric generator

- The steam flows over and turns the turbine

- The electric generator rotates and produces electricity.

1.10.2 Combined Heat and Power

- Most biomass-fired steam turbine plants are located at industrial sites that have a steady supply of biomass available.

- These include factories that make sugar and/or ethanol from sugarcane at pulp and paper mills.

- At these sites, waste heat from the steam turbine can be recovered and used for meeting industrial heat needs further enhancing the economic attractiveness of such plants.

- Conventional thermoelectric stations convert only about one-third of the fuel energy into electricity.

- The rest is lost as heat.

- CHP provides more efficient production of electricity, where more than four-fifths of the fuel's energy is converted into usable energy, resulting in both economics and environmental benefits.

- Cogeneration is the consecutive (simultaneous) production and exploitation of two energy sources, electrical (or mechanical) and thermal, from a system utilizing the same fuel.

- CHP could be applied to industry in West Africa where there is simultaneous demand for electricity and heat.

1.11 Biomass Gasification

- **Biomass gasification**, or production gas from biomass under restricted air supply for the generation of producer gas.

- Producer gas is a mixture of gases : 18 % - 22 % Carbon Monoxide (CO), 8 % - 12 % hydrogen (H_2) 8 % - 12 % carbon dioxide (CO_2), 2% - 4% methane

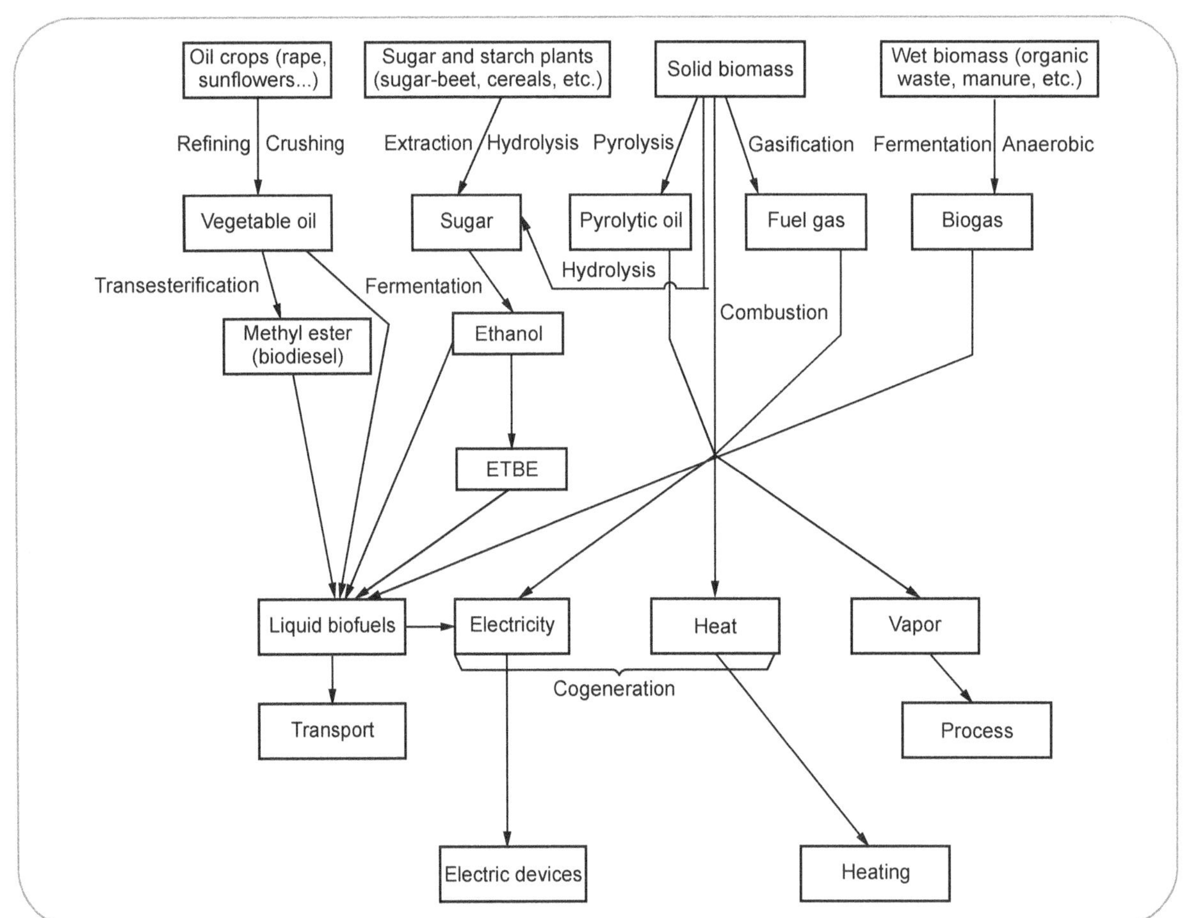

Fig. 1.10.1 Biomass energy conversion

(CH_4) and 45 % - 50 % nitrogen (N_2) making up the rest.

- Producing gas from biomass consists of the following main reactions, which occur inside a biomass gasifier.

  - **Drying :** Biomass fuels usually contain 10 % - 35 % moisture. When biomass is heated to about 100 °C, the moisture is converted into steam.

  - **Pyrolysis :** After drying, as heating continues, the biomass undergoes pyrolysis. Pyrolysis involves burning biomass completely without supplying any oxygen solids, liquids, and gases. Charcoal is the solid part, tar is the liquid part, and flue gases make up the gaseous part.

  - **Oxidation :** Air is introduced into the gasifier after the decomposition process. During oxidation, which takes place at about 700 - 1400 ° C, charcoal, or the solid carbonized fuel, reacts with the oxygen in the air to produce carbon dioxide and heat.

$$C + O_2 \rightarrow CO_2 + heat$$

Reduction : At higher temperatures and under reducing conditions, that is when not enough oxygen is available, the following reactions take place forming carbon dioxide, hydrogen, and methane.

$$C + CO_2 \rightarrow 2\,CO$$

$$C + H_2O \rightarrow CO + H_2$$

$$CO + H_2O \rightarrow CO_2 + H_2$$

$$C + 2\,H_2 \rightarrow CH_4$$

1.11.1 Advantages of Biomass Gasification Technologies

- **Mature technology :** Biomass gasifier technology is a mature technology and gasifiers are available in several designs and capacities to suit different requirements.

- **Small and modular :** The technology is suitable and economical for small, decentralized applications, typically with capacities smaller than a megawatt.

- **Flexible operation :** A gasifier based power system, unlike those based on other renewable sources such as the sun and wind, can generate electricity when required and also whenever required.

- **Economically viable :** For small - scale systems, the cost of power generation by biomass gasification technology is far more reasonable that of conventional diesel based power generation.

- **Socio-economically beneficial :** Biomass gasifier based systems generate employment for local people.

- **Mitigate climate change :** Biomass is a CO_2 neutral fuel and, therefore, unlike fossil fuels such as diesel does not contribute to net CO_2 emissions, which makes biomass based power generation systems an attractive option in mitigating the adverse effects of climate change.

1.12 Factors Affecting Biogas Production

The rate of production of biogas depends on the following factors :

1) Temperature and pressure
2) Solid concentration and loading rate
3) Retention period
4) pH value
5) Nutrients composition
6) Toxic substances
7) Digester size and shape
8) Stirring agitation of the content of digestion

1.13 : Grades of Energy

- In physics, **energy** is the quantitative property that must be transferred to an object in order to perform work on, or to heat, the object.

- Energy is a conserved quantity; the law of conservation of energy states that energy can be converted in form, but not created or destroyed.

- The SI unit of energy is the joule (J), which is the energy transferred to an object by the work of moving it a distance of 1 metre against a force of 1 Newton.

- Common forms of energy include
  - the kinetic energy of a moving object,
  - the potential energy stored by an object's position in a force field (gravitational, electric or magnetic),

- ○ the elastic energy stored by stretching solid objects,
- ○ the chemical energy released when a fuel burns,
- ○ the radiant energy carried by light,
- ○ the thermal energy due to an object's temperature, etc.

Some forms of energy (that an object or system can have as a measurable property)	
Type of energy	**Description**
Mechanical	The sum of macroscopic translational and rotational kinetic and potential energies.
Electric	Potential energy due to or stored in electric fields.
Magnetic	Potential energy due to or stored in magnetic fields.
Gravitational	Potential energy due to or stored in gravitational fields.
Chemical	Potential energy due to chemical bonds.
Ionization	Potential energy that binds an electron to its atom or molecule.
Nuclear	Potential energy that binds nucleons to form the atomic nucleus (and nuclear reactions).
Sound wave	Kinetic and potential energy in a fluid due to a sound propagated wave. (a particular form of mechanical wave)
Radiant	Potential energy stored in the fields of propagated by electromagnetic radiation, including light.
Thermal	Kinetic energy of the microscopic motion of particles, a form of disordered equivalent of mechanical energy.

- It is important to assign the quality to the energy.
- Based on the thermodynamic concepts, an energy source can be called as high-grade or low-grade, depending on the ease with which it can be converted into other forms.

- High grade form of energy are highly organised in nature and conversion of high grade energy to low grade energy is not desirable.
- However, there may be some conversion to low grade energy as work is converted into other useful form. This is because of dissipation of heat due to friction (example : mechanical work → Electricity, some losses are there due to the friction in bearing of machineries).
- Low grade energy such as heat due to combustion, fission, fusion reactions as well as internal energies are highly random in nature. Conversion of such form of energy into high grade energy (Heat→Work) is of interest. This is due to the high quality of organised form of energy obtained from low quality energy.
- Thus electrical energy is called a high-grade energy, as it is very easy to convert almost all of it into other energy forms such as thermal energy (say by using an electrical heater).
- Whereas, it is not possible to convert thermal energy completely into electrical energy (typical efficiencies of thermal power plants are around 30 %), hence thermal energy is called a low-grade energy.
- Naturally, high-grade energy sources are more expensive compared to low-grade energy sources.

1.14 : Basic Terms

i) Specifications :

- Specification basically is an act of specifying.
- It is a detailed description of requirement, dimensions, materials, etc as of a proposed building, machine, bridge, etc.
- Also it is an act of making specific.
- For example : Specification of vehicle or specifications of engine.

ii) Input and output :

- Input is nothing but anything which is taken in to operate any system or process.
- Example : For a motor input is electrical energy (current) which is converted into mechanical energy (rotation of shaft), for a vehicle input is a fuel, etc.

- Output is defined as the amount of energy, work, goods or services produced by a machine, industry or an individual in a period.

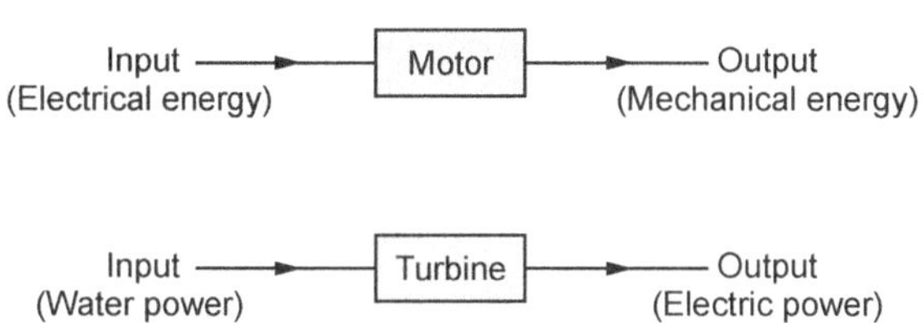

Fig. 1.14.1 Input and Output

- Example : For a turbine input is water power (hydraulic energy) and output is electric energy.

iii) Efficiency :

- Efficiency is defined as the comparison of what is actually produced with what can achieved with the consumption of resources.

- It signifies a level of performance that describes using the least amount of input to achieve the highest amount of output.

- Basically it is the state or quality of being efficient.

- Mathematically, efficiency is the ratio of output power of system to the input power of system. It is denoted by 'η'.

- It is always expressed in percentage (%) and it is always less than 1 (100 %). Mathematically,

$$\eta = \frac{\text{Output power}}{\text{Input power}} \times 100$$

- If for a system η = 80 % means system is 80 % effective or there is loss of 20 % in the system.

- The relation between input, output and losses can be given as,

$$\text{Input} = \text{Output} + \text{Losses}$$

or $$\text{Losses} = \text{Input} - \text{Output}$$

Efficiency of Termal Power Plant

- We know that, in thermal power plant work is done by the turbine and power is generated.

- But, out of this generated power, some power is wasted to run the pump in the system.

- Let,

$$W_T = \text{Workdone by the trubine in J/kg}$$

$$W_P = \text{Workdone on the pump in J/kg}$$

$$\text{H.S.} = \text{Heat supplied by the boiler in J/kg}$$

- For thermal power plant input is heat supplied by the boiler [Here we are considering for 1 kg of steam so we writing work as J/kg]

- Here, output is net work done of the system i.e.

$$\text{Net workdone} = \frac{\text{Turbinework}}{\text{(Power generated)}} - \frac{\text{Pumpwork}}{\text{(Power consumed)}}$$

$\therefore$
$$\boxed{W_{net} = W_T - W_P}$$

- It means, for thermal power plant efficiency is given as,

$$\eta = \frac{\text{Output}}{\text{Input}} \times 100 = \frac{W_{net}}{\text{Heat supplied}} \times 100$$

or
$$\boxed{\eta = \left(\frac{W_T - W_P}{\text{H.S.}}\right) \times 100} \qquad \dots (1.14.1)$$

1.15 Solved Examples

Ex. 1.1 : *In a thermal power plant the work done by the steam turbine is 900 J/kg. The work consumed by the pump is 50 J/kg. The heat supplied by the boiler to the system is 2800 J/kg. Find : i) Net workdone ii) Efficiency of the plant.*

Sol. : Given data : W_T = 900 J/kg, W_P = 50 J/kg, H.S. = 2800 J/kg

To find : i) W_{net} ii) η

$$\text{Net workdone} = W_T - W_P = 900 - 50$$

$\therefore$ W_{net} = **850 J/kg**   ... **Ans.**

Efficiency of plant is,

$$\eta = \left(\frac{W_{net}}{\text{H.S.}}\right) \times 100 = \frac{850}{2800} \times 100$$

$\therefore$ η = **30.35 %**   ... **Ans.**

Ex. 1.2 : *For a thermal power plant efficiency is 38 %. The work done by the turbine is 1 kJ/kg and the amount of heat supplied by the boiler is 2.2 KJ/kg. Find the work consumed by the pump and other equipments.*

Sol. : Given data : W_T = 1 kJ/kg = 1×10^3 J/kg, H.S. = 2.2 kJ/kg = 2.2×10^3 J/kg, η = 38 %

To find : W_P

Efficiency of plant is,

$$\eta = \left(\frac{W_{net}}{H.S.}\right) \times 100 = \left(\frac{W_T - W_P}{H.S.}\right) \times 100$$

$$\therefore \quad 38 = \left(\frac{1 \times 10^3 - W_P}{2.2 \times 10^3}\right) \times 100$$

$$\therefore \quad 1 \times 10^3 - W_P = 83.6$$

$$\therefore \quad \mathbf{W_P} = 1 \times 10^3 - 83.86 = \mathbf{164 \ J/kg} \quad \text{... Ans.}$$

Ex. 1.3 : *In a steam power plant the heat supplied by boiler is 2900 J/kg. For this amount of heat supplied, the turbine work is 900 J/kg and pump work is 100 J/kg. Find efficiency of the plant.*
If it is required to increase efficiency by 3 % for the same amount of heat supplied and turbine work then what will be the pump work ?

Sol. : Given data :

Case I :   H.S. = 2900 J/kg, W_T = 900 J/kg,
         W_P = 100 J/kg

Case II : η_{new} = η_{old} + 3 %, H.S. = 2900 J/kg,
         W_T = 900 J/kg

To find :   i) η_{old} ii) $(W_P)_{new}$

Efficiency of plant is,

$$\eta_{old} = \left(\frac{W_{net}}{H.S.}\right) \times 100 = \left(\frac{W_T - W_P}{H.S.}\right) \times 100$$

$$\therefore \quad \eta_{old} = \left(\frac{900 - 100}{2900}\right) \times 100 = \mathbf{27.58 \ \%} \text{ ... Ans.}$$

If efficiency is increased by 5 %

$$\eta_{new} = \eta_{old} + 3 \% = 27.58 + 3 = 30.58 \%$$

$$\therefore \quad \eta_{new} = \left(\frac{W_T - (W_P)_{new}}{H.S.}\right) \times 100$$

$$\therefore \quad 30.58 = \left(\frac{900 - (W_P)_{new}}{2900}\right) \times 100$$

$$\therefore \quad \mathbf{(W_P)_{new}} = \mathbf{13.18 \ J/kg} \quad \text{... Ans.}$$

Ex. 1.4 : *A steam power plant uses an impulse type turbine. The work done of the turbine is 1200 J/kg when the heat supplied by steam generator is 4000 J/kg. Find efficiency of plant when :*
a) Pump work is neglected
b) Pump work is 75 J/kg

Sol. : Given data : W_T = 1200 J/kg, H.S. = 4000 J/kg, W_{P1} = 0, W_{P2} = 75 J/kg

To find :   Efficiency η

Efficiency of plant is,

$$\eta = \left(\frac{W_{net}}{H.S.}\right) \times 100 = \left(\frac{W_T - W_P}{H.S.}\right) \times 100$$

$$\therefore \quad \eta_1 = \left(\frac{W_T - W_{P1}}{H.S.}\right) \times 100$$

$$= \left(\frac{1200 - 0}{4000}\right) \times 100$$

$$\therefore \quad \eta_1 = \mathbf{30 \ \%} \quad \text{...Ans.}$$

For second case,

$$\eta_2 = \left(\frac{W_T - W_{P2}}{H.S.}\right) \times 100$$

$$= \left(\frac{1200 - 75}{4000}\right) \times 100$$

$$\therefore \quad \eta_2 = \mathbf{29.375 \ \%} \quad \text{... Ans.}$$

Review Questions

1. *Explain the importance of electricity. What are the types of available energy sources ?*
2. *What are the types of power plants ?*
3. *Explain thermal power plant with block diagram. State its advantages and disadvantages.*
4. *Write short note on :*
   i) Hydroelectric power plant
   ii) Nuclear power plant.
5. *State the advantages and disadvantages of nuclear power plant.*

6. *Draw a neat sketch of hydroelectric power plant and nuclear power plant.*

7. *Write short note on :*

   i) Solar energy system   ii) Wind energy system.

8. *Explain the working of hybrid power system. State its advantages and disadvantages.*

9. *State the advantages and disadvantages of tidal power plant.*

10. *With neat sketch explain tidal energy.*

11. *What is geothermal energy ? Explain its set up with neat sketch.*

12. *Explain hydrogen energy. State it's advantages and disadvantages.*

13. *Write short note on biomass energy.*

14. *What are the grades of energy ?*

15. *Explain biomass conversion technologies.*

1.16   University Questions with Answers

Dec. - 2004

Q.1   *Draw a neat sketch of nuclear power plant. Why are these plants less in India ?*
(Refer section 1.4)   **[6]**

May - 2005

Q.2   *Draw a neat labelled diagram of a nuclear power plant.* **(Refer section 1.4)**   **[3]**

May - 2006

Q.3   *Draw a neat sketch of nuclear power plant and explain its working.* **(Refer section 1.4)**   **[8]**

May - 2007

Q.4   *Draw schematic sketch of steam power plant.* **(Refer section 1.2)**   **[4]**

Q.5   *Draw a neat labelled diagram of a nuclear power plant.* **(Refer section 1.4)**   **[6]**

Dec. - 2007

Q.6   *Explain the working of steam power plant with neat sketch.* **(Refer section 1.2)**   **[6]**

Dec. - 2008

Q.7   *What are the advantages of wind energy ? Explain a simple modern wind mill with a neat sketch.* **(Refer section 1.5.2)**   **[6]**

May - 2009

Q.8   *Draw a neat labelled diagram of a nuclear power plant.* **(Refer section 1.4)**   **[4]**

Q.9   *Explain hydroelectric power plant with neat sketch.* **(Refer section 1.3)**   **[6]**

Dec. - 2009

Q.10   *Draw a neat labelled diagram of a nuclear power plant.* **(Refer section 1.4)**   **[4]**

Q.11   *Compare thermal and nuclear power plants on any four parameters.* **(Refer section 1.4)**   **[4]**

Q.12   *Explain working of solar-wind hybrid power plant with sketch.* **(Refer section 1.5.3)**   **[8]**

May - 2010

Q.13   *Describe thermal power plant with block diagram.* **(Refer section 1.2)**   **[9]**

Q.14   *What is hybrid power plant ? State its advantages. Explain use of solar energy with block diagram.* **(Refer sections 1.5.3 and 1.5.1)**   **[9]**

Dec. - 2010

Q.15   *State advantages and disadvantages of thermal power plant.* **(Refer section 1.2)**   **[6]**

Q.16   *State advantages and disadvantages of hydro-power plant.* **(Refer section 1.3)**   **[6]**

Q.17   *Draw a neat labelled diagram of a nuclear power plant.* **(Refer section 1.4)**   **[5]**

Q.18   *Explain use of solar energy for any one application.* **(Refer section 1.5.1)**   **[5]**

May - 2011

Q.19   *Draw and explain thermal power plant.* **(Refer section 1.2)**   **[6]**

Q.20   *State advantages and disadvantages of hydro-power plant.* **(Refer section 1.3)**   **[6]**

Q.21   *Explain any one hybrid power plant with block diagram.* **(Refer section 1.5.3)**   **[6]**

Dec.- 2011

Q.22 *State the advantages and disadvantages of nuclear power plants.***(Refer section 1.4)**   **[4]**

May - 2013

Q.23 *Write a note on hydro-electric power plant.* **(Refer section 1.3)**   **[6]**

Dec. - 2013

Q.24 *Explain working of thermal power plant with neat sketch.* **(Refer section 1.2)**   **[6]**

Q.25 *Draw a layout of nuclear power plant and explain the energy extraction.* **(Refer section 1.4)**   **[4]**

May - 2014

Q.26 *Explain working of thermal power plant with neat sketch.* **(Refer section 1.2)**   **[7]**

Q.27 *Draw a layout of wind power plant and explain the energy extraction (transfer). State its limitations.* **(Refer section 1.5.2)**   **[4]**

Dec. - 2014

Q.28 *With neat sketches explain nuclear power plant.* **(Refer section 1.4)**   **[6]**

May - 2015

Q.29 *Draw a layout of solar power plant. State the limitations of the plant.* **(Refer section 1.5.1)**   **[4]**

Q.30 *Draw a layout of hydro-electric power plant and explain the energy extraction (energy conversion) process.* **(Refer section 1.3)**   **[4]**

Dec. - 2015

Q.31 *Draw a layout of solar power plant. State the limitations of the plant.* **(Refer section 1.5.1)**   **[4]**

Q.32 *Draw a layout of hydro-electric power plant and explain the energy extraction (energy conversion) process.***(Refer section 1.3)**   **[4]**

May - 2016

Q.33 *Draw a layout of solar power plant. State the limitations of the plant.* **(Refer section 1.5.1)**   **[4]**

Dec. - 2016

Q.34 *Draw a layout of hydro-electric power plant and explain the energy extraction (energy conversion) process.***(Refer section 1.3)**   **[4]**

Dec. - 2016

Q.35 *Draw a block diagram of :*   **[6]**

1) *Wind power plant, and* **(Refer section 1.5.2)**

2) *Solar power plant* **(Refer section 1.5.1)**

May - 2017

Q.36 *Draw a sketch of wind power plant. Explain energy transfer (extraction) in the power plant and state its limitations.* **(Refer section 1.5.2)**   **[6]**

Q.37 *Draw a sketch of hydro-electric power plant. Explain energy transfer (extraction) in the power plant and state its limitations.* **(Refer section 1.3)**   **[6]**

Dec. - 2017

Q.38 *List various conventional and non - conventional energy resources ? Draw block diagram of solar power plant. Explain energy extraction (transfer) in the plant.* **(Refer sections 1.1.1 and 1.5.1)**   **[6]**

May - 2018

Q.39 *What are the types of energy resources ? Draw block diagram of wind power plant. Explain energy extraction (transfer) in the plant.* **(Refer sections 1.1.1 and 1.5.2)**   **[6]**

Dec. - 2018

Q.40 *Explain working of wind power plant. State its limitations.* **(Refer section 1.5.2)**   **[6]**

Q.41 *Draw block diagrams of nuclear power plant and hydro-electric power plant.* **(Refer section 1.4)**   **[6]**

May - 2019

Q.42 *Draw layout of hydroelectric power plant and explain the energy conversion proess and its limitations.* **(Refer section 1.3)**   **[6]**

❑❑❑

UNIT - I

2   Energy Conversion Devices

Syllabus

Introduction of pump, compressor, turbines, wind mills etc. (Simple numerical on power and efficiency calculations) .

Contents

Mind Map - Energy Conversion Devices

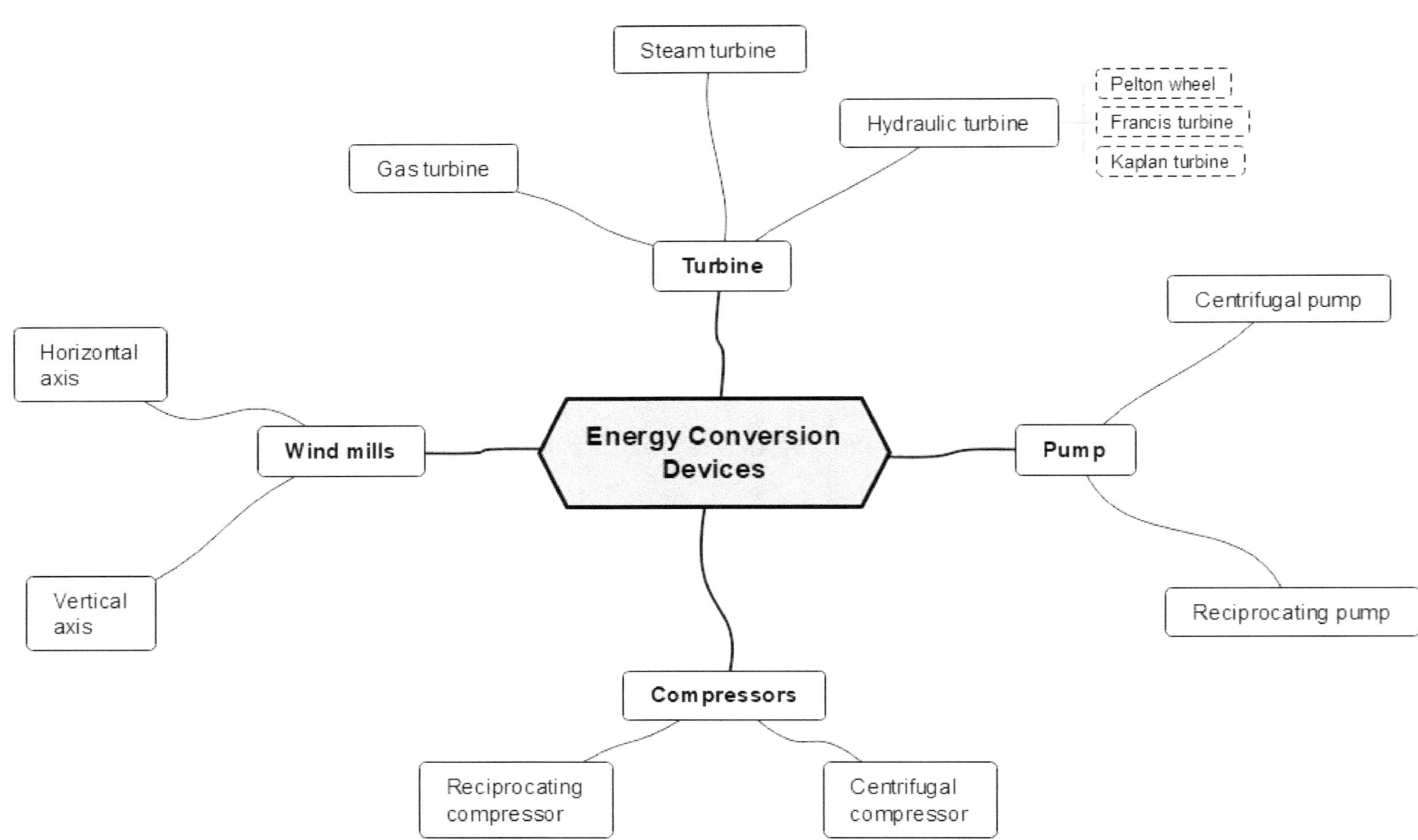

2.1 Introduction

- In various engineering applications, one form of energy is converted into another form.
- Usually, the heat is converted into work or work is converted into heat.
- The systems or devices which are used for converting energy are called as **energy conversion devices.**
- The energy converting devices are divided into two categories :

> 1. Power producing devices
> 2. Power absorbing devices

1. Power producing devices

- Power producing devices operates on a cycle and produces mechanical power output on the expense of supplied energy.
- These devices generally absorb thermal energy from the source and rejects some part of it to a sink, while doing so it produces work as per the laws of thermodynamics.
- **For example :** Boilers, Steam and Gas tubines, I.C. engines, Hydraulic turbines, etc.

2. Power absorbing devices

- Power absorbing devices uses work energy from outside source to increase the energy of working fluid during a cycle.
- **For example :** Pumps, Compressors, Blowers, Fans, Refrigerators, Air conditioners, etc.

TURBINES

2.2 Turbines   SPPU : May-06, Dec.-06, 07

- **Hydraulic machine** is a prime mover which convert the energy of flowing water (hydraulic energy) into the mechanical energy.
- This mechanical energy is used to run an electric generator which is directly connected to the shaft of hydraulic machine. From this generator, electric power can be transmitted over long distances.
- The hydraulic machines which convert the hydraulic energy into mechanical energy are called as turbines whereas the hydraulic machines which convert the mechanical energy into hydraulic energy are called **pumps**.
- The hydraulic turbines are also called as **water turbines** because the fluid used is water.
- Generally, water turbine consists of a wheel called as *runner* or *rotor* having number of specially designed vanes or blades. The water with large amount of hydraulic energy strikes the runner and causes it to rotate. Thus, mechanical energy is supplied to the generator coupled to runner which generates electrical energy.
- The electric power which is obtained from the hydraulic energy is called as **hydro-electric power**.
- Generation of hydro-eletric power is the cheapest as compared to other power gererating methods.
- The conmonly used water turbines are Pelton turbine, Francis turbine and Kaplan turbine. Similarly, the commonly used pumps are centrifugal pumps and reciprocating pumps.

2.2.1 Classification of Hydraulic Turbines

Hydraulic turbines are classified on the basis of following :

1. **According to the action of water flowing through the turbine runners**
   - a) Impulse turbine
   - b) Reaction turbine

2. **According to the direction of flow of water in the turbine**
   - a) Tangential flow turbine
   - b) Radial flow turbine
   - c) Axial flow turbine
   - d) Mixed flow turbine

3. **According to the head and quantity of water required**
   - a) High head turbine
   - b) Medium head turbine
   - c) Low head turbine

1. Impulse turbine

- In these turbines, all the available energy of water is converted into kinetic energy or velocity head by passing it through a nozzle provided at the end of penstock.

- This high velocity water is impinged on a series of buckets of the runner thus causing it to revolve.

- A casing is provided on the runner to prevent splashing and to guide the water discharged from the buckets to the tail race.

- Commonly used impulse turbines are **Pelton wheel,** Girard turbine, Banki turbine, Jonval turbine, etc.

2. Reaction turbine

- In these turbines, at the entrance to the runner, only a part of the available energy of water is converted into kinetic energy and substantial part is in the form of pressure energy.

- The pressure at the inlet of turbine is much higher than the pressure at the outlet and varies throughout the passage of water through the turbine.

- In these turbines, the runner is completely enclosed in air-tight casing and the passage is full of water throughout the operation.

- The pressure difference between inlet and outlet of the runner is called as *reaction pressure* and hence these turbines are called **reaction turbines.**

- Commonly used reaction turbines are **Francis turbine, Kaplan turbine,** Thomson turbine, Propeller turbine, etc.

2.2.2 Pelton Wheel (Impulse Turbine)

SPPU : May-06, Dec.-07

- This is the only impulse type of hydraulic turbine which is now commonly used.

- It was first developed by Lester A. Pelton an American Engineer in 1880.

- This turbine is most suitable for operating under high heads.

- An arrangement of a typical Pelton wheel is shown in Fig. 2.2.1.

- The main elements of a Pelton wheel are :
    - ○ Runner and buckets
    - ○ Casing
    - ○ Nozzle and spear assembly

- The water from reservoir flows through the penstock to the nozzle which converts pressure energy into kinetic energy.

- This high velocity jet (water) strikes the buckets or vanes on the outer periphery of runner.

- The buckets have a shape of double semi-ellipsoidal cups and each bucket is divided into two

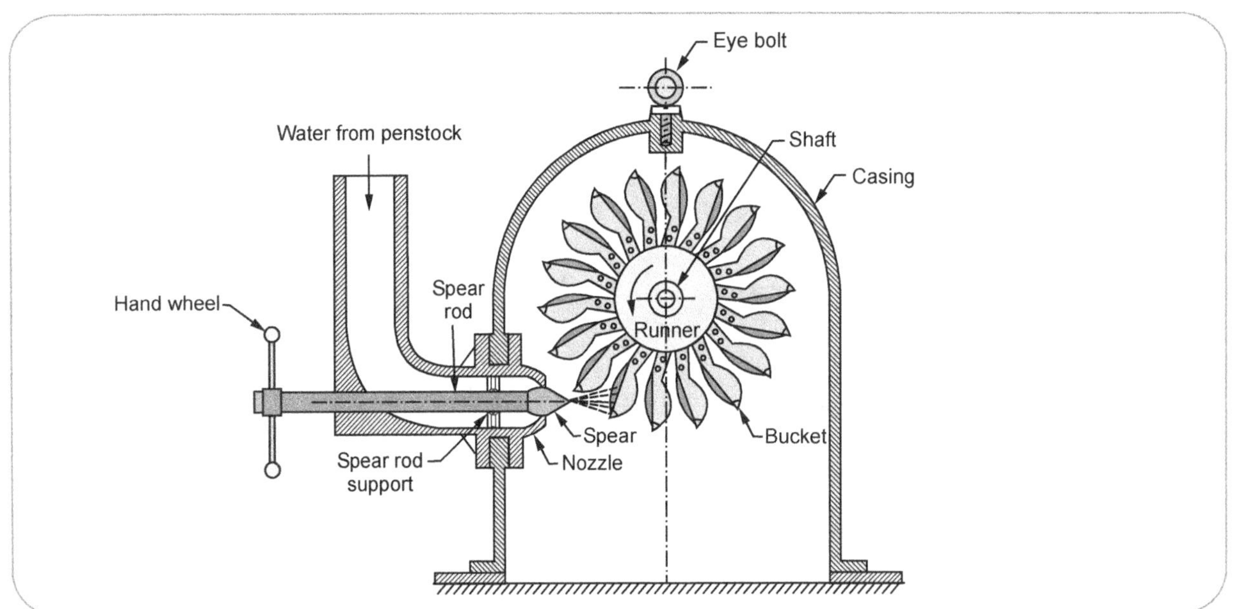

Fig. 2.2.1 : Pelton wheel

symmetrical parts by a sharp-edged ridge called as **splitter.** Refer Fig. 2.2.2.

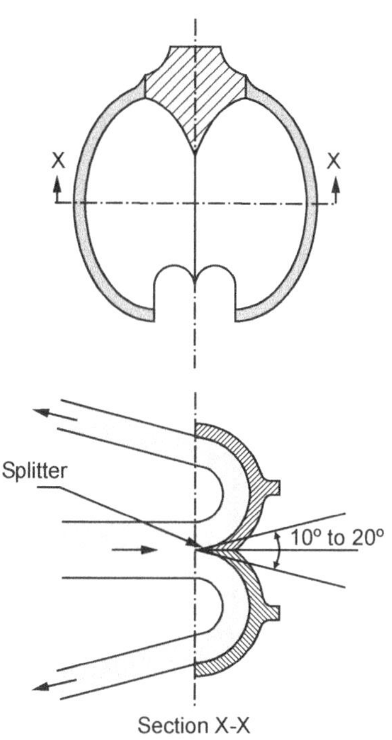

Fig. 2.2.2 : Shape of bucket

- The jet of water impinges on the splitter which divides the jet into two equal portions, each of which flowing round the smooth inner surface and leaves at its outer edge.

- Due to double cup-shaped buckets the axial thrust neutralize each other, being equal and opposite and the wheel bearings are not subjected to axial or end thrust.

- The back of the bucket is so shaped that as it swings downwards into the jet, no water is wasted by splashing.

- The buckets are generally made of cast iron, cast steel, bronze or stainless steel.

- The casing has no hydraulic function. It is provided only to prevent splashing of water, to lead the water to the tail race and it also act as a safeguard against accidents.

- In order to control the quantity of water striking the runner, the nozzle fitted at the end of penstock is provided with a spear which is fixed to the end of a rod.

- It may be operated manually by a wheel or automatically by a governor.

2.2.3 Francis Turbine (Reaction Turbine)

SPPU : May-06

- Francis turbine is a mixed flow type of reaction turbine.

- It was invented by James B. Francis an American Engineer in 1849.

- In these turbines water enters the runner radially at the outer periphery and leaves axially at its centre.

- The main components of Francis turbine are :
    - Spiral or scroll casing
    - Runner and vanes
    - Guide mechanism
    - Draft tube

- The water from the pen stock enters a spiral casing which completely surrounds the runner.

- The casing is provided for an even distribution of water around the circumference of the turbine runner and maintains approximately constant velocity for the water so distributed.

- Fig. 2.2.3 shows the arrangement of Francis turbine.

- To keep the volocity of water constant throughout its path around the runner, the cross-sectional area of the cassing is gradually decreased. Generally, the casing is made of cast steel, steel, concrete, etc.

- From the scroll casing the water passes through a stay ring which consists of an upper and lower ring held together by series of fixed vanes called as **stay vanes.**

- The stay ring which is made of cast steel or cast iron directs the water from scroll casing to the guide vanes.

- From the speed ring water passes through a series of guide vanes provided around the periphery of the turbine runner.

- The guide vanes are of aerofoil shape and are used to regulate the quantity of water supplied to the runner and to direct the water on the runner at an appropriate angle.

- The guide vanes are operated either by means of a wheel or automatically by a governor.

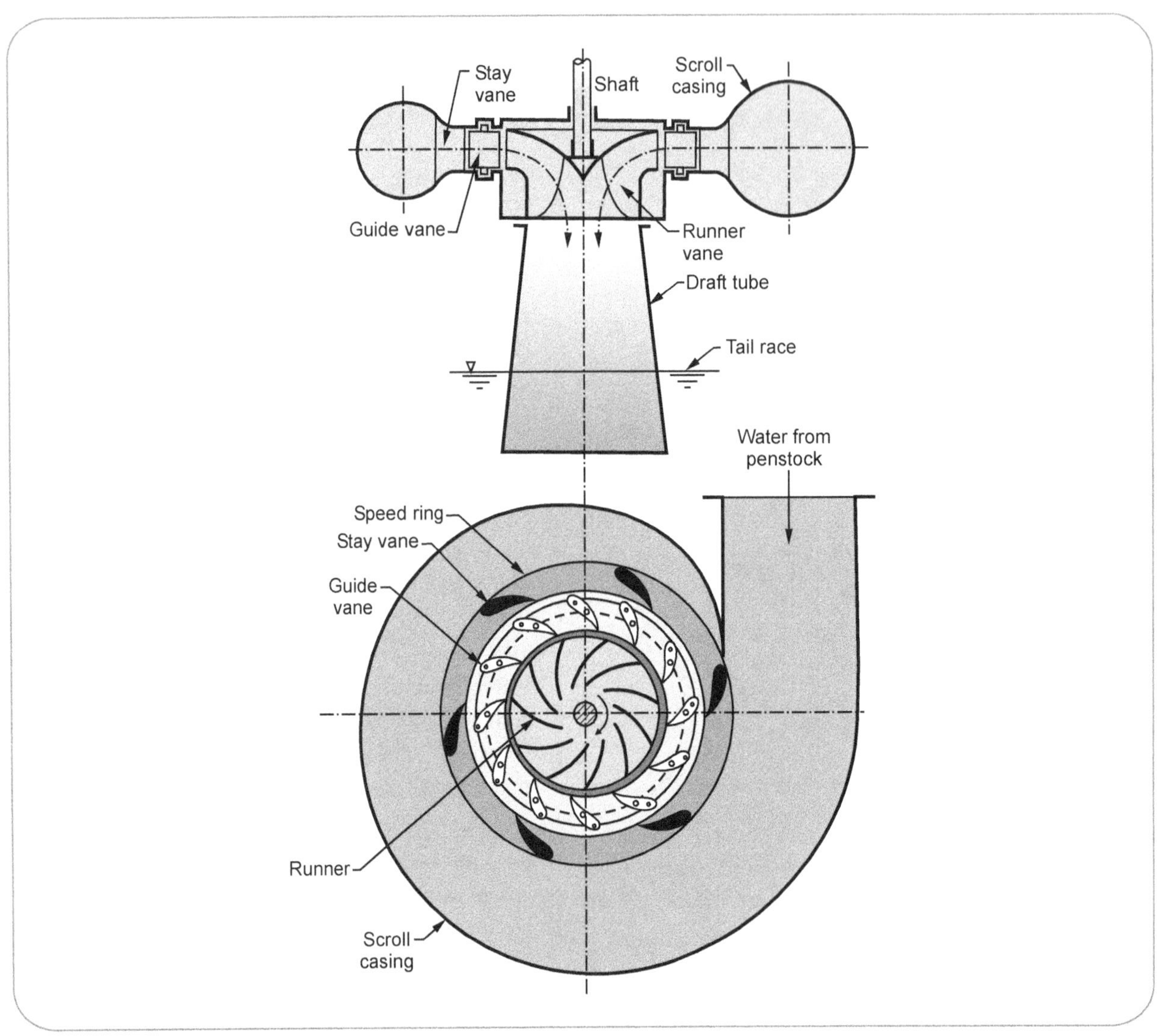

Fig. 2.2.3 : Francis turbine

- The vanes are so shaped that water enters the runner radially at the outer periphery and leaves axially at the inner periphery.

- The change in direction of water produces a circumferential force on the runner which makes it to rotate and produces useful output.

- After passing through the runner, the water flows to the tail race through a draft tube.

- A draft tube is a pipe or passage of gradually increasing cross-sectional area which connects the runner exit and tail race.

- It is always submerged below the level of water in the tail race.

2.2.4 Kaplan Turbine (Reaction Turbine)

SPPU : Dec.-06

- Kaplan turbine is an axial flow turbine which is suitable for low heads and hence it requires large quantity of water to develop large amount of power.

- It was developed by V. Kaplan an Australian Engineer.

- It operates in an entirely closed conduit from the head race to tail race.

- Fig. 2.2.4 shows the arrangement of Kaplan turbine.

- The main elements of Kaplan turbine are similar to Francis turbine.

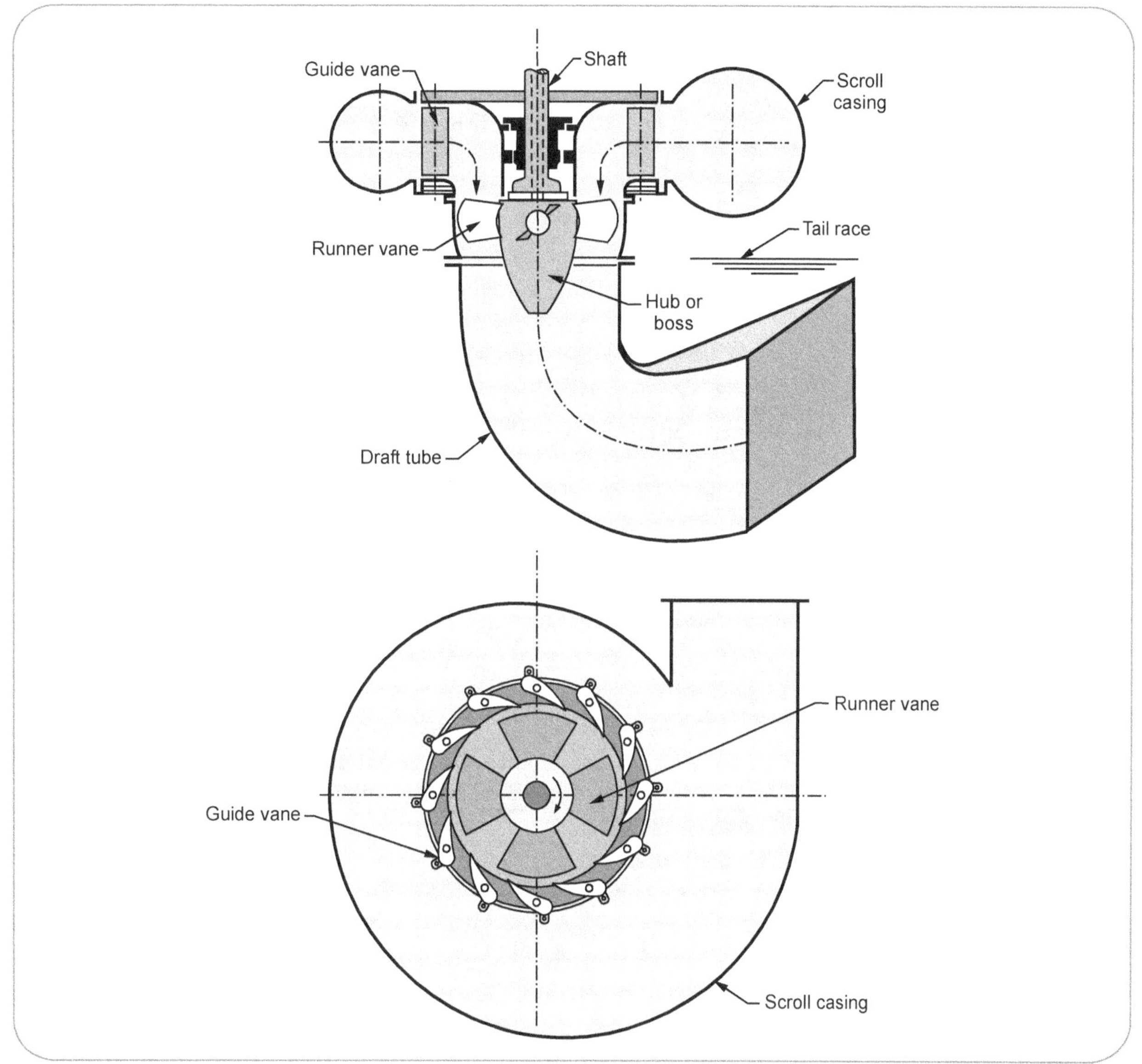

Fig. 2.2.4 : Kaplan turbine

- In these turbines, water turns between the guide vanes and runner at right-angle into the axial direction and then passes through the runner.

- The vanes attached to the hub or boss are so shaped that the water flows axially through the runner.

- The main advantage of Kaplan turbine is that, its runner blades can be turned about their own axis for adjusting the angle of inclination.

- The working principle of Kaplan turbine is similar to Francis turbine.

2.2.5 Comparison between Impulse Turbine and Reaction Hydraulic Turbine

Sr. No.	Impulse turbine	Reaction turbine
1.	The available water energy is converted into kinetic energy by using a nozzle.	Only some part of available energy is converted into kinetic energy in the guide vanes.

2.	During the operation, pressure remains constant.	After entering at the runner, water undergoes chage in pressure and velocity.
3.	Draft tube is not required for the operation.	Draft tube is required for the operation.
4.	Water from the nozzles comes out in the form of jet which impinges on the buckets of runner.	Water is guided by the guide blades to flow over the vanes of aerofoil shape.
5.	Water does not fill the bucket fully hence the air has easy access to the buckets.	Water completely fills all the passages between the blades.
6.	It is always installed above the tail race.	It may be installed above or below the tail race.
7.	These turbines are suitable for high heads.	These turbines are suitable for low heads.
8.	Power is developed due to change in kinetic energy.	Power is developed due to change in pressure head.

2.3 Steam Turbines

SPPU : Dec.-03,06,09,12,16, May-04,09,10

- A steam turbine is a mechanical device that extracts thermal energy from pressurized steam and converts it into rotary motion.

- Steam turbine depends directly upon dynamic action of the steam.

- As per the Newton's second law of motion, the rate of change of momentum is caused in the steam by allowing high velocity jet of steam to pass over curved plate (blade) and the steam will impart a force to the blade.

- If the blade is free then it will rotate in the direction of applied force.

- The steam from the boiler is expanded in the nozzle where due to fall in pressure of steam, thermal energy of steam is converted into kinetic energy of steam.

- This high velocity jet of steam is impinged on the blades mounted on a shaft. Refer Fig. 2.3.1.

- The change in flow direction of steam causes a force to be exerted on the blades and due to rotation of these blades power is developed.

2.3.1 Types of Steam Turbines

SPPU : May-09, Dec.-09, 12

Steam turbines are mainly divided into two groups :

1. Impulse turbine
2. Reaction turbine

1. Impulse Turbine

SPPU : May-10, Dec.-16

- In these turbines, the steam comes out at a very high vleocity through the fixed nozzle and

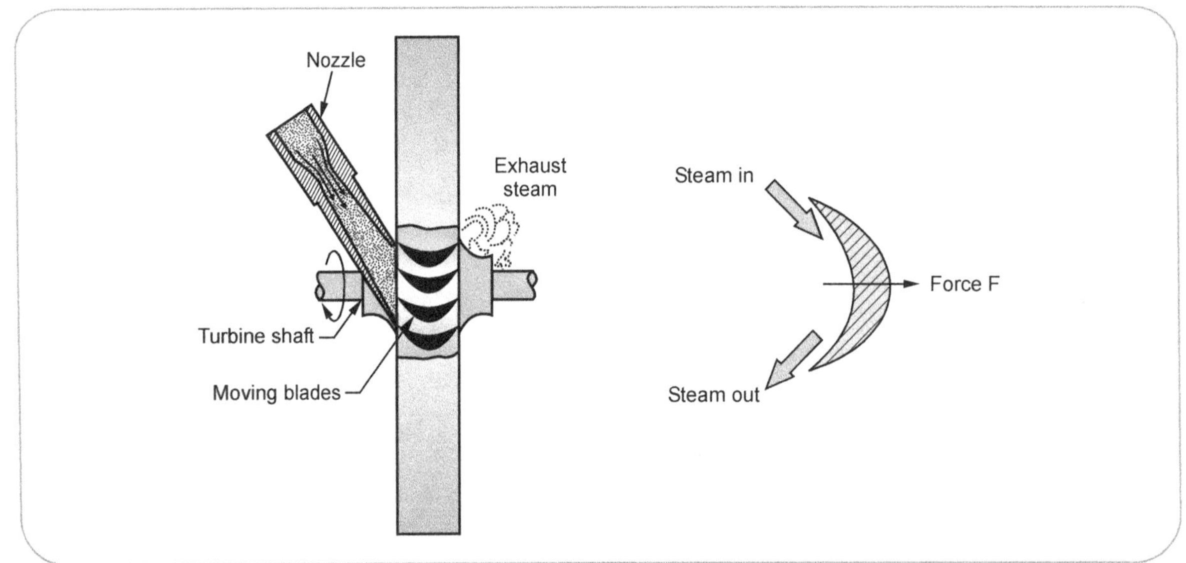

Fig. 2.3.1 Working principle of steam turbine

impinges on the blades fixed on the periphery of a rotor. Refer Fig. 2.3.2 (a).

- The blades change the direction of the steam flow without changing its pressure.

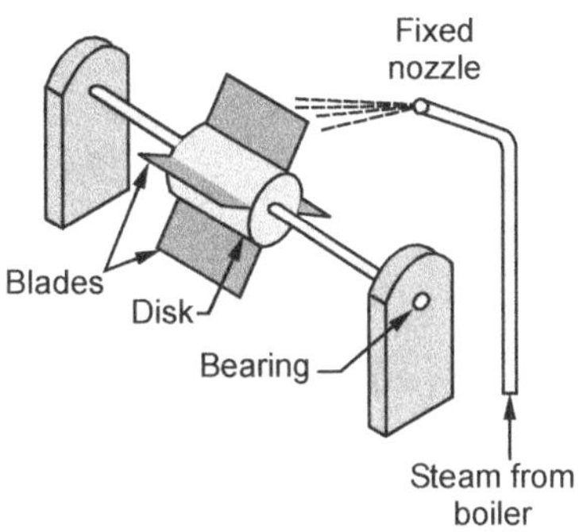

(a) Working principle

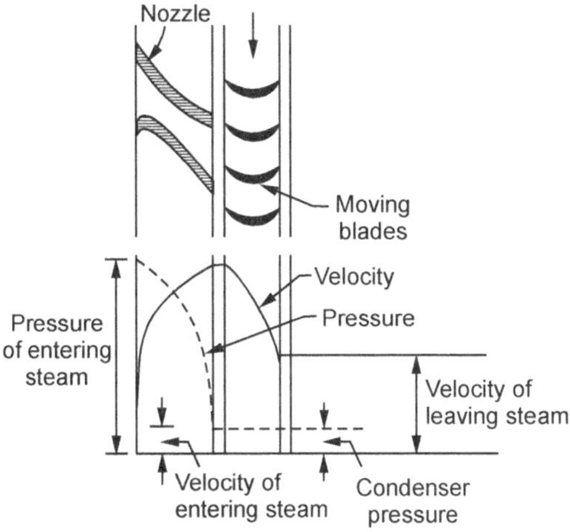

(b) Arrangement of blades

Fig. 2.3.2 Impulse turbine

- The resulting force due to the change in momentum causes the rotation of the turbine shaft.

- Fig. 2.3.2 (b) shows the blade arrangement for impulse turbine and the variation of pressure and velocity of steam passing through the turbine.

- Examples of impulse turbine are De-Laval turbine, Rateau turbine, Curties turbine.

- It is important to note that, in case of impulse turbine the shape of blades in *profile type*.

2. Reaction Turbine

SPPU : May-04, Dec.- 16

- In these turbines, the high pressure steam from the boiler is passed through the nozzles as shown in Fig. 2.3.3 (a).

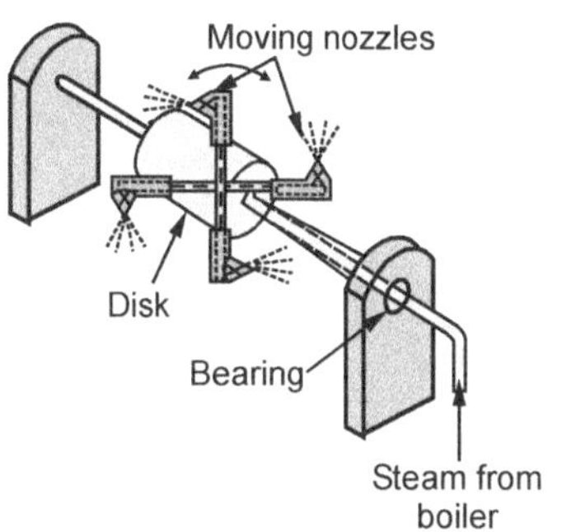

(a) Working principle

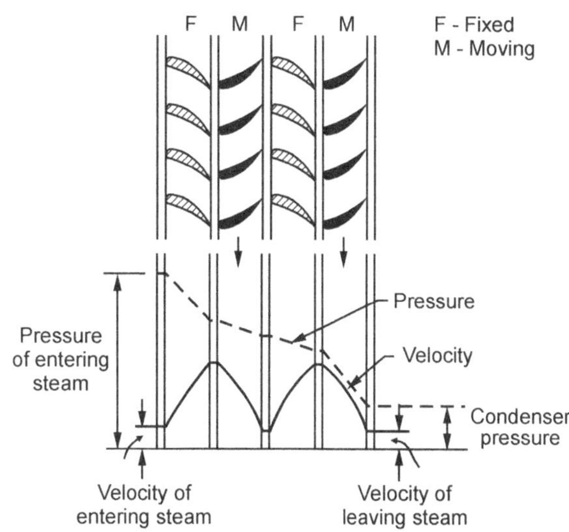

(b) Arrangement of blades

Fig. 2.3.3 Reaction turbine

- When the steam comes out through these nozzles, the velocity of the steam increases relative to the rotating disc.

- This results in reacting force of the steam on nozzle which gives rotating motion to the disc and shaft.

- In these turbines steam expands both in fixed and moving blades continuously when the steam passes over them.

- Hence, pressure drop occurs gradually and continuously over both moving and fixed blades.

- For example, Parson's reaction turbine.

- Fig. 2.3.3 (b) shows the blade arrangement for reaction turbine and the variation of pressure and velocity of steam passing through the turbine.

- It is important to note that, in case of reaction turbine the shape of blades is *aerofoil*.

2.3.2 Advantages of Steam Turbines

Steam turbines offer following advantages :

- Steam turbine is highly simplified in operation and construction.

- The thermal efficiency of steam turbines is much higher.

- It is compact and it has low weight to power ratio.

- It can operate at high speeds and greater range of speeds is possible.

- Due to absence of reciprocating parts, the vibrations and noise are greatly minimized.

- Steam turbine can take considerable over load.

- It can be designed in sizes ranging from a few kW to 1000 MW in a single unit.

- In steam turbines there is no condensation loss.

- Life of steam turbine is high.

- Initial cost, maintenance cost and installation cost are low.

2.3.3 Comparison between Impulse and Reaction Steam Turbine   SPPU : Dec.-03, 06

Sr. No.	Impulse turbine	Reaction turbine
1.	In these turbines, pressure drops only in nozzle and not in moving blades channel.	In these turbines, pressure drops in nozzle as well as in moving blades channel.
2.	The blades are of profile shape.	The blades are of aerofoil shape.
3.	Blade channel area is constant.	Blade channel area is varying.
4.	By using these turbines much power cannot be developed.	Much power can be developed by using these turbines.
5.	These turbines occupy less space for same power.	These tubines occupy more space for same power.
6.	Velocity of steam is slightly higher.	Velocity of steam is lower.
7.	Blade manufacturing is simple and less costly.	Blade manufacturing is difficult hence costly.
8.	Efficiency of these turbines is low.	Efficiency of these turbines is high.

2.4 Gas Turbines   SPPU : Dec.-03,04,06, May-04,07,08

- The working principle of gas turbines is almost similar to steam turbines; only the working fluid is gas instead of steam.

- Gas turbines represent the most satisfactory way of producing large quantities of power in a compact and self-contained unit.

- Sometimes, the thermal efficiency of gas turbines is increased by using a heat exchanger.

2.4.1 Classification of Gas Turbines

Gas turbines may be classified as follows :

> 1. Constant pressure combustion gas turbine
>   a) Open cycle constant pressure gas turbine
>   b) Closed cycle constant pressure gas turbine
> 2. Constant volume combustion gas turbine

1. Open cycle constant pressure gas turbine
          SPPU : Dec.-04, May-07

- An open cycle constant pressure gas turbine is shown in Fig. 2.4.1.

- It consists of an air compressor, combustion chamber and gas turbine.

- In these turbines, an air compressor and gas turbine are mounted on a common shaft.

- During the operation, fresh air is drawn into the compressor from the atmosphere and it is compressed to high pressure.

- This high pressure air is supplied to the combustion chamber where heat is added by injecting the fuel.

- Hence, the formed gases are allowed to expand in the turbine and it performs mechanical work by driving the turbine shaft.

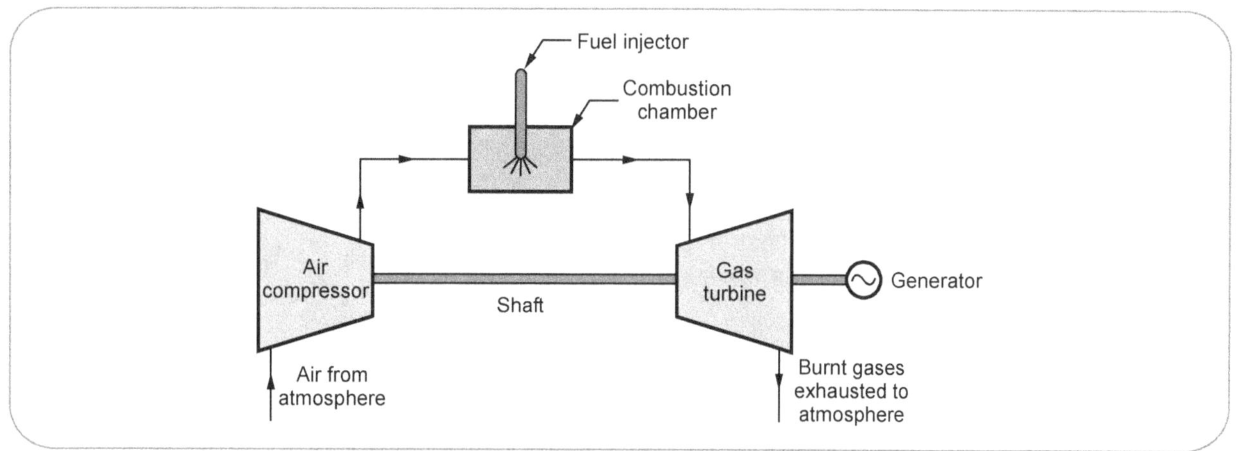

Fig. 2.4.1 : Open cycle constant pressure gas turbine

- Some part of this developed power is used to run the compressor and remaining is considered as useful work.

- After expansion, the gases are exhausted to the atmosphere.

- The working fluid (mixture of air and fuel) must be continuously replaced beacuse it is exhausted to the atmosphere.

2. Closed cycle constant pressure gas turbine

- A closed cycle constant pressure gas turbine is shown in Fig. 2.4.2.

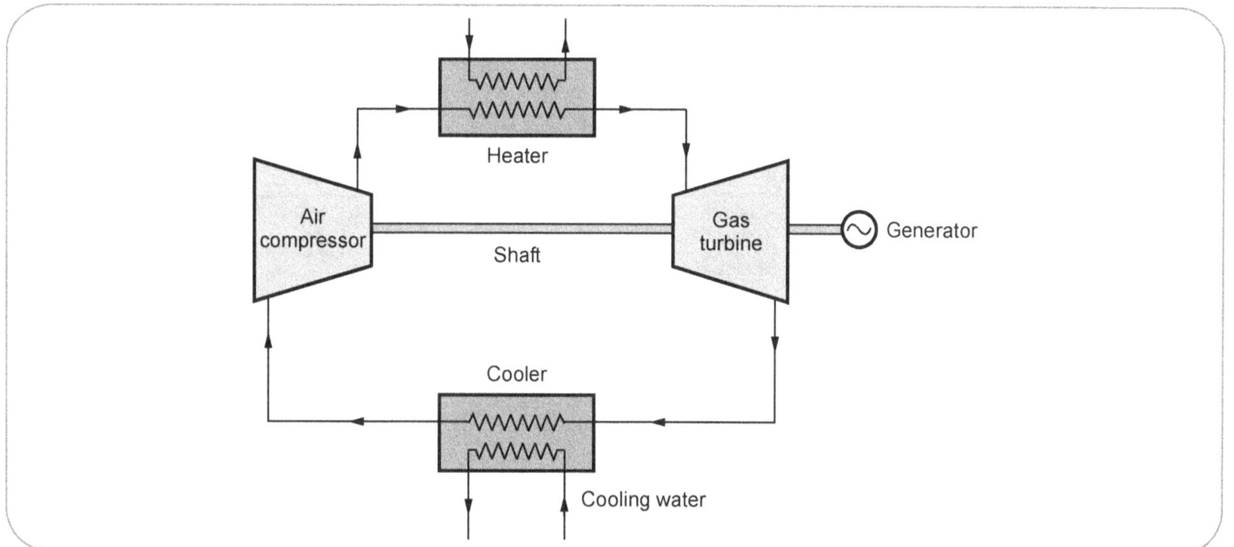

Fig. 2.4.2 : Closed cycle constant pressure gas turbine

- In addition to air compressor combustion chamber and gas turbine, it consists of heater and cooler.

- During the operation, the fluid is compressed in the compressor and the compressed fluid coming out of the compressor is heated in the heater by an external source and at constant pressure.

- This fluid expands in the turbine and power is developed.

- The expanded fluid coming out of the turbine is cooled to its original temperature in the cooler using external cooling source before fed to the compressor.

- Hence, the working fluid is continuously circulated in the circuit.

Advantages of closed cycle gas turbine

- A high pressure can be maintained throughout the cycle with consequent reduction in volume of air.

- As the combustion is external, any type of fuel can be used.

- The working medium remains free from pollution of products.

- As the working circuit is closed, there is no loss of working medium.

- Thermal efficiency of these type of turbines is high.

3. Constant volume combustion gas turbine

- A constant volume combustion gas turbine is shown in Fig. 2.4.3.

- The working principle of these turbines is almost similar to open cycle gas turbine at constant pressure.

- During the operation, the atmospheric air is compressed in the compressor and this compressed air is supplied to the combustion chamber through the inlet valve.

- When inlet valve is closed, at that time fuel is supplied to the combustion chamber with the help of fuel pump.

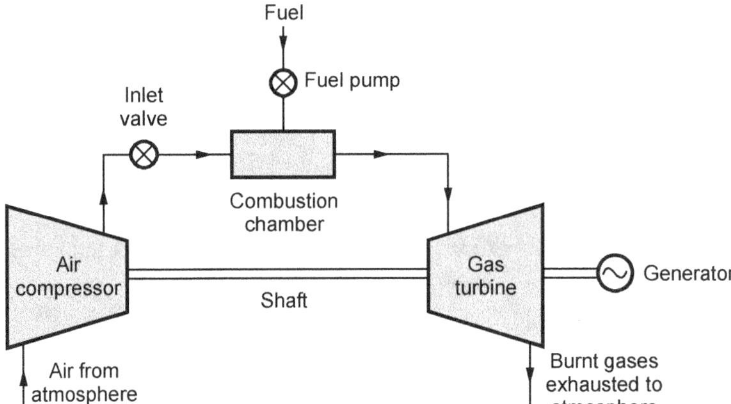

Fig. 2.4.3 : Constant volume combustion gas turbine

- This air-fuel mixture is ignited and combustion takes place at constant volume.

- These high pressure gases are expanded in the turbine and power is developed.

- After expansion the gases are exhausted to atmosphere and velocities of gas are not constant.

2.4.2 Comparision between Open Cycle and Closed Cycle Gas Turbines

Sr. No.	Open cycle gas turbine	Closed cycle gas turbine
1.	Maximum operating pressure of these turbines is less.	Maximum operating pressure of these turbines is high.
2.	In these turbines, working fluid is air.	Working fluid is gas like helium, argon, air-fuel mixture, etc.
3.	For given output size of plant is larger.	For given output size of plant is smaller.
4.	Due to contaminated gases corrosion of blades takes place.	As there is indirect heating in the cycle, there is no corrosion of blades.
5.	Commonly used fuels are kerosene, petrol, oil, etc.	It can use any type of fuel for its operation.
6.	Cooling medium is not required.	Large amount of cooling water is required.
7.	Weight and size of plant is small.	Weight and size of plant is high.
8.	Low thermal efficiency.	High thermal efficiency.

2.4.3 Comparison between Gas Turbines and Steam Turbines

Sr. No.	Gas turbine	Steam turbine
1.	For the operation of these turbines, boiler is not required.	Boiler is required for the operation of these turbines.
2.	For given output less space is required.	It requires more space for given output.
3.	Starting and power generation time is less.	Start and power generation time is more.
4.	Weight to power ratio is less.	Weight to power ratio is high.
5.	It is suitable for part loads and peak loads.	It is suitable only for part loads.
6.	It does not require cooling water.	It requires large amount cooling water.
7.	Thermal efficiency is low.	Thermal efficiency is high.
8.	Capital cost and running cost is low.	Capital cost and running cost is high.

2.4.4 Applications of Gas Turbines

SPPU : Dec.-04, 06, May-07

Gas turbines are used in the following fields :

i) **Turbojet engine :** Energy turbojet and turbopropeller engine has a gas turbine.

ii) **Marine applications :** Gas turbines are mostly used in marine applications because of absence of boiler.

iii) **Railway engines :** The gas turbines are now also used in railway engines.

iv) **Supercharging :** In supercharging small gas turbine run by the hot flue gases drives supercharger for aviation gasoline engines.

v) **Electric power generation :** As gas turbines require less amount of water, now they are more popular then the steam turbines.

vi) **Industrial applications :** Gas turbines are also used in steel industries, chemical industries, oil industries, etc.

PUMPS

2.5 Pumps

- Pump is a mechanical device which when connected in a pipeline converts the mechanical energy supplied to it into the hydraulic energy and transfers it to the liquid.

- Hence, pump increases the energy of flowing fluid. Almost all the pumps increases the pressure energy of the liquid.

- The main difference between turbine and pump is that, in case of turbine flow of fluid takes place from high pressure to low pressure whereas in case of pump flow of fluid takes place from low pressure to high pressure.

2.5.1 Classification of Pumps

Generally pumps are divided into two major groups :

1. Positive-displacement pumps
2. Rotodynamic pumps (Dynamic pressure pumps)

1. **Positive-displacement pumps :** In these pumps, the fluid is sucked and actually displaced or pushed due to the thrust exerted on it by a moving member (piston) which results in lifting the liquid to the desired height. The commonly used positive-displacement pump is a *reciprocating pump.*

2. **Rotodynamic or dynamic pressure pumps :** These pumps have a rotating element (impeller) through which as the liquid passes its angular momentum changes hence the pressure energy of liquid is increased. These pumps do not push the liquid as in case of positive-displacement pump. The commonly used rotodynamic pump is a *centrifugal pump.*

2.6 Reciprocating Pumps

- Reciprocating pump is a positive-displacement type pump in which liquid is displaced by a piston-cylinder arrangement which is driven by crank and connecting rod mechanism.

- Reciprocating pump is suitable for small capacities and high heads.

• Reciprocating pumps can be classified as follows :

> 1. Single acting pump
> 2. Double acting pump

2.6.1 Construction of Reciprocating Pump

Fig. 2.6.1 shows the arrangement of single acting reciprocating pump. The various components of reciprocating pump are as follows :

> 1. Suction pipe and delivery pipe
> 2. Piston and cylinder
> 3. Crank and connecting rod mechanism
> 4. Sump (Reservoir)

1. **Suction pipe and delivery pipe :** Suction pipe is connected to the sump of the pump and through this pipe liquid is sucked into the pump. Delivery pipe is connected to the discharge end and it carries liquid at high pressure through some height. It also consists of a non-return valve.

2. **Piston and cylinder :** The piston reciprocates inside the cylinder. The connecting rod transfers rotary motion of the crank into reciprocating motion of the crank into reciprocating motion of piston.

3. **Crank and connecting rod mechanism :** The crank is mounted on a crankshaft and it is driven by either I.C. engine or electric motor. The crank is connected to the piston by connecting rod hence the rotary motion of crank is converted into reciprocating motion of piston.

4. **Sump :** It is the reservoir through which the liquid is pumped into the system.

2.6.2 Single Acting Reciprocating Pump

SPPU : Dec.-07, 11, 12, 16, May-11

• Fig. 2.6.1 shows the arrangement of single acting reciprocating pump which has the same construction as discussed above.

• Initially the crank is at IDC and starts rotating in clockwise direction. When the crank rotates, the piston moves towards right side and on the left side of the piston vacuum is created. This vacuum opens the suction valve and the liquid will be forced from sump to the left side of the piston. When the crank

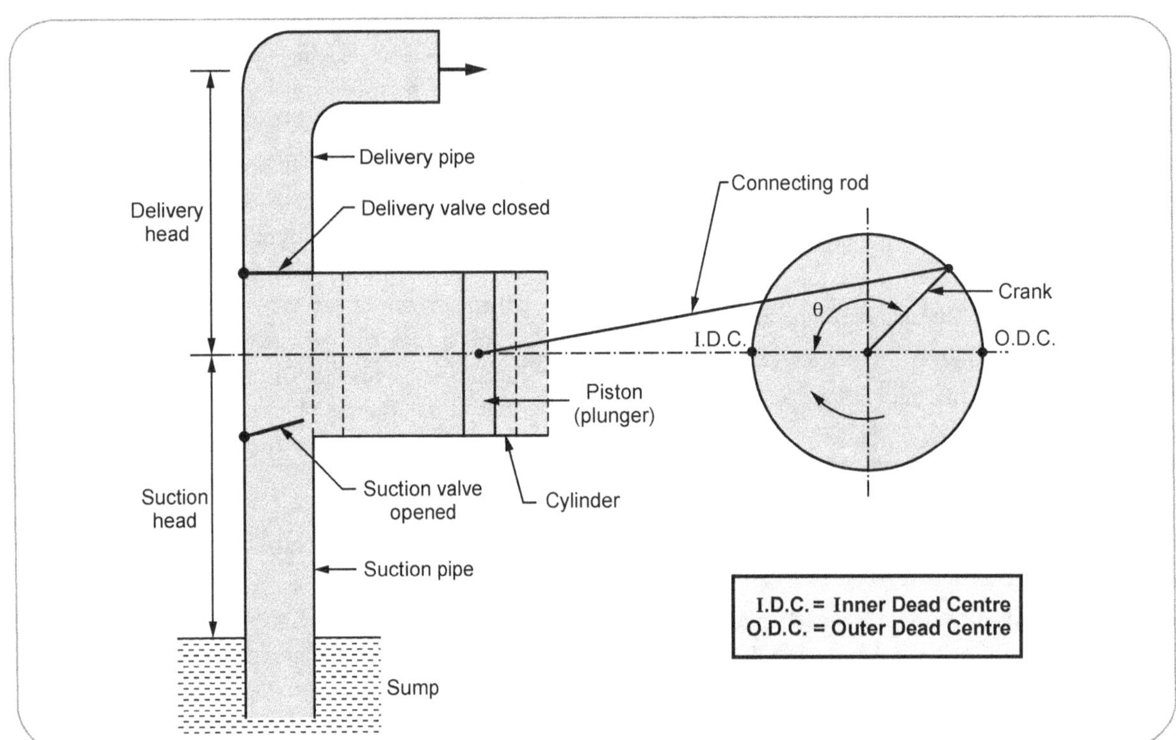

Fig. 2.6.1 Single acting reciprocating pump

reaches to ODC, the piston is on extreme right side the suction stroke is completed.

- At the end of suction stroke, the cylinder is full of liquid. When the crank rotates from ODC to IDC in clockwise direction, the liquid will be compressed and high pressure will be built in the cylinder. Because of high pressure delivery valve opens and liquid is delivered through the delivery pipe.

- At the end of delivery stroke, the crank is at IDC and piston is on extreme left position.

2.6.3 Applications of Reciprocating Pumps

SPPU : Dec.-11, 16

Reciprocating pumps are used in following applications :

- It is used as a feed water pump in boilers, hydraulic jacks, kerosene pumps, hand operated pumps, etc.

- It is used in industries and agriculture field.

- It is also used in service stations for pressure washing.

2.7 Centrifugal Pumps   **SPPU : Dec.-16, May-17**

- The hydraulic machines which convert mechanical energy into hydraulic energy are called as pumps. The hydraulic energy is generally in the form of pressure energy.

- If the mechanical energy is converted into pressure by using centrifugal force acting on the fluid, the hydraulic machine is known as **centrifugal pump**.

- It works on the principle of forced vortex flow which means that when a certain mass of liquid is rotated, the rise in pressure of the rotating liquid takes place.

- The main parts of a centrifugal pump are as follows (Refer Fig. 2.7.1) :

> 1. Impeller   2. Casing
> 3. Suction pipe   4. Delivery pipe

1. **Impeller :** It is a rotating part of a centrifugal pump. It consists of backward curved vanes. It is mounted on a shaft which is connected to the shaft of an electric motor.

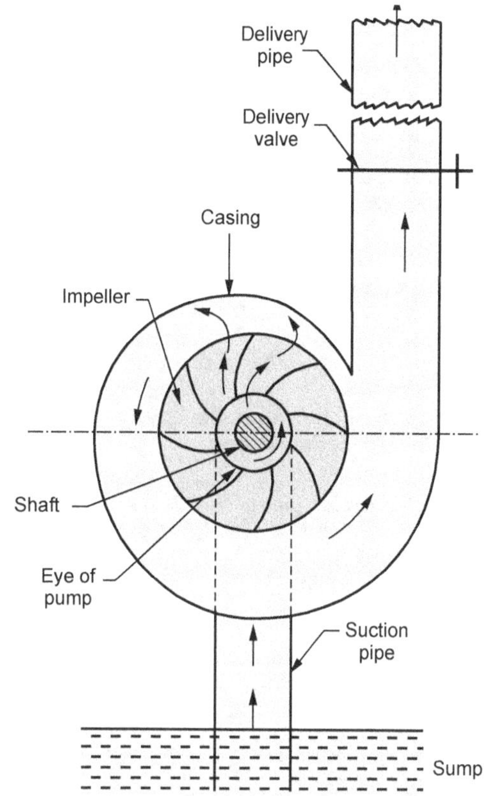

Fig. 2.7.1 Centrifugal pump

2. **Casing :** It is an air-tight passage surrounding the impeller. It is designed in such a way that the kinetic energy of the water discharged at the outlet of impeller is converted into pressure energy before the water leaves the casing and enters the delivery section. The casing is of spiral type in which the area of flow increases gradually. This increase in area reduces the velocity of flow and increases the pressure of the liquid flowing through the casing.

3. **Suction pipe :** It is a pipe whose one end is connected to the inlet of the pump and other end dips into the water sump. A non-return valve is fitted at the lower end of the suction pipe. It opens only in upward direction. To remove dust, dirt, etc. a strainer (filter) is also fitted at the lower end of the suction pipe.

4. **Delivery pipe :** It is a pipe whose one end is connected to the outlet of the pump and other end delivers the water at a desired height. To control the flow of liquid a delivery valve is connected to this pipe.

2.7.1 Applications of Centrifugal Pump

SPPU : Dec.-16

- It is used in domestic water supply.
- It is used in energy and oil refineries, power plants.
- It is used in building services for pressure boosting, fire protection sprinkler systems, air conditioners, etc.
- It is used in boiler feed applications, waste water management, irrigation, etc.
- It is also used in chemical and process industries.

2.7.2 Comparison between Centrifugal Pump and Reciprocating Pump

Sr. No.	Centrifugal pump	Reciprocating pump
1.	The flow rate is continuous and smooth.	The flow rate is fluctuating.
2.	It can supply large quantity of liquid.	It can supply small quantity of liquid only.
3.	These pumps run at high speed.	These pumps run at low speed.
4.	The working of these pumps is smooth and without much noise.	The working these pumps is complicated and with much noise.
5.	Cost of centrifugal pump is low.	Cost of reciprocating pump is high.
6.	The cost of maintenance and installation is also low.	The cost of maintenance and installation is also high.
7.	Efficiency of these pumps is high.	Efficiency of these pumps is low.
8.	It requires smaller floor area.	It requires larger floor area.

COMPRESSORS

2.8 Air Compressors

SPPU : Dec.-05,07,08,10,11,12, May-04,05,07,09,10,11,12,14,17

- In various industrial applications compressed air is required. A device or machine providing air at high pressure is called as **air compressor.**

- An air compressor takes in air at atmospheric pressure, compresses it and delivers the high pressure air to a storage vessel called as **receiver.**
- From receiver compressed air may be conveyed by the pipe-line to a place where the supply of compressed air is required.

Applications of compressed air　**SPPU : Dec.-08, 10**

The compressed air is used in various industrial applications some of them are as follows :

- It is used to operate pneumatic drills, air motors, hammers, riveting and nut tightening, etc.
- Cleaning of workshops and automobiles.
- Supercharging of I.C. engines and in gas turbine power plants.
- In paint industries for spray painting.
- In refrigeration and air-conditioning industry.
- In construction roads dams, tunnels, bridges, etc.
- For spraying fuel in high speed diesel engines.
- For driving mining machinery.
- For conveying sand materials, concrete, etc.
- For operating air brakes.
- In paper industries and printing machinery.

2.8.1 Classification of Air Compressors

SPPU : May-04, 07, 09

Air compressors may be classifed as follows :

1. According to the type of motion

a) **Reciprocating air compressors :** In these compressors air is compressed by the reciprocating action of piston in a cylinder. It provides high pressure air with intermittent discharge.

b) **Rotary air compressors :** These compressors have rotating element to compress the air. It provides the air at low pressure but in large and continuous quantity. These compressors are further classified as follows :

i) **Positive displacement compressors :** These compressors have two sets of mutually engaging surfaces. Air is trapped between these lobes and squeezing action takes place. For example : Roots blower, vane blower, etc.

ii) Non-positive displacement compressors : In these compressors kinetic energy is converted into pressure energy by the arrangement of impeller and fixed rings. For example : Centrifugal compressor, axial flow compressor, etc.

2. According to the number of stages

a) Single stage compressor : In these compressors, air is sucked and compressed in a single cylinder.

b) Multi-stage compressor : In these compressors, air is sucked and compressed in two or more cylinders. Generally, number of stages is equal to the number of cylinders.

3. According to the working position (side) of piston

a) Single acting compressors : In these compressors, air is sucked and compressed on one side of piston.

b) Double acting compressors : When suction and delivery of air takes place on both sides of piston then it is called as double acting compressor. It handles double the air than single acting air compressor.

4. According to the discharge pressure

a) Low pressure compressors : Delivery pressure is less than 10 bar.

b) Medium pressure compressors : Delivery pressure is between 10 to 80 bar.

c) High pressure compressors : Delivery pressure is greater than 80 bar.

5. According to the capacity of compressor

a) Low capacity compressors : Discharge capacity is less than $0.15 \text{ m}^3/\text{s}$.

b) Medium capacity compressors : Discharge capacity is between 0.15 to $5 \text{ m}^3/\text{s}$.

c) High capacity compressors : Discharge capacity is greater than $5 \text{ m}^3/\text{s}$.

2.8.2 Reciprocating Compressors

SPPU : May-05, 10, 11, 12, 14, 17, Dec.-05, 07, 08, 10, 11

- Fig. 2.8.1 shows the principal parts of reciprocating air compressor, which is a single stage and single acting type.

- In these compressors, crank is connected to prime mover or electric motor.

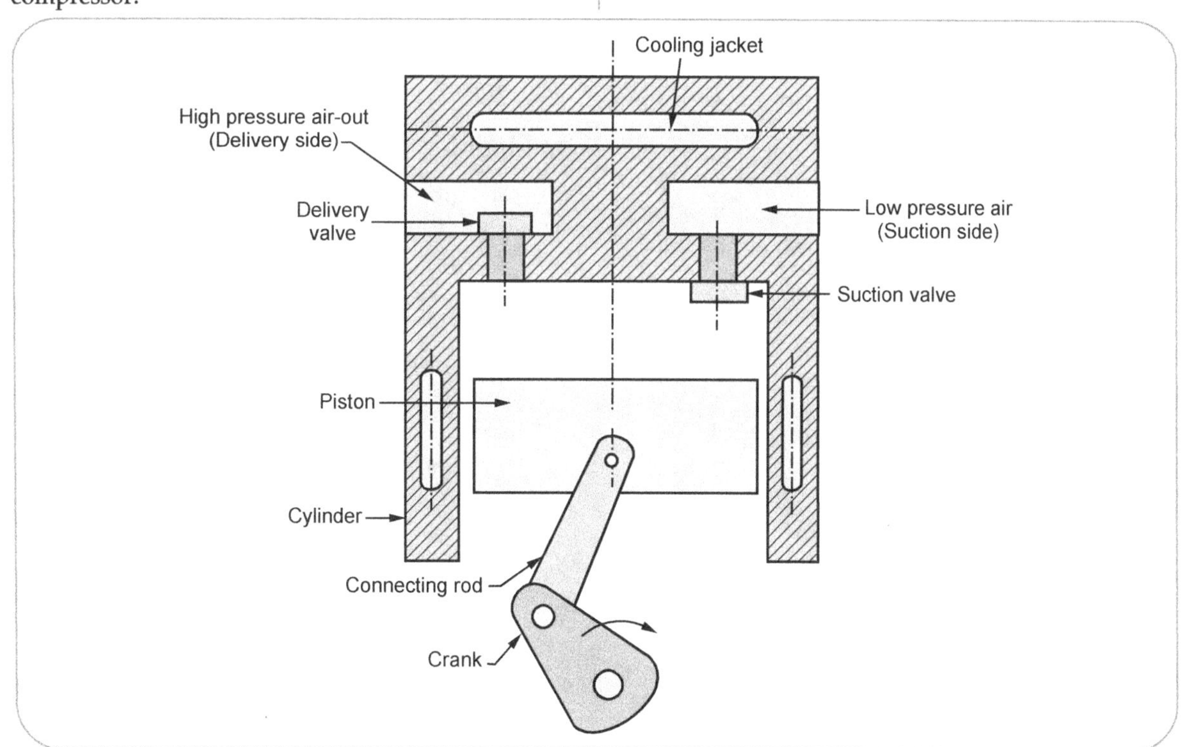

Fig. 2.8.1 Reciprocating air compressor

- The inlet and outlet valves are opened and closed by the pressure difference on both sides of valves.

- The operation of reciprocating compressor is divided in two strokes i.e. suction stroke and delivery stroke.

- During suction stroke, piston moves downwards due to which pressure in the cylinder falls below atmospheric and inlet valve opens and air is sucked.

- During delivery stroke, piston moves upwards with compression of air in the cylinder. At that time, both inlet and delivery valves are closed.

- At the end of this stroke pressure increases above the receiver pressure, hence delivery valve opens and air is discharged to the receiver. (Receiver is a vessel which acts as a storage tank).

- These pumps are widely used in industries, agriculture. They are suitable for high pressure heads with low flow rates.

2.8.3 Terminology of Air Compressor

SPPU : May-09, Dec.-12

i) **Free air delivery (F.A.D)** : It is the volume of air delivered under the conditions of pressure and temperature existing at compressor intake. Generally it is considered as 1.01325 bar pressure and 15 °C temperature.

ii) **Power capacity** : It is the quantity of the free air actually delivered by the compressor in m^3/min or m^3/s.

iii) **Pressure ratio** : It is defined as the ratio of outlet or discharge pressure to inlet or suction pressure.

iv) **Shaft power/Brake power** : It is the power required to drive the compressor or it is the power delivered to the compressor shaft.

2.9 Centrifugal Compressor   **SPPU : Dec.-16**

- The centrifugal compressors are used to supply large quantities of air but with low pressure ratio. It can develop pressure ratio upto 4 with single stage.

- A centrifugal compressor consists of a rotating impeller, a diffuser and a casing as shown in Fig. 2.9.1.

i) Impeller :

○ The impeller consists of a disc on which radial blades are attached. It can run at speeds of 20,000 to 30,000 rpm.

○ The air enters the eye of the compressor at low velocity and atmospheric pressure. The air moves radially outwards passing through the impeller and it is guided by the impeller vanes.

○ The impeller increases the momentum of air which results in rise in temperature as well as pressure of the air. (Refer Fig. 2.9.2)

○ The main function of impeller is to impart the energy to the air by rotating the blades. The rotation of impeller causes a static pressure rise alongwith increase in kinetic energy of the air.

ii) Diffuser :

○ The diffuser is an important part of the compressor which surrounds the impeller and provides diverging passages for air flow. This increases the pressure of air.

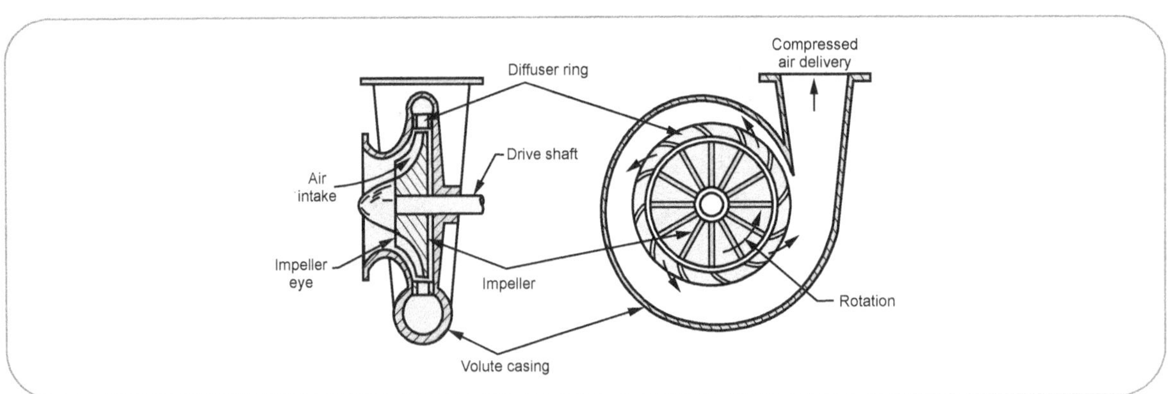

Fig. 2.9.1 : Centrifugal compressor

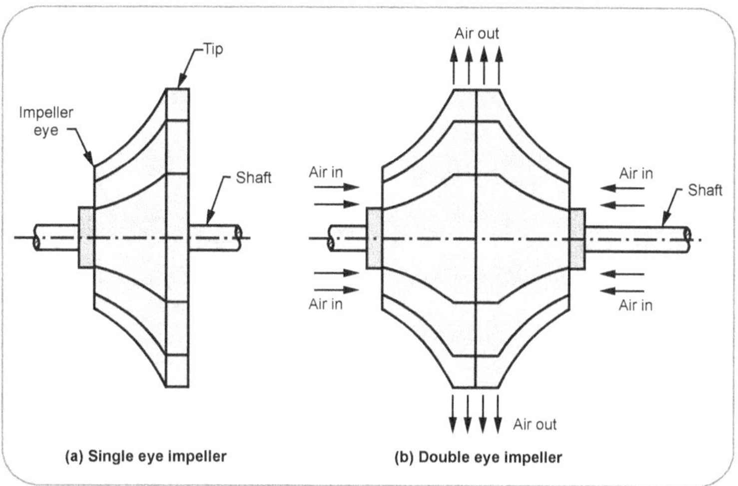

Fig. 2.9.2 : Types of impeller

○ The air leaving the impeller blades enters the diffuser where its velocity is reduced by providing more cross-sectional area for the air flow.

○ It causes to convert the kinetic energy into pressure energy by diffusion. Almost half of the total pressure rise is achieved in impeller and the remaining half in the diffuser.

iii) Volute casing :

○ The air coming out of the diffuser is collected in the casing and taken out from the outlet of the compressor.

○ The increased cross-sectional area of volute casing also causes small rise in pressure with reduction in the kinetic energy.

2.9.1 Advantages, Disadvantages and Applications of Centrifugal Compressor

Advantages

• Centrifugal compressors are simple in design and operation.

• They can operate efficiently over a wide range of mass flow at any particular speed.

• Centrifugal compressors are less expensive.

• It occupies less length as compared to axial compressors.

• Contaminated atmosphere does not affect the performance of the compressor.

Disadvantages

• It has low maximum efficiency.

• It needs large frontal area for given mass flow rates.

• It is not suitable for multistaging due to large losses in between the stages.

Applications

• These compressors are used in supercharging of I.C. engines.

• They are also used in jet engines.

WIND MILLS

2.10 Wind Mills

• A wind mill is a machine for wind energy conversion. A wind turbine converts the kinetic energy of the winds motion to mechanical energy transmitted by the shaft. A generator further converts it to electrical energy, there by generating electricity. There are two broad classifications of wind mills.

1. Horizontal axis type

2. Vertical axis type

2.10.1 Horizontal Axis Wind Mill

• The following are different types

1. Horizontal axis using two aerodynamic blades

2. Horizontal axis propeller type using single blade

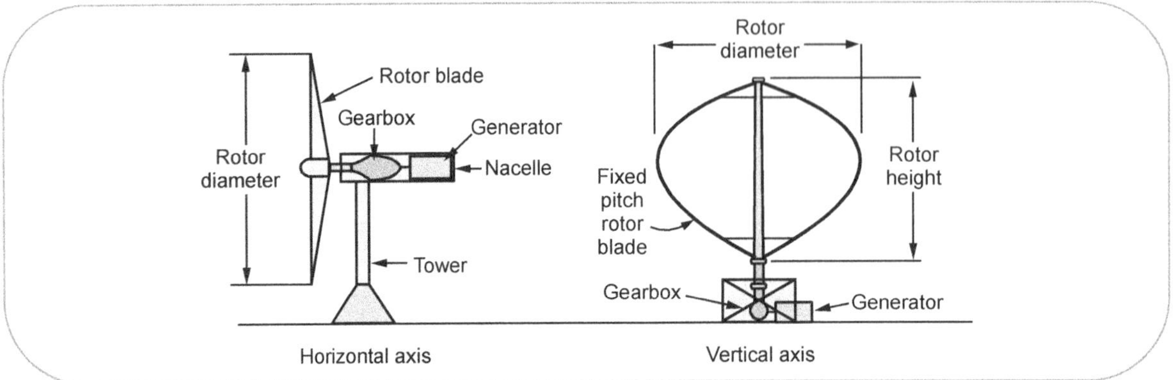

Fig. 2.10.1 : Wind turbine configurations

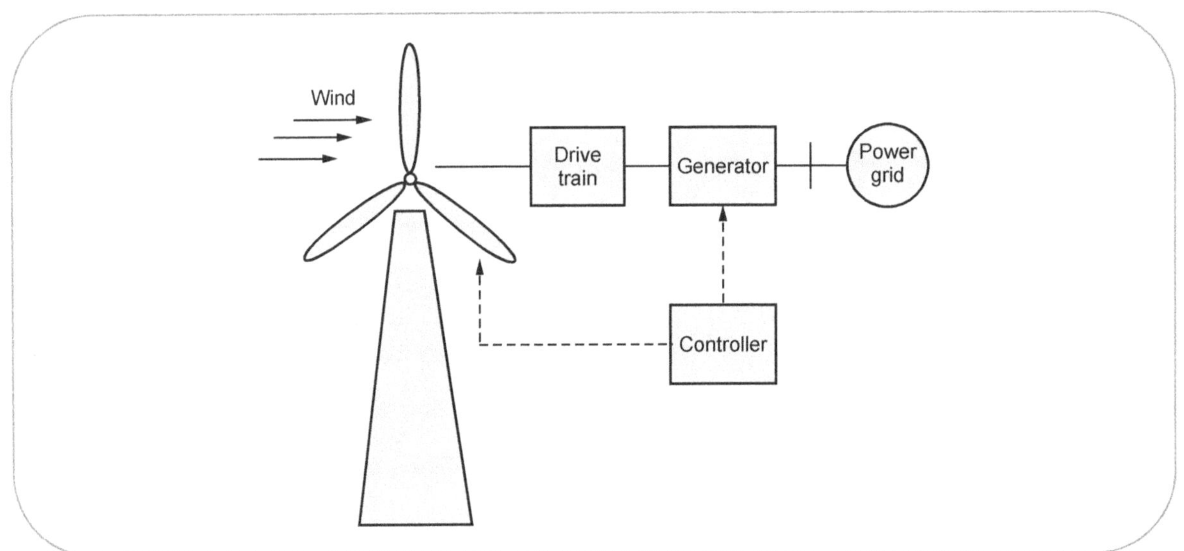

Fig. 2.10.2

3. Horizontal axis multi blade type

4. Horizontal axis wind mill Dutch type

5. Sail type

Horizontal axis type wind mills

A horizontal axis wind mill has its blades rotating on axis parallel to the ground.

1. Horizontal axis using two aerodynamic blades :

- In this type of design, rotor drives a generator through a step up gear box.
- The blade rotor is usually designed to be oriented downwind of the tower.

- The components are mounted on a bed plate which is attached on a pintle at the top of the tower.
- The rotor blades are continuously flexed by unsteady aerodynamic, gravitational and inertia loads, when the machine is in operation.

- If the blades are made of metal, flexing reduces their fatigue life. With rotor the tower is also subjected to above loads, which may cause serious damage.

- If the vibrational modes of the rotor happen to coincide with one of the natural mode of vibration of the tower, the system may shake itself to pieces.

- Because of the high cost of the blade rotors with more than two blades are not recommended.

- Rotors with more than two, say 3 or 4 blades would have slightly, higher power coefficient.

2. Horizontal axis propeller type using single blade :

- In this arrangement, a long blade is mounted on a rigid hub induction generator and gear box.

- If extremely long blades (above 60 m) are mounted on rigid hub, large blade root bending moments may occur due to tower shadow, gravity and sudden shifts in wind directions.

- To reduce rotor cost, use of low cost counter weight is recommended which balances long blade centrifugally.

3. Horizontal axis multi blade type :

- This type of design for multi blades made from sheet metal or aluminium.

- The rotors have high strength to weight rations and have been known to survive hours of freewheeling operation in 60 km/hr winds.

- They have good power coefficient, high starting torque and added advantage of simplicity and low cost.

4. Horizontal axis wind mill Dutch type :

- It is one of the oldest designs. The blade surface from an array of wooden slats which 'feather' at high wind speeds.

5. Sail type :

- It is of recent origin. The blade surface is made from cloth, nylon or plastics arranged as mast and pole or sail wings.

- There is also variation in number of sails used.

- The horizontal axis types generally have better performance.

- They have been used for various applications, but the two major areas of interest are electric power generation and pumping water.

2.11 Power and Efficiency Calculations

(A) Water turbines

- The important efficiencies of a water turbine are as follows :

> (a) Hydraulic efficiency
> (b) Mechanical efficiency
> (c) Overall efficiency

a) Hydraulic efficiency (η_h) :

- It is defined as the ratio of power developed by the runner of a turbine to the power supplied by the water at the entry of a turbine.

- The power at the entry of the turbine is more and it goes on decreasing as the water flows over the vanes of the turbine beacause the vanes are not so smooth. It is given as,

$$\eta_h = \frac{\text{Power developed by the runner}}{\text{Power supplied at the entry of turbine}}$$

$$\eta_h = \frac{\text{Runner Power (RP)}}{\text{Water Power (WP)}} \qquad \ldots (2.11.1)$$

b) Mechanical efficiency (η_m) :

- It is defined as the ratio of the power available at the shaft of a turbine to the power developed by the runner.
- Due to mechanical losses the power available at the shaft of a turbine is less than the power developed by the runner. It is given by,

$$\eta_m = \frac{\text{Power available at the turbine shaft}}{\text{Power developed by the runner}}$$

$$\therefore \qquad \eta_m = \frac{\text{Shaft Power (SP)}}{\text{Runner Power (RP)}} \qquad \ldots (2.11.2)$$

c) Overall efficiency (η_o) :

- It is defined as the ratio of the power available at the turbine shaft to the power supplied by the water at the entry of turbine.
- It is given by,

$$\eta_o = \frac{\text{Power available at the turbine shaft}}{\text{Power supplied at the entry of turbine}}$$

$$\eta_o = \frac{\text{Shaft Power (SP)}}{\text{Water Power (WP)}} \qquad \ldots (2.11.3)$$

Multiplying and dividing by Runner Power (RP)

$$\therefore \qquad \eta_o = \frac{\text{SP}}{\text{WP}} \times \frac{\text{RP}}{\text{RP}} = \frac{\text{SP}}{\text{RP}} \times \frac{\text{RP}}{\text{WP}}$$

$$\therefore \qquad \eta_o = \eta_m \times \eta_h \qquad \ldots (2.11.4)$$

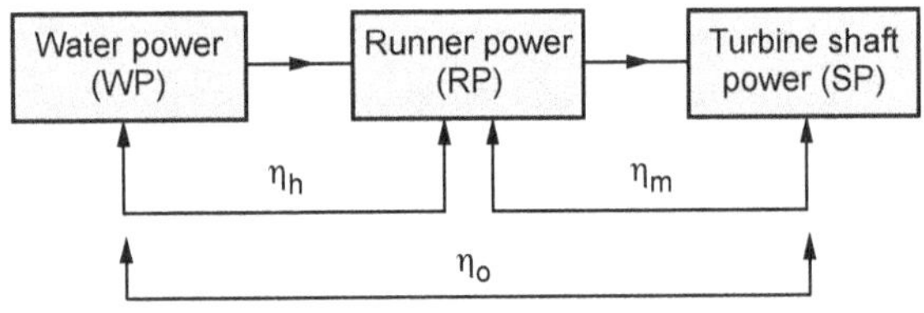

Fig. 2.11.1 : Efficiencies of water turbines

Ex. 2.11.1 : *A pelton wheel works under a water power of 1.3 MW. The hydraulic efficiency of turbine is 95 %. Find runner power developed. If mechanical efficiency of turbine is 90 % find overall efficiency and shaft power of turbine.*

Sol. : Given data :

$$WP = 1.3 \text{ MW} = 1.3 \times 10^6 \text{ Watts,}$$

$$\eta_h = 95 \text{ \%, } \eta_m = 90 \text{ \%}$$

To find : i) RP   ii) η_o iii) SP

Hydraulic efficiency is,

$$\eta_h = \left(\frac{\text{RP}}{\text{WP}}\right) \times 100$$

$$\therefore \qquad 95 = \left(\frac{\text{RP}}{1.3 \times 10^6}\right) \times 100$$

$$\therefore \qquad \mathbf{RP = 1.235 \times 10^6 \text{ W} = 1.235 \text{ MW} \quad ...Ans.}$$

Overall efficiency is,

$$\eta_o = \eta_h \times \eta_m = 0.95 \times 0.9$$

$$\therefore \qquad \mathbf{\eta_o = 0.855 = 85.5 \text{ \%}} \qquad \text{...Ans.}$$

Overall efficiency is given by,

$$\eta_o = \left(\frac{\text{SP}}{\text{WP}}\right) \times 100$$

$$\therefore \qquad 85.5 = \left(\frac{\text{SP}}{1.3 \times 10^6}\right) \times 100$$

$$\therefore \qquad \mathbf{SP = 1.1115 \times 10^6 \text{ W}}$$

$$\mathbf{= 1.1115 \text{ MW}} \qquad \text{...Ans.}$$

Ex. 2.11.2 : *A pelton turbine receives water from penstock and develops power of 10,000 kW. If overall efficiency is 88 % find water power supplied to turbine. Also find mechanical efficiency if runner power of wheel is 11 MW.*

Sol. : Given data :

$$SP = 10,000 \text{ kW} = 10 \times 10^6 \text{ Watts,}$$

$$\eta_o = 88 \text{ \%,}$$

$$RP = 11 \text{ MW} = 11 \times 10^6 \text{ Watts}$$

To find : i) WP   ii) η_m

Overall efficiency is,

$$\eta_o = \left(\frac{\text{SP}}{\text{WP}}\right) \times 100$$

$$\therefore \quad 88 = \left(\frac{10\times10^6}{WP}\right)\times100$$

$$\therefore \quad WP = 11.3636\times10^6 \ W$$

$$= 11.3636 \ MW \qquad ...Ans.$$

Mechanical efficiency is,

$$\eta_m = \left(\frac{SP}{RP}\right)\times100 = \left(\frac{10\times10^6}{11\times10^6}\right)\times100$$

$$\eta_m = 90.9 \ \% \qquad ... Ans.$$

Ex. 2.11.3 : *In an Impulse water turbine water power is 150 kW. Out of this 85 % power is used to drive the runner. If mechanical losses (in bearing and all) are 20 % , find overall efficiency and shaft power of turbine.*

Sol. : Given data :

$$WP = 150 \ kW = 150\times10^3 \ Watts$$

$$RP = 85 \ \% \ WP$$

$$\therefore \quad \frac{RP}{WP} = 85 \ \% = \eta_h$$

Mechanical losses = 20%

$$\therefore \quad \text{Mechanical efficiency } \eta_m = 100 - 20 = 80 \ \%$$

To find : i) Overall efficiency η_o ii) SP

Overall efficiency is,

$$\eta_o = \eta_h \times \eta_m = 0.85 \times 0.8$$

$$\therefore \quad \eta_o = 0.68 = 68 \ \% \qquad ... Ans.$$

It is also given as,

$$\eta_o = \left(\frac{SP}{WP}\right)\times100$$

$$\therefore \quad 68 = \left(\frac{SP}{150\times10^3}\right)\times100$$

$$\therefore \quad SP = 102\times10^3 \ watts = 102 \ kW \ \ ... Ans.$$

(B) Centrifugal pumps

- Basically, pump is a power consuming and a turbine is power producing device. So the efficiency terms are opposite of turbine terms.

- The efficiencies of a centrifugal pump are as follows :

> (a) Manometric efficiency
> (b) Mechanical efficiency
> (c) Overall efficiency

a) Manometric efficiency (η_{ma}) :

- It is defined as the ratio of actual power delivered by the pump (water power) to the power delivered by impeller (Impeller power).

$$\eta_{ma} = \left(\frac{\text{Water Power (WP)}}{\text{Impeller Power (IP)}}\right)\times100$$

b) Mechanical efficiency (η_m) :

- It is defined as the ratio of the power delivered by the impeller (Impeller power) to the power input to the pump shaft (Shaft power).

$$\eta_m = \left(\frac{\text{Impeller Power (IP)}}{\text{Shaft Power (SP)}}\right)\times100$$

c) Overall efficiency (η_o) :

- It is defined as the ratio of the output power of the pump (water power) to the input power of pump shaft (SP).

$$\eta_o = \left(\frac{\text{Water Power (WP)}}{\text{Shaft Power (SP)}}\right)\times100$$

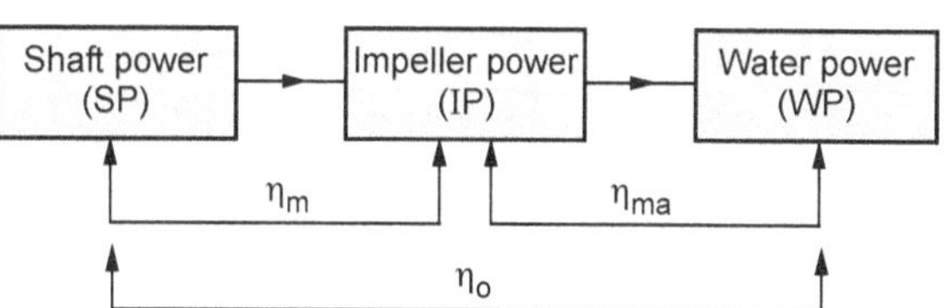

Fig. 2.11.2 : Efficiencies of centrifugal pump

Ex. 2.11.4 : *In a centrifugal pump work done by impeller is 346 kW and manometric efficiency is 60%. Find power delivered by the pump. If mechanical efficiency is 75%, find input power of pump shaft.*

Sol. : Given data :

$IP = 346 \ kW = 346 \times 10^3 \ Watts, \ \eta_{ma} = 60 \ \%,$
$\eta_m = 75 \ \%$

To find : i) WP ii) SP

Manometric efficiency is,

$$\eta_{ma} = \left(\frac{\text{Water power}}{\text{Impeller power}}\right) \times 100$$

$$\therefore \quad 60 = \left(\frac{\text{WP}}{346 \times 10^3}\right) \times 100$$

$$\therefore \quad \text{WP} = 207.6 \times 10^3 \text{ Watts} = 207.6 \text{ kW}$$

...Ans.

Mechanical efficiency is,

$$\eta_m = \left(\frac{\text{Impeller power}}{\text{Shaft power}}\right) \times 100$$

$$\therefore \quad 75 = \left(\frac{346 \times 10^3}{\text{SP}}\right) \times 100$$

$$\therefore \quad \text{SP} = 461.333 \times 10^3 \text{ Watts}$$

$$= 461.333 \text{ kW} \quad \text{...Ans.}$$

Ex.2.11.5 : *For a centrifugal pump, power required to drive the pump is 15 kW. The overall efficiency of pump is 62 %. Find output power of the pump. Also, find mechanical efficiency of pump if manometric efficiency is 75 %.*

Sol. : Given data :

SP = 15 kW = 15×10^3 Watts, η_0 = 62 %, η_{ma} = 75 %

To Find : i) WP ii) η_m

Overall efficiency of pump is,

$$\eta_0 = \left(\frac{\text{WP}}{\text{SP}}\right) \times 100$$

$$\therefore \quad 62 = \left(\frac{\text{WP}}{15 \times 10^3}\right) \times 100$$

$$\therefore \quad \text{WP} = 9300 \text{ Watts} = 9.3 \text{ kW} \quad \text{...Ans.}$$

It is also given as,

$$\eta_0 = \eta_m \times \eta_{ma} \quad \therefore 0.62 = \eta_m \times 0.75$$

$$\therefore \quad \eta_m = \frac{0.62}{0.75} = 0.8266 = 82.66 \% \quad \text{...Ans.}$$

(C) Wind mills

The electric power generated from wind mill is given by,

$$P = \frac{\pi}{2} r^2 \times V^3 \times \rho \times \eta \qquad \text{...(2.11.5)}$$

where,   P = Power generated in watts

     r = Radius or length of blade in m

     V = Wind speed in m/sec

     ρ = Density of air in kg/m^3

     η = Efficiency factor in %

Ex. 2.11.6 : *An offshore wind turbine has radius of 80 m at a wind speed of 15 m/s. If density of air is 1.2 kg/m^3 and efficiency factor is 40 % find power generated by wind mill.*

Sol. : Given data :

$$r = 80 \text{ m}, V = 15 \text{ m/s}, \rho = 1.2 \text{ kg}/\text{m}^3,$$

$$\eta = 40 \% = 0.4$$

To find : Power P

Power developed by wind mill is,

$$P = \frac{\pi}{2} \times r^2 \times V^3 \times \rho \times \eta$$

$$= \frac{\pi}{2} \times 80^2 \times 15^3 \times 1.2 \times 0.4$$

$$\therefore \quad P = 16.286 \times 10^6 \text{ Watts}$$

$$= 16.286 \text{ MW (Megawatts)} \quad \text{... Ans.}$$

(D) Reciprocating compressors

The important efficiencies of a reciprocating compressor are as follows :

> (a) Polytropic efficiency
> (b) Isothermal efficiency
> (c) Mechanical efficiency

a) Polytropic efficiency (η_p)

- It is the ratio of polytropic work supplied to the actual work supplied.

$$\eta_p = \frac{\text{Polytropic work supplied (WD}_p)}{\text{Actual work supplied (WD}_a)} \times 100$$

b) Isothermal efficiency (η_{iso})

• It is the ratio of isothermal work to actual work.

$$\eta_{iso} = \frac{\text{Isothermal work or Ideal work (WD}_i)}{\text{Actual work supplied (WD}_a)} \times 100$$

c) Mechanical efficiency (η_m)

• It is the ratio of indicated power to the brake power of compressor.

$$\eta_m = \frac{\text{Indicated power (IP)}}{\text{Brake power (BP)}} \times 100$$

Ex. 2.11.7 : *In a single acting single cylinder air compressor the following data is available :*
Indicated power = 5.8 kW
Brake power = 6.25 kW
Work supplied actual = 6.2 kW
Polytropic work = 6 kW
Isothermal work = 5.5 kW
Calculate : a) Isothermal efficiency   b) Mechanical efficiency c) Polytropic efficiency

Ans. : Given :

IP = 5.8 kW, BP = 6.25 kW, WD_a = 6.2 kW,

WD_p = 6 kW, WD_i = 5.5 kW

To find : a) η_{iso} b) η_m c) η_p

Isothermal efficiency is,

$$\eta_{iso} = \left(\frac{WD_i}{WD_a}\right) \times 100 = \left(\frac{5.5}{6.2}\right) \times 100$$

$\therefore$ η_{iso} = **88.709 %**   **... Ans.**

Mechanical efficiency is,

$$\eta_m = \left(\frac{IP}{BP}\right) \times 100 = \left(\frac{5.8}{6.25}\right) \times 100$$

$\therefore$ η_m = **92.5 %**   **... Ans.**

Polytropic efficiency is,

$$\eta_p = \left(\frac{WD_i}{WD_a}\right) \times 100 = \left(\frac{6}{6.2}\right) \times 100$$

$\therefore$ η_p = **96.77 %**   **... Ans.**

Review Questions

1. What do you mean by power producing and power absorbing devices?
2. What are the different power producing devices?
3. What is steam turbine? What are its types?
4. Explain impulse and reaction turbine with neat sketch.
5. What are the advantages of steam turbine?
6. Compare impulse and reaction steam turbine. (at least 6 points).
7. Explain the classification of gas turbines.
8. Explain the open cycle and closed cycle gas turbine (constant pressure type).
9. Compare open cycle and closed cycle gas turbines.
10. Compare gas turbine and steam turbine.
11. Explain the applications of gas turbines.
12. Classify hydraulic turbines.
13. Explain following types of hydraulic turbines :
    i) Kaplan turbine   ii) Francis turbine
    iii) Pelton wheel
14. Compare impulse and reaction hydraulic turbines.
15. What do you mean by pump? Classify it.
16. Explain the working of single acting reciprocating pump with neat sketch.
17. How double acting reciprocating pump differs from single acting?
18. What are the applications of compressed air?
19. How compressors are classified? Explain.
20. Explain any one rotary compressor with neat sketch.
21. Explain the construction and working of reciprocating air compressor.
22. What is wind will ? What are the types of wind mill ? Explain.

❑❑❑

Notes

Unit - II
Introduction to Thermal Engineering

3

Thermodynamics

Syllabus

Laws of thermodynamics, heat engine, heat pump, refrigerator (simple numerical).

Contents

Mind Map - Thermodynamics

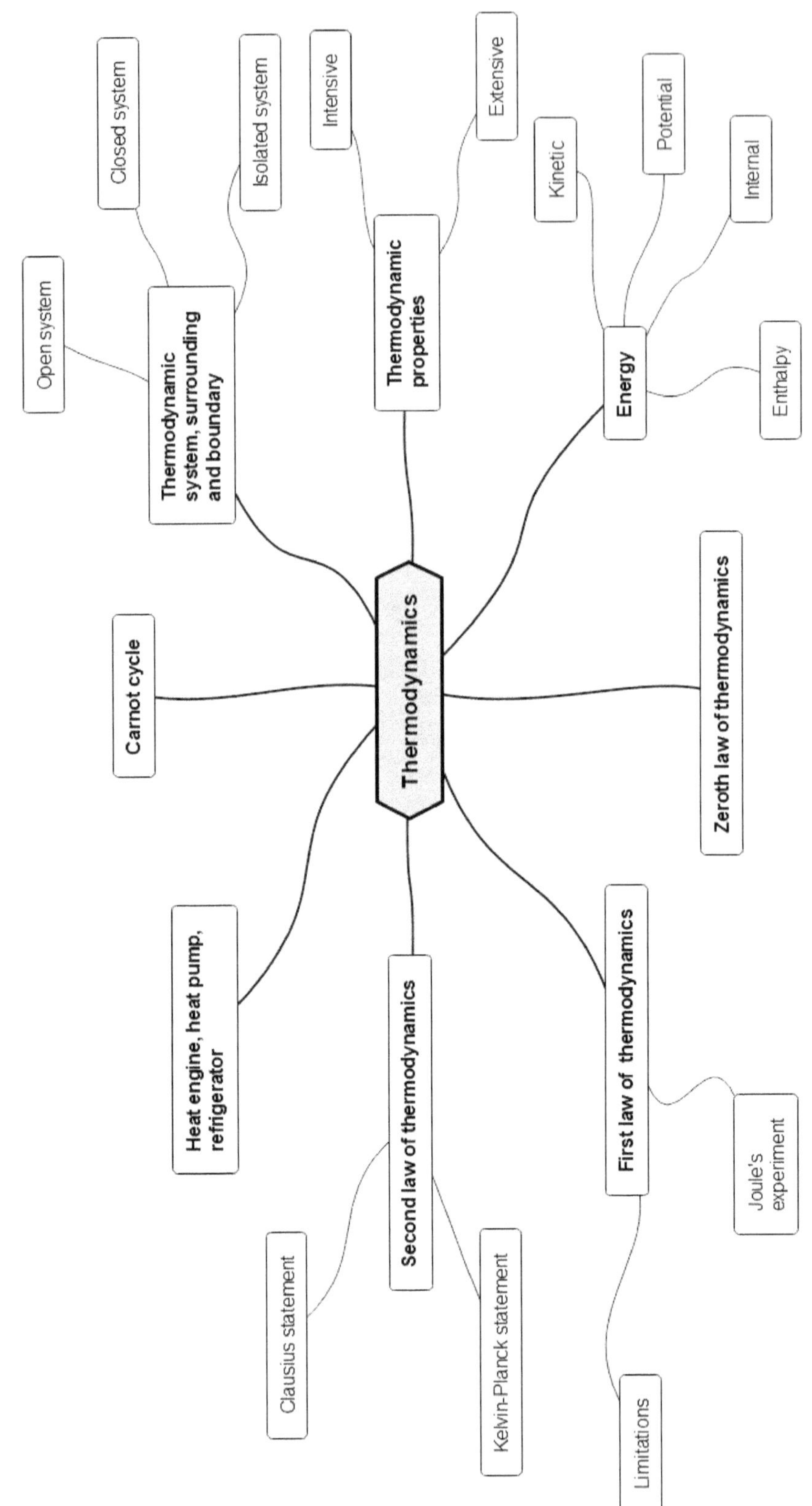

3.1 : Introduction

- **Thermodynamics** is the science which deals with energy interaction (heat and work) and the effect of energy interaction on the properties of a system.

- The study of thermodynamics was started with the need to convert heat into mechanical work and to produce desired changes in a system by means of exchange of heat or work.

- The concept of thermodynamics is used in domestic as well as industrial applications.

- In domestic applications it is used in referigerators, air-conditioning systems, ovens, pressure cooker, pump, computer, TV, etc. whereas in industrial applications it is used in the design of I.C. engines, jet engines, power plants, turbines, etc.

- Thermodynamics is basically concerned with four laws known as zeroth law, first law, second law and third law of thermodynamics.

1. **Zeroth law :** This law deals with the thermal equilibrium and establishes a concept of temperature.

2. **First law :** This law deals with the law of conservation of energy and the concept of internal energy.

3. **Second law :** It deals with the direction of flow of heat and the limit of converting heat into work. It also introduces the concept of entropy.

4. **Third law :** It deals with the absolute zero of entropy.

3.2 :   Thermodynamic System, Surroundings and Boundary

SPPU : Dec.-09,10, 16, 17, May-17, 18, 19

- A thermodynamic **system** is defined as a region in space or a quantity of matter upon which particular attention is concentrated for the study of work and heat transfer and their conversions.

- Everything outside the system which has direct effect on the behaviour of sytem is called as **surroundings.** Refer Fig. 3.2.1.

- The system is separated from the surroudings by an envelope which is called as **boundary.**

- The boundary may be real physical surface or some imaginary surface. It may be fixed or moving.

- When a system and surroundings are put together, then it is called as a **universe.**

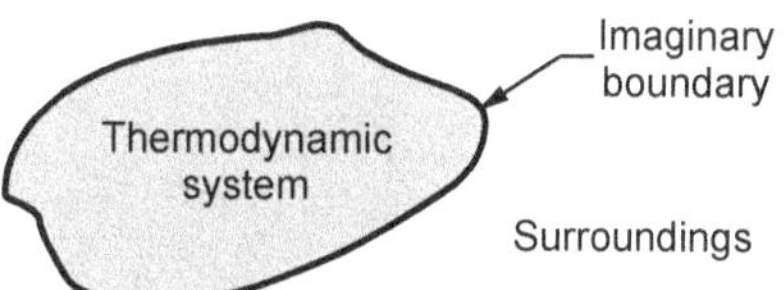

Fig. 3.2.1 Thermodynamic system, surroundings and boundary

- In a thermodynamic system, it is not necessary that the volume or shape should remain fixed.

3.2.1 Types of Thermodynamic Systems

SPPU : Dec.-03,08,09,12,13,17,18, May-04,13,14,17,18,19

According to the *mass* and *energy* transfer between the system and surroundings, the thermodynamic systems are classified as follows :

1. Closed system or non-flow system.
2. Open system or flow system.
3. Isolated system.

1. Closed system or non-flow system

- A system is called as closed system when the mass within the boundary of the system remains constant and only the energy (heat and work) may transfer across the boundary.

- For example, consider a piston cylinder arrangement of an engine as shown in Fig. 3.2.2 in which gas is enclosed between the piston and cylinder.

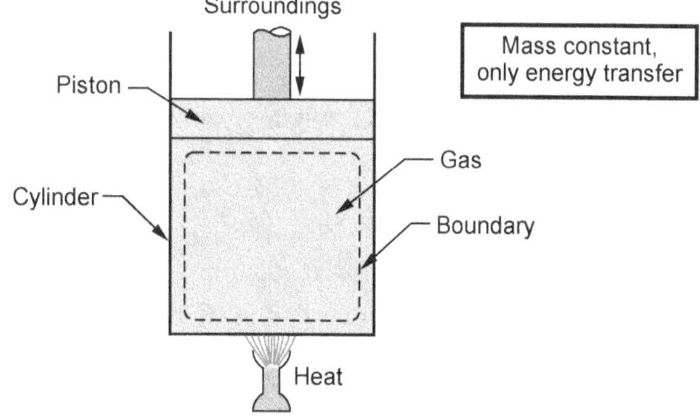

Fig. 3.2.2 Closed system

- The mass of the gas is constant within the system eventhough the gas may be expanded or compressed, as it is closed from all sides.

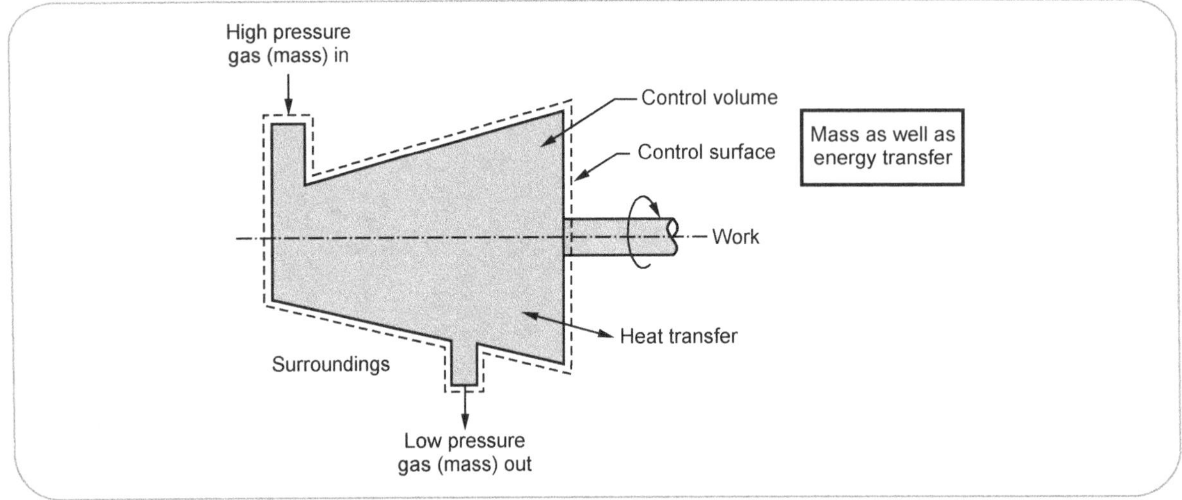

Fig. 3.2.3 Open system

- Hence, there is only transfer of energy (heat and work) takes place across the boundary of system.

- **Other examples** are pressure cooker, hot water stored in a tank, cup of tea, etc.

2. Open system or flow system

- If mass as well as energy (heat and work) transfers across the boundary of system, then it is called as **open system.**

- For example, consider a turbine in which both mass as well as energy transfer takes place. Refer Fig. 3.2.3.

- The fixed region through which mass and energy flow takes place is called as *control volume* and the surface of control volume is called as *control surface.*

- **Other examples** are I.C. engines, air compressors, etc.

3. Isolated system

- When there is no transfer of mass as well as energy across the boundary of system, it is called as **isolated system.**

- For example, consider an enclosed vessel containing gas as shown in Fig. 3.2.4.

- Another example of isolated system is thermo flask.

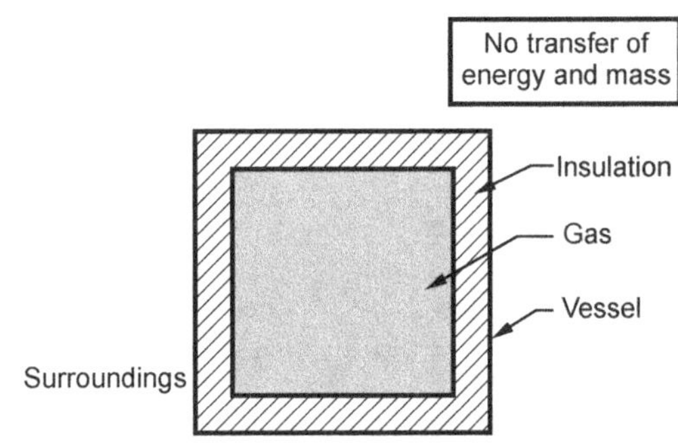

Fig. 3.2.4 Isolated system

3.3 : Thermodynamic Properties

SPPU : Dec.-18

- Any characteristic of the substance which can be measured or observed is known as **property** of the substance. For example, pressure, temperature, volume, etc.

- The thermodynamic properties are divided in following two categories :

  1. Intensive properties.
  2. Extensive properties.

1. **Intensive properties :** The properties which are independent of mass of the system are called as **intensive properties.** For example, pressure, temperature, etc.

2. **Extensive properties :** The properties which are dependent of mass of the system are called as **extensive properties.** For example, volume, energy, enthalpy, etc.

3.4 : Energy and Forms of Energy

SPPU : Dec.-06, 09, 10, 12

- Energy is defined as capacity to do work. It is a scalar quantity that cannot be observed directly but can be recorded and found by indirect measurements.

- Energy input or removal from a substance causes change in properties of substance.

- Any one form of energy can be converted into any other form.

- Energy can also be stored in different forms. The different forms are as follows :

> 1. Potential energy
> 2. Kinetic energy
> 3. Internal energy

1. **Potential energy :** Potential energy of a system is energy stored in the system due to its position in a gravitational field. Let 'z' be the distance of mass 'm' from the reference (datum). If 'F' is the force acting on the mass, then

$$d(P.E.) = F \cdot dz$$

$$\therefore \quad d(P.E.) = m \cdot g \cdot dz \quad \ldots (\because F = mg)$$

$$\therefore \quad \int_1^2 d(P.E.) = m \cdot g \int_1^2 dz$$

$$\therefore (P.E.)_2 - (P.E.)_1 = m \cdot g \, (z_2 - z_1)$$

$$\therefore \quad \Delta P.E. = m \cdot g \, (z_2 - z_1), \text{ J or N-m} \quad \ldots (3.4.1)$$

2. **Kinetic energy :** Kinetic energy of a system is the energy which arises from the motion of a mass. If mass 'm' is moving through a distance 'dx' then the workdone is $'F \cdot dx'$.

$$\therefore \quad d(K.E.) = F \cdot dx = m \cdot a \cdot dx \quad \ldots (\because F = ma)$$

$$\therefore \quad d(K.E.) = m \cdot \frac{dV}{dt} \cdot dx \quad \ldots \left(\because a = \frac{dV}{dt} \right)$$

$$\therefore \quad d(K.E.) = m \cdot \frac{dx}{dt} \cdot dV = m \cdot V \cdot dV \ldots \left(\because V = \frac{dx}{dt} \right)$$

$$\int_1^2 d(K.E.) = m \int_1^2 V \cdot dV$$

$$(K.E.)_2 - (K.E.)_1 = \frac{m}{2} \, (V_2^2 - V_1^2), \text{ J or N-m}$$

$$\therefore \quad \Delta K.E. = \frac{1}{2} m \, (V_2^2 - V_1^2), \text{ J or N-m}$$

$$\ldots (3.4.2)$$

3. **Internal energy :**

- It is the energy arising in the form of motions of molecules and atoms. It is the sum of the kinetic energy of the individual atoms or molecules. It is denoted by 'U'.

- The total energy of a system is the summation of potential energy, kinetic energy and internal energy, i.e.

Total energy, E = P.E. + K.E. + U

$$\therefore \quad E = m \cdot g \cdot z + \frac{1}{2} m \cdot V^2 + U \quad \ldots (3.4.3)$$

3.4.1 Enthalpy

SPPU : May-12, Dec.-12

- Enthalpy is defined as the energy which is the algebraic sum of internal energy and flow work.

  Mathematically,

$$\boxed{h = U + p \cdot V} \quad \ldots (3.4.4)$$

where,   h = Enthalpy in J

     U = Internal energy in J

     $p \cdot V$ = Flow work in J

- Enthalpy is also called as **total heat** or **heat content.**

- The above equation (3.4.4) is applicable for unit mass gas hence **enthalpy (h) is an extensive property.**

3.5 : Zeroth Law of Thermodynamics

SPPU : May-07,10,11, Dec.-08

- Zeroth law of thermodynamics states that *"if two systems are in thermal equilibrium with a third system, then they are in thermal equilibrium with each other."*

- For understanding of the statement, consider three bodies A, B and C which are at different temperatures.

- If A and C are brought into good contact, then the energy in the form of heat will transfer from high temperature body to low temperature body.

- After some time bodies A and C are in thermal equilibrium. Refer Fig. 3.5.1.

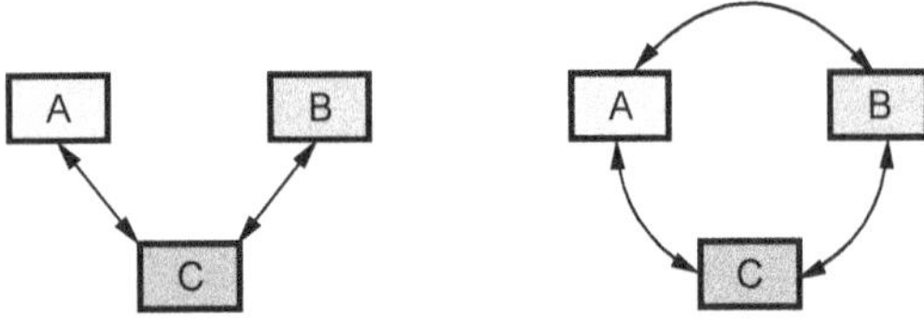

Fig. 3.5.1 Zeroth law of thermodynamics

- Now, if B and C are brought in good contact, then after some time they are also in thermal equilibrium.

- Hence, bodies A and B are also in thermal equilibrium.

> **Note :**
> - When a system is isolated from its surroundings then after some time it reaches a steady state.
> - The steady state implies a state of balance or thermodynamic equilibrium.
> - It means in thermodynamic equilibrium, values of all thermodynamic properties of a system remains constant or the system satisfies the condition of mechanical, thermal and chemical equilibrium.
> - A system is said to be in **mechanical equilibrium** when there are no unbalanced forces (pressure forces) in the system from its surroundings.
> - If the temperature of the system does not change with respect to time and it has the same value at all points then the system is said to be in **thermal equilibrium.**

3.6 : Joule's Experiment

SPPU : Dec.-05, 08, May-10

- The English scientist J. P. Joule conducted experiments which were the first step in the quantitative analysis of thermodynamics and it led to the **first law of thermodynamics.**

- In this experiment, consider a closed system with known mass of water contained in an insulated vessel having a paddle wheel as shown in Fig. 3.6.1.

- The work is supplied across the boundary of the system by a falling weight W.

- The system formed by the liquid in a container is a closed system which is initially at temperature T_1.

- The weight is allowed to fall through a height 'Z' which causes rotation of paddle wheel.

- The work input to the paddle wheel is measured by the fall of weight (potential energy).

- Then the system was placed in the water bath so that the heat was transferred from the fluid to the water until the original state of fluid was re-established. (indicated by thermometer). Refer Fig. 3.6.1.

- In this way, the system was undergone through a cycle.

- It was observed that, the amount of heat rejected by the fluid to the water was directly proportional to the increase of energy of the water bath which could be found by measuring the increase in temperature of the water bath.

- By this experiment, Joule concluded that the net work input W to the system was proportional to the

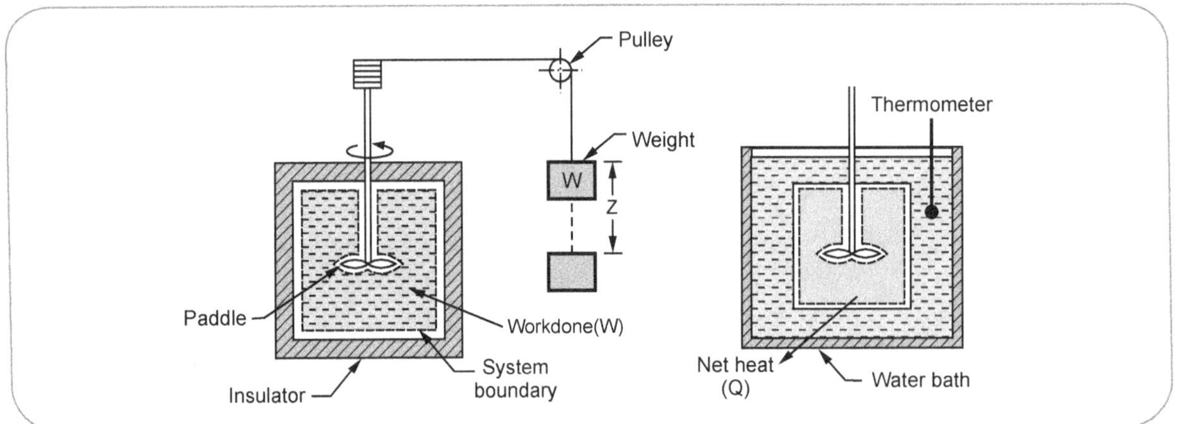

Fig. 3.6.1 Joule's experiment

net heat Q transferred to the system at the end of a cycle. Refer Fig. 3.6.2.

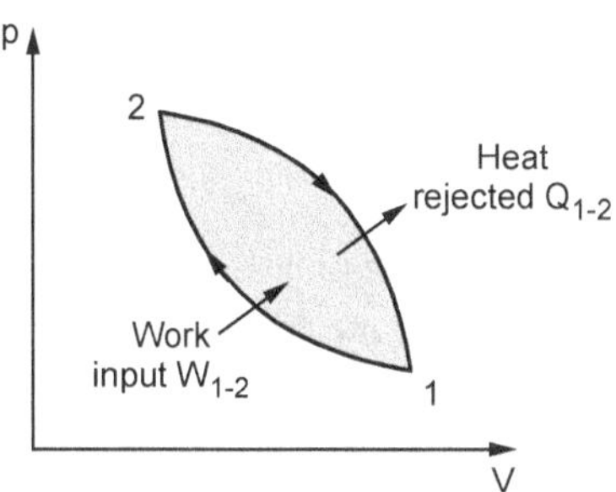

Fig. 3.6.2 Representation of Joule's experiment on p-V diagram

- For cyclic integral $\oint$ sign is used, hence

$$\oint dQ \propto \oint dW$$

$$\therefore \qquad \oint dQ = \frac{1}{J} \oint dW$$

$$\therefore \qquad \oint dW = J \oint dQ \qquad \qquad ...(3.6.1)$$

Where, J is a constant and it is called as **mechanical equivalent of heat.**

- In S.I. system of units 1 N-m = 1 J.

- The equation (3.6.1) indicates that, the algebraic sum of the work and heat interactions during a cycle is zero.

3.6.1 : First Law of Thermodynamics
SPPU : Dec.-05, 07, 16, 17, May-08, 13, 14, 17

The statement of first law of thermodynamics in different forms can be stated as follows :

a) When system undergoes a cyclic change, then the algebraic sum of work delivered to the surrounding is directly proportional to the algebraic sum of heat taken from the surrounding i.e.

$$\oint dW \propto \oint dQ$$

b) Both heat and work are mutually convertible one into another i.e. the energy can neither be created nor destroyed but converted from one form to another.

c) There is no any machine which is capable of producing work without expenditure of energy.

3.7 : Corollaries of First Law of Thermodynamics

The first law of thermodynamics has the various important consequences which are stated in the form of corollaries as follows :

3.7.1 Corollary 1 (For a Process)
SPPU : Dec.-03, May-09

> i) Corollary 1 (For a process)
> ii) Corollary 2 (For an isolated system)
> iii) Corollary 3 (PMM-I)

- The property of a closed system is such that a change in total energy (ΔE) is equal to the difference between the heat supplied (Q) and the workdone (W) during the change of state. [From statement (a) of first law of thermodynamics].

$$\therefore \qquad \Delta E = Q - W \qquad \qquad ...(3.7.1)$$

Proof

- Consider a cyclic change of a system as shown in Fig. 3.7.1.

- Let, the system proceeds from state 1 to 2 along the path A and then brought back to state 1 along the either path B or C.

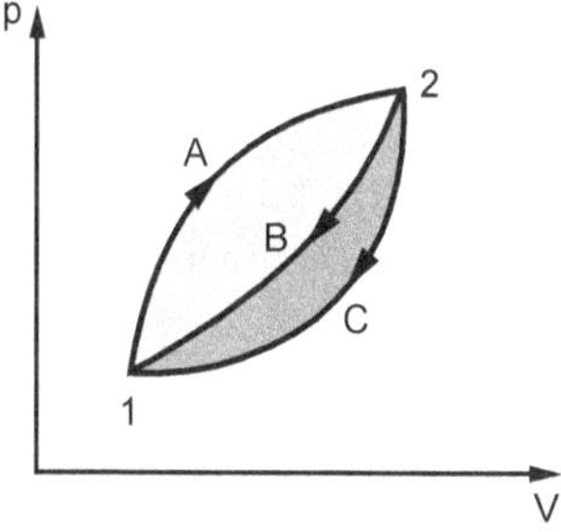

Fig. 3.7.1 Corollary 1

- If the system completes the cycle through the path 1-A-2-B-1 then,

$$\int\limits_{1A}^{2} (dQ - dW) + \int\limits_{2B}^{1} (dQ - dW) = 0$$

$$... \text{[From equation (3.6.1)]} ... (i)$$

- If the system completes the cycle through the path 1-A-2-C-1 then,

$$\int\limits_{1A}^{2} (dQ - dW) + \int\limits_{2C}^{1} (dQ - dW) = 0 \qquad \dots \text{(ii)}$$

- Now subtracting equation (ii) form equation (i), we get,

$$\int\limits_{2B}^{1} (dQ - dW) - \int\limits_{2C}^{1} (dQ - dW) = 0$$

$$\therefore \qquad \int\limits_{2B}^{1} (dQ - dW) = \int\limits_{2C}^{1} (dQ - dW) \qquad \dots \text{(iii)}$$

- It means that, when system changes from states 1 to 2, the quantity $\int\limits_{2}^{1} (dQ - dW)$ is constant whether the system proceeds through the path B or C.

- Hence, $\int\limits_{1}^{2} (dQ - dW)$ is function of the initial and final states of the sytem and it does not depend on the path of the process.

- Therefore, the quantity $\int\limits_{1}^{2} (dQ - dW)$ is a **property** because it is **point function.**

- The energy of the system which includes all forms of energy is called as **total energy** (E) of the system. It is given by,

$$E = Q - W$$

or $\qquad \Delta E = dQ - dW \qquad \dots \text{(3.7.2)}$

- But, total energy includes internal energy (U), kinetic energy (K.E.) and potential energy (P.E.).

$\therefore$ Total energy, E = Internal energy (U) + K.E. + P.E.

or $\qquad \Delta E = \Delta U + \Delta(K.E.) + \Delta(P.E.)$

Substituting the value of ΔE from equation (3.7.2),

$$\therefore \qquad \Delta E = dQ - dW = \Delta U + \Delta(K.E.) + \Delta(P.E.)$$

- If change in K.E. and P.E. are neglected then,

$$\therefore \qquad \boxed{\Delta U = U_2 - U_1 = dQ - dW} \qquad \dots \text{(3.7.3)}$$

3.7.2 Corollary 2 (For an Isolated System)

- For a closed system the internal energy remains unchanged if the sytem is isolated from its surroundings. [From statement (b) of first law of thermodynamics].

- For isolated system,

$$dQ = 0, \ dW = 0$$

$$\therefore \qquad \Delta U = U_2 - U_1 = 0$$

$$\therefore \qquad U_2 = U_1 \qquad \dots \text{(3.7.4)}$$

3.7.3 Corollary 3 (PMM-I)

- A **perpetual motion machine** of first kind (PMM-I) is impossible to construct. [From statement (c) of first law of thermodynamics].

- Fig. 3.7.2 shows PMM-I which delivers work continuously without any input.

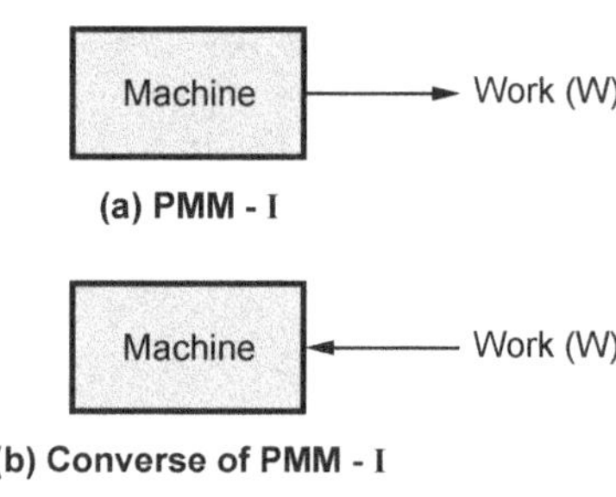

Fig. 3.7.2 Representation of PMM-I

- In this kind of machine the interaction with the surrounding is the delivery of work.

- It is impossible to have such a device as it violates first law of thermodynamics.

- **Converse of PMM-I** is also true i.e. there is no any machine which is capable of absorbing work without expenditure of energy.

3.8 : Limitations of Law of Thermodynamics
SPPU : Dec.-05,07,11,17,18 May-09,13,14,17,18,19

- We know that, first law of thermodynamics states that, when a closed system undergoes a cyclic process then the cyclic integral of the work is proportional to the cyclic integral of heat i.e.

$$\oint dW = \oint dQ$$

- Also, first law of thermodynamics relates to the principle of conservation of energy i.e. if one form of energy disappears then it appears in another form in equal amount or energy is neither created nor destroyed.

- According to the first law, the energy transfer can take place in either direction and it does not specify the direction of the energy transfer.

- We know that, non-violation of the first law by a proposed cycle does not ensure that the cycle will occur; this drawback has led to the formulation of the second law of thermodynamics.

- Hence, a cycle can occur if it satisfies both the first and second laws of thermodynamics.

- Second law states whether it is possible for energy transfer to proceed in a particular direction or not and the reverse direction is impossible.

- The limitations of first law of thermodynamics can be understood with the help of following examples :

  o An ice-cream when kept open to atmosphere absorbs heat and gets melted but liquid ice-cream does not solidify without doing any work on it.

  o In case of Joule's experiment, the fall of weight causes rotation of paddle and it increases the temperature of water (work transforms into heat), but increase in temperature of water does not rotate the paddle and lift the weight (means heat can not be transformed into work).

  o When a running vehicle is stopped by applying brakes, the internal energy of brakes increases by an amount equal to the decrease in kinetic energy, but hot brakes after losing their internal energy to wheels does not cause them to rotate.

- In all the above examples, the energy transfer or the process can proceed along a particular direction but it is impossible in the opposite direction even if the first law is satisfied.

- It means the first law of thermodynamics is necessary but not sufficient for a process to take place. Hence, it is necessary to study the laws which give more information about the transformation of heat into useful work.

3.9 : Basic Definitions SPPU : May-13, Dec.-13,17

To understand the second law of thermodynamics following basic definitions should be clear :

> 1. Heat or thermal reservoir
> 2. Heat source and heat sink
> 3. Heat engine
> 4. Heat pump

1. Heat or Thermal reservoir

- Heat or thermal reservoir is a sufficiently large system in stable equilibrium to which and from which finite amount of heat can be transferred without any change in its temperature.

- For example, large bodies of water like lakes, rivers, etc and the atmosphere.

2. Heat sink and heat source

- A heat reservoir which supplies heat to a system or high temperature reservoir is called as **heat source.**

- A heat reservoir which absorbs heat from a system or low temperature reservoir is called as **heat sink.**

3. Heat engine

- Heat engine is a thermodynamic system or device operating in a cycle to which net heat is transferred and from which net work is delivered.

- For example, steam engine, steam turbine, I.C. engine, etc.

4. Heat pump

- Heat pump is a thermodynamic system or device operating in a cycle which removes heat from low temperature body and delivers to high temperature body.

- For example, refrigerator, air-conditioner, etc.

3.10 : Heat Engine and Heat Pump
SPPU : Dec.-09,11,12,13,16 May-09,12,13,14,17,18

Heat Engine

- Heat engine is a thermodynamic system operating in a cycle to which net heat is transferred and from which net work is delivered.

- Fig. 3.10.1 shows a heat engine using a cyclic process.

- In this process, steam is generated from water in the boiler and it is used to develop useful work in the turbine. The steam coming out of the turbine is condensed in the condenser and again fed back to the boiler by using pump to complete the cycle.

- All the four elements of the system, boiler, turbine, condenser and pump form a heat engine.

- The input to the system is heat supplied to the boiler whereas net output from the system is work done by the turbine.

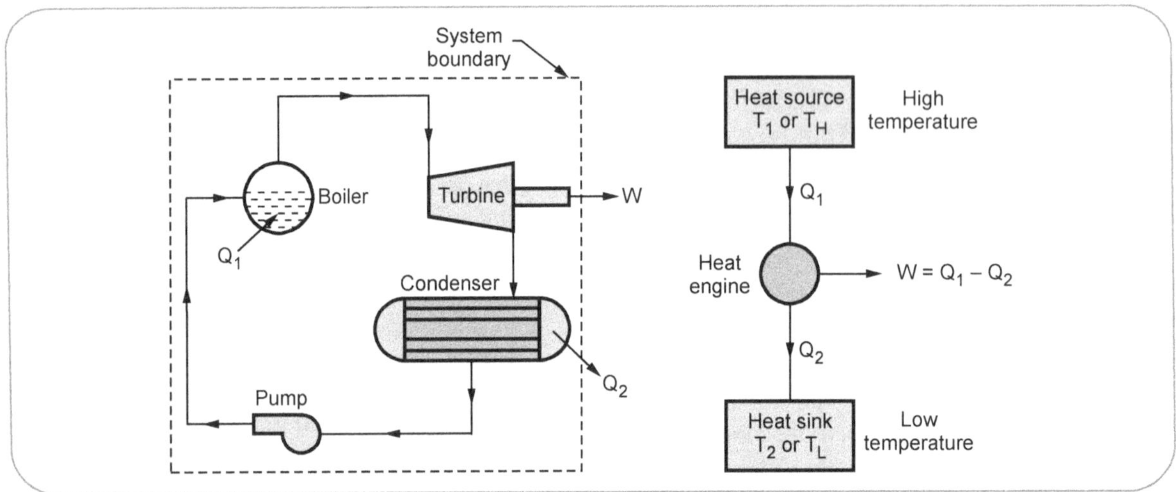

Fig. 3.10.1 Heat engine

- If Q_1 is the heat supplied to the boiler and Q_2 is the heat rejected from the condenser then net heat interaction between the system and surroundings is $(Q_1 - Q_2)$. Refer Fig. 3.10.1.

- From first law of thermodynamics,

$$W = Q_1 - Q_2$$

- The efficiency (measure of performance) of heat engine is,

$$\eta = \frac{\text{Output work}}{\text{Input work}} = \frac{W}{Q_1} = \frac{Q_1 - Q_2}{Q_1}$$

$$\therefore \quad \boxed{\eta = 1 - \frac{Q_2}{Q_1}} \qquad \ldots (3.10.1)$$

- *Hence, efficiency of heat engine is always less than unity.*

Heat Pump

- Heat pump is a thermodynamic system operating in a cycle which removes heat from low temperature body and delivers to high temperature body.

- Heat is pumped out so as to cool houses in summer season.

- Fig. 3.10.2 shows the heat pump using a cyclic process.

- In a domestic refrigerator, the energy from it is delivered at low temperature to the atmosphere which is at high temperature by using heat pump. Hence, maintaining a low temperature in the refrigerator.

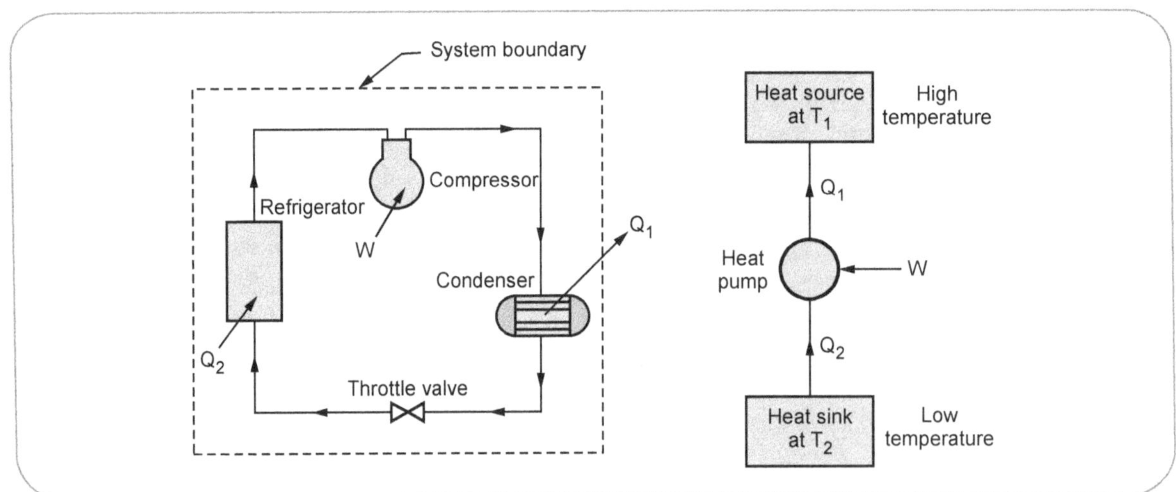

Fig. 3.10.2 Heat pump

- All the four elements of the system, compressor, condenser, throttle valve and referigerator form a heat pump.

- In a refrigerator main purpose is maintain a low temperature by removing heat coming in and discharge the same to the atmosphere which is at high temperature.

- From first law of thermodynamics,

$$W = Q_1 - Q_2$$

3.10.1 : Coefficient of Performance (C.O.P.)

SPPU : Dec.-09, 10, 12, 13, 16, 17, May-10, 13, 17

- In case of heat pump, the efficiency is replaced by a term Coefficient of Performance (C.O.P.).

- The defination of C.O.P. depends on the use of system whether like a heat pump or like a refrigerator.

- If Q_2 is heat added to the system and Q_1 is heat rejected from a system then the net output is $(Q_1 - Q_2)$.

Now,

$$(COP)_{refrigerator} = \frac{\text{Desired output}}{\text{Energy input}} = \frac{Q_2}{W}$$

$$\therefore \quad (COP)_{refrigerator} = \frac{Q_2}{Q_1 - Q_2} \quad ... (3.10.2)$$

$$\text{and } (COP)_{heat\ pump} = \frac{\text{Desired output}}{\text{Energy input}} = \frac{Q_1}{W}$$

$$\therefore \quad (COP)_{heat\ pump} = \frac{Q_1}{Q_1 - Q_2} \quad ... (3.10.3)$$

- Subtracting equation (3.10.3) and (3.10.2),

$$\left(\frac{Q_1}{Q_1 - Q_2}\right) - \left(\frac{Q_2}{Q_1 - Q_2}\right) = \frac{Q_1 - Q_2}{Q_1 - Q_2} = 1$$

$$\therefore \quad (COP)_{heat\ pump} - (COP)_{refrigerator} = 1$$

$$... (3.10.4)$$

Note : $(COP)_{heat\ pump}$ *is the reciprocal of the efficiency of heat engine, for the same sources.*

3.11 : Second Law of Thermodynamics

SPPU : May-09,10,11,12,17,19, Dec.-11,13,18

Based on the limitations of first law of thermodynamics many statements of second law are available. Some of them are as follows :

> 1. Kelvin-Planck statement
> 2. Clausius statement

1. Kelvin-Planck Statement   **May-13**

- It states that, *it is impossible to construct an engine working in a cyclic process whose sole effect is the conversion of all the heat energy supplied to it into an equivalent amount of work.*

- It means only part of heat is converted into work and remaining part is rejected to low temperature reservoir. Refer Fig. 3.11.1.

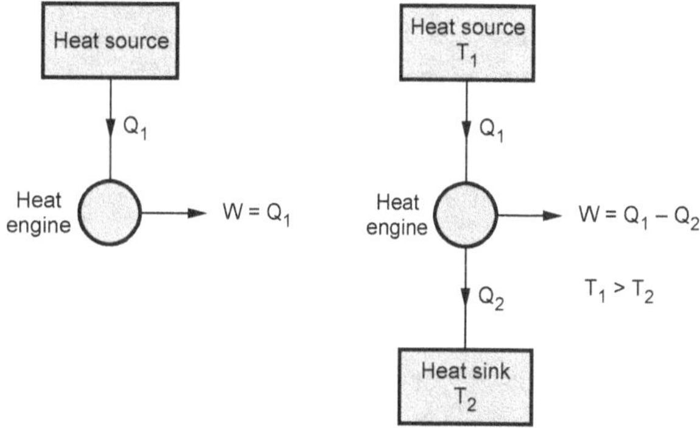

Fig. 3.11.1 Kelvin-Planck statement

- Therefore, to operate heat engine at least two thermal reservoirs at different temperatures are required.

2. Clausius Statement

- It states that, *it is impossible to construct a heat pump working in a cyclic process whose sole effect is the transfer of energy in the form of heat from a body at low temperature to a body at high temperature.*

- It means, energy in the form of work must be added or supplied to heat pump for transferring the heat from cold body to hot body. Refer Fig. 3.11.2.

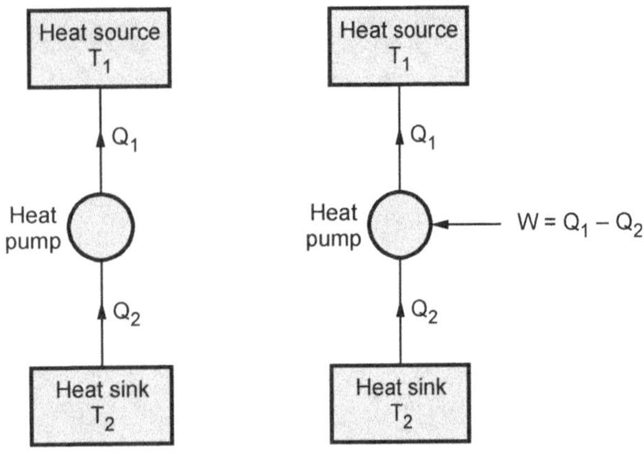

Fig. 3.11.2 Clausius statement

3.11.1 Perpetual Motion Machines (P.M.M.)

SPPU : Dec.-09

- Any machine or device that violates either the first law or second law of thermodynamics is called as perpetual motion machine (P.M.M.).

- There are following two kinds of P.M.M.

---

1. Perpetual motion machine of first kind (P.M.M. – I)
2. Perpetual motion machine of second kind (P.M.M. – II)

---

1. P.M.M. – I

- A machine or device that violates the first law of thermodynamics is called as P.M.M. – I.

- These machines will give continuous work without receiving energy from other system.

- But, it is impossible to construct P.M.M. – I i.e. no machine is capable of producing work without expenditure of energy.

2. P.M.M. – II

- A machine or device that violates the second low of thermodynamics is called as P.M.M. – II.

- It is a machine which converts internal energy from the surrounding atmosphere into work and gives 100 % efficiency.

- P.M.M. – II satisfies first law but violates second law.

- As per the second law, it is impossible to have a P.M.M. – II which just converts the internal energy of the atmosphere into useful work.

3.12 : Carnot Cycle

- In 1824 a French scientist **Sadi Carnot** originated a classifical reversible cycle in the study of engines which is termed as **Carnot cycle.**

- The following **assumptions** are made in the operation of carnot cycle :
  - During the motion there is no friction between the piston and cylinder.
  - The walls of cylinder and piston are perfectly insulated.
  - The cylinder head of the cylinder is arranged in such a way that it can be a perfect heat conductor or perfect heat insulator, as per the requirement.
  - The heat transfer does not affect the temperature of the source and sink.
  - The cycle can operate on any substance, but here it is a gas.

3.12.1 Processes in Carnot Cycle **SPPU : May-11**

- The Carnot cycle consists of four reversible processes, out of which two are frictionless isothermals and other two are frictionless adiabatics.

- Fig. 3.12.1 shows proposed reversible engine working between heat source and heat sink.

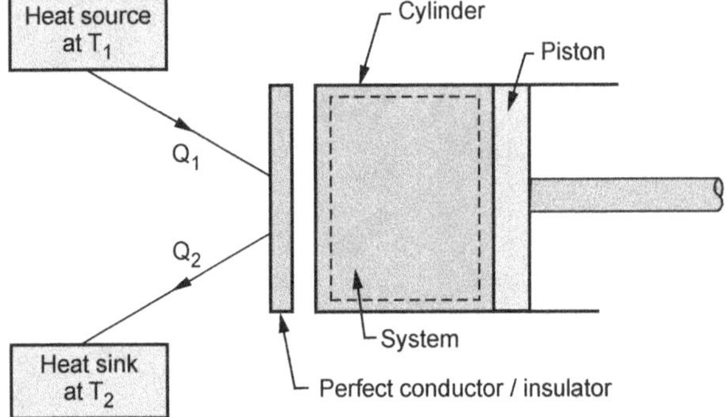

Fig. 3.12.1 Carnot engine

- The working of the Carnot engine is represented on p-V and T-S (Temperature-Entropy) diagram.

- The working of the cycle and sequence of operations are described as follows :

i) **Isothermal expansion process (1-2) :** During the begining of a stroke, the cylinder head is made perfect conductor and heat Q_1 is supplied to the system at temperature T_1. The gas expands isothermally at temperature T_1 from 1 to 2.

ii) Reversible adiabatic expansion process (2-3) : During this process, the system is allowed to expand adiabatically so that its temperatuure becomes law i.e. heat sink temperature T_2 (on T-S diagram it is T_3).

iii) Isothermal compression process (3-4) : In this process, perfect conductor is removed and heat sink is brought in contact with the cylinder head. During this process, gas is compressed isothermally from 3 to 4. Here work is done on the gas.

iv) Reversible adiabatic compression process (4-1) : Here, the perfect conductor is brought in contact with the cylinder head and the gas is compressed adiabatically. During the process, the temperature is raised from T_4 to T_1 on T-S diagram.

3.12.2 Efficiency of Carnot Engine

`SPPU : Dec.-13`

- Thermal efficiency of Carnot engine is defined as the ratio of work output to the heat input.

$$\eta = \frac{\text{Work output (W)}}{\text{Heat input }(Q_1)}$$

But, work output $W = Q_1 - Q_2$

$$\therefore \quad \eta = \frac{W}{Q_1} = \frac{Q_1 - Q_2}{Q_1}$$

$$\therefore \quad \boxed{\eta = 1 - \frac{Q_2}{Q_1}} \qquad \dots (3.12.1)$$

- The equations of various processes of Carnot cycle are as follows :

For process (1 - 2) :
For isothermal process,

$$\Delta U = U_2 - U_1 = 0 \quad \dots (\because T_2 = T_1 \text{ on cycle})$$

$$\therefore \quad Q_1 = W_{1\text{-}2} = m R T_1 \, ln\left(\frac{V_2}{V_1}\right) \qquad \dots (i)$$

For process (3 - 4) :
For isothermal process,

$$\Delta U = U_4 - U_3 = 0 \quad \dots (\because T_3 = T_4 \text{ on cycle})$$

$$\therefore \quad Q_2 = W_{3-4} = m R T_2 \, ln\left(\frac{V_4}{V_3}\right)$$

$$\text{or} \quad Q_2 = W_{3-4} = - m R T_2 \, ln\left(\frac{V_3}{V_4}\right) \qquad \dots (ii)$$

Negative sign indicates that the work is done on the gas or heat is rejected to the surrounding, this is the significance of negative sign.

Substituting the equation (i) and (ii) in equation (3.12.1),

$$\therefore \quad \eta = 1 - \frac{Q_2}{Q_1} = 1 - \frac{m R T_2 \, ln\left(\dfrac{V_3}{V_4}\right)}{m R T_1 \, ln\left(\dfrac{V_2}{V_1}\right)}$$

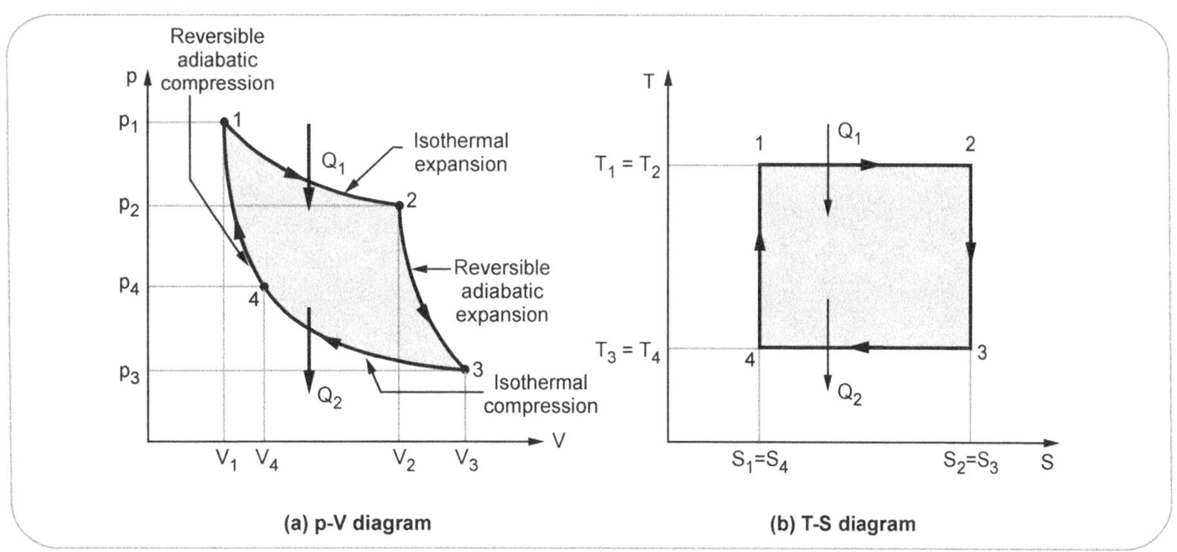

(a) p-V diagram   (b) T-S diagram

Fig. 3.12.2 Representation of Carnot cycle

or
$$\eta = 1 - \left(\frac{T_2}{T_1}\right) \frac{ln\left(\frac{V_3}{V_4}\right)}{ln\left(\frac{V_2}{V_1}\right)} \qquad \text{... (iii)}$$

For process (2-3) and (4-1) :

For reversible adiabatic process,

$$p V^\gamma = C$$

$$\frac{T_2}{T_3} = \left(\frac{V_3}{V_2}\right)^{\gamma-1} \quad \text{and} \quad \frac{T_1}{T_4} = \left(\frac{V_4}{V_1}\right)^{\gamma-1}$$

But, on cycle $T_3 = T_4 = T_2$ (Heat sink temperature)

and on cycle $T_1 = T_2 = T_1$ (Heat source temperature)

$$\therefore \qquad \frac{T_1}{T_2} = \left(\frac{V_3}{V_2}\right)^{\gamma-1} = \left(\frac{V_4}{V_1}\right)^{\gamma-1}$$

$$\therefore \qquad \frac{V_3}{V_2} = \frac{V_4}{V_1}$$

or
$$\frac{V_3}{V_4} = \frac{V_2}{V_1} \qquad \text{... (iv)}$$

Substituting equation (iv) in equation (iii), we get

$$\therefore \qquad \boxed{\eta = 1 - \frac{T_2}{T_1}} \qquad \text{... (3.12.2)}$$

Equating equation (3.12.1) and (3.12.2),

$$1 - \frac{Q_2}{Q_1} = 1 - \frac{T_2}{T_1}$$

$$\therefore \qquad \frac{Q_2}{Q_1} = \frac{T_2}{T_1}$$

or
$$\frac{Q_1}{T_1} = \frac{Q_2}{T_2} = \text{Constant}$$

or
$$\boxed{\frac{Q}{T} = \text{Constant}} \qquad \text{... (3.12.3)}$$

From above equation (3.12.3) it is clear that *heat transfer from a heat reservoir is proportional to its temperature* and this called as **Carnot principle.**

Important Points of Carnot Cycle

- Carnot heat engine is a hypothetical device.
- The efficiency of Carnot cycle depends on the absolute temperatures of heat source (T_1) and heat sink (T_2).

- Carnot cycle cannot be used in practical engine due to following reasons :
  - For isothermal process, the piston should move slowly to allow time for heat transfer so that the temperature remains constant.
  - It is not possible to perform a frictionless process.
  - An interchangeble cylinder head makes the process impractical.
  - As compared to other engines, Carnot engine has the maximum efficiency while operating between given temperature limits.

3.12.3 Carnot Heat Pump and Refrigerator

- In Carnot cycle it is possible to reverse the process individually and continue the cycle in reverse order.
- During the reverse process, all the energy transfers related with the process are reversed in direction but their magnitude remains same. Refer Fig. 3.12.3.

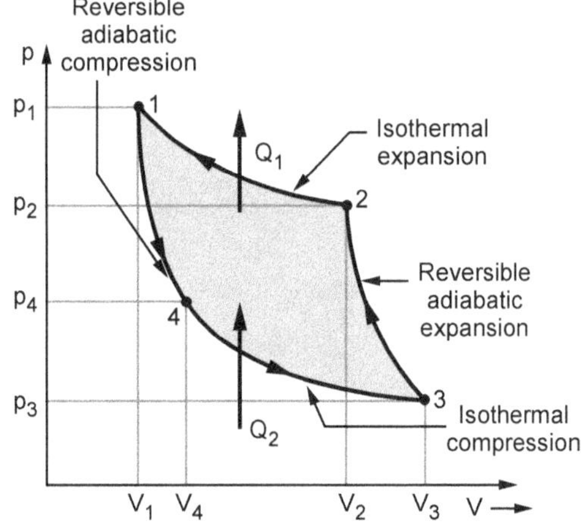

Fig. 3.12.3 Carnot refrigerator and heat pump

- Therefore, we can construct hypothetical device which absorbs heat Q_2 from heat reservoir at temperature T_2 and rejects the heat Q_1 to high temperature reservoir T_1. During this external work W is supplied to the system from the surrounding.
- We know that COP of refrigerator and heat pump is given by,

$$(\text{COP})_{\text{refrigerator}} = \frac{Q_2}{Q_1 - Q_2}$$

$$\text{... [From equation (3.10.2)]}$$

and $\qquad (COP)_{\text{heat pump}} = \dfrac{Q_1}{Q_1 - Q_2}$

$\qquad\qquad\qquad$ ... [From equation (3.10.3)]

But, $\qquad\qquad \dfrac{Q}{T} = \text{Constant C}$

$\therefore \qquad\qquad\qquad Q = T\,C$

or $\qquad\qquad Q_1 = T_1\,C \ \text{ and } \ Q_2 = T_2\,C$

and

$$(COP)_{\text{refrigerator}} = \dfrac{Q_2}{Q_1 - Q_2} = \dfrac{T_2}{T_1 - T_2}$$

$$(COP)_{\text{heat pump}} = \dfrac{Q_1}{Q_1 - Q_2} = \dfrac{T_1}{T_1 - T_2} \quad \text{... (3.12.4)}$$

Note : *Heat reservoir of temperature T_2 is called as evaporator and heat reservoir of temperature T_1 is called as condenser.*

3.13 : Solved Examples

Ex. 3.13.1 : *The COP of a refrigerator is 6, when it maintains the temperature of $-3\,°C$ in the evaporator. Determine the condenser temperature and refrigerating effect if the power required to run the refrigerator is 7.5 kW.*

Sol. : Given data : $(COP)_{\text{ref.}} = 6,$
$T_2 = -3\,°C = -3 + 273 = 270 \ K, \ W = 7.5 \ kW$

To find : i) Condenser temperature T_1 and
$\qquad\qquad$ ii) Refrigerating effect.

Step - 1 : **Calculate the condenser temperature T_1**
COP of refrigerator is given by,

$$(COP)_{\text{ref.}} = \dfrac{T_2}{T_1 - T_2}$$

$\therefore \qquad\qquad 6 = \dfrac{270}{T_1 - 270}$

$\therefore \qquad\qquad T_1 = \mathbf{315\ K} \qquad\qquad \text{... Ans.}$

Step - 2 : **Calculate the refrigerating effect**
Power required to run the refrigerator

$= 7.5 \ kW = 7500 \ W = 7500 \ J/sec$

We know that,

$$(COP)_{\text{ref.}} = \dfrac{Q_2}{W} = \dfrac{\text{Work output}}{\text{Work input}}$$

$$(COP)_{\text{ref.}} = \dfrac{\text{Refrigerating effect}}{\text{Input power}}$$

$\therefore \qquad\quad 6 = \dfrac{\text{Refrigerating effect}}{7500}$

$\therefore$ Refrigerating effect = 45000 J/sec.

$\qquad\qquad\qquad = 45000 \ W = \mathbf{45\ kW} \qquad \text{... Ans.}$

Ex. 3.13.2 : *The efficiency of a Carnot engine rejecting heat to a cooling pond at 28 °C is 30 %. If the cooling pond receives 1050 kJ/min. What is the power developed by the cycle in kW. Also find the temperature of the source.*

Sol. : Given data : $T_2 = 28\,°C = 28 + 273 = 301 \ K,$
$\eta_{\text{Carnot}} = 0.3, \ Q_2 = 1050 \ kJ/min.$

To find : i) Power developed W $\quad$ ii) T_1

Step - 1 : **Calculate the source temperature T_1**
Refer Fig. 3.13.1.

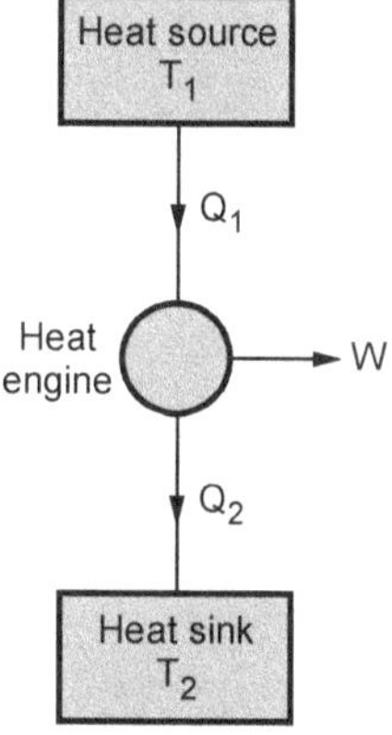

Fig. 3.13.1

We know that,

$$\eta_{\text{Carnot}} = 1 - \dfrac{T_2}{T_1}$$

$\therefore \qquad\quad 0.3 = 1 - \dfrac{301}{T_1}$

$\therefore \qquad\quad T_1 = \mathbf{430\ K} \qquad\qquad \text{... Ans.}$

Step - 2 : **Calculate the heat rejecting from heat source and power**
Carnot efficiency is also given by,

$$\eta_{\text{Carnot}} = 1 - \dfrac{Q_2}{Q_1}$$

$$\therefore \qquad 0.3 = 1 - \frac{1050}{Q_1}$$

$$\therefore \qquad Q_1 = 1500 \ \text{kJ/min.}$$

Power developed by the cycle is,

$$\text{Power} = W = Q_1 - Q_2 = 1500 - 1050$$

$$\therefore \qquad W = 450\,\text{kJ/min} = 7.5\,\text{kJ/sec.}$$

$$\therefore \qquad \mathbf{W = 7.5 \ kW} \qquad\qquad \textbf{... Ans.}$$

Ex. 3.13.3 : *A heat pump is used to maintain an auditorium hall at 25 °C, when the atmospheric temperature is 10 °C. The heat leakage from the hall is 1500 kJ/min. Calculate the power required to run the actual heat pump, if the COP of the actual heat pump is 30 % of the COP of the Carnot heat pump working between the same temperature limits.*

Sol. :

Given data : $T_1 = 25\ °\text{C} = 25 + 273 = 298$ K,

$$T_2 = 10\ °\text{C} = 10 + 273 = 283\ \text{K}$$

$$Q_1 = 1500 \ \text{kJ/min} = 25 \ \text{kJ/sec} = 25 \ \text{kW}$$

$$(\text{COP})_{\text{actual}} = 0.3\,(\text{COP})_{\text{Carnot}}$$

To find : Power

Step - 1 : Calculate the power required to run the actual heat pump
Refer Fig. 3.13.2.

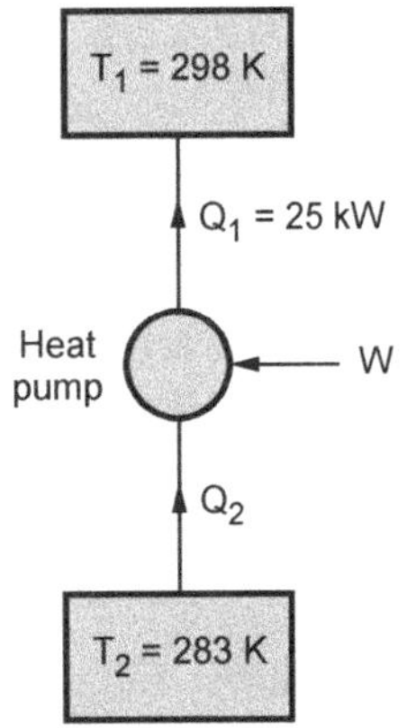

Fig. 3.13.2

$$(\text{COP})_{\text{Carnot hp}} = \frac{T_1}{T_1 - T_2} = \frac{298}{298 - 283} = 19.867$$

$$\therefore (\text{COP})_{\text{actual hp}} = 0.3 \times (\text{COP})_{\text{Carnot hp}} = 0.3 \times 19.867$$

$$= 5.96$$

$$(\text{COP})_{\text{heat pump}} = \frac{Q_1}{Q_1 - Q_2} = \frac{Q_1}{W}$$

$$\therefore \qquad 5.96 = \frac{25}{W}$$

$$\mathbf{W = 4.195 \ kW} \qquad\qquad \textbf{... Ans.}$$

Ex. 3.13.4 : *A reversible heat engine operates on Carnot cycle between source and sink temperatures of 225 °C and 25 °C. If the heat engine receives 40 kW from the source, find the net workdone, heat rejected to sink and efficiency of the engine.*

Sol. :

Given data : $T_1 = 225\ °\text{C} = 225 + 273 = 498$ K

$$T_2 = 25\ °\text{C} = 25 + 273 = 298 \ \text{K}$$

$$Q_1 = 40 \ \text{kW}$$

To find : W, Q_2 and η_{Carnot}.

Step - 1 : Calculate the efficiency of heat engine
Refer Fig. 3.13.3.

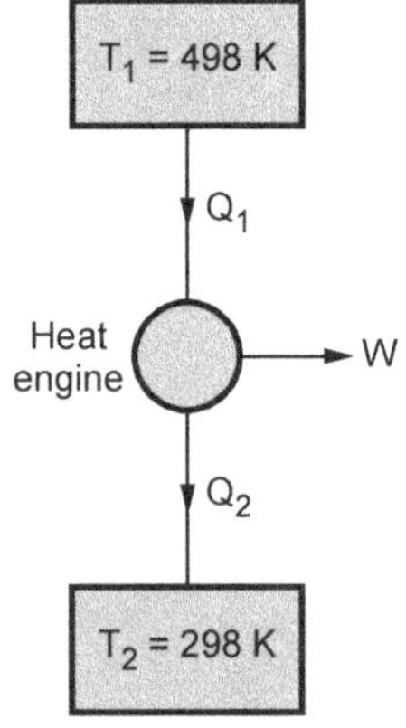

Fig. 3.13.3

Efficiency of Carnot cycle is,

$$\eta_{\text{Carnot}} = 1 - \frac{T_2}{T_1} = 1 - \frac{298}{498}$$

$$\therefore \qquad \eta_{\text{Carnot}} = 0.4 = 40 \ \% \qquad\qquad \textbf{... Ans.}$$

Step - 2 : Calculate the net workdone and heat rejected to the sink

We know that,

$$\eta_{Carnot} = \frac{Q_1 - Q_2}{Q_1} = \frac{W}{Q_1}$$

$$\therefore \quad 0.4 = \frac{W}{40}$$

$$\therefore \quad \mathbf{W = 16 \ kW} \qquad \text{... Ans.}$$

But, $\quad W = Q_1 - Q_2$

$$\therefore \quad 16 = 40 - Q_2$$

$$\therefore \quad \mathbf{Q_2 = 24 \ kW} \qquad \text{... Ans.}$$

Ex. 3.13.5 : *A Carnot engine operates between source and sink temperatures of 230 °C and 30 °C. If the engine receives 450 kJ of heat from the source, determine i) Workdone   ii) Heat rejected   iii) η_{Carnot} iv) Also find COP if it operates as a heat pump and when it operates as a refrigerator?*

Sol. :

Given data : $T_1 = 230$ °C $= 230 + 273 = 503$ K

$$T_2 = 30 \text{ °C} = 30 + 273 = 303 \text{ K}$$

$$Q_1 = 450 \text{ kJ}$$

To find :   i) W   ii) Q_2 iii) η_{Carnot} iv) $(COP)_{hp}$ and $(COP)_{ref.}$

Step - 1 : Calculate the efficiency and workdone of cycle

We know that,

$$\eta_{Carnot} = 1 - \frac{T_2}{T_1} = 1 - \frac{303}{503}$$

$$\therefore \quad \eta_{Carnot} = 0.398 = \mathbf{39.8 \ \%} \qquad \text{... Ans.}$$

Efficiency is also given by,

$$\eta_{Carnot} = \frac{Q_1 - Q_2}{Q_1} = \frac{W}{Q_1}$$

$$0.398 = \frac{W}{450}$$

$$\therefore \quad \mathbf{W = 179.1 \ kJ} \qquad \text{... Ans.}$$

But, $\quad W = Q_1 - Q_2$

$$\therefore \quad 179.1 = 450 - Q_2$$

$$\therefore \quad \mathbf{Q_2 = 270.9 \ kJ} \qquad \text{... Ans.}$$

Step - 3 : Calculate the $(COP)_{hp}$ and $(COP)_{ref.}$

When the engine operates like heat pump,

$$(COP)_{hp} = \frac{Q_1}{W} = \frac{Q_1}{Q_1 - Q_2} = \frac{450}{450 - 270.9}$$

$$\therefore \quad \mathbf{(COP)_{hp} = 2.515} \qquad \text{... Ans.}$$

When it operates like refrigerator,

$$(COP)_{ref.} = \frac{Q_2}{W} = \frac{Q_2}{Q_1 - Q_2} = \frac{270.9}{450 - 270.9}$$

$$\therefore \quad \mathbf{(COP)_{ref.} = 1.515} \qquad \text{... Ans.}$$

Ex. 3.13.6 : *An invertor claims to have developed a refrigeration unit which maintains at − 5 °C in the refrigerator which is in a room of surrounding temperature 28 °C. It has COP of 9.5. Check whether his claim is right or not ?*

Sol. :

Given data : $T_2 = -5$ °C $= -5 + 273 = 268$ K

$$T_1 = 28 \text{ °C} = 28 + 273 = 301 \text{ K}$$

$$(COP)_{inv.} = 9.5$$

To find :   Check whether inventor's claim is right or not.

Step - 1 : Calculate the $(COP)_{ref.}$ and check it

Refer Fig. 3.13.4.

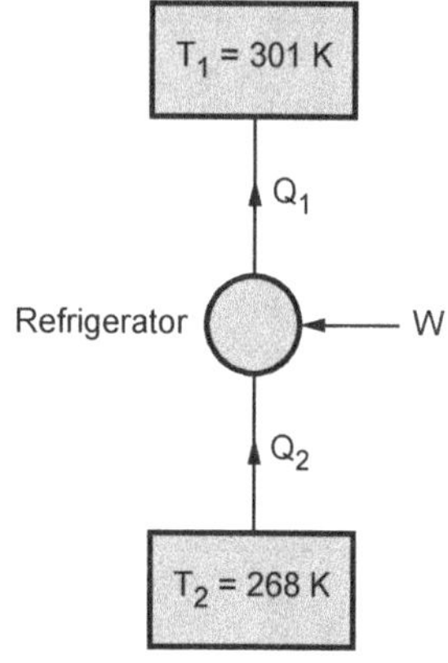

Fig. 3.13.4

We know that,

$$(COP)_{ref.} = \frac{Q_2}{Q_1 - Q_2} = \frac{T_2}{T_1 - T_2}$$

$$\therefore \quad (COP)_{ref.} = \frac{268}{301 - 268} = 8.121$$

$$\therefore \quad (COP)_{ref.} < (COP)_{inv.}$$

It means for given temperature range inventor's claim is invalid.   **... Ans.**

Ex. 3.13.7 : *A fish freezing plant of 100 tons capacity is to be maintained at – 40 °C for which the outside atmospheric temperature is 30 °C. The actual COP of the refrigeration system is 10 % of the theoretical Carnot pump working between the same temperature limits. Calculate the power required to run the plant. Take, 1 ton of refrigeration = 3.5 kW*

Sol. :

Given data : $T_1 = 30 °C = 30 + 273 = 303 \text{ K}$

$$T_2 = -40 °C = -40 + 273 = 233 \text{ K}$$

$$(COP)_A = 0.1 \, (COP)_T$$

$$Q_1 = 100 \text{ tons} = 100 \times 3.5 = 350 \text{ kW}$$

To find : Power (W)

Step - 1 : Calculate the power required to run the plant

Refer Fig. 3.13.5.

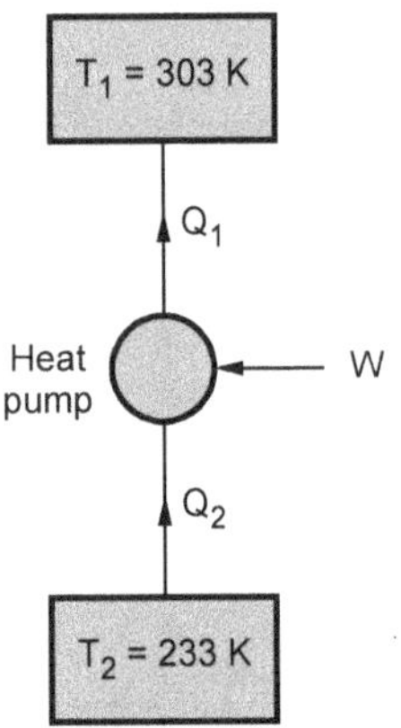

Fig. 3.13.5

COP of Carnot or theoretical heat pump is given by,

$$(COP)_T = \frac{Q_1}{Q_1 - Q_2} = \frac{T_1}{T_1 - T_2}$$

$$\therefore \quad (COP)_T = \frac{303}{303 - 233} = 4.33$$

$$(COP)_A = 0.1 \, (COP)_T \quad\quad \text{... (Given)}$$

$$\therefore \quad (COP)_A = 0.1 \times 4.33 = 0.433$$

Now, $\quad (COP)_A = \dfrac{Q_1}{Q_1 - Q_2} = \dfrac{Q_1}{W}$

$$\therefore \quad 0.433 = \frac{350}{W}$$

$$\therefore \quad \mathbf{W = 808.31 \text{ kW}} \quad\quad \text{... Ans.}$$

Ex. 3.13.8 : *The power input required for a grinding mill is 30 MJ/min. A heat source at 420 °C is available for supplying the energy and the surrounding is at 30 °C. If the actual engine is 25 % as efficient as a Carnot engine working between the same temperature limits. Calculate the energy supplied by the source per sec.*

Sol. :   **Given data :** Power input to grinding mill

$$W = 30 \text{ MJ/min} = \frac{30 \times 10^3}{60} \frac{\text{kJ}}{\text{sec.}} = 500 \text{ kW}$$

$$T_1 = 420 °C = 420 + 273 = 693 \text{ K}$$

$$T_2 = 30 °C = 30 + 273 = 303 \text{ K}$$

$$(\eta)_A = 0.25 \, (\eta)_T$$

To find : Q_1

Step - 1 : Calculate the energy supplied by the source

Refer Fig. 3.13.6.

Efficiency of Carnot engine is,

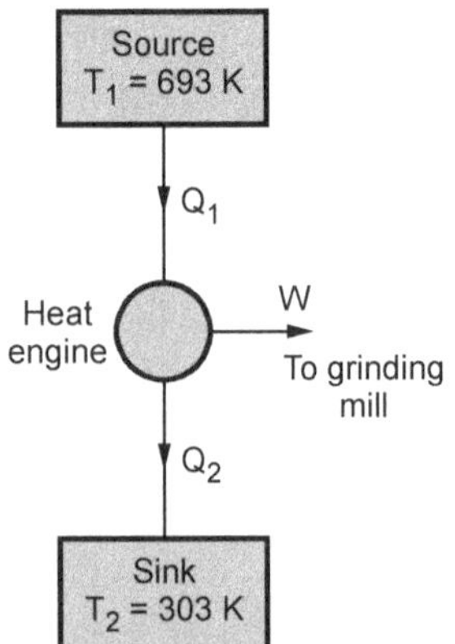

Fig. 3.13.6

$$\eta_{Carnot} = \frac{Q_1 - Q_2}{Q_1} = \frac{T_1 - T_2}{T_1}$$

$$\therefore \quad \eta_{Carnot} = \frac{693 - 303}{693} = 0.563 = 56.3 \%$$

But, $\quad \eta_A = 0.25 \, \eta_{Carnot} = 0.25 \times 0.563$

$$\therefore \quad \eta_A = 0.141$$

Now, $\quad \eta_A = \frac{Q_1 - Q_2}{Q_1} = \frac{W}{Q_1}$

$$\therefore \quad 0.141 = \frac{500}{Q_1}$$

$$\therefore \quad Q_1 = 3546.1 \text{ kJ/sec.} = 3546.1 \text{ kW} \quad \textbf{... Ans.}$$

Ex. 3.13.9 : *An inventor claims to have invented refrigeration machine operating between −23 °C and 27 °C. It consumes 1 kW electrical power and gives 20000 kJ of refrigerating effect in one hour. Comment on his claim.*

Sol. :

Given data : $T_1 = 27 \text{ °C} = 27 + 273 = 300 \text{ K}$

$$T_2 = -23 \text{ °C} = -23 + 273 = 250 \text{ K}$$

$$W = 1 \text{ kW,}$$

$$Q_2 = 20000 \text{ kJ/hr}$$

$$= \frac{20000}{3600} \frac{\text{kJ}}{\text{sec.}} = 5.55 \text{ kW}$$

To find : Check whether inventor's claim is right or not.

Step - 1 : Calculate the actual and theoretical COP of refrigerator

$$(COP)_T = (COP)_{Carnot} = \frac{Q_2}{Q_1 - Q_2}$$

$$= \frac{T_2}{T_1 - T_2} = \frac{250}{300 - 250}$$

$$\therefore \quad (COP)_{Carnot} = 5$$

$$(COP)_{Actual} = \frac{Q_2}{Q_1 - Q_2} = \frac{Q_2}{W} = \frac{5.55}{1}$$

$$\therefore \quad (COP)_{Actual} = 5.55$$

It means COP of Carnot refrigerator is less than the COP of actual refrigerator. Hence, inventor's claim is invalid.   **... Ans.**

Ex. 3.13.10 : *The Carnot engine is operating between source and sink temperature T_1 and T_2 respectively. It has efficiency of 30 %. If the sink temperature is reduced by 25 °C then the efficiency is increased to 35 %. Find the temperatures T_1 and T_2.*

Sol. :

Given data : $\eta = 0.3 = 1 - \dfrac{T_2}{T_1}$

$$\eta' = 0.35 \quad \text{when} \quad T_2' = T_2 - 25$$

and $\quad T_1' = T_1$

To find : T_1 and T_2.

Step - 1 : Calculate the source and sink temperatures T_1 and T_2

It is given that,

$$\eta = 0.3 = 1 - \frac{T_2}{T_1} \qquad \text{... (i)}$$

and $\quad \eta' = 0.35 = 1 - \dfrac{T_2'}{T_1'}$

$$\therefore \quad 0.35 = 1 - \frac{(T_2 - 25)}{T_1} \qquad \text{... (ii)}$$

Dividing equation (i) and (ii),

$$\frac{\eta'}{\eta} = \frac{0.35}{0.3} = \frac{1 - \dfrac{(T_2 - 25)}{T_1}}{1 - \left(\dfrac{T_2}{T_1}\right)} = \frac{T_1 - T_2 + 25}{T_1 - T_2}$$

$$\therefore \quad 1.167 = \frac{T_1 - T_2 + 25}{T_1 - T_2}$$

$$\therefore \quad 1.167 \, T_1 - 1.167 \, T_2 = T_1 - T_2 + 25$$

$$\therefore \quad 0.167 \, T_1 = 0.167 \, T_2 + 25$$

$$\therefore \quad T_1 = T_2 + 149.7 \qquad \text{... (iii)}$$

Substituting this value in equation (i),

$$\therefore \quad 0.3 = 1 - \frac{T_2}{T_1} = 1 - \frac{T_2}{(T_2 + 149.7)}$$

$\therefore \qquad \dfrac{T_2}{T_2 + 149.7} = 1 - 0.3 = 0.7$

$\therefore \qquad \mathbf{T_2 = 349.3\ K} \qquad \qquad \text{... Ans.}$

and $\qquad T_1 = T_2 + 149.7 = 349.3 + 149.7$

$\therefore \qquad \mathbf{T_1 = 499\ K} \qquad \qquad \text{... Ans.}$

Ex. 3.13.11 : *A reversible engine is supplied with heat from two constant temperature sources at 900 K and 600 K and rejects heat to a constant temperature sink at 300 K. If the engine executes a number of complete cycles while developing 80 kW and rejecting 3500 kJ of heat per min. Determine the heat supplied by each source per minute and efficiency of the engine.*

Sol. :

Given data : Work developed $W = 80$ kW,
$T_{1A} = 900$ K, $T_{1B} = 600$ K,

$\qquad T_2 = 300$ K,

Heat rejected $Q_2 = 3500\,\text{kJ/min}$

$\therefore \qquad Q_2 = \dfrac{3500}{60}\ \dfrac{\text{kJ}}{\text{sec.}} = 58.33\ \text{kW}$

To find : i) Heat supplied by each source Q_{1A} and Q_{1B}.

ii) η of engine.

Step - 1 : Calculate the heat supplied by each source

Refer Fig. 3.13.7.

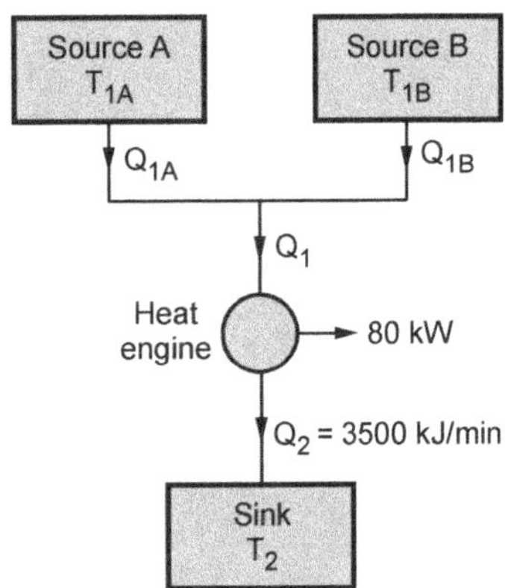

Fig. 3.13.7

We know that,

$\qquad W = Q_1 - Q_2$

$\therefore \qquad 80 = Q_1 - 58.33$

$\therefore \qquad Q_1 = 138.33\ \text{kW}$

If Q_{1A} and Q_{1B} are the heat supplied by source A and B then

$\qquad Q_1 = Q_{1A} + Q_{1B}$

$\therefore \qquad Q_{1B} = Q_1 - Q_{1A}$

$\therefore \qquad Q_{1B} = 138.33 - Q_{1A} \qquad \text{... (i)}$

As the engine is reversible, the required condition is,

$$\left(\dfrac{Q}{T}\right)_{\text{supplied}} = \left(\dfrac{Q}{T}\right)_{\text{rejected}}$$

$\therefore \qquad \dfrac{Q_{1A}}{Q_{1A}} + \dfrac{Q_{1B}}{T_{1B}} = \dfrac{Q_2}{T_2}$

$\dfrac{Q_{1A}}{900} + \dfrac{(138.33 - Q_{1A})}{600} = \dfrac{58.33}{300}$

$\therefore \quad \dfrac{Q_{1A}}{900} + \dfrac{138.33}{600} - \dfrac{Q_{1A}}{600} = \dfrac{58.33}{300}$

$\dfrac{138.33}{600} - \dfrac{58.33}{300} = \dfrac{Q_{1A}}{600} - \dfrac{Q_{1A}}{900}$

$0.0316 = 5.56 \times 10^{-4}\ Q_{1A}$

$\therefore \qquad \mathbf{Q_{1A} = 56.834\ kW} \qquad \text{... Ans.}$

and $\qquad Q_{1B} = 138.33 - 56.834$

$\therefore \qquad \mathbf{Q_{1B} = 81.496\ kW} \qquad \text{... Ans.}$

Step - 2 : Calculate the efficiency of the engine

Efficiency of the engine is,

$\qquad \eta = \dfrac{\text{Work done}}{\text{Heat supplied}} = \dfrac{W}{Q_1}$

$\therefore \qquad \eta = \dfrac{80}{138.33} = 0.5783$

$\therefore \qquad \mathbf{\eta = 57.83\ \%} \qquad \text{... Ans.}$

Ex. 3.13.12 : *A reversible engine working in a cycle takes 4800 kJ/min of heat from a source of 800 K and develops 22 kW power. The engine rejects heat to two reservoirs at 300 K and 350 K. Calculate the heat rejected to each sink.*

Sol. :

Given data : Q_1 = 4800 kJ/min = 80 kW

$$T_1 = 800 \text{ K}, \quad W = 22 \text{ kW}$$

$$T_{2A} = 300 \text{ K} \quad \text{and} \quad T_{2B} = 360 \text{ K}$$

To find : Q_{2A} and Q_{2B}.

Step - 1 : Calculate the heat rejected to each sink

Refer Fig. 3.13.8.

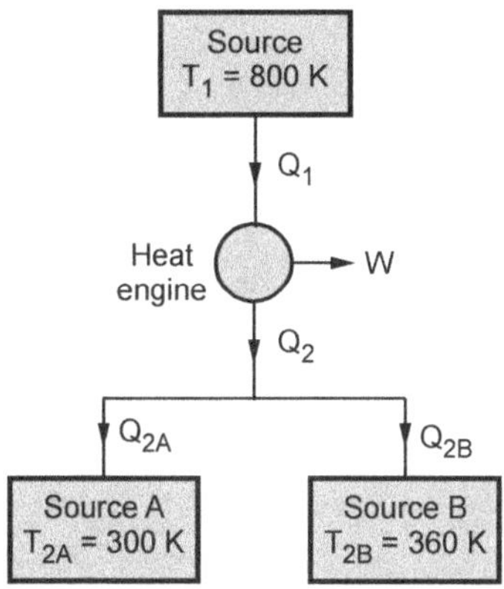

Fig. 3.13.8

We know that,

$$W = Q_1 - Q_2$$

$$\therefore \quad 22 = 80 - Q_2$$

$$\therefore \quad Q_2 = 58 \text{ kW}$$

If Q_{2A} and Q_{2B} are the heat rejected to sink A and B then

$$Q_2 = Q_{2A} + Q_{2B}$$

$$\therefore \quad 58 = Q_{2A} + Q_{2B}$$

$$\therefore \quad Q_{2B} = 58 - Q_{2A} \qquad \text{... (i)}$$

As the engine is reversible, the required condition is,

$$\left(\frac{Q}{T}\right)_{\text{supplied}} = \left(\frac{Q}{T}\right)_{\text{rejected}}$$

$$\therefore \quad \frac{Q_1}{T_1} = \frac{Q_{2A}}{T_{2A}} + \frac{Q_{2A}}{T_{2B}}$$

$$\therefore \quad \frac{80}{800} = \frac{Q_{2A}}{300} + \frac{(58 - Q_{2A})}{360}$$

$$\therefore \quad 0.1 = 3.333 \times 10^{-3}\, Q_{2A} + 2.78 \times 10^{-3}\,(58 - Q_{2A})$$

$$\therefore \quad 0.1 = 3.333 \times 10^{-3}\, Q_{2A} + 0.161 - 2.78 \times 10^{-3}\, Q_{2A}$$

$$\therefore \quad -0.061 = 5.53 \times 10^{-4}\, Q_{2A}$$

$$\therefore \quad \mathbf{Q_{2A} = -110.31 \text{ kW}} \qquad \text{... Ans.}$$

and $Q_{2B} = 58 - Q_{2A} = 58 - (-110.31)$

$$\therefore \quad \mathbf{Q_{2B} = 168.31 \text{ kW}} \qquad \text{... Ans.}$$

Ex. 3.13.13 : *A reversible heat engine operates between two reservoirs at temperature of 600 °C and 40 °C. The engine drives a reversible refrigerator, which operates between reservoir at temperature 40 °C and −20 °C. The heat transfer to the heat engine is 2000 kJ and the net work output of the combined engine refrigerator plant is 360 kJ.*

Sol. :

Given data : T_1 = 600 °C = 600 + 273 = 873 K,

$$T_2 = 40 \text{ °C} = 40 + 273 = 313 \text{ K},$$

$$T_3 = -20 \text{ °C} = -20 + 273 = 253 \text{ K},$$

$$T_4 = T_2 = 313 \text{ K}, \quad Q_1 = 2000 \text{ kJ},$$

$$W_1 = 360 \text{ kJ}$$

To find : i) Heat transfer to the refrigerator and reservoir and

ii) Heat transfer to the refrigerator and reservoir at

$$\eta_A = 0.4\,\eta_T \quad \text{and} \quad (\text{COP})_A = 0.4\,(\text{COP})_T$$

Step - 1 : Calculate heat transfer to the refrigerator and reservoir

Refer Fig. 3.13.9.

$$W = W_1 + W_2 \qquad \text{... (i)}$$

But, $\eta_T = 1 - \dfrac{T_2}{T_1}$

$$\therefore \quad \eta_T = 1 - \frac{313}{873}$$

$$\therefore \quad \eta_T = 0.6415$$

Now, $\eta_T = \dfrac{W}{Q_1}$

$$\therefore \quad 0.6415 = \frac{W}{2000}$$

$$\therefore \quad W = 1283 \text{ kJ}$$

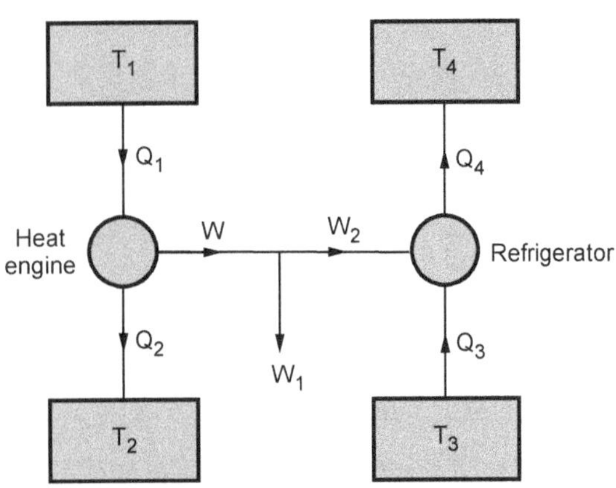

Fig. 3.13.9

Substituting in equation (i),

$$1283 = 360 + W_2$$

$\therefore$ $\mathbf{W_2 = 923 \ kJ}$ **... Ans.**

But, $\eta_T = \dfrac{W}{Q_1} = \dfrac{Q_1 - Q_2}{Q_1} = 1 - \dfrac{Q_2}{Q_1}$

$\therefore$ $0.6415 = 1 - \dfrac{Q_2}{2000}$

$\therefore$ $Q_2 = 717 \ kJ$

We know that,

$$(COP)_{ref.} = \dfrac{Q_3}{W_2} = \dfrac{Q_3}{Q_4 - Q_3} = \dfrac{T_3}{T_4 - T_3}$$

$\therefore$ $(COP)_{ref.} = \dfrac{253}{313 - 253} = 4.22$

But, $(COP)_{ref.} = \dfrac{Q_3}{W_2}$

$\therefore$ $4.22 = \dfrac{Q_3}{923}$

$\therefore$ $Q_3 = 3895.06 \ kJ$

and $W_2 = Q_4 - Q_3$

$\therefore$ $923 = Q_4 - 3895.06$

$\therefore$ $Q_4 = 4818.06 \ kJ$

$\therefore$ Total heat transfer to the reservoir $= Q_2 + Q_4$

$$= = 717 + 4818.06 = \mathbf{5535.06 \ kJ} \ \text{... Ans.}$$

Step - 2 : Calculate heat transfer to the refrigerator and reservoir at 40 % of their maximum values

$$\eta_A = 0.4 \ \eta_T = 0.4 \times 0.6415 = 0.2566$$

Now, $\eta_A = 1 - \dfrac{T_2}{T_1} = 1 - \dfrac{Q_2}{Q_1}$

$\therefore$ $0.2566 = 1 - \dfrac{Q_2}{2000}$

$\therefore$ $\mathbf{Q_2 = 1486.8 \ kJ}$ **... Ans.**

and $\eta_A = \dfrac{W}{Q_1}$

$\therefore$ $0.2566 = \dfrac{W}{2000}$ $\therefore W = 513.2 \ kJ$

But, $W = W_1 + W_2$

$\therefore$ $513.2 = 360 + W_2$

$\therefore$ $W_2 = 153.2 \ kJ$

Similarly,

$$(COP)_A = 0.4 \ (COP)_T = 0.4 \times 4.22 = 1.688$$

Now, $(COP)_A = \dfrac{Q_3}{W_2}$ $\therefore 1.688 = \dfrac{Q_3}{153.2}$

$\therefore$ $Q_3 = 258.6 \ kJ$

But, $W_2 = Q_4 - Q_3$

     $153.2 = Q_4 - 258.6$

$\therefore$ $Q_4 = 411.8 \ kJ$

$\therefore$ Total heat transfer to the reservoir $= Q_2 + Q_4$

$$= 1486.8 + 411.8 = \mathbf{1898.6 \ kJ} \ \text{... Ans.}$$

Ex. 3.13.14 : *Two Carnot engines work in series between the source and sink temperatures of 600 K and 400 K. If both engines develop same power determine the intermediate temperature.*

Sol. :

Given data : $T_1 = 600 \ K$, $T_2 = 400 \ K$,   W is same.

To find : T_3, (Refer Fig. 3.13.10).

Step - 1 : Calculate the intermediate temperature of engine

Efficiency of engines H.E.1 and H.E.2 is,

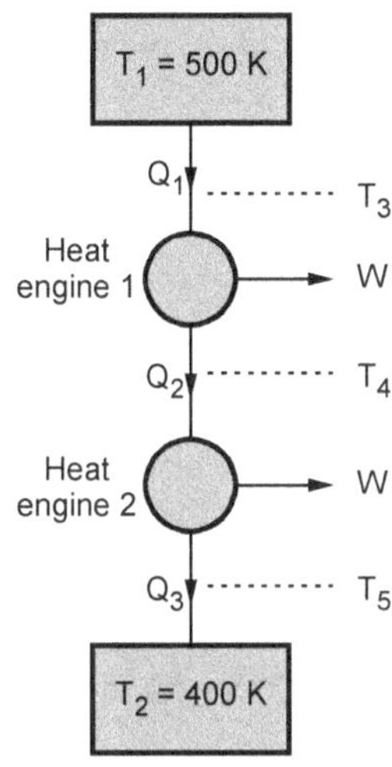

Fig. 3.13.10

$$\eta_1 = \frac{W}{Q_1} = \frac{T_3 - T_4}{T_3} = \frac{Q_1 - Q_2}{Q_1}$$

and $$\eta_2 = \frac{W}{Q_2} = \frac{T_4 - T_5}{T_4} = \frac{Q_2 - Q_3}{Q_2}$$

or $$\eta_1 = \frac{T_3 - T_4}{T_3} = \frac{W}{Q_2 + W} = \frac{W}{Q_1} \qquad \text{... (i)}$$

and $$\eta_2 = \frac{T_4 - T_5}{T_4} = \frac{W}{Q_3 + W} = \frac{W}{Q_2} \qquad \text{... (ii)}$$

From equation (i), we can write

$$W = (Q_2 + W)\left(\frac{T_3 - T_4}{T_3}\right)$$

$$\therefore \qquad W = Q_2\left(\frac{T_3 - T_4}{T_3}\right) + W\left(\frac{T_3 - T_4}{T_3}\right)$$

$$\therefore \qquad W - W\left(\frac{T_3 - T_4}{T_3}\right) = Q_2\left(\frac{T_3 - T_4}{T_3}\right)$$

$$\therefore \qquad W\left[1 - \left(\frac{T_3 - T_4}{T_3}\right)\right] = Q_2\left(\frac{T_3 - T_4}{T_3}\right)$$

$$\therefore \qquad W\left(\frac{T_4}{T_3}\right) = Q_2\left(\frac{T_3 - T_4}{T_3}\right)$$

$$\therefore \qquad W = Q_2\left(\frac{T_3 - T_4}{T_4}\right) \qquad \text{.. (iii)}$$

From equation (ii), we can write

$$W = Q_2\left(\frac{T_4 - T_5}{T_4}\right) \qquad \text{... (iv)}$$

Equating equation (iii) and (iv), we get,

$$\frac{T_3 - T_4}{T_4} = \frac{T_4 - T_5}{T_4}$$

or $$T_3 - T_4 = T_4 - T_5$$

But, $$T_3 = T_1 = 600 \text{ K}$$

and $$T_5 = T_2 = 400 \text{ K}$$

$$\therefore \qquad 600 - T_4 = T_4 - 400$$

$$\therefore \qquad 1000 = 2\,T_4$$

$$\therefore \qquad \mathbf{T_4 = 500\ K} \qquad \textbf{... Ans.}$$

Ex. 3.13.15 : *Which is more effective way to increase efficiency of Carnot engine, either to increase T_1 keeping T_2 constant, or to decrease T_2 keeping T_1 constant.*

Sol. : We know that, the efficiency of Carnot cycle working between the temperature limits T_1 and T_2 is given by,

$$\eta = \frac{T_1 - T_2}{T_1}$$

Consider that the source temperature is increased by 'ΔT' maintaining the sink temperature T_2 constant then,

$$\eta_1 = \frac{(T_1 + \Delta T) - T_2}{T_1 + \Delta T}$$

$$= \frac{T_1 + \Delta T - T_2}{T_1 + \Delta T} \qquad \text{...(3.13.1)}$$

Now consider that the sink temperature is decreased by 'ΔT' maintaining source temperature T_1 constant then,

$$\eta_2 = \frac{T_1 - (T_2 - \Delta T)}{T_1}$$

$$= \frac{T_1 + \Delta T - T_2}{T_1} \qquad \text{...(3.13.2)}$$

Comparing equations (3.13.1) and (3.13.2) it is obvious that the numerators of both the equations are same and denominator of equation (3.13.2) is less than the denominator of equation (3.13.1) as $T_1 < (T_1 - \Delta T)$.

$$\therefore \qquad \eta_2 > \eta_1$$

$\therefore$ In order to increase the efficiency of Carnot cycle, **decreasing of T_2 will be more effective.** **... Ans.**

Ex. 3.13.16 : *The overall volume expansion ratio of a Carnot cycle is 16. The cycle works between the temperature limit of*
320 °C and 40 °C. Determine :
i) Volume ratio of isothermal and adiabatic process
ii) Thermal efficiency of the cycle.
Assume working fluid is air and γ = 1.4.

Sol. :

Given data : Expansion ratio $\dfrac{V_3}{V_1} = 16$, $\gamma = 1.4$

$T_1 = 320$ °C $= 320 + 273 = 593$ K $= T_2$

$T_3 = 40$ °C $= 40 + 273 = 313$ K $= T_4$

To find : i) $\dfrac{V_3}{V_2}$ and $\dfrac{V_2}{V_1}$ ii) η

Step - 1 : Calculate the expansion ratio of isothermal and adiabatic processes
Refer Fig. 3.13.11.

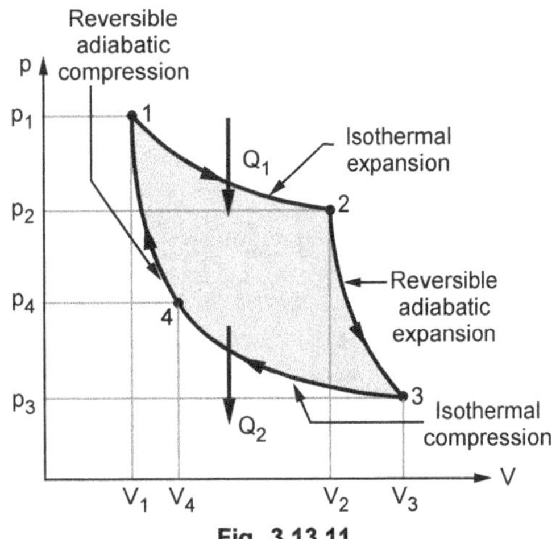

Fig. 3.13.11

For adiabatic process (2 - 3),

$$\frac{T_2}{T_3} = \left(\frac{V_3}{V_2}\right)^{\gamma-1}$$

$\therefore$ $$\frac{V_3}{V_2} = \left(\frac{T_2}{T_3}\right)^{\left(\frac{1}{\gamma-1}\right)} = \left(\frac{593}{313}\right)^{\left(\frac{1}{1.4-1}\right)}$$

$\therefore$ $$\frac{V_3}{V_2} = 4.94 \qquad \text{... Ans.}$$

For isothermal process (3-4), we can write

$$\frac{V_2}{V_1} = \frac{V_2}{V_3} \times \frac{V_3}{V_1} = \frac{1}{4.94} \times 16$$

$$\frac{V_2}{V_1} = 3.24 \qquad \text{... Ans.}$$

Step - 2 : Calculate the efficiency of the cycle
Thermal efficiency of the Carnot cycle is,

$$\eta = 1 - \frac{T_3}{T_2} = 1 - \frac{313}{593}$$

$\therefore$ $$\eta = 0.5278 = 52.78 \% \qquad \text{... Ans.}$$

3.14 : Solved University Examples

Ex. 3.14.1 : *A household refrigerator with a C.O.P. of 1.8 removes heat from the refrigerated space at a rate of 90 kJ/min.*
Determine : 1) The electric power consumed by the refrigerator.
2) The rate of heat transfer to the kitchen air.

SPPU : May-09, Marks 6, Dec.-17, Marks 5

Sol. :

Given data : $(COP)_{ref.} = 1.8$,

Heat removed $Q_2 = 90 \dfrac{kJ}{min} = \dfrac{90}{60} \dfrac{kJ}{sec.} = 1.5$ kW

To find : i) W ii) Q_1

Step - 1 : Calculate the power consumed by the refrigerator
Refer Fig. 3.14.1.

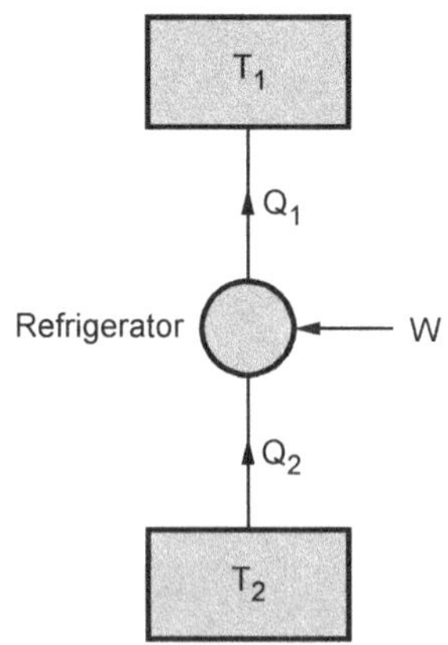

Fig. 3.14.1

We know that,

$$(COP)_{ref.} = \frac{Q_2}{Q_1 - Q_2} = \frac{Q_2}{W}$$

$$\therefore \quad 1.8 = \frac{1.5}{W}$$

$$\therefore \quad W = 0.833 \text{ kW} \quad \quad \text{... Ans.}$$

Step - 2 : Calculate the rate of heat transfer to the kitchen air

We know that,

$$W = Q_1 - Q_2$$

$$\therefore \quad 0.833 = Q_1 - 1.5$$

$$\therefore \quad Q_1 = 2.333 \text{ kW} = 140 \text{ kJ/min.} \quad \text{... Ans.}$$

Ex. 3.14.2 : *Heat Pump is used to maintain house at 23 °C. The house is losing heat to outside air through walls at 60,000 kJ/hr. While energy generated in house by various appliances is 4,000 kJ/hr. For a COP of 1.5, find required power input in kW, supplied to the heat pump.*

SPPU : Dec.-10, May-18, Marks 6

Sol. :

Given data : $T_1 = 23°C = 23 + 273 = 296$ K
Heat lost = 60000 kJ/hr, Heat generated = 4000 kJ/hr.,
COP = 1.5

To find : Power input in kW.

Step - 1 : Calculate the power input to the heat pump
Refer Fig. 3.14.2.

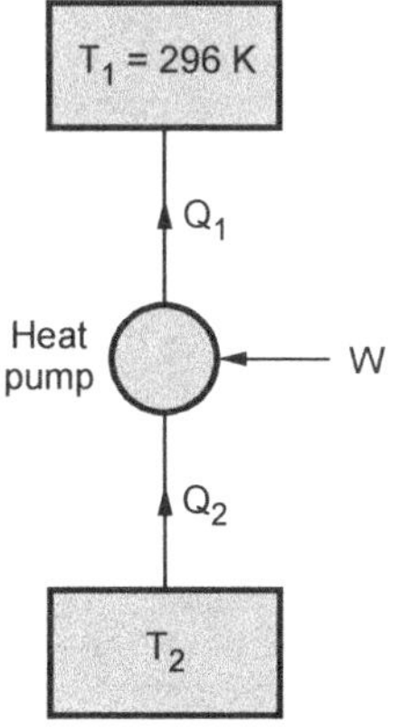

Fig. 3.14.2

Heat delivered by the heat pump is,

$$Q_1 = \text{Heat lost} - \text{Heat generated}$$

$$\therefore \quad Q_1 = 60000 - 4000$$

$$\therefore \quad Q_1 = 56000 \text{ kJ/hr.}$$

$(COP)_{\text{heat pump}}$ is given by,

$$(COP)_{hp} = \frac{Q_1}{Q_1 - Q_2} = \frac{Q_1}{W}$$

$$\therefore \quad 1.5 = \frac{56000}{W}$$

$$\therefore \quad W = 37333.33 \text{ kJ/hr} = \frac{37333.33}{3600} \text{ kJ/sec.}$$

$$\therefore \quad W = 10.37 \text{ kW} \quad \quad \text{... Ans.}$$

Ex. 3.14.3 : *A cold storage is to be maintained at − 5 °C while the surroundings are at 35 °C. The heat leakage from the surroundings into the cold storage is estimated to be 29 kW. The actual C.O.P. of the refrigeration plant is one-third of an ideal plant working between the same temperatures. Find the power required to drive the plant.*

SPPU : Dec.-11, Marks 6

Sol. :

Given data : $T_1 = 35 °C = 35 + 273 = 308$ K

$$T_2 = −5 °C = −5 + 273 = 268 \text{ K}$$

$$(COP)_A = \frac{1}{3}(COP)_T = 0.333 (COP)_T$$

$$Q_1 = 29 \text{ kW}$$

To find : Power (W)

Step - 1 : Calculate the power required to drive the plant
Refer Fig. 3.14.3.

COP of Carnot or theoretical heat pump is given by,

$$(COP)_T = \frac{Q_1}{Q_1 - Q_2} = \frac{T_1}{T_1 - T_2}$$

$$\therefore \quad (COP)_T = \frac{308}{308 - 268} = 7.7$$

$$(COP)_A = 0.333 (COP)_T \quad \quad \text{... (Given)}$$

$$\therefore \quad (COP)_A = 0.333 \times 7.7 = 2.566$$

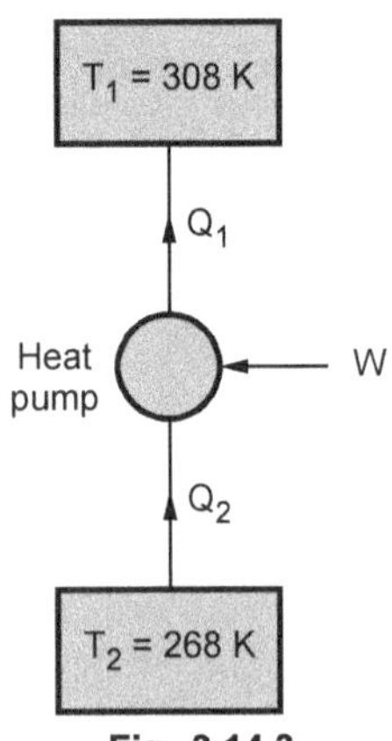

Fig. 3.14.3

Now, $(COP)_A = \dfrac{Q_1}{Q_1 - Q_2} = \dfrac{Q_1}{W}$

$\therefore \qquad 2.566 = \dfrac{29}{W}$

$\therefore \qquad\qquad W = 11.298 \text{ kW} \qquad\qquad \text{... Ans.}$

Ex. 3.14.4 : *A heat engine operates between source and sink temperatures of 235 °C and 30 °C respectively. If heat engine receives 35 kW from the source, find the net work done by the engine, the heat rejected to the sink by the engine and the efficiency of engine. Draw the sketch of system.*

SPPU : May.-13, 19, Marks 5

Sol. :

Given data : $T_1 = 235\ ^\circ C = 235 + 273 = 508 \text{ K}$

$\qquad\qquad\quad T_2 = 30\ ^\circ C = 30 + 273 = 303 \text{ K}$

$\qquad\qquad\quad Q_1 = 35 \text{ kW}$

To find : W, Q_2 and η_{Carnot}

Step - 1 : **Calculate the efficiency of heat engine**

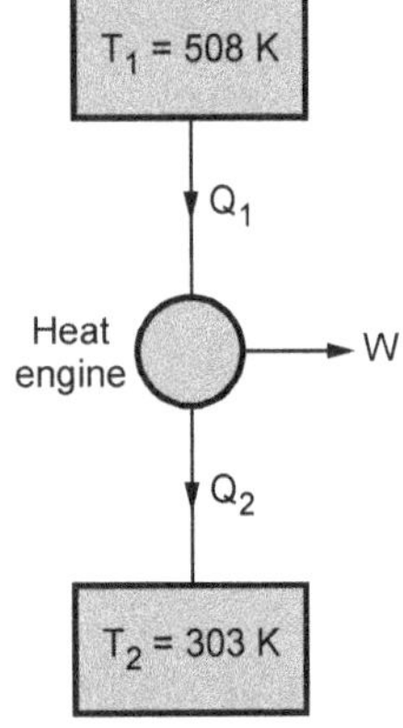

Fig. 3.14.4

Refer Fig. 3.14.4.

Efficiency of Carnot cycle is,

$$\eta_{Carnot} = 1 - \frac{T_2}{T_1} = 1 - \frac{303}{508}$$

$\therefore \qquad \eta_{Carnot} = 0.4035 = 40.35\ \% \qquad\qquad \text{... Ans.}$

Step - 2 : **Calculate the net workdone and heat rejected to the sink**

We know that,

$$\eta_{Carnot} = \frac{Q_1 - Q_2}{Q_1} = \frac{W}{Q_1}$$

$\therefore \qquad 0.4035 = \dfrac{W}{35}$

$\therefore \qquad\qquad W = 14.1225 \text{ kW} \qquad\qquad \text{... Ans.}$

But, $\qquad\qquad W = Q_1 - Q_2$

$\therefore \qquad 14.1225 = 35 - Q_2$

$\therefore \qquad\qquad Q_2 = 20.8775 \text{ kW} \qquad\qquad \text{... Ans.}$

Ex. 3.14.5 : *An engine develops 80 kW of work output when heat is supplied at the rate of 240 kW. Find the efficiency of the engine and heat rejected to atmosphere. Draw the sketch of system.*

SPPU : Dec.-13, Marks 5

Sol. : Given data : $W = 80 \text{ kW}$, $Q_1 = 240 \text{ kW}$

To find : η and Q_2

Step - 1 : **Calculate the efficiency and heat rejected to atmosphere.**

As per first law of thermodynamics,

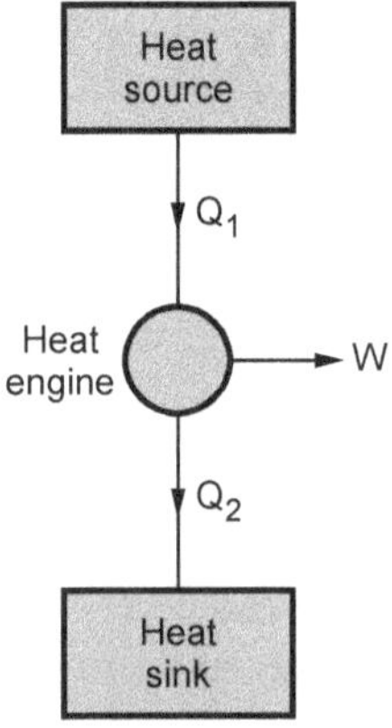

Fig. 3.14.5

$$W = Q_1 - Q_2$$

$$80 = 240 - Q_2$$

$$\therefore \quad Q_2 = 160 \text{ kW} \qquad \text{... Ans.}$$

Efficiency of heat engine is,

$$\eta = \frac{Q_1 - Q_2}{Q_1} = \frac{W}{Q_1}$$

$$\therefore \quad \eta = \frac{80}{240} = 0.3333 = \textbf{33.33 \%} \qquad \text{... Ans.}$$

Ex. 3.14.6 : *A refrigerator with COP of 1.5 absorbs heat from food compartment at the rate of 360 kJ/min. Draw the sketch of system and find*
1) Power consumed by the refrigerator and
2) The amount of heat rejected to surrounding.

SPPU : May-14, Marks 5

Sol. : Given data : $(COP)_{ref.} = 1.5$, Heat removed $Q_2 = 360 \dfrac{kJ}{min} = \dfrac{360}{60} \dfrac{kJ}{sec.} = 6$ kW

To find : i) W ii) Q_1

Step - 1 : Calculate the power consumed by the refrigerator
Refer Fig. 3.14.6.

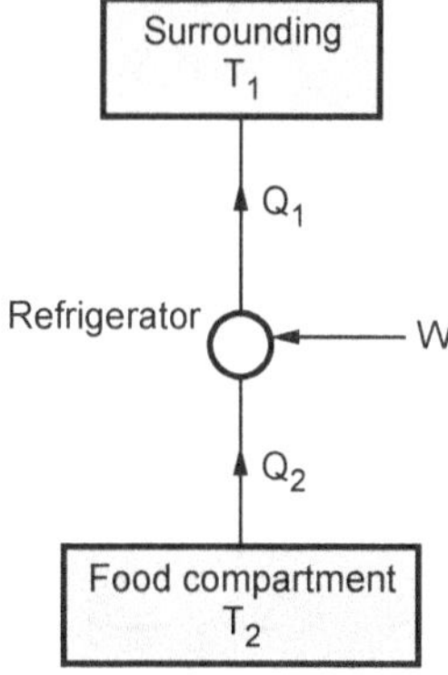

Fig. 3.14.6

We know that,

$$(COP)_{ref.} = \frac{Q_2}{Q_1 - Q_2} = \frac{Q_2}{W}$$

$$\therefore \quad 1.8 = \frac{6}{W}$$

$$\therefore \quad \textbf{W = 3.3333 kW} \qquad \text{... Ans.}$$

Step - 2 : Calculate the amount of heat rejected to surrounding.
We know that,

$$W = Q_1 - Q_2$$

$$\therefore \quad 3.3333 = Q_1 - 6$$

$$\therefore \quad \textbf{Q}_1 = \textbf{9.3333 kW}$$

$$= \textbf{559.998 kJ/min.} \qquad \text{... Ans.}$$

Ex. 3.14.7 : *A refrigeration system is used to maintain a cold storage at 4 degree C. The heat leakage from surrounding into the cold storage is estimated to be 1800 kJ/min. If COP of the refrigeration system is 1.5. Find :*
(1) The amount of heat rejected to the surrounding and
(2) Power required to drive the refrigeration system. Draw the sketch of system. SPPU : May-15, Marks 5

Sol. : Given data : $T_2 = 4\,°C = 4 + 273 = 277$ K,
$Q_2 = 1800$ kJ/min $= \dfrac{1800}{60}$ kJ/sec $= 30$ kW,
$(C.O.P.)_{ref} = 1.5$

To find : i) Q_1 ii) W

Step - 1 : Calculate the heat rejected to surroundings
Refer Fig. 3.14.7.

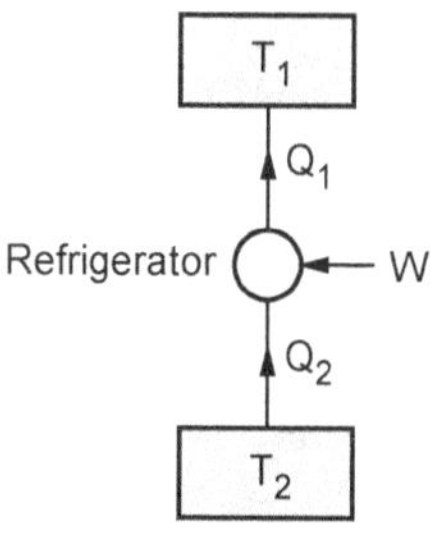

Fig. 3.14.7

$$(C.O.P.)_{ref} = \frac{Q_1}{Q_1 - Q_2}$$

$$\therefore \quad 1.5 = \frac{Q_1}{Q_1 - 30}$$

$$\therefore \quad \textbf{Q}_1 = \textbf{90 kW} \qquad \text{... Ans.}$$

Step - 2 : Calculate the power required to drive the refrigeration system

We know that,

$$W = Q_1 - Q_2 = 90 - 30$$

$$\therefore \qquad W = 60 \text{ kW} \qquad \qquad \text{... Ans.}$$

Ex. 3.14.8 : *A heat pump is used to maintain the house at 24 degree C. The house is losing the heat at the rate of 1800 kJ/min to the surrounding. The heat pump is driven by an electric motor of power rating 12 kW. Find :*
(i) The amount of heat absorbed from surrounding
(ii) COP of the heat pump
Draw the sketch of the system.

SPPU : Dec.-15, May-16, Marks 5

Sol. : $T_1 = 24 \ °C = 24 + 273 = 297$ K,

$Q_1 = 1800$ kJ/min $= \dfrac{1800}{60} = 30$ kJ/sec. $= 30$ kW,
$W = 12$ kW.

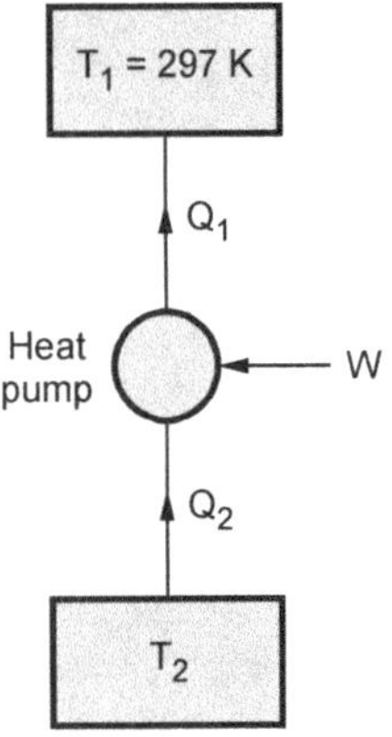

Fig. 3.14.8

To find : i) Q_2 ii) COP

Step 1 : Calculate the amount of heat absorbed and COP.

$$(\text{COP})_{\text{heat pump}} = \frac{Q_1}{Q_1 - Q_2} = \frac{Q_1}{W}$$

$$\therefore (\text{COP})_{\text{heat pump}} = \frac{30}{12} = 2.5 \qquad \text{... Ans.}$$

But, $\qquad W = Q_1 - Q_2 \qquad \therefore \ 12 = 30 - Q_2$

$$\therefore \qquad Q_2 = 18 \text{ kW} \qquad \qquad \text{... Ans.}$$

Ex. 3.14.9 : *A reversible heat engine develops 30 kW of work output with efficiency of 30%. Find the heat supplied to the engine and heat rejected from the engine. If the engine is reversed to act as refrigerator with same rate of energy transfer, find its COP.*

SPPU : Dec.-16, 18, Marks 5

Sol. :

Given data : $W = 30$ kW, $\eta = 30 \% = 0.30$

To find : Q_1, Q_2, $(\text{COP})_{\text{HP}}$

Step 1 :   Calculate heat supplied and heat rejected from the engine

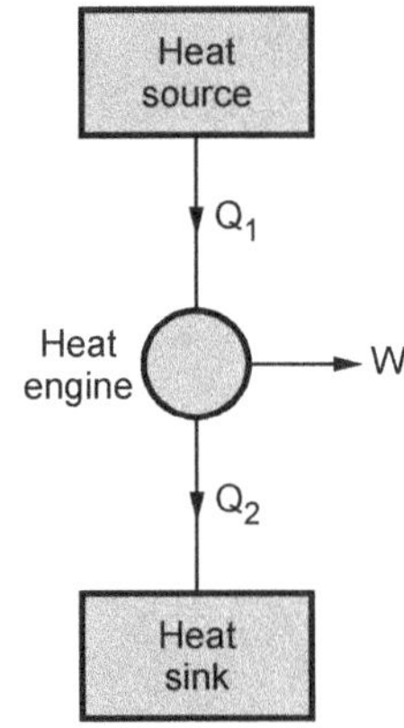

Fig. 3.14.9 (a)

Efficiency of heat engine is,

$$\eta = \frac{W}{Q_1} \quad \therefore \quad 0.30 = \frac{30}{Q_1}$$

$$\therefore \qquad Q_1 = 100 \text{ kW} \qquad \qquad \text{...Ans.}$$

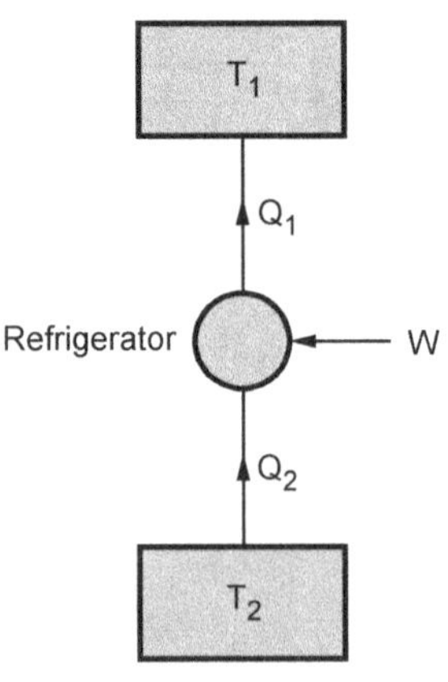

Fig. 3.14.9 (b)

Also, $\quad \eta = \dfrac{Q_1 - Q_2}{Q_1} \quad \therefore \ 0.30 = \dfrac{100 - Q_2}{100}$

$\therefore \qquad \mathbf{Q_2 = 70 \ kW}$ **...Ans.**

Step 2 : Calculate COP if the engine acts as refrigerator (COP)

$(COP)_{ref}$ is given by,

$$(COP)_{ref} = \dfrac{Q_2}{Q_1 - Q_2} = \dfrac{70}{100 - 70}$$

$\therefore \qquad \mathbf{(COP)_{ref} = 2.33}$ **...Ans.**

Ex. 3.14.10 : *A fish freezing plant is to be maintained at – 10 degree C. If power required to drive the plant is 30 kW and COP of refrigeration system is 3. Find :*
i) heat sucked (absorbed) from the freezing plant and
ii) heat rejected to the surrounding.
Draw sketch of the system.   **SPPU : May-17, Marks 5**

Sol. :   **Given data :** $T_2 = -10\ ^\circ C = -10 + 273 = 263$ K $W_{net} = 30$ kW, COP = 3
To find : i) Q_1 ii) Q_2

Step 1 :   Calculate heat absorbed (Q_1)
We know that,

$$COP = \dfrac{Q_1}{W_{net}}$$

$\therefore \qquad Q_1 = COP \times W_{net}$

$\therefore \qquad Q_1 = 3 \times 30 = \mathbf{90\ kW}$ **... Ans.**

Step 2 :   Calculate heat rejected to the surrounding (Q_2)

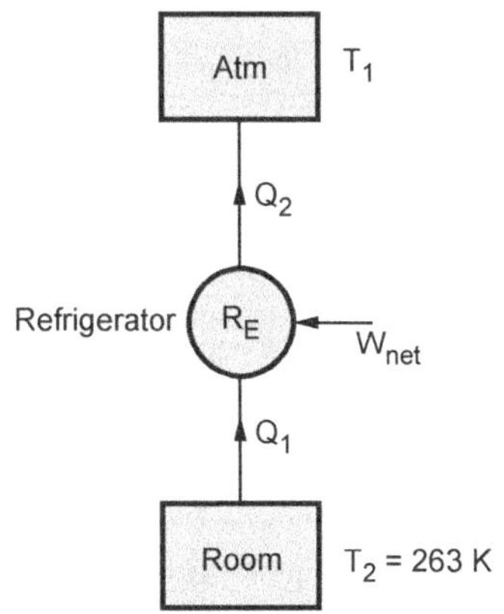

Fig. 3.14.10

$\qquad W_{net} = Q_2 - Q_1$

$\therefore \qquad Q_2 = W_{net} + Q_1$

$\therefore \qquad Q_2 = 30 + 90 = \mathbf{120\ kW}$ **... Ans.**

Review Questions

1. State whether the following systems are closed/open/isolated systems. Give reasons.
   i) Air compressor   ii) Pressure cooker
   iii) Thermas flask
2. Explain the various types of thermodynamic system with suitable examples for each.
3. Explain in brief Joule's experiment.
4. State first law of thermodynamics.
5. What are the limitations of first law of thermodynamics ?
6. What is total energy ? Explain different forms of energies that constitute total energy.
7. State and explain zeroth law of thermodynamics.
8. Prove that internal energy is a property of a system.
9. Show that the efficiency of heat engine is always less than unity.
10. Give Kelvin-Planck and Clausius statement of second law of thermodynamics.
11. Define with example : System, Surrounding, isolated system.
12. Draw sketch of heat engine and refrigerator using source and sink concepts. Also state the relations for efficiency and C.O.P.
13. What is PMM - II ?
14. Explain Carnot cycle using P-V diagram.

3.15 University Questions with Answers

Dec. - 2011

Q.1   Define : a) Heat engine   b) Heat pump
     (Refer section 3.10)   **[4]**

Q.2   Explain the significance of first and second law of thermodynamics by giving suitable examples.
     (Refer sections 3.8 and 3.11)   **[6]**

May - 2013

Q.3 *Define thermodynamic system. Explain its types with examples.* **(Refer section 3.2.1)** [4]

Q.4 *Explain the various types of thermodynamic system with suitable examples for each.* **(Refer section 3.2.1)** [6]

Q.5 *State various statements and limitations of first law of thermodynamics.* **(Refer sections 3.6.1 and 3.8)** [4]

Q.6 *Define : Heat source, Heat sink, Thermal efficency and Coefficient of performance.* **(Refer sections 3.9, 3.10 and 3.10.1)** [4]

Q.7 *Explain second law of thermodynamics for heat engine.* **(Refer section 3.11)** [4]

Dec. - 2013

Q.8 *Explain with example. i) Closed System ii) Open System* **(Refer section 3.2.1)** [4]

Q.9 *Define : i) Heat source ii) heat sink* **(Refer section 3.9)** [2]

Q.10 *Define : COP; Coefficient of Performance.* **(Refer section 3.10.1)** [1]

Q.11 *State and explain second law of thermodynamics.* **(Refer section 3.11)** [4]

Q.12 *Define : Thermal Efficiency* **(Refer section 3.12.2)** [1]

May - 2014

Q.13 *Explain the various types of thermodynamic system with suitable examples for each.* **(Refer section 3.2.1)** [2]

Q.14 *State whether the following systems are closed/open/isolated systems. Give reasons.*
i) Air compressor   ii) Pressure cooker
iii) Thermas flask **(Refer section 3.2.1)** [2]

Q.15 *State various statements and limitations of first law of thermodynamics.* **(Refer sections 3.6.1 and 3.8)** [4]

Q.16 *Define : a) Heat engine   b) Heat pump* **(Refer section 3.10)** [4]

Dec. - 2014

Q.17 *Define intensive and extensive properties with examples.* **(Refer section 3.3)** [4]

Q.18 *Explain first law of thermodynamics and give its limitations.* **(Refer sections 3.6.1 and 3.8)** [4]

Q.19 *Explain various types of thermodynamic system.* **(Refer section 3.2.1)** [4]

May - 2015

Q.20 *Explain the following terms :*
(1) Zeroth law of thermodynamics
(Refer section 3.5)
(2) Intensive properties **(Refer section 3.3)**
(3) Open system **(Refer section 3.2)**
(4) Heat engine **(Refer section 3.10)** [4]

Q.21 *Explain "Kelvin-Planck and Clausius" statement of second law of thermodynamics.* **(Refer section 3.11)** [4]

Q.22 *Draw a sketch of heat pump and refrigerator using heat source and sink concept.Prove that :*
$(C.O.P.)_{Heat\ Pump} = 1 + (C.O.P.)_{Refrigerator}$
(Refer section 3.10) [4]

Dec. - 2015

Q.23 *Explain the following terms*
(i) Zeroth law of thermodynamics
(Refer section 3.5)
(ii) Extensive properties **(Refer section 3.3)**
(iii) Closed system **(Refer section 3.2.1)**
(iv) Heat engine. **(Refer section 3.9)** [4]

Q.24 *Explain "Kelvin-Planck and Clausius" statement of second law of thermodynamics.* **(Refer section 3.11)** [4]

Q.25 *Draw a sketch of heat pump and refrigerator using heat source and sink concept.Prove that :*
$(C.O.P.)_{Heat\ Pump} = 1 + (C.O.P.)_{Refrigerator}$
(Refer section 3.10) [4]

May - 2016

Q.26 *Explain the following terms*
(i) Zeroth law of thermodynamics
(Refer section 3.5)

(ii) Extensive properties **(Refer section 3.3)**

(iii) Closed system **(Refer section 3.2.1)**

(iv) Heat engine. **(Refer section 3.9)**   **[4]**

Q.27 *Explain "Kelvin-Planck and Clausius" statement of second law of thermodynamics.*
(Refer section 3.11)   **[4]**

Q.28 *Draw a sketch of heat pump and refrigerator using heat source and sink concept.Prove that :*
(C.O.P.)$_{Heat\ Pump}$ = 1 + (C.O.P.)$_{Refrigerator}$
(Refer section 3.10)   **[4]**

Dec. - 2016

Q.29 *Explain the concept of heat engine and refrigerator with neat sketch.*
(Refer section 3.10)   **[5]**

Q.30 *Explain the following :*
1) Any two statements of first law of thermodynamics **(Refer section 3.6.1)**
2) COP of Heat Pump and COP of Refrigerator **(Refer section 3.10.1)**
3) System, surrounding and boundary.
(Refer section 3.2)   **[4]**

May - 2017

Q.31 *State any two statements and limitations of first law of thermodynamics.*
(Refer sections 3.6.1 and 3.8)   **[4]**

Q.32 *Draw schematic sketches of : Isolated system*
(Refer section 3.2.1)   **[1]**

Q.33 *Explain the following :*   **[4]**

i) System, surrounding and boundary
(Refer section 3.2)

ii) Kelvin Plank's statement of second law of thermodynamics. **(Refer section 3.11)**

Q.34 *Draw sketches of heat pump and refrigerator system. Derive the relation between COP of heat pump and COP of refrigerator.*
(Refer sections 3.10 and 3.10.1)   **[4]**

Dec. - 2017

Q.35 *What is thermodynamic system ? Explain various types of thermodynamic systems with example.* **(Refer sections 3.2 and 3.2.1)**   **[4]**

Q.36 *State any two statements and discuss any two limitations of first law of thermodynamics.*
(Refer sections 3.6.1 and 3.8)   **[4]**

Q.37 *Define the following : Heat pump and COP of heat pump, Refrigerator and COP of refrigerator.* **(Refer sections 3.9 and 3.10.1)** **[4]**

May 2018

Q.38 *What is thermodynamic system ? Explain various types of thermodynamic systems with example.* **(Refer sections 3.2 and 3.2.1)**   **[4]**

Q.39 *State any two statements and discuss any two limitations of first law of thermodynamics.*
(Refer section 3.6.1)   **[4]**

Q.40 *Define and explain the following devices with sketch; Heat engine and refrigerator.*
(Refer section 3.10)   **[4]**

Dec. - 2018

Q.41 *State and explain two statements of second law of thermodynamics.* **(Refer section 3.11)**   **[4]**

Q.42 *Discuss limitations of first law of thermodynamics with two examples.*
(Refer section 3.8)   **[4]**

Q.43 *Explain the following : i) Open System and Isolated System. ii) Intensive properties and Extensive properties.*
(Refer sections 3.2.1 and 3.3)   **[4]**

May 2019

Q.44 *Explain following terms : (i) System, sourrounding and Boundary (ii) Kelvin Plank Statement of second low of thermodynamics.*
(Refer sections 3.2 and 3.11)   **[4]**

Q.45 *State any two statements and limitations of first law of thermodynamics.*
(Refer sections 3.6.1 and 3.8)   **[4]**

Q.46 *With neat sketch explain open system closed system and isolated system.*
(Refer section 3.2.1)   **[4]**

❑❑❑

Notes

UNIT - II

4

Heat Transfer

Syllabus

Laws of thermodynamics, heat engine, heat pump, refrigerator (simple numerical) Modes of heat transfer: conduction, convection and radiation, Fourier's law, Newton's law of cooling, Stefan Boltzmann's law. (Simple numerical) Two stroke and Four stroke engines (Petrol, Diesel and CNG engines). Steam generators.

Contents

Mind Map - Heat Transfer

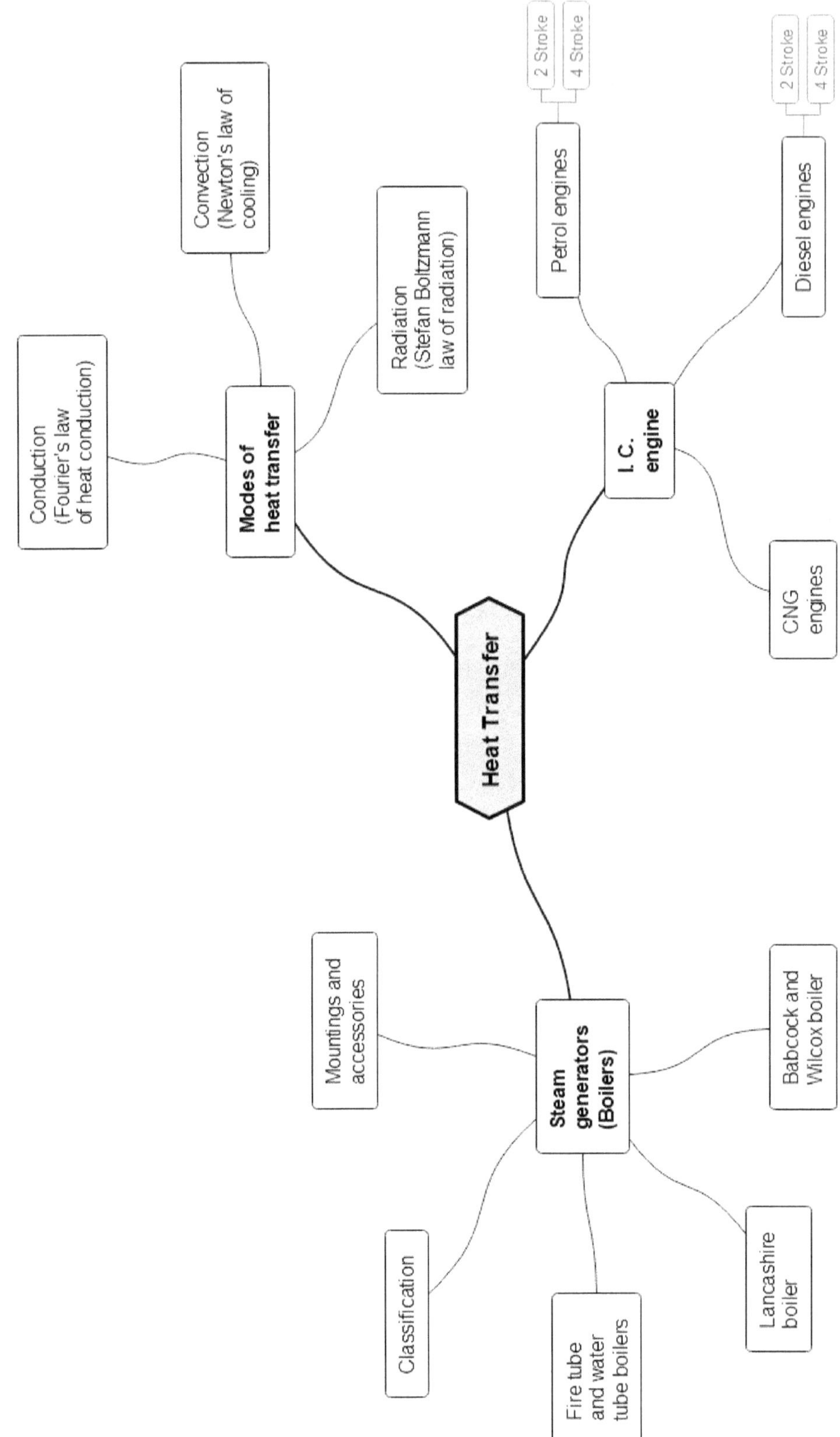

4.1 : Introduction

- **Heat** is the mechanism by which energy is transferred across a boundary between systems, without transfer of mass because of the difference in temperature of the two systems.

- The heat or energy transfer takes place from higher temperature to lower temperature.

- Heat is a transient quantity which can be identified only as it crosses the boundary of a system.

- It is important to note that, heat and temperature are two different terms. The relation between them is that, a temperature difference between two systems is required for heat to be transferred.

- The basic **unit of heat** in S.I. unit system is **Joule.**

- The heat transfer phenomenon plays an important role in many industrial and domestic applications. Some of the important applications are listed below :

  ○ I.C. engines are cooled in order to function properly.

  ○ Houses and buildings using air-conditioning systems.

  ○ The heating and cooling required in food processing plants.

  ○ Melting of metals, plastics, etc. during manufacturing process.

  ○ Cooling of electronic components in computers, T.V., laptops, etc.

  ○ The steam generating equipments like boilers, etc.

  ○ Cooling of motors, generators, transformers, etc.

4.2 : Modes of Heat Transfer

- When temperature difference or gradient exists, heat is always transferred in the direction of lower temperature. This heat can be transferred by three modes :

> 1. Conduction
> 2. Convection
> 3. Radiation

1. Conduction

- Conduction heat through a substance is because of exchange of energy between the molecules by direct interactions.

- It is recognized as the transfer of heat within a substance from high temperature to low temperature region.

- **For example :** When solid rod of metal is heated from one end, then its other end will also get hot.

2. Convection

- When a fluid flows over a hot body, heat will be transferred from hot body to flowing fluid.

- Hence, convection is associated with transfer of heat due to flowing fluid.

- **For example :** When the wind is blowing, the flowing air carries heat away from our body and we feel cool.

3. Radiation

- Conduction and convection modes of heat transfer requires some medium for heat transfer, but in radiation mode there is no need of any medium.

- Radiation can take place in space also, from high temperature body to low temperature body in the form of electromagnetic waves.

- **For example :** During winter season, we feel warm while standing in front of electric heater or fire.

> **Note :** *In all heat transfer modes, a temperature difference must exist to cause heat flow and heat always flows in the direction of lower temperature.*

4.3 : Conduction Mode of Heat Transfer

- It is already discussed that, conduction of heat through a substance is due to exchange of energy between the molecules by direct interactions.

- Conduction occurs in solids, liquids and gases also.

- The conduction phenomena is greatly influenced by the flow of free electrons.

4.3.1 Fourier's Law of Heat Conduction

SPPU : May-06, Dec.-09

- The important condition for conduction to take place is the presense of temperature difference or gradient.

- The relationship between the rate of heat flow and temperature gradient can be derived from Fourier's law.

- It states that *rate of heat flow by conduction per unit area normal to the direction of heat flow in any direction directly proportional to the temperature gradient present in that direction.*

$\therefore$
$$\frac{Q}{A} \propto \frac{dT}{dx}$$

$\therefore$
$$\boxed{\frac{Q}{A} = -K \frac{dT}{dx}} \qquad \text{... (4.3.1)}$$

Where, $\quad Q = $ Rate of heat flow in J/s or Watt,

$\quad A = $ Area normal to the direction of heat flow in m^2,

$\quad \dfrac{dT}{dx} = $ Temperature gradient in the direction of heat flow in $^\circ$C/m or K/m

(Here X-direction),

$\quad K = $ Constant of proportionality and it is called as **thermal conductivity**

of the material in $\dfrac{W}{m\text{-}K}$ or $\dfrac{W}{m\text{-}^\circ C}$

- The negative sign in the above equation indicates that, the heat always flows in the direction of lower temperature. Hence, the temperature gradient in the direction of heat flow is negative.

4.3.2 Heat Conduction through a Slab or Thick Wall

- Consider a wall or slab of surface area A and thickness be as shown in Fig. 4.3.1.
- Let, Q be the rate of heat flow in the direction as shown and the wall faces are maintained at temperature T_1 and T_2 respectively.
- According to the Fourier's law of heat conduction

$$\frac{Q}{A} = -K \frac{dT}{dx} \qquad \text{... [From equation (4.3.1)]}$$

$\therefore \quad Q\,dx = -K\,A\,d\,T$

Integrating between the boundary conditions,

at $\qquad\qquad x = 0, \quad T = T_1$

and $\qquad\qquad x = b, \quad T = T_2$

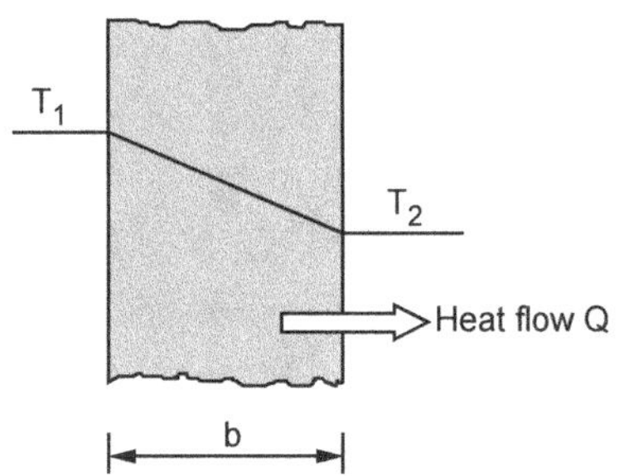

Fig. 4.3.1 : Heat conduction through a slab

$\therefore \qquad Q\displaystyle\int_{0}^{b} dx = -KA \int_{T_1}^{T_2} dT$

$\therefore \qquad Q[x]_0^b = -KA[T]_{T_1}^{T_2}$

$\therefore \qquad Q{\cdot}b = -KA\,(T_2 - T_1)$

$$\therefore \qquad Q = \frac{KA}{b}(T_1 - T_2)$$

... (4.3.2)

4.3.3 Electrical Analogy of Heat Conduction

SPPU : Dec.-07, 09, May-09

- The heat flow calculations through composite sections like wall of boiler house, insulated steam pipe, etc. can be easily worked out by using a technique called as **electrical analogy.**

- The process of heat flow by conduction through a substance can be proved to be analogous with the process of flow of electrons through the conductor as follows :

Sr. No.	Quantity	Analogous parameters	
		Electrical system	**Heat conduction**
1.	Quantity that flows	Electrons	Heat
2.	Rate of flow	Current I in Amp.	Heat transfer Q in Watts.
3.	Driving force	Potential difference V in volts.	Temperature difference ΔT in ºC or K.
4.	Responsible factors	i) Resistivity ρ, ii) Length of conductor L, iii) Cross-sectional area of conductor A.	i) Thermal conductivity K, ii) Width of slab b, iii) Area normal to the direction of heat flow.

- According to the ohm's law,

$$V = I \cdot R \qquad or \qquad I = \frac{V}{R}$$

... (a)

We know that,

$$R = \frac{\rho \cdot L}{A} \qquad or \qquad R = f(\rho, L, A)$$

So equation (a) becomes,

$$I = \frac{V}{\left(\dfrac{\rho \cdot L}{A}\right)}$$

... (b)

- From equation (4.3.2) it is clear that rate of heat flow in conduction depends on K, A and b.

- A function that defines three parameters K, A and b will be an analogous term to the electrical resistance which is called as **thermal resistance** $(\mathbf{R_{th}})$. Refer Fig. 4.3.2.

We can write,

$$R_{Th} = f(K, b, A)$$

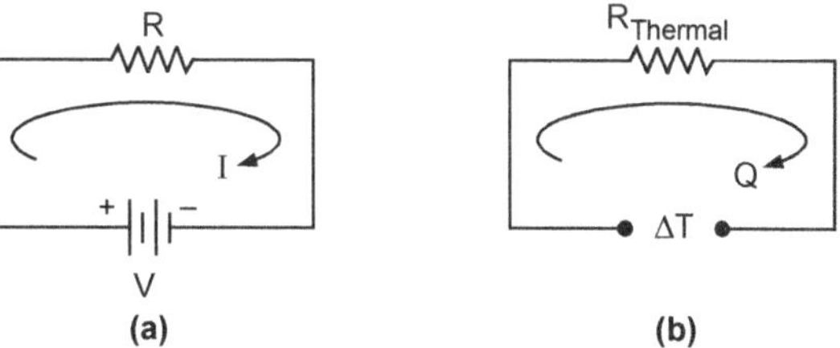

Fig. 4.3.2 : Electrical analogy of heat conduction

From equation (4.2) we can write,

$$Q = \frac{KA}{b}(T_1 - T_2) = \frac{KA}{b}\Delta T$$

$$\therefore \qquad Q = \frac{\Delta T}{\left(\dfrac{b}{KA}\right)} = \frac{\Delta T}{R_{Th}}$$

... (4.3.3)

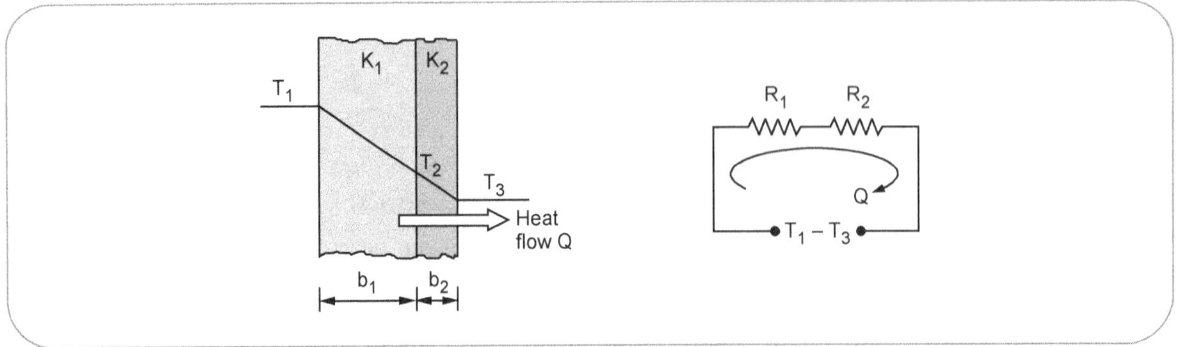

Fig. 4.3.3 : Composite wall in series

- By comparing equations (a), (b) and (4.3.3) it is clear that, the process of heat flow is analogous with the process of flow of current, and $R_{Th} = b/KA$.

4.3.4 Heat Conduction through Composite Wall

SPPU : May-09, 10, Dec.-09

Composite Wall in Series

- Consider a composite wall of surface area A and thickness b_1 and b_2 respectively. Refer Fig. 4.3.3.

- Let, Q be the rate of heat flow in the direction as shown and the wall faces are maintained at temperature T_1, T_2 and T_3.

- The composite wall is made of two different materials hence their thermal conductivities are also different. Let, K_1 and K_2 be the thermal conductivity of materials.

- We know that, for resistors in series

$$R_{eq} = R_{Th} = R_1 + R_2$$

But, $\quad R_1 = \dfrac{b_1}{K_1 A} \quad$ and $\quad R_2 = \dfrac{b_2}{K_2 A}$

$$\text{... [From equation (4.3.3)]}$$

$$\therefore \quad R_{Th} = \dfrac{b_1}{K_1 A} + \dfrac{b_2}{K_2 A}$$

- Rate of heat flow is,

$$\therefore \quad \boxed{Q = \dfrac{\Delta T}{R_{Th}} = \dfrac{(T_1 - T_3)}{R_{Th}}} \quad \text{... (4.3.4)}$$

Composite wall in Parallel

- Consider a composite wall of same thickness b as shown in Fig. 4.3.4.

- Let the wall faces are maintained at temperatures T_1 and T_2.

- We know that for resistors in parallel,

$$\dfrac{1}{R_{eq}} = \dfrac{1}{R_{Th}} = \dfrac{1}{R_1} + \dfrac{1}{R_2}$$

$$\therefore \quad R_{Th} = \dfrac{R_1 R_2}{R_1 + R_2}$$

But, $\quad R_1 = \dfrac{b}{K_1 A_1} \quad$ and $\quad R_2 = \dfrac{b}{K_2 A_2}$

- Rate of heat flow is,

$$\therefore \quad \boxed{Q = \dfrac{\Delta T}{R_{Th}} = \dfrac{(T_1 - T_2)}{R_{Th}}} \quad \text{... (4.3.5)}$$

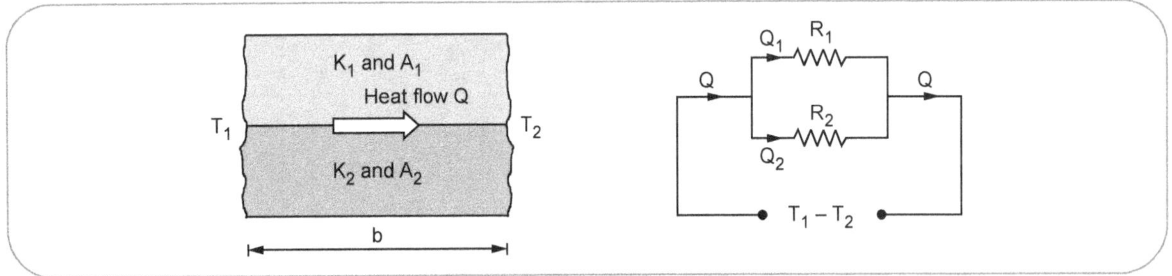

Fig. 4.3.4 : Composite wall in parallel

4.3.5 Heat Conduction through Infinitely Long Cylinder   `SPPU : May-04, 08, Dec.-05`

- Consider an infinitely long hollow cylinder with inner and outer radii as r_1 and r_2. The temperatures of the inner and outer surfaces are maintained at T_1 and T_2 respectively.

- Let, L be the length of cylinder and K be the thermal conductivity of cylinder material.

- As the length of cylinder is infinite, neglect the temperature gradient along the axis and consider heat flow only in radial direction. Refer Fig. 4.3.5.

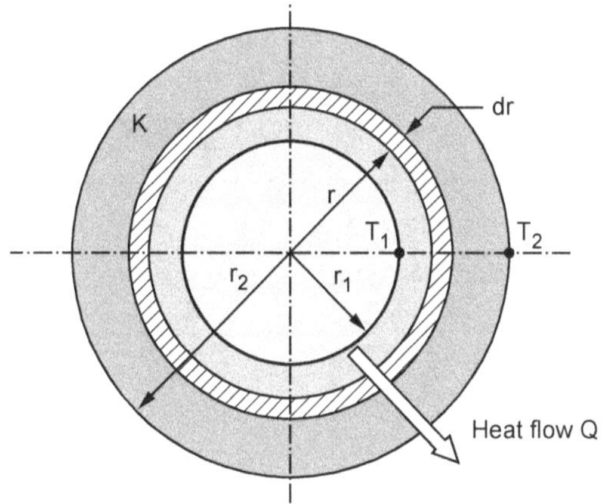

Fig. 4.3.5 : Heat conduction through hollow cylinder

- Consider an elemental ring of thickness 'dr' at a distance 'r' from the centre. The surface area of the ring is $A = 2\pi r L$.

From Fourier's law of heat conduction,

$$\frac{Q}{A} = -K \frac{dT}{dx}$$

Here,

$$\frac{Q}{A} = -K \frac{dT}{dr}$$

$$\therefore \quad \frac{Q}{2\pi r L} \cdot dr = -K \, dT \qquad \dots (\because A = 2\pi r L)$$

Integrating between the boundary conditions,

at $\quad r = r_1, \quad T = T_1$

and at $\quad r = r_2, \quad T = T_2$

$$\therefore \quad \frac{Q}{2\pi L} \int_{r_1}^{r_2} \frac{dr}{r} = -K \int_{T_1}^{T_2} dT$$

$$\therefore \quad \frac{Q}{2\pi L} \left[\log (r)\right]_{r_1}^{r_2} = -K \left[T\right]_{T_1}^{T_2}$$

$$\therefore \quad \frac{Q}{2\pi L} \cdot \log\left(\frac{r_2}{r_1}\right) = K \left[T_1 - T_2\right]$$

$$\therefore \quad \boxed{Q = \frac{2\pi L K (T_1 - T_2)}{\log\left(\frac{r_2}{r_1}\right)}} \qquad \dots (4.3.6)$$

or $\qquad Q = \dfrac{(T_1 - T_2)}{\left(\dfrac{\log (r_2/r_1)}{2\pi L K}\right)} = \dfrac{\Delta T}{R_{Th}}$

$$\therefore \quad R_{Th} = \frac{\log (r_2/r_1)}{2\pi L K}$$

4.3.6 Heat Conduction through Hollow Sphere   `SPPU : May-05, Dec.-05`

- Consider a hollow sphere with inner and outer radii r_1 and r_2. The temperatures of the inner and outer surfaces are, maintained at T_1 and T_2 respectively. Refer Fig. 4.3.6.

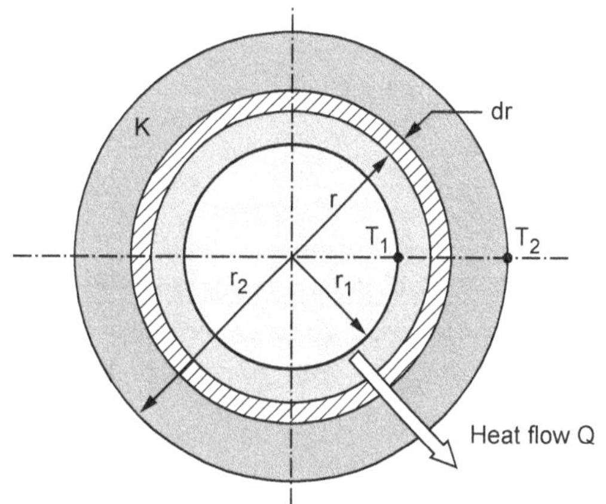

Fig. 4.3.6 : Heat conduction through hollow sphere

- Let, K be the thermal conductivity of sphere material and heat flow only in radial direction.

- Consider an elemental ring of thickness 'dr' at radius 'r' from the centre. The surface area of the ring is $A = 4\pi r^2$.

From Fourier's law of heat conduction,

$$\frac{Q}{A} = -K \frac{dT}{dx}$$

Here,

$$\frac{Q}{A} = -K \frac{dT}{dr}$$

$$\therefore \quad \frac{Q}{4\pi r^2} \cdot dr = -K\, dT \qquad \dots (\because A = 4\pi r^2)$$

Integrating between the boundary conditions,

at $\quad r = r_1, \quad T = T_1$

and $\quad r = r_2, \quad T = T_2$

$$\therefore \quad \frac{Q}{4\pi} \int_{r_1}^{r_2} \frac{dr}{r^2} = -K \int_{T_1}^{T_2} dT$$

$$\therefore \quad \frac{Q}{4\pi}\left[-\frac{1}{r}\right]_{r_1}^{r_2} = -K\,[T]_{T_1}^{T_2}$$

$$\therefore \quad \frac{Q}{4\pi}\left[\frac{1}{r_1} - \frac{1}{r_2}\right] = K\,(T_1 - T_2)$$

$$\therefore \quad \frac{Q}{4\pi}\frac{(r_2 - r_1)}{r_1\, r_2} = K\,(T_1 - T_2)$$

$$\therefore \qquad \boxed{Q = \frac{4\pi K \cdot r_1\, r_2\,(T_1 - T_2)}{(r_2 - r_1)}} \qquad \dots (4.3.7)$$

$$\text{or} \qquad Q = \frac{(T_1 - T_2)}{\dfrac{(r_2 - r_1)}{4\pi K \cdot r_1 r_2}} = \frac{\Delta T}{R_{Th}}$$

$$\therefore \qquad R_{Th} = \frac{(r_2 - r_1)}{4\pi K \cdot r_1 r_2}$$

4.4 : Convection Mode of Heat Transfer

- It is already discussed that, convection mode of heat transfer is associated with transfer of heat due to flowing fluid.

- In this process, the flow of heat is primarily due to the movement of fluid molecules.

- According to the physical behaviour of the fluid involved in convection and the force causing the flow of fluid over the surface, the convection process is classified as follows :

> 1. Forced convection
> 2. Natural convection

1. Forced convection

- When the fluid over a hot or cold surface under external pressure, then the flow is called as **forced flow** and heat transfer is known as **forced convection.**

- The force which acts on the fluid to cause the motion of fluid is due to fan, blower, etc.

- In this type, due to forced flow high rate of heat transfer is obtained.

- **For example :** To obtain the flow of flue gases in the boiler fans are used.

2. Natural Convection

- In this type, the motion of fluid molecules takes place not because of any external forces.

- It is due to inherent forces like gravitational force, buoyancy force, etc.

- In this case, rate of heat transfer is low.

- **For example :** Cooling of tea put in a saucer.

4.4.1 Newton's Law of Cooling

SPPU : Dec.-08, 09, May-10

- Rate of heat flow in convection process is found by using Newton's law of cooling.

- It states that *the rate of heat transfer is directly proportional to surface area and the temperature difference between the surface and flowing fluid in the direction of heat flow.*

$$\therefore \qquad Q \propto A\,(T - T_f)$$

$$\therefore \qquad \boxed{Q = h\,A\,(T - T_f)} \qquad \dots (4.4.1)$$

Where, Q = Rate of heat flow in J/s or or Watt,

$\quad A$ = Surface are in m^2,

$\quad T$ = Temperature of the surface in °C or K,

$\quad T_f$ = Temperature of surrounding fluid in °C or K,

$\quad h$ = Constant of proportionality and it is called as **convective heat transfer** coefficient in $\dfrac{W}{m^2 \text{-°C}}$ or $\dfrac{W}{m^2 \text{-K}}$.

- Rearranging the equation (4.4.1),

$$Q = \frac{(T - T_f)}{\dfrac{1}{hA}} = \frac{\Delta T}{R_{Th}}$$

$$\therefore \qquad R_{Th} = \frac{1}{hA} \qquad \dots (4.4.2)$$

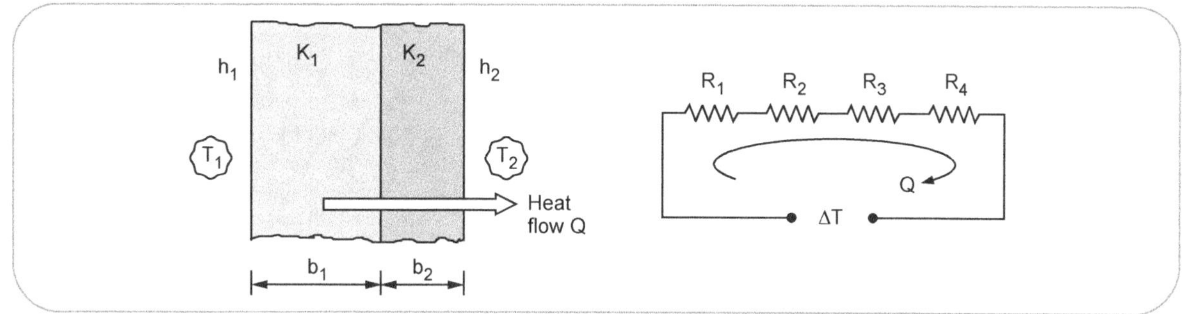

Fig. 4.4.1 : Overall heat transfer coefficient for composite slab

4.4.2 Overall Heat Transfer Coefficient

- Overall heat transfer coefficient is the term equivalent to the convective heat transfer coefficient (h) which can be used to find the rate of heat flow through the composite structures.

- To find the rate of heat flow, the equation similar to Newton's law cooling is used i.e.,

$$\therefore \quad \boxed{Q = U A (T - T_f) = U A (\Delta T)} \quad \dots (4.4.3)$$

Where $\quad U$ = Overall heat transfer coefficient in

$$W/m^2 \text{-K or } W/m^2 \text{-}^\circ C$$

- From electrical analogy, we know that

$$Q = \frac{\Delta T}{R_{eq}} = \frac{\Delta T}{\Sigma R} \quad \dots (4.4.4)$$

- Comparing the equations (4.4.3) and (4.4.4),

$$\Sigma R = \frac{1}{UA} \text{ or } UA = \frac{1}{\Sigma R} \quad \dots (4.4.5)$$

- For various cases the value of U can be obtained as follows :

1. Composite slab

- Consider a composite slab having thickness b_1 and b_2 with thermal conductivity of materials K_1 and K_2. Refer Fig. 4.4.1.

- Let, T_1 and T_2 be the surface temperatures and h_1 and h_2 be the convective heat transfer coefficients of respective surfaces.

- In this case, composite slab is equivalent to four resistors in series.

- Out of four resistors, R_1 and R_4 represents convective resistance whereas, R_2 and R_3 represents conductive resistance.

- As per equation (4.4.5),

$$UA = \frac{1}{\Sigma R} = \frac{1}{R_1 + R_2 + R_3 + R_4}$$

$$\therefore \quad UA = \frac{1}{\dfrac{1}{h_1 A_1} + \dfrac{b_1}{K_1 A_1} + \dfrac{b_2}{K_2 A_2} + \dfrac{1}{h_2 A_2}}$$

$$\dots \text{[From equations (4.4.2) and (4.3.3)]}$$

If, $\quad A = A_1 = A_2$

$$\therefore \quad \boxed{U = \frac{1}{\dfrac{1}{h_1} + \dfrac{b_1}{K_1} + \dfrac{b_2}{K_2} + \dfrac{1}{h_2}}} \quad \dots (4.4.6)$$

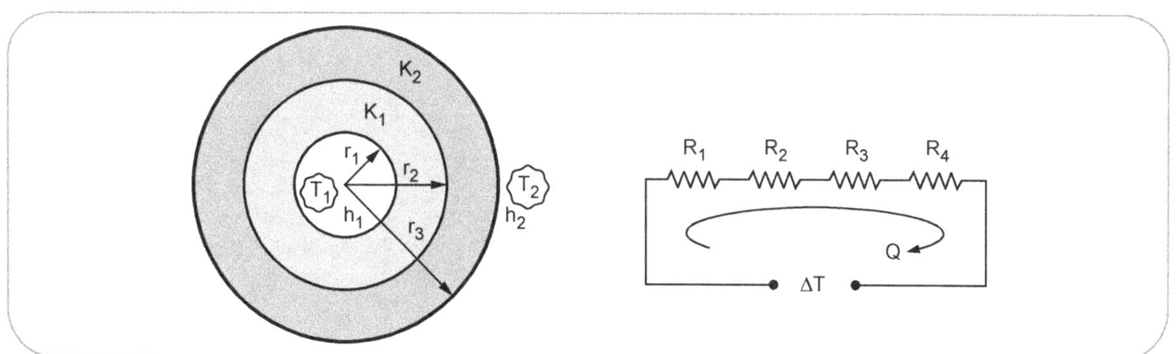

Fig. 4.4.2 : Overall heat transfer coefficient for composite cylinder

2. Composite cylinder

Consider a composite cylinder having radii r_1, r_2 and r_3 with thermal conductivity of materials K_1 and K_2. Refer Fig. 4.4.2.

Let, T_1 and T_2 be the inner and outer temperatures and h_1 and h_2 be the convective heat transfer coefficients of respective surface. Consider L as length of cylinder.

From equation (4.4.5),

$$UA = \frac{1}{\Sigma R} = \frac{1}{R_1 + R_2 + R_3 + R_4}$$

$\therefore \qquad UA = \dfrac{1}{\dfrac{1}{h_1 A_1} + R_1 + R_3 + \dfrac{1}{h_2 A_2}} \qquad \text{... [From equation (4.4.2)]}$

$\therefore \qquad UA = \dfrac{1}{\dfrac{1}{h_1 \cdot 2\pi r_1 L} + \dfrac{\log\left(\dfrac{r_2}{r_1}\right)}{2\pi L K_1} + \dfrac{\log\left(\dfrac{r_3}{r_2}\right)}{2\pi L K_2} + \dfrac{1}{h_2 \cdot 2\pi r_2 L}} \qquad \text{... (4.4.7)}$

4.5 : Radiation Mode of Heat Transfer

- It is already discussed that, in radiation mode of heat transfer there is no need of any medium.

- The heat transfer from one body to another body without any transmitting medium is called as **radiation** mode of heat transfer.

- This mode of heat transfer is associated with electromagnetic waves.

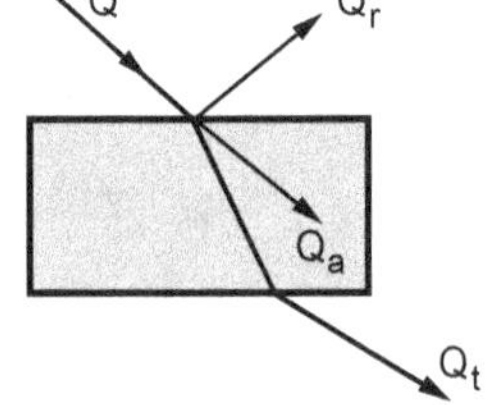

Fig. 4.5.1 : Radiation mode of heat transfer

- The radiant energy falling on a body is partly reflected, partly transmitted and partly absorbed. Refer Fig. 4.5.1.

- Let, Q is the incident energy falling on a body, Q_a is the energy absorbed, Q_t is the energy transmitted and Q_r is the energy reflected.

- Hence, total incident energy Q is,

$$Q = Q_a + Q_r + Q_t$$

or $\qquad 1 = \dfrac{Q_a}{Q} + \dfrac{Q_r}{Q} + \dfrac{Q_t}{Q}$

$\therefore \qquad 1 = \alpha + \gamma + \tau$

or $\qquad \alpha + \gamma + \tau = 1 \qquad\qquad \text{... (4.5.1)}$

Where, $\qquad \alpha$ = Coefficient of absorptivity

$\qquad\qquad \gamma$ = Coefficient or reflectivity

$\qquad\qquad \tau$ = Coefficient of transmittivity

1. Coefficient of Absorptivity :

It is defined as the ratio of amount of energy absorbed (Q_a) to the amount of energy incident on the body (Q). It is denoted by α.

$$\therefore \qquad \alpha = \frac{Q_a}{Q}$$

2. Coefficient of Reflectivity :

It is defined as the ratio of amount of energy reflected (Q_r) to the amount of energy incident on the body (Q). It is denoted by γ.

$$\therefore \qquad \gamma = \frac{Q_r}{Q}$$

3. Coefficient of Transmittivity :

It is defined as the ratio of amount of energy transmitted (Q_t) to the amount of energy incident on the body (Q). It is denoted by τ.

$$\therefore \qquad \tau = \frac{Q_t}{Q}$$

- A body which absorbs all the incident radiation is called as **black body** irrespective of its colour.
- For a black body, $\alpha = 1$ and $\gamma = \tau = 0$.

4.5.1 Stefan Boltzmann Law of Radiation

SPPU : May 10

- The Stefan Boltzmann law is used to find the emussive power of a black body.
- It states that *the emussive power of a black body is directly proportional to the fourth power of its absolute temperature.*

$$\therefore \qquad E_b \propto T^4$$

$$\therefore \qquad \boxed{E_b = \sigma \cdot T^4} \qquad \dots (4.5.2)$$

Where, $\qquad E_b$ = Emussive power of a black body per unit area in W/m^2,

$\qquad\qquad T$ = Absolute temperature in K,

$\qquad\qquad \sigma$ = Constant of proportionality and it is called as Stefan Boltzmann's constant in $\dfrac{W}{m^2 \text{-}K^4}$

- The value of σ is constant which is,

$$\sigma = 5.67 \times 10^{-8} / W\, m^2 \text{-} K^4$$

Emussive power of a black body (E_b) is defined as the rate at which the radiant flux (heat per unit area) is emitted from a surface at a certain temperature. Its unit is W/m^2.

4.6 : List of Formulae

1) Heat flux, $\dfrac{Q}{A} = -K \dfrac{dT}{dx}$

2) For thick wall or slab, $Q = \dfrac{KA}{b}(\Delta T)$

3) For composite wall or slab, $Q = \dfrac{\Delta T}{\Sigma R}$

4) For infinitely long hollow cylinder, $Q = \dfrac{2\pi L K (\Delta T)}{\log\left(\dfrac{r_2}{r_1}\right)}$

5) For hollow sphere, $Q = \dfrac{4\pi K r_1 r_2 (\Delta T)}{(r_2 - r_1)}$

6) For convection, $Q = hA(\Delta T)$

7) For overall heat transfer coefficient, $Q = UA(\Delta T)$

8) For radiation, $\alpha + \gamma + \tau = 1$

9) Emissive power, $E_b = \sigma T^4$

$\qquad$ where, $\sigma = 5.67 \times 10^{-8}\ W/m^2 \text{-} K^4$

4.7 : Solved Examples

Ex. 4.7.1 : *The glass windows of a room has total area of 10 m^2 and glass is 4 mm thick. Calculate quantity of heat leaving from room through glass, when inside surface of window is at 25 °C and outside surface is at 10 °C. The value of thermal conductivity for a glass is 0.84 W/mK.*

SPPU : Dec.-09, Marks 4

Sol : **Given data :** $A = 10\,m^2$,

$b = 4\,mm = 4 \times 10^{-3}\,m$, $\qquad T_1 = 25\,°C$, $\qquad T_2 = 10\,°C$,

$K = 0.84\,W/mK$

To find : Heat loss, Q

Step - 1 : Calculate the heat leaving from the window

From Fig. 4.7.1,

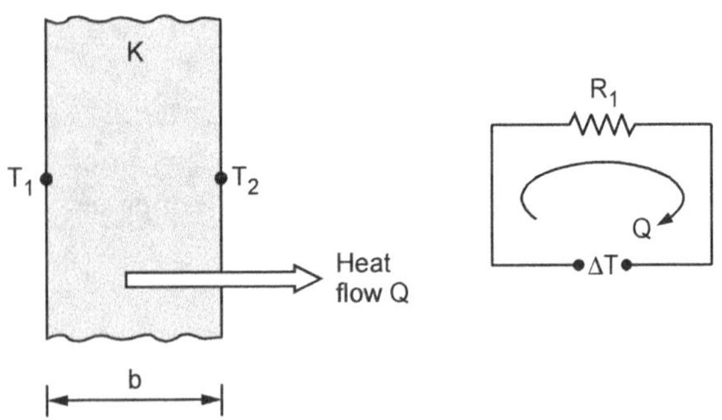

Fig. 4.7.1

$$R = \frac{b}{KA} = \frac{4\times 10^{-3}}{0.84\times 10} = 4.762\times 10^{-4} \text{ K/W}$$

We know that,

$$Q = \frac{\Delta T}{\Sigma R} = \frac{(25-10)}{4.762\times 10^{-4}}$$

$$\therefore \qquad \mathbf{Q = 31.499\times 10^3 \ W} \qquad \textbf{... Ans.}$$

Ex. 4.7.2 : *Calculate rate of heat transfer by convection between roof of area $20\times 20\ m^2$ and ambient air, if roof temperature is 10 °C and air temperature is 40 °C. Assume average heat transfer coefficient for convection as 10 $W/m^2 K$. Comment about heat flow.* **SPPU : May-10, Marks 3**

Sol. : Given data : $A = 20 \times 20\,\text{m}^2 = 400\ \text{m}^2$, $T = 10\,^\circ C$
$T_f = 40\,^\circ C$, $h = 10\ W/m^2 K$

To find : Rate of heat transfer, Q

Step - 1 : Calculate the rate of heat transfer from the roof
From Newton's law of cooling,

$$Q = h\,A\,(\Delta T) = h\,A\,(T - T_f)$$

$$\text{... [From equation (4.4.1)]}$$

$$\therefore \qquad Q = 10\times 400\times (10-40)$$

$$\therefore \qquad \mathbf{Q = -120\times 10^3 \ W} \qquad \textbf{... Ans.}$$

Negative sign indicates that, the direction of heat flow is opposite to the direction of x. It means, heat flow from roof to the surrounding. **... Ans.**

Ex. 4.7.3 : *A wire 1.5 mm in diameter and 150 mm long is submerged in fluid. An electric current is passed through wire and is increased until the fluid reaches 100 °C. Under the condition if convective heat transfer coefficient is 4500 $W/m^2\ °C$, find how much electrical power must be supplied to wire to maintain wire surface at 120 °C?* **SPPU : Dec.-10, Marks 6**

Sol. : Given data : Wire diameter d = 1.5 mm
$= 1.5 \times 10^{-3}\,\text{m}$, Length of wire L = 150 mm = 0.15 m,
Temperature of wire surface $T = 120\ ^\circ C$,

Temperature of fluid $T_f = 100\,^\circ C$,

Convective heat transfer coefficient $h = 4500\ W/m^2\ ^\circ C$

To find : Electrical power (Q)

Step - 1 : Calculate the amount of electrical power
We know that,

Supplied electrical power = Heat loss by convection (Q)

But, $\qquad Q = hA\,(T - T_f)$

$$= h \times \pi dL \times (T - T_f) \qquad (\because A = \pi dL)$$

$$\therefore \qquad Q = 4500\times \pi\times 1.5\times 10^{-3} \times 0.15\times (120 - 100)$$

$$\therefore \qquad \mathbf{Q = 63.617 \ W} \qquad \textbf{... Ans.}$$

Ex. 4.7.4 : *A 60 W incandescent lamp has coil surface temperature of 2500 K and room temperature of 300 K. Estimate the surface area of the coil.* **SPPU : Dec.-09, Marks 4**

Sol. : Given data : $T_1 = 2500\ K$, $T_2 = 300\ K$,
Q = 60 W

To find : Surface area, A

Step - 1 : Calculate the surface area of the coil
From Stefan Boltzmann law,

$$E = \sigma T^4$$

$$\text{or} \qquad \frac{Q}{A} = \sigma T^4 = \sigma\left(T_1^4 - T_2^4\right)$$

$$\therefore \qquad \frac{60}{A} = 5.67\times 10^{-8}\,[(2500)^4 - (300)^4]$$

$$\text{... } (\because \sigma = 5.67\times 10^{-8})$$

$$\therefore \qquad \mathbf{A = 2.7095\times 10^{-5} \ m^2} \qquad \textbf{... Ans.}$$

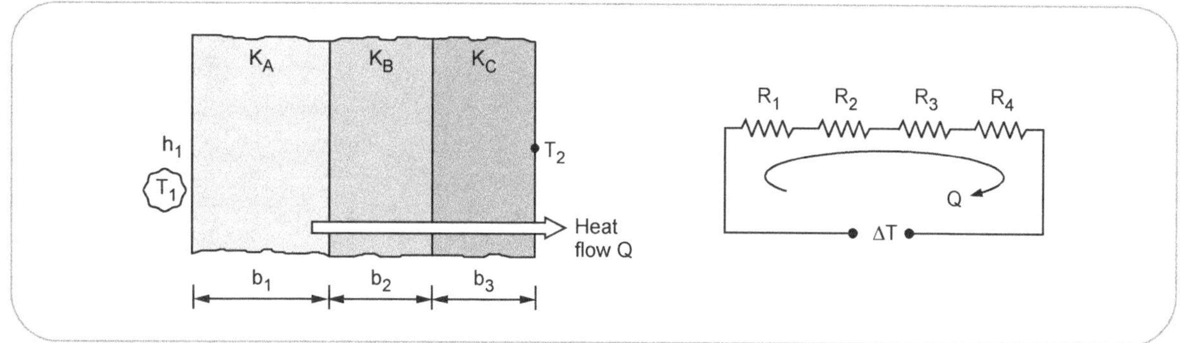

Fig. 4.7.2

Ex. 4.7.5 : *Calculate rate of heat transfer per m^2 through wall of 200 mm thick inner layer of 'A', a central layer of 'B' 100 mm thick and a outer layer of 'C' 100 mm thick. Temperature of gas in the furnace is 1670 °C with $h_{in} = 74$ W/m^2 ° C and outside surface temperature of 'C' is 70 °C.*
Given : $K_A = 1.25$ $W/m° C$, $K_B = 0.074$ $W/m° C$
$K_C = 0.55$ $W/m ° C$
Assume steady state, 1-D flow of heat.

SPPU : Dec.-10, Marks 6

Sol. : **Given data :** $T_1 = 1670 ° C$, $T_2 = 70 ° C$,
$b_1 = 200$ mm $= 0.2$ m, $b_2 = 100$ mm $= 0.1$ m,
$b_3 = 100$ mm $= 0.1$ m, $h_1 = 74$ W/m^2 ° C,
$K_A = 1.25$ W/m ° C, $K_B = 0.074$ W/m° C,
$K_C = 0.55$ W/m ° C

To find : Q/A

Step - 1 : Calculate the amount of heat transfer per unit area
From Fig. 4.7.2,

$$R_1 = \frac{1}{h_1 A} = \frac{1}{74 \times A} = \text{K/W}$$

$$R_2 = \frac{b_1}{K_A A} = \frac{0.2}{1.25 \times A} = \frac{0.16}{A} \text{K/W}$$

$$R_3 = \frac{b_2}{K_B A} = \frac{0.1}{0.074 \times A} = \frac{1.351}{A} \text{K/W}$$

$$R_4 = \frac{b_3}{K_C A} = \frac{0.1}{0.55 \times A} = \frac{0.182}{A} \text{K/W}$$

and $$\Sigma R = R_1 + R_2 + R_3 + R_4$$

$$= \frac{1}{74 A} + \frac{0.16}{A} + \frac{1.351}{A} + \frac{0.182}{A}$$

$\therefore$ $$\Sigma R = \frac{1.706}{A} \text{K/W}$$

We know that,

$$Q = \frac{\Delta T}{\Sigma R} = \frac{(T_1 - T_2)}{\Sigma R} = \frac{(1670 - 70)}{\left(\frac{1.706}{A}\right)}$$

$\therefore$ $$\frac{Q}{A} = 937.87 \, W/m^2 \qquad \text{... Ans.}$$

Ex. 4.7.6 : *A composite wall consists of 1.5 mm thick steel sheet and 10 mm plywood sheet separated by 2 mm thick glasswool in between. Calculate the rate of heat flow per m^2 if the temperatures on the steel sheet and plywood sides are 25 °C and − 15 °C respectively. Also calculate interface temperatures.*
K for steel = 23.23 W/mK
K for plywood = 0.052 W/mK
K for glasswool = 0.014 W/mK

SPPU : Dec.-08, Marks 8

Sol. : **Given data :** $b_1 = 1.5$ mm $= 0.0015$ m,
$b_2 = 2$ mm $= 0.002$ mm, $b_3 = 10$ mm $= 0.01$ m,
$K_1 = 23.23$ W/mK, $K_2 = 0.014$ W/mK, $K_3 = 0.052$ W/mK,
$T_1 = 25 ° C$, $T_2 = -15 ° C$

To find : i) Q ii) T_3 and T_4

Step - 1 : Calculate the rate of heat transfer
From Fig. 4.7.3,

$$R_1 = \frac{b_1}{K_1 A} = \frac{0.0015}{23.23 \times 1} = 6.457 \times 10^{-5} \text{ K/W}$$

$$R_2 = \frac{b_2}{K_2 A} = \frac{0.002}{0.014 \times 1} = 0.1428 \text{ K/W}$$

and $$R_3 = \frac{b_3}{K_3 A} = \frac{0.01}{0.052 \times 1} = 0.1923 \text{ K/W}$$

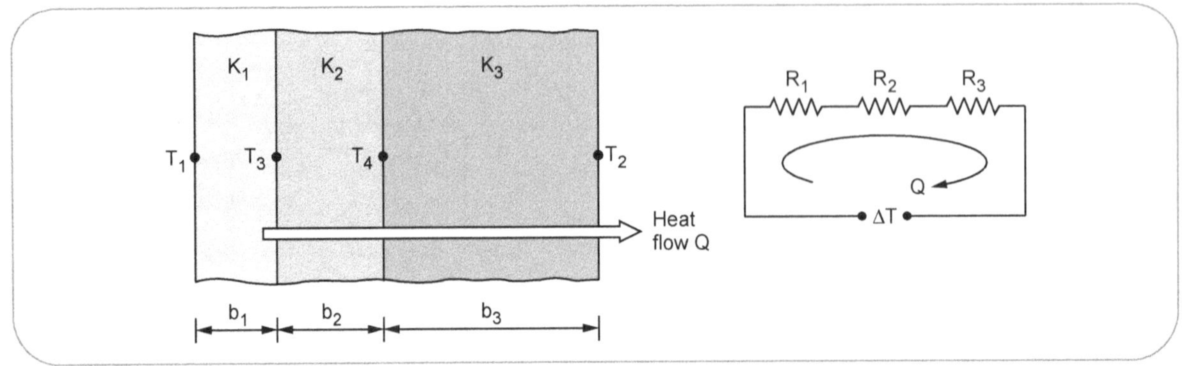

Fig. 4.7.3

Now, $\quad \Sigma R = R_1 + R_2 + R_3$

$$= 6.457 \times 10^{-5} + 0.1428 + 0.1923$$

$\therefore \quad \Sigma R = 0.3352 \ K/W$

We know that,

$\therefore \quad Q = \dfrac{\Delta T}{\Sigma R} = \dfrac{(T_1 - T_2)}{\Sigma R} = \left[\dfrac{25 - (-15)}{0.3352}\right]$

$\therefore \quad \mathbf{Q = 119.332 \ W} \qquad \text{... Ans.}$

Step - 2 : Calculate the interface temperatures

Now, $\quad Q = \dfrac{\Delta T}{\Sigma R} = \dfrac{(T_1 - T_3)}{R_1}$

$\therefore \quad 119.332 = \dfrac{(25 - T_3)}{6.457 \times 10^{-5}}$

$\therefore \quad \mathbf{T_3 = 24.99 \ °C} \qquad \text{... Ans.}$

Similarly, $\quad Q = \dfrac{\Delta T}{\Sigma R} = \dfrac{(T_4 - T_2)}{R_3}$

$\therefore \quad 119.332 = \dfrac{[T_4 - (-15)]}{0.1923}$

$\therefore \quad \mathbf{T_4 = 7.947 \ °C} \qquad \text{... Ans.}$

Ex. 4.7.7 : *A furnace wall is made of 75 mm thick fire clay brick and 6.4 mm thick mild steel plate. The inside surface of the brick at 647 °C and outside air temperature is 27 °C. Determine the heat loss per m^2 area of the furnace wall and outside surface temperature of the steel plate. Take conductivity of brick = 1.1 W/m°C, conductivity of steel = 39 W/m°C and outside heat transfer coefficient = 68 W/m² °C.*

SPPU : Dec.-05, Marks 7

Sol. : Given data : $b_1 = 75$ mm $= 0.075$ m ,
$b_2 = 6.4$ mm $= 0.0064$ m, $T_1 = 647\,°C$, $T_3 = 27\,°C$,
$K_1 = 1.1 \ W/m\,°C$, $K_2 = 39 \ W/m\,°C$, $h_2 = 68 \ W/m^2\,°C$

To find : i) Q/A \quad ii) T_2

Step - 1 : Calculate heat loss from the furnace wall
From Fig. 4.7.4,

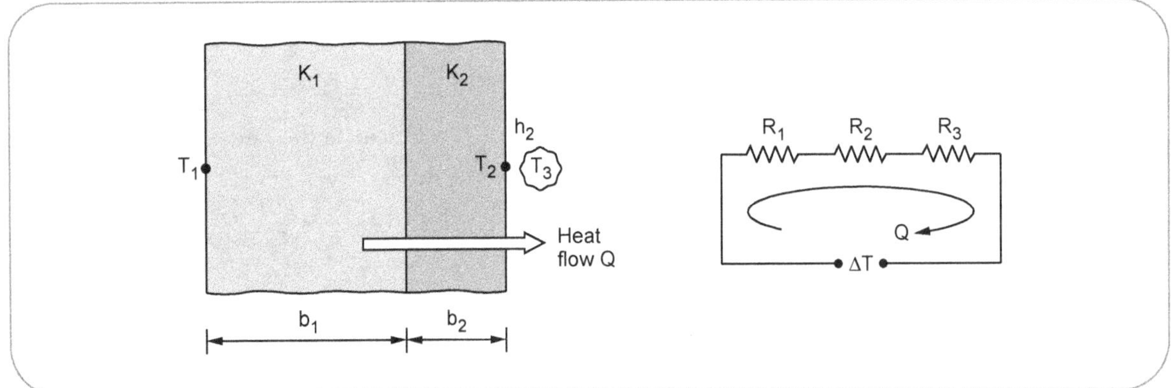

Fig. 4.7.4

$$R_1 = \frac{b_1}{K_1 A} = \frac{0.075}{1.1 \times A} = \frac{0.0682}{A} \text{ K/W}$$

$$R_2 = \frac{b_2}{K_2 A} = \frac{0.0064}{39 \times A} = \frac{1.641 \times 10^{-4}}{A} \text{ K/W}$$

and

$$R_3 = \frac{1}{h_2 A} = \frac{1}{68 \times A} = \frac{0.0147}{A} \text{ K/W}$$

Now,

$$\Sigma R = R_1 + R_2 + R_3$$

$$= \frac{0.0682}{A} + \frac{1.641 \times 10^{-4}}{A} + \frac{0.0147}{A}$$

$$= \frac{0.083}{A} \text{ K/W}$$

We know that,

$$Q = \frac{\Delta T}{\Sigma R} = \frac{(T_1 - T_3)}{\Sigma R} = \frac{(647 - 27)}{\left(\dfrac{0.083}{A}\right)}$$

$$\therefore \quad \frac{Q}{A} = 7469.879 \ \text{W/m}^2 \qquad \text{... Ans.}$$

Step - 2 : **Calculate the temperature at the interface**
Let, T_2 be the temperature on the surface of steel plate.

Now,

$$Q = \frac{\Delta T}{\Sigma R}$$

$$= \frac{(T_1 - T_2)}{R_1 + R_2} = \frac{(647 - T_2)}{\left(\dfrac{0.0682}{A} + \dfrac{1.641 \times 10^{-4}}{A}\right)}$$

$$\therefore \quad \frac{Q}{A} = 7469.879 = \frac{(647 - T_2)}{0.0683}$$

$$\therefore \qquad T_2 = 136.073 \ ^\circ\text{C} \qquad \text{... Ans.}$$

Ex. 4.7.8 : *Hot air temperature of 60 °C flowing through a steel pipe of 10 cm diameter. The pipe is covered with two layers of different insulating materials of the thickness 5 cm and 3 cm and their corresponding thermal conductivities are 0.23 and 0.37 W/m° K. The atmosphere is at 25 °C. Find the rate of heat loss from a 50 m length of pipe. Neglect the resistance of the steel pipe.*

Sol. : **Given data :** $T_1 = 60\ ^\circ\text{C}$, $T_2 = 25\ ^\circ\text{C}$,
$r_1 = 5 \text{ cm} = 0.05 \text{ m}$, $r_2 = r_1 + t_1 = 5 + 5 = 10 \text{ cm} = 0.1 \text{ m}$,
$K_1 = 0.23 \text{ W/m-}^\circ\text{K}$, $r_3 = r_2 + t_2 = 10 + 3 = 13 \text{ cm} = 0.13 \text{ m}$,
$K_2 = 0.37 \text{ W/m-}^\circ\text{K}$, $L = 50 \text{ m}$

To find : Heat loss Q

Step - 1 : **Calculate the rate of heat loss from a pipe**
From Fig. 4.7.5,

$$R_1 = \frac{\log\left(\dfrac{r_2}{r_1}\right)}{2\pi L\, K_1} = \frac{\log\left(\dfrac{0.1}{0.05}\right)}{2\pi \times 50 \times 0.23}$$

$$\therefore \qquad R_1 = 9.5928 \times 10^{-3} \text{ K/W}$$

$$\text{... [Take 'ln' on calculator]}$$

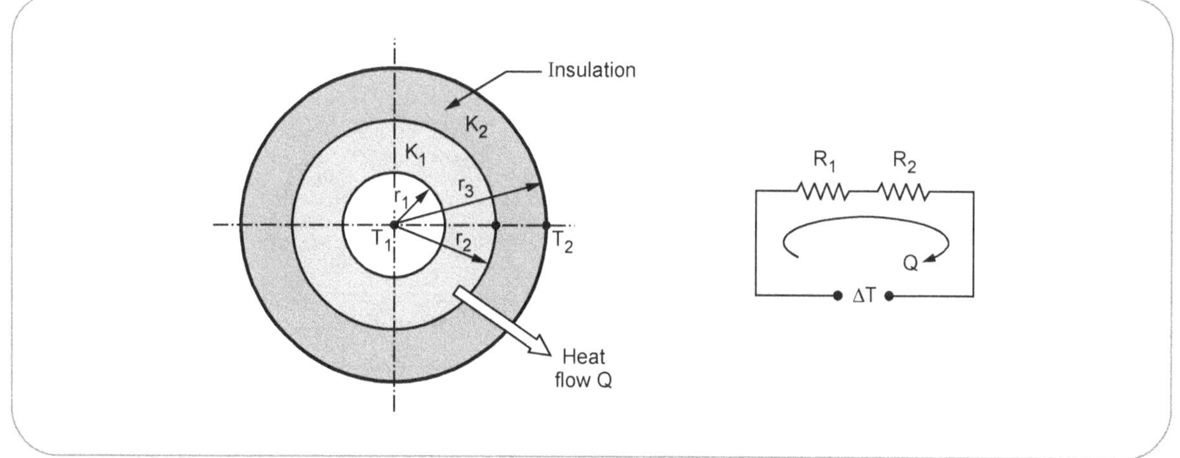

Fig. 4.7.5

and
$$R_2 = \frac{\log\left(\dfrac{r_3}{r_2}\right)}{2\pi\, L\, K_2} = \frac{\log\left(\dfrac{0.13}{0.1}\right)}{2\pi\times 50\times 0.37}$$

$$= 2.2571\times 10^{-3}\ K/W$$

Now,
$$\Sigma R = R_1 + R_2$$

$$= 9.5928\times 10^{-3} + 2.2571\times 10^{-3}$$

$$\therefore \qquad \Sigma R = 0.0118\ K/W$$

We know that,

$$Q = \frac{\Delta T}{\Sigma R} = \frac{(T_1 - T_2)}{\Sigma R} = \frac{(60-25)}{0.0118}$$

$$\therefore \qquad \mathbf{Q = 2966.101\ W} \qquad\qquad \textbf{... Ans.}$$

Ex. 4.7.9 : *Determine heat flux across a wall of an air conditioned hall maintained at 22 °C. The thickness of wall is 20 cm and it's thermal conductivity is 0.7 W/m-K. Heat transfer coefficient on each side of the wall is 10 W/m^2-K. Take outside air temperature as 44 °C.*   **SPPU : Dec.-04, Marks 6**

Sol. : Given data : $T_1 = 22\,^\circ C$, $T_2 = 44\,^\circ C$, $K = 0.7\,W/m\text{-}K$, $h_1 = h_2 = 10\ W/m^2\text{-}K$

To find : Heat flux, Q/A

Step - 1 : **Calculate heat flux across a wall**
From Fig. 4.7.6,

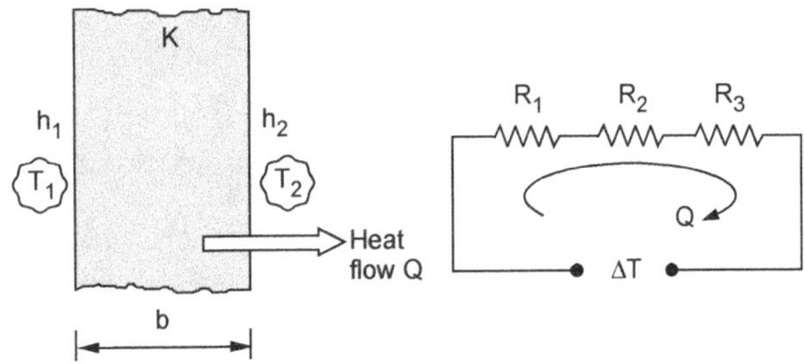

Fig. 4.7.6

$$R_1 = \frac{1}{h_1 A} = \frac{1}{10\,A} = \frac{0.1}{A}\ K/W$$

$$R_2 = \frac{b}{KA} = \frac{0.2}{0.7\,A} = \frac{0.2857}{A}\ K/W$$

and
$$R_3 = \frac{1}{h_2\,A} = \frac{1}{10\,A} = \frac{0.1}{A}\ K/W$$

Now,
$$\Sigma R = R_1 + R_2 + R_3$$

$$= \frac{0.1}{A} + \frac{0.2857}{A} + \frac{0.1}{A} = \frac{0.4857}{A}\ K/W$$

We know that,

$$Q = \frac{\Delta T}{\Sigma R} = \frac{(T_1 - T_2)}{\Sigma R} = \frac{(22-44)}{\left(\dfrac{0.4857}{A}\right)}$$

$$\frac{Q}{A} = -\,45.295\ W/m^2 \qquad\qquad \textbf{... Ans.}$$

Negative sign indicates that, heat transfer is in opposite direction of x.

Ex. 4.7.10 : *A steam pipe having 4 cm outer diameter is to be insulated by two layers of insulation each 2 cm thick. The material M_1 has conductivity K and the material M_2 has conductivity 3 K. Assuming that the inner and outer surface temperatures of composite insulation to be used are T_1 and T_2. Find i) What arrangement would give less heat loss rate; M_1 near the pipe surface and M_2 as outer layer or vice versa ? ii) What is the percentage reduction ?*   **SPPU : May-05, Marks 8**

Sol. : Given data : $r_1 = 2\ cm = 0.02\ m$, $r_2 = r_1 + t = 2 + 2 = 4\ cm = 0.04\ m$, $K_1 = K$, $K_2 = 3\,K$, $r_3 = r_2 + t = 4 + 2 = 6\ cm = 0.06\ m$

To find : i) Possible arrangement
ii) Percentage reduction in heat.

Step - 1 : **Obtain the most suitable arrangement of pipe**

Case - I : *When layer of material M_1 is near to the surface*
From Fig. 4.7.7,

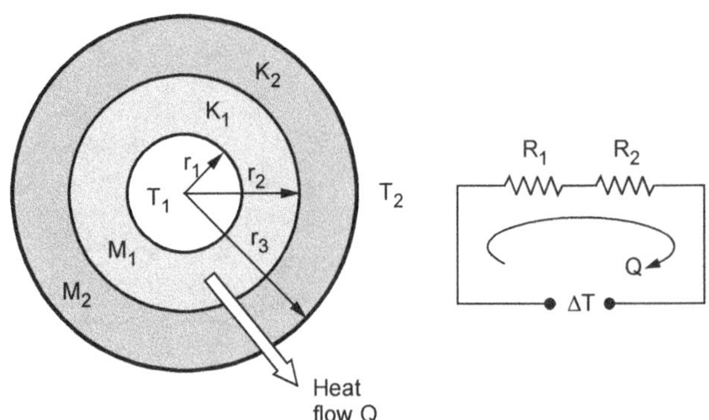

Fig. 4.7.7

$$R_1 = \frac{\log\left(\dfrac{r_2}{r_1}\right)}{2\pi L K_1} = \frac{\log\left(\dfrac{0.04}{0.02}\right)}{2\pi \times L \times K}$$

$$\therefore \qquad R_1 = \frac{0.1103}{LK} \ K/W$$

$$\dots \text{[Take 'ln' on calculator]}$$

$$\text{and} \qquad R_2 = \frac{\log (r_3/r_2)}{2\pi L K_2}$$

$$= \frac{\log\left(\dfrac{0.06}{0.04}\right)}{2\pi L \times 3K} = \frac{0.0215}{LK} \ K/W$$

Now, $\qquad \Sigma R = R_1 + R_2$

$$= \frac{0.1103}{LK} + \frac{0.0215}{LK} = \frac{0.1318}{LK} \ K/W$$

We know that,

$$Q_1 = \frac{\Delta T}{\Sigma R} = \frac{\Delta T}{\left(\dfrac{0.1318}{LK}\right)}$$

$$= 7.587 \ LK(\Delta T), \ W \qquad \dots \text{(a)}$$

Case - II : *When layer of material M_2 is near to the surface*

Here, $K_2 = K_1$ and $K_1 = K_2$

$$R_1 = \frac{\log\left(\dfrac{r_2}{r_1}\right)}{2\pi L K_1}$$

$$= \frac{\log\left(\dfrac{0.04}{0.02}\right)}{2\pi \times L \times 3K} = \frac{0.0367}{LK} \ K/W$$

$$\text{and} \qquad R_2 = \frac{\log\left(\dfrac{r_3}{r_2}\right)}{2\pi L K_2} = \frac{\log\left(\dfrac{0.06}{0.04}\right)}{2\pi \times L \times K} = \frac{0.0645}{LK} \ K/W$$

Now, $\qquad \Sigma R = R_1 + R_2$

$$= \frac{0.0367}{LK} + \frac{0.0645}{LK} = \frac{0.1012}{LK} \ K/W$$

We know that,

$$Q_2 = \frac{\Delta T}{\Sigma R} = \frac{(\Delta T)}{\left(\dfrac{0.1012}{LK}\right)}$$

$$= 9.8814 \ LK \ (\Delta T), \ W \qquad \dots \text{(b)}$$

Comparing equations (a) and (b) it is clear that, in first case heat transfer rate is less hence, **material M_1 is near to the surface.**   **... Ans.**

Step - 2 : **Calculate percentage reduction in heat transfer**

Percentage reduction in Q is given by,

$$\% \text{ reduction in Q} = \left(\frac{Q_2 - Q_1}{Q_2}\right) \times 100$$

$$= \left(\frac{9.8814 \ LK \ (\Delta T) - 7.587 \ LK \ (\Delta T)}{9.8814 \ LK \ (\Delta T)}\right) \times 100$$

$$\therefore \quad \% \text{ reduction in Q} = 23.219 \ \% \qquad \text{... Ans.}$$

Ex. 4.7.11 : *A spherical vessle (K = 40 W/m-K) of inner and outer radii of 490 mm and 500 mm respectively contains liquid nitrogen of 80 K. Rockwool (K = 0.065 W/mK) of thickness 100 mm is lagged round the vessel. Calculate rate of heat transfer from outside air of 40 °C into the liquid nitrogen. Assume inside surface temperature of the vessel as 80 K and convective heat ransfer coefficient on the outer most surface as 10 W/m² K.*

SPPU : Dec.-04, Marks 4

Sol. : Given data : $K_1 = 40 \ W/m\text{-}K$, $\quad r_1 = 0.49 \ m$,

$r_2 = 0.5 \ m$, $\ K_2 = 0.065 \ W/m\text{-}K$, $\ r_3 = r_2 + t = r_2 = 0.5 \ m$,

$T_1 = 80 \ K$, $T_2 = 40 \ °C = 40 + 273 = 313 \ K$,

$h_2 = 10 \ W/m^2 K$

To find : Rate of heat transfer Q

Step - 1 : **Calculate the rate of heat transfer from vessel**

From Fig. 4.7.8,

$$R_1 = \frac{(r_2 - r_1)}{4\pi K_1 r_1 r_2} = \frac{(0.5 - 0.49)}{4\pi \times 40 \times 0.49 \times 0.5}$$

$$= 8.121 \times 10^{-5} \ K/W$$

$$R_2 = \frac{(r_3 - r_2)}{4\pi K_2 r_2 r_3}$$

$$= \frac{(0.6 - 0.5)}{4\pi \times 0.065 \times 0.5 \times 0.6} = 0.4081 \ K/W$$

$$\text{and} \qquad R_3 = \frac{1}{h_2 A} = \frac{1}{h_2 4\pi r_3^2}$$

$$= \frac{1}{10 \times 4\pi \times (0.6)^2} = 0.0221 \ K/W$$

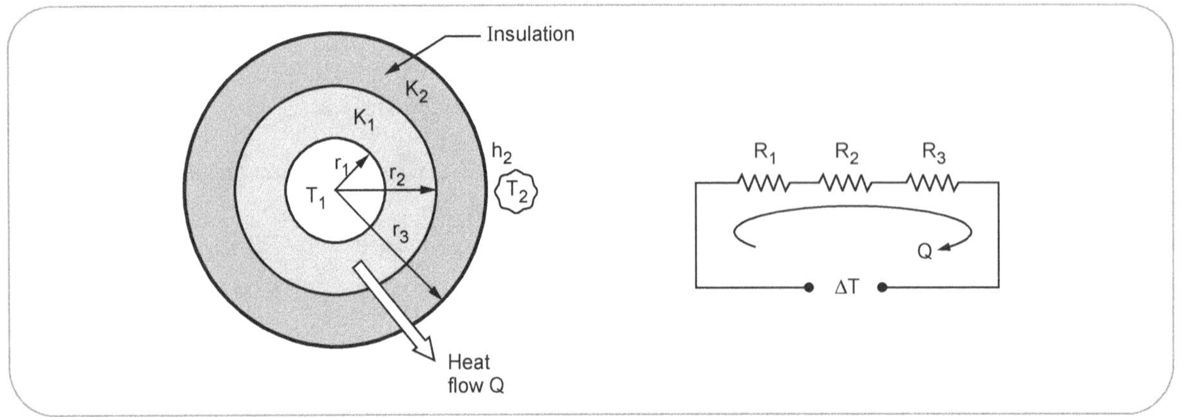

Fig. 4.7.8

Now, $\quad \Sigma R = R_1 + R_2 + R_3$

$$= 8.121 \times 10^{-5} + 0.4081 + 0.0221$$

$$= 0.4302 \ K/W$$

We know that, $Q = \dfrac{\Delta T}{\Sigma R} = \dfrac{(T_1 - T_2)}{\Sigma R} = \dfrac{(80 - 313)}{0.4302}$

$\therefore \qquad \mathbf{Q = -541.608 \ W} \qquad$ **... Ans.**

Negative sign indicates that, heat transfer is in opposite direction of r.

I.C. Engines

4.8 : Internal Combustion (I.C.) Engines

- Any machine which derives heat energy from the combustion of fuel (chemical energy of fuel) and convert part of this energy into mechanical work is called as **heat engine.**

- These heat engines may be externally combustion type or internally combustion type.

- In case of external combustion engines, the combustion of fuel takes place outside the cylinder.

  For example : Steam engines, steam and gas turbines, air engines, etc.

- In case of internal combustion engines, the combustion of fuel in the presence of air takes place inside the cylinder and produced gases act on piston to develop the power.

  For example : Petrol engines, diesel engines, gas engines, etc.

- Internal combustion (I.C.) engines are commonly used in vehicles, locomotives and in various industrial applications.

4.8.1 Classification of I.C. Engines

SPPU : Dec.-09, 11, May-12

I.C. engines are classified as follows :

1. According to the cycle of operation

a) Four stroke engines : In these engines cycle is completed in two revolution of crank.

b) Two stroke engines : In these engines cycle is completed in one revolution of crank.

2. According to the nature of thermodynamic cycle

a) Constant volume or Otto cycle.

b) Constant pressure or diesel cycle.

c) Partly Otto and partly diesel cycle or dual cycle.

3. According to the method of ignition

a) **Spark Ignition (S.I.) engines :** In spark ignition or **petrol engines,** air-fuel mixture supplied by carburettor is ignited by producing spark.

b) **Compression Ignition (C.I) engines :** In compression ignition or **diesel engines** air is compressed to higher pressure and in fine atomised form fuel is injected in combustion chamber. Hence air-fuel mixture is ignited without producing spark.

4. According to the number of cylinders

a) Single cylinder engines

b) Multi-cylinder engines

5. According to the type of cooling system

a) **Air cooled engines :** For cooling purpose air is used.

b) **Water cooled engines :** Water jackets are provided around the cylinder to cool the engine.

6. According to the speed of engine

a) Low speed engines

b) Medium speed engines

c) High speed engines

7. According to the lubrication method

a) Wet sump lubrication

b) Dry sump lubrication

c) Pressure lubrication

8. According to the fuel used

a) Petrol engines

b) Diesel engines

c) Gas engines

9. According to the field of operation

a) Automotive engines

b) Stationary engines

c) Locomotive engines

d) Marine engines

10. According to the arrangement of cylinder (Refer Fig. 4.8.1)

a) **Vertical engines :** In these engines, cylinder is in vertical position.

b) **Horizontal engines :** In these engines, cylinder is in horizontal position.

c) **V-engines :** In these engines, two cylinders are arranged at an angle connected to a common crank.

d) **Radial engines :** In these engines 4 to 6 cylinders are placed radially and equally speaced and connected to a common crank.

e) **Opposed cylinder engines :** In these engines, cylinders are arranged on opposite side of a common crank.

f) **Opposed piston engines :** In these engines, single cylinder houses two pistons, each of which drives separate crank.

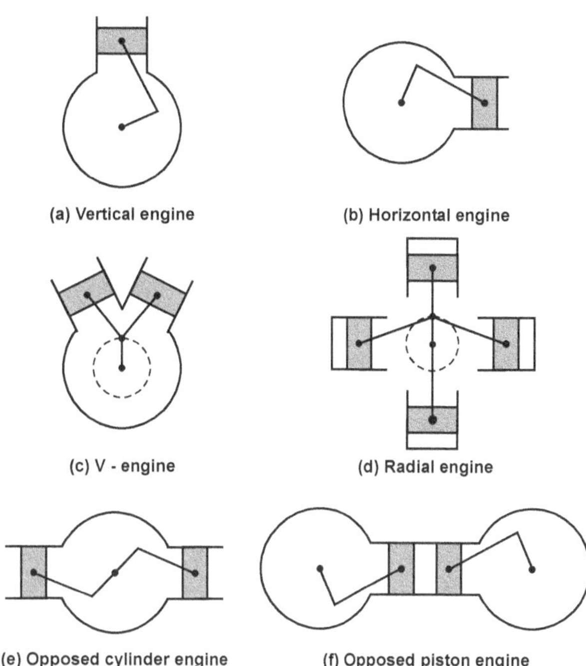

Fig. 4.8.1 Arrangement of cylinders in I.C. engine

4.8.2 I.C. Engine and its Components

Fig. 4.8.2 shows reciprocating I.C. engine and its components. The main components of this 4-stroke petrol engine are as follows :

1. Cylinder	2. Piston
3. Crank and crankshaft	4. Connecting rod
5. Spark plug	6. Inlet and exhaust valve
7. Crank pin	8. Gudgeon pin
9. Cylinder head	10. Piston rings
11. Suction and exhaust manifold	
12. Carburetor	

1. **Cylinder :** It is the main body of an I.C. engine in which piston reciprocates to develop power. It has to withstand very high pressures (upto 75 bar) and temperatures (upto 2400 °C), because the the combustion takes place in the cylinder. It may be air cooled or water cooled. It is generally made of cast iron or alloy steel.

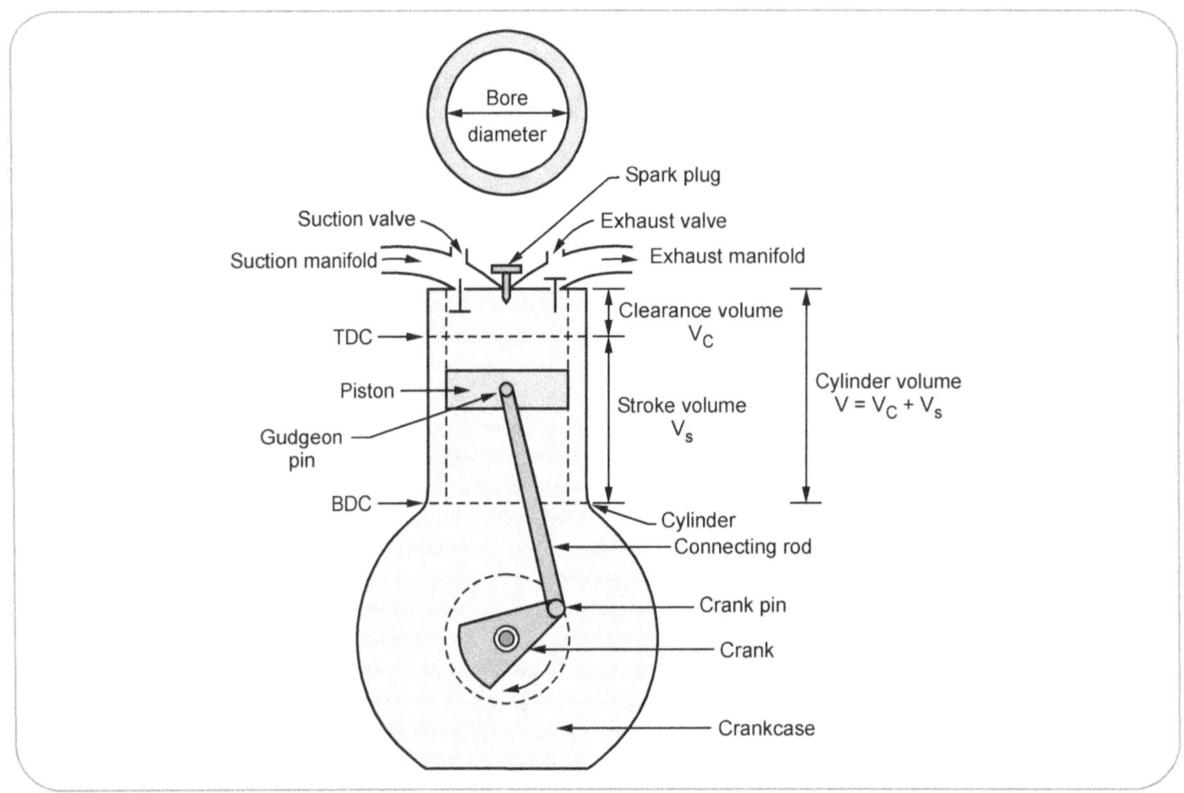

Fig. 4.8.2 Four stroke I.C. engine (petrol) and its components

2. **Piston :** It is used to compress the charge (air-fuel mixture) during compression stroke and to transmit the gas force to the connecting rod and then crank during power stroke.

3. **Crank and crankshaft :** Crank is the integral part of the crankshaft and crankshaft is the backbone of the engine, crankshaft is supported in main bearings and it has a heavy wheel called as **flywheel** to control the fluctuation of torque.

4. **Connecting rod :** It is a rod of circular or rectangular cross section. Its small end is connected to the piston and big end is connected to the crank. It converts reciprocating motion of pistion into rotary motion of crankshaft.

5. **Spark plug :** It is used to initiate the spark for the combustion of air-fuel mixture in petrol engines.

6. **Inlet and exhaust valve :** Inlet valve controls the admission of the charge during suction stroke and exhaust valve is used for the removal of exhaust gases after doing work on the pistion.

7. **Crank pin and Gudgeon pin :** Crank pin is used to connect big end of connecting rod to the crank and gudgeon pin is used to connect small end of connecting rod to the piston.

8. **Cylinder head :** It is used for closing the one end of the cylinder. It houses inlet and exhaust valves through which charge is taken inside the cylinder and burned gases are exhausted to the atmosphere from the cylinder.

9. **Piston rings :** Piston rings are housed in the circumferential grooves provided on the outer surface of the piston. It gives tight fit between the pistion and cylinder hence prevents the leakage of high pressure gases.

10. **Suction and exhaust manifold :** Suction manifold is the passage which carries the charge from carburettor to the engine where as exhaust manifold is the passage which carries the exhaust gases from the exhaust valve to the atmosphere.

11. **Carburettor :** It is used to supply the uniform air-fuel mixture to the cylinder of petrol engine through the intake manifold. It is controlled by a throttle valve.

4.8.3 Terminology used in I.C. Engines

In addition to the above components of I.C. engine, certain standard terminology used in I.C. engine is as follows :

> 1. Bore
> 2. Stroke
> 3. Top Dead Centre (T.D.C.)
> 4. Bottom Dead Centre (B.D.C.)
> 5. Clearance volume
> 6. Swept volume
> 7. Compression ratio

1. **Bore :** It is the inside diameter of the cylinder.

2. **Stroke :** It is the maximum distance travelled by the piston in the cylinder in one direction. It is generally twice the radius of crank.

3. **Top Dead Centre (T.D.C) :** It is the extreme position of the piston at the top of the cylinder. For horizontal engines it is called as inner dead centre (I.D.C).

4. **Bottom Dead Centre (B.D.C) :** It is the extreme position of the piston at the bottom of the cylinder. For horizontal engines it is called as outer dead centre (O.D.C).

5. **Clearance volume :** It is the volume contained in the cylinder above the top of the piston when it is at T.D.C. It is denoted by V_c.

6. **Swept volume :** It is the volume swept by the piston in moving between TDC and BDC. It is denoted by V_s.

7. **Compression ratio :** It is the ratio of the volume when the piston is at BDC to the volume when the piston is at TDC. It is denoted by r.

4.8.4 Four Stroke Petrol Engines (S.I. Engines)

SPPU : Dec.-07, 08, 13, May-11, 13, 17

- In a four stroke cycle engine, the cycle is completed in *two revolutions* of crankshaft and four strokes are performed.

> 1. Suction stroke   2. Compression stroke
> 3. Expansion/Power stroke   4. Exhaust stroke

- Each stroke consists of 180° of crankshaft rotation, hence the cycle consists of 720° of crankshaft rotation for one cycle.

- Four stroke cycle engine consists of following strokes (Refer Fig. 4.8.3) :

1. Suction stroke

- This stroke starts when the piston is at TDC and about to move downwards.

- *During this stroke inlet valve remains open and exhaust valve remains closed.*

- Due to low pressure created by downward moving piston, the charge (air-fuel mixture) is drawn into the cylinder.

- At the end of this stroke the inlet valve closes.

2. Compression stroke

- During this stroke, the compression of fresh sucked charge takes place by the return stroke (BDC to TDC) of piston.

- *During this stroke, both inlet and exhaust valve remains closed.*

- Just before completion of compression stroke a charge is ignited by a spark plug and combustion takes place.

3. Expansion or power stroke

- The products of combustion exerts pressure on the piston hence forcing it to move downwards i.e. TDC to BDC.

- *During this stroke, both inlet and exhaust valve remains closed.*

- Power is developed during expansion of gases hence this stroke is also called as **power stroke.**

- During power stroke, both temperature and pressure decreases.

4. Exhaust stroke

- *At the end of power stroke the exhaust valve opens but the inlet valve remains closed.*

- During this stroke piston moves from BDC to TDC and burnt gases inside the cylinder are exhausted.

- At the end of this stroke exhaust valve closes but some residual gases remain in the cylinder.

- In this way, again the inlet valve opens and the new cycle starts.

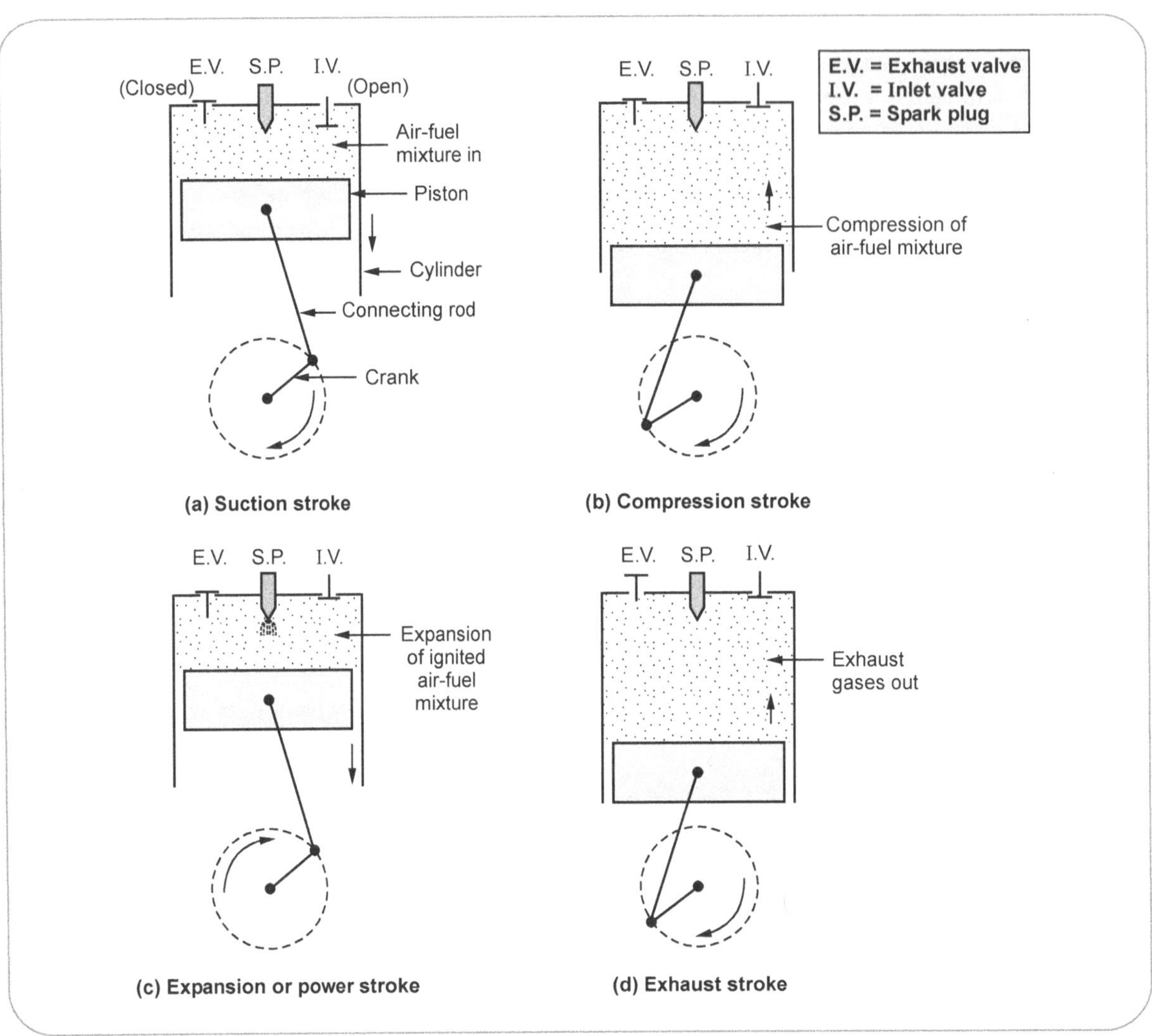

Fig. 4.8.3 Four stroke petrol (S.I.) engine

Note : *Each cycle of four stroke engine completes the above strokes in two crank rotations. One revolution of crankshaft occurs during the suction and compression strokes and the other revolution during the expansion and exhaust stroke.*

4.8.5 Four Stroke Diesel Engines (C.I. Engines)

SPPU : May-06, 08

The working of four stroke diesel engine is similar to four stroke petrol engine, except the following two major points :

○ In C.I. engines the air is only drawn during the suction stroke instead of air-fuel mixture drawn in S.I. engines.

○ In C.I. engines the fuel is injected by using injector into the cylinder and the high pressure and temperature mixture is burnt. Hence there is no need of spark plug.

• Diesel engine cycle consists of following four strokes (Refer Fig. 4.8.4) :

1. Suction stroke	2. Compression stroke
3. Expansion/Power stroke	4. Exhaust stroke.

1. Suction stroke

• During this stroke air is drawn into the cylinder and piston moves from TDC to BDC.

• The exhaust valve remains closed during this stroke.

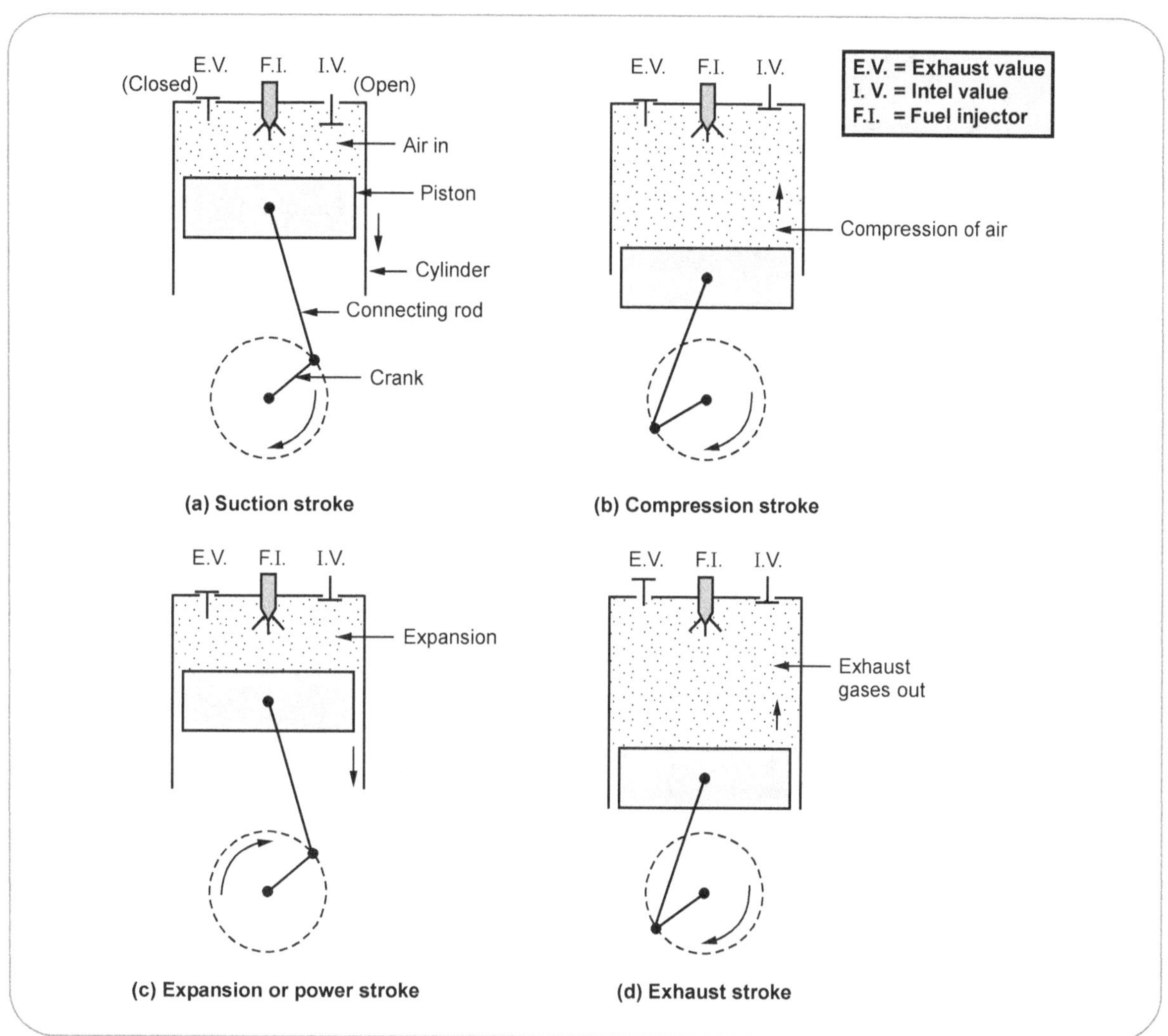

Fig. 4.8.4 Four stroke diesel (C.I.) engine

2. Compression stroke

- During this stroke the piston moves from BDC to TDC and both the valves remain closed.

- The air sucked during the suction stroke is compressed hence there is rise in the pressure and temperature of the air.

- Before the end of this stroke, a fine spray of fuel is injected into the compressed air which is at high temperature and hence combustion takes place.

3. Expansion or power stroke

- During this stroke both the valves remain closed and the piston moves from TDC to BDC.

- The heat energy released by the combustion of fuel results in rise in pressure of the gases which drives the piston downwards. Hence, power is developed.

4. Exhaust stroke

- In this stroke exhaust valve is opened and inlet valve remains closed and the piston moves from BDC to TDC.

- The upward movement of the piston forces the burnt gases out of the cylinder through the exhaust valve.

4.8.6 Comparison between Petrol Engine and Diesel Engine `SPPU : Dec.-03, 05, May-09`

The comparison between four stroke petrol and four stroke diesel engines is as follows :

Sr. No.	Petrol (S.I.) engines	Diesel (C.I.) engines
1.	Petrol engine works on Otto cycle.	Diesel engine works on diesel cycle.
2.	In these engines, air-fuel mixture is sucked during suction stroke.	In these engines, only air is sucked during suction stroke.
3.	Fuel supply is controlled by throttle valve in the carburettor.	Fuel supply is controlled by fuel pump.
4.	For ignition, spark plug is required.	Spark plug is not required.
5.	Compression ratio is low (about 6 to 12).	Compression ratio is high (about 14 to 22).
6.	Due to low compression ratio, efficiency of these engines is low.	Due to high compression ratio, efficiency of these engines is high.
7.	Due to low compression ratio, starting of these engines is easy.	Due to high compression ratio, starting of these engines is difficult.
8.	Petrol engine is light in weight.	Diesel engine is heavier due to high pressure.
9.	Due to light weight these engines can rotate at high speed.	Due to heavy weight these engines cannot rotate at high speed.
10.	The operation of these engines is silent.	The operation of these engines is noisy.
11.	Running cost is more since it uses costilier fuel.	Running cost is low since it uses cheaper fuel.
12.	Initial cost is low.	Initial cost is high.
13.	These engines are used in light duty vehicles like motor cycles, scooters, cars, etc.	These engines are used in heavy duty vehicles like buses, trucks, etc.

4.8.7 Two Stroke Petrol Engines (S.I. Engines)

`SPPU : May-05, 07, 10, Dec.-05, 10, 12`

- In four stroke engines, there is one power stroke in one cycle i.e. in two revolutions of crankshaft.

- In two stroke engines the suction and exhaust strokes are eliminated. Also there are no inlet and exhaust valves, instead of that these engines have inlet and exhaust ports.

- A piston of these engines is given a specific crown shape which helps to prevent the loss of fresh incoming charge which can be escaped alongwith the gases through exhaust port.

- A two stroke petrol engine consists of following strokes (Refer Fig. 4.8.5) :

> 1. First stroke
> 2. Second stroke

1. First stroke

- Initially the piston is at BDC. The arrangement of the ports is such that the piston performs two operations simultaneously.

- When the piston starts rising from BDC it closes the transfer port and exhaust port and the already existing charge is compressed.

- At that time, vacuum is created in the crankcase which is gas tight.

- As soon as the inlet port is uncovered the fresh air is sucked in the crankcase and the charging is continued untill the crankcase is filled.

- At the end of this first stroke piston reaches TDC.

2. Second stroke

- In this stroke, piston moves from TDC to BDC.

- Before the completion of compression stroke, the compressed charge is ignited using spark plug and the gas pressure is exerted on the crown of the piston.

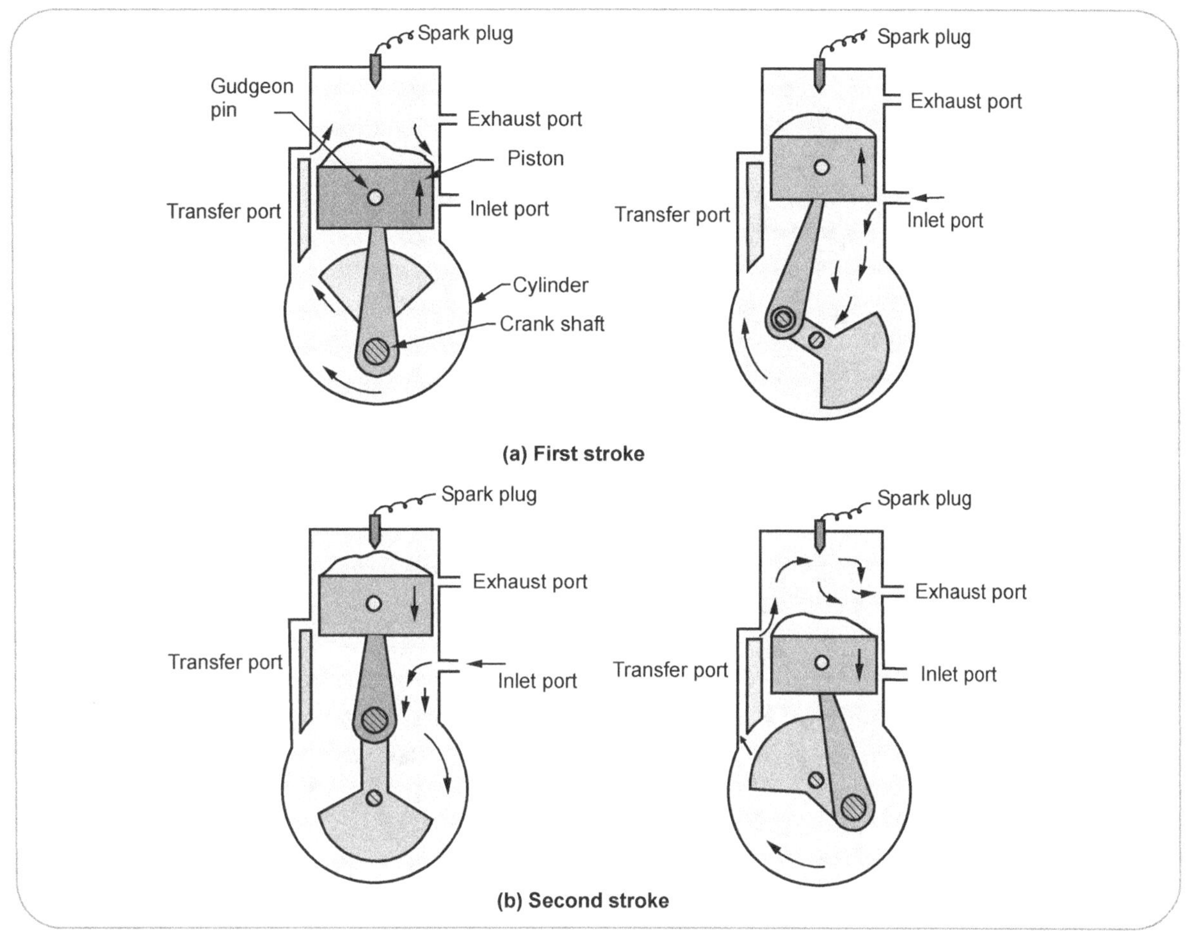

Fig. 4.8.5 Two stroke petrol (S.I.) engine

- This pressure forces the piston in downward direction which produces some useful work.

- The downward movement of piston closes the inlet port and compresses the charge already sucked in the crankcase. This fresh charge is transferred on the top side of the piston through transfer port.

- At that time, exhaust port opens and burnt gases are escaped.

4.8.8 Two Stroke Diesel Engines

- We know that, in two stroke engines the suction and exhaust strokes are eliminated. Also, there is no inlet and exhaust valves, instead of that these engines have inlet and exhaust ports.

- A two stroke diesel engine consists of following strokes :

    1) First stroke   2) Second stroke

- In the Fig. 4.8.6 (a) the piston is shown at Top Dead Centre (TDC).

- In the upward motion of the piston, suction is created in the crank-case and air enters the closed crankcase through the inlet port which open inside due to suction.

- Therefore, for full 180° of crank rotation, the suction will continue below the piston in crankcase.

- Above the piston, compression will start only after both the ports have been covered by the piston, thus trapping air above it.

- When the piston moves upwards towards TDC, the air above it will be compressed.

- Just before the end of compression fuel is injected and at TDC the fuel auto ignites due to heat of compression.

- On the downward stroke, the high pressure product of combustion expand and the air below the piston

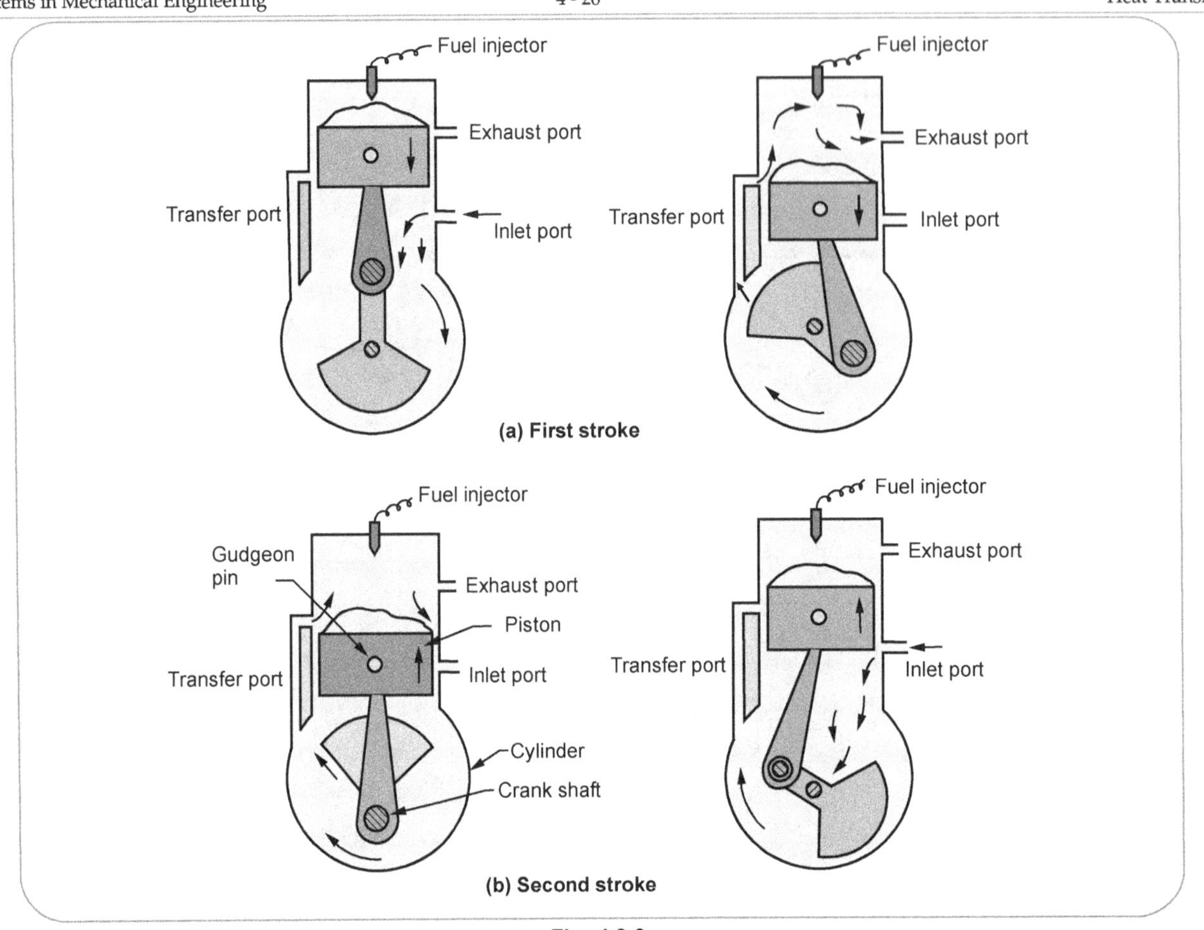

Fig. 4.8.6

compresses. Because of positive pressure due to compression in the crankcase, the exhaust valve closes.

- This continues till the piston uncovers the exhaust port when products of combustion are released out to the atmosphere.

4.8.9 Comparison between Two Stroke and Four Stroke Engines

SPPU : Dec.-04, 08, 11, 13, May-10, 14, 17

Sr. No.	Two stroke engine	Four stroke engine
1.	The cycle is completed in one revolution of crankshaft.	The cycle is completed in two revolutions of crankshaft.
2.	Power stroke is obtained in every revolution of crankshaft.	Power stroke is obtained in every two revolutions of crankshaft.
3.	Due to one power stroke in one revolution, power produced for same size of engine is more.	Due to one power stroke for two revolutions, power produced for same size of engine is small.
4.	In these engines ports are present.	Instead of ports these engines contains valves.
5.	Due to simplicity and light weight, its initial cost is low.	Due to complication and heavy weight, its initial cost is high.
6.	The piston is having crown shape or dome shape.	The crown of the piston is flat.
7.	Low thermal efficiency.	High thermal efficiency.
8.	It consumes more amount of lubricating oil.	It consumes less amount of lubricating oil.
9.	During the operation it produces more noise.	It produces less noise during the operation.
10.	These type of engines are used in scooters, mopeds, etc.	These type of engines are used in cars, buses, trucks, etc.

4.9 : CNG (Compressed Natural Gas) Engines

- CNG engine uses compressed natural gas to power the car. CNG is a substitute for gas and diesel fuel, and is considered to be much cheaper and cleaner than gas or diesel.

- The engine functions the same way as a gasoline engine. The fuel-air mixture is compressed and ignited by a spark plug.

- The natural gas is stored in a fuel tank, or cylinder, typically at the back of the vehicle.

- A CNG fuel system transfers high-pressure natural gas from the fuel tank to the engine.

- The pressure is then reduced to a level compatible with the engine fuel injection system, through which the fuel is introduced into the intake manifold or combustion chamber.

- As an efficient and environmentally-friendly alternative to gas powered cars, the CNG engine has become increasingly popular, and many people choose to convert their cars using professional mechanics.

- It is made by compressing natural gas (which is mainly composed of methane (CH_4)), by about 75 %.

- It is stored and distributed in hard containers, at a normal pressure of 200 - 220 bar (20 - 22 MPa), usually in cylindrical or spherical shapes to maintain equal pressure on the walls of the containers.

Why Use Compressed Natural Gas ?

- As many have noted, compressed natural gas is marginally cheaper than ordinary gasoline or diesel.

- In addition, the CNG engine is considered to be more environmentally friendly. There are considerably less pollutants associated with compressed natural gas being ignited, and studies show that it gives off 40 percent less greenhouse gas.

- On the downside, a CNG engine will usually get fewer miles to the full tank than a regular gas engine, and you may also struggle to find a suitable engine and tank conversion kit, which means that the price of converting your vehicle can negate the savings from the cheaper fuel alternatives.

The CNG Engine

- The CNG engine uses a second fuel tank which has to be attached to the car, and is usually placed in the trunk (or other place where there is suitable room).

- This tank is usually very large, as it has to keep the gas used compressed.

- The driver can then decide which of the fuels they wish to use by simply pressing a switch on the dashboard. This means that the car can alternate between the different tanks, drawing fuel from either.

How the CNG Engine Powers the Car

- Once the driver selects the CNG tank, the compressed gas in the tank is pulled through a series of highly pressurized lines until it reaches the regulator.

- Inside the regulator, the pressure on the gas is lessened until it matches the amount needed by the fuel injection system of the car's engine.

- Once the gas has reached an acceptable pressure, the solenoid valve allows the gas to move into the fuel injection system and from there into the engine.

- Just as with gasoline, once the engine has received the gas, it is ignited in the combustion chamber, and this provides the energy to power the car forward.

4.9.1   Components of a Natural Gas Vehicle

The key components of a natural gas vehicle are as follows :

(i) **Battery :** The battery provides electricity to start the engine and power vehicle electronics/accessories.

(ii) **Electronic control module (ECM) :** The ECM controls the fuel mixture, ignition timing, and emissions system; monitors the operation of the vehicle; safeguards the engine from abuse; and detects and troubleshoots problems.

(iii) **Exhaust system :** The exhaust system channels the exhaust gases from the engine out through the tailpipe.

(iv) Fuel filler : This is a filler or "nozzle" used to add fuel to the tank.

(v) Fuel injection system : This system introduces fuel into the engine's combustion chambers for ignition.

(vi) Fuel line : A metal tube or flexible hose (or a combination of these) transfers fuel from the tank to the engine's fuel injection system.

(vii) Fuel tank (compressed natural gas) : Stores compressed natural gas on board the vehicle until it's needed by the engine.

(viii) High pressure regulator : Reduces and regulates the pressure of the fuel exiting the tank, lowering it to an acceptable level required by the engine's fuel injection system.

(ix) Internal combustion engine (spark-ignited) : In this configuration, fuel is injected into either the intake manifold or the combustion chamber, where it is combined with air, and the air/fuel mixture is ignited by the spark from a spark plug.

(x) Manual shut off : Allows the vehicle operator or mechanic to manually shut off the fuel supply.

(xi) Natural gas fuel filter : Traps dirt and other particles to prevent them from clogging critical fuel system components, such as fuel injectors.

(xii) Transmission : The transmission transfers mechanical power from the engine and/or electric traction motor to drive the wheels.

Advantages of CNG :

- **CNG is economic :** Natural gas is significantly less expensive than gasoline.

- **Environment friendly :** CNG is more eco-friendly than gasoline. Natural gas produces far fewer harmful emissions and hydrocarbons than gasoline.

- **Clean engine :** Using CNG makes the engine cleaner and more efficient. Unlike gasoline, CNG minimizes harmful carbon deposits when combusted. This results to a cleaner and more efficient engine as well as longer lasting spark plugs.

- **Ease of use and flexibility :** Car runs on both CNG and gasoline. So its option for gasoline.

- **Lubrication :** It requires less lubrication.

- **CNG is Safer :** CNG is safer than petrol and diesel. CNG gas tanks are safer and stronger.

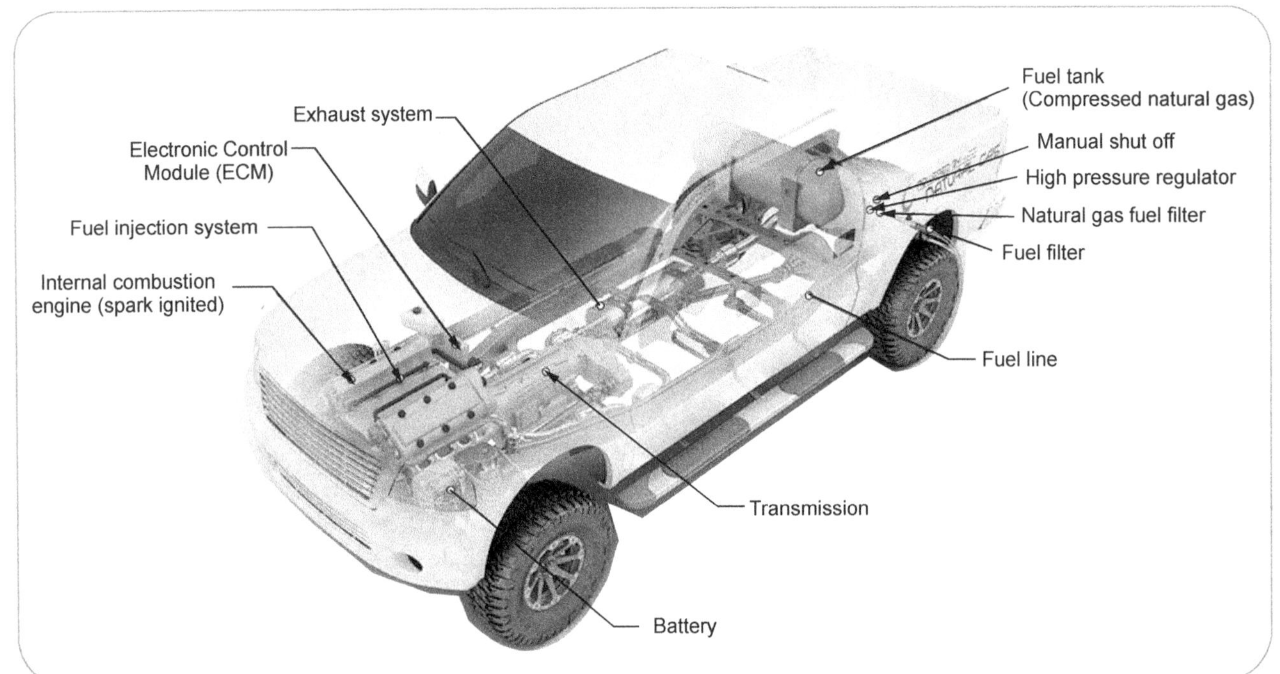

Fig. 4.8.7 CNG Vehicle

Disadvantages of CNG :

- CNG tanks require storage space. You may need to sacrifice some of the space in the trunk (for cars), truck bed (for pickup trucks) or behind the back seat (for SUVs).

- The CNG cylinder can be heavy, the added weight of the tank is offset by the reduced weight of a gasoline fuel.

- CNG filling stations have limited availability. CNG conversion is most practical and convenient for people living in areas with easy access to CNG filling stations -or- considers installing a Home CNG Fueling Unit.

- CNG gas kit prices are between ₹ 20000 to ₹ 30000. But it can be recovered by fuel savings.

4.10 : Boilers (Steam Generators)

- **Boiler** or **steam generator** is a closed pressure vessel used for generation of steam under high pressure.

- This steam is mainly used for power generation, process heating and space heating purposes.

- A boiler is commonly made of steel in which the chemical energy of fuel is converted into heat by the combustion process. This heat energy is transferred to water so as to produce steam.

- The design of boilers is very complicated and depends on the type of fuel used and its power (capacity).

4.10.1 Classification of Boilers

> **SPPU : Dec.-09, 10, 12, May-11, 12**

The boilers may be classified as follows :

1. According to the relative position of water and flue gases

a) **Water tube boilers :** In these boilers, the water passes through the tubes and flue gases pass through the external surface of the tubes.

   For example : Babcock-Wilcox, La-mont, Benson and Package boilers.

b) **Fire tube boilers :** In these boilers, the flue gases are passed through the tubes and the tubes are surrounded by the water.

For example : Cochran, Lancashire and Locomotive boilers.

2. According to the method of furnace

a) **Externally fired boilers :** In these boilers, the furnace is placed outside the boiler shell.

   For example : All water tube boilers.

b) **Internally fired boilers :** In these boilers, the furnace is placed inside the boiler shell.

   For example : All fire tube boilers.

3. According to the method of water circulation

a) **Natural circulation boilers :** In these boilers, the water is circulated by natural convection which is set up due to heating of water. Due to temperature gradient water flows from high density region to low density region.

   For example : Babcock and Wilcox boiler.

b) **Forced circulation boilers :** In these boilers, the water is circulated by using a pump driven by motor.

   For example : La-mont and Benson boilers.

4. According to the use

a) **Stationary boilers :** These type of boilers are commonly used in industries, power plants, etc. for power generation.

   For example : Lancashire and Babcock-Wilcox boilers.

b) **Mobile boilers :** These boilers are used in ships, locomotives, etc. because they continuously move from one place to another place.

   For example : Locomotive boilers.

5. According to the axis of shell

a) **Vertical boilers :** In these boilers the axis of shell is vertical.

   For example : Cochran boilers.

b) **Horizontal boilers :** In these boilers, the axis of shell is horizontal.

   For example : Lancashire and Locomotive boilers.

6. According to the pressure of steam generated

a) **Low pressure boilers :** When the pressure of generated steam is below 25 bar, then it is called as low pressure boiler.

For example : Cochran, Lancashire and Locomotive boilers.

b) **High pressure boilers :** When the pressure of generated steam is above 25 bar and upto 160 bar then it is called as high pressure boiler.

For example : La-mont and Loeffler boilers.

7. According to the heat source

a) Heat is generated due to combustion of solid, liquid and gaseous fuels.

b) Waste heat from other processes or nuclear energy.

4.10.2 Requirements of a Good Boiler

A good boiler must possess the following qualities :

- The boiler should be capable to generate steam at the desired pressure and quantity in minimum possible time with minimum fuel consumption.

- The boiler should be light in weight and it should occupy less floor area.

- The boiler must be able to meet the fluctuating demands without fluctuations in pressure.

- All the boiler parts should be easily accessible or approachable for cleaning and inspecation.

- The boiler should be leak proof.

- It should have simple installation with minimum time and manpower.

- It should start quickly.

- There should be no deposition of mud and foreign materials on heated surface because it affect the efficiency of boiler.

- The design of boiler should allow high heat transfer rates with minimum pressure drop by providing high velocity of water and flue gases.

- The initial cost, installation cost and maintenance cost of boiler should be minimum.

- The boiler should confirm to the safety regulations as per the *Indian Boiler Act (IBA) 1923.*

4.10.3 Package Boilers `SPPU : Dec.-10, May-11`

- Package boilers are widely used in the pharmaceutical industries, food industries and ceramic industries.

- These boilers require less fuel and electricity for their operation.

- If the oxygen concentration in the flue gases is reduced then these boilers operate more efficiently.

- But, insufficient oxygen causes incomplete combustion which results in increased smoke emission.

- The typical package boiler is either *water tube* or *fire tube* with a capacity of 5 to 25 tonns/hr steam generation capacity.

- The most commonly used fuels for these boilers are heavy oil, light oil or gas.

Fire tube package boiler `SPPU : May-13`

- A fire tube boiler is a boiler in which hot gases from a fire pass through one or more tubes running through a sealed container of water.

- The heat of these gases is transferred through the walls of the tubes by thermal conduction hence heating the water and ultimately creating the steam. Refer Fig. 4.9.1.

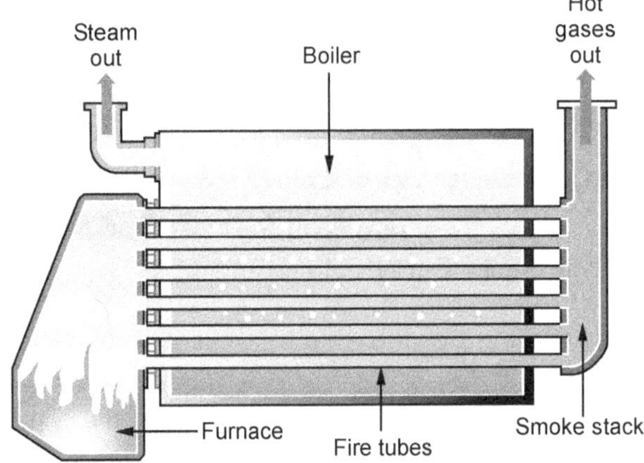

Fig. 4.9.1 Fire tube package boiler

- These type of boilers are commonly used in locomotives, marine applications, rockets, etc.

- Cornish boiler, Lancashire boiler, Locomotive boiler, Vertical boiler are the types of fire tube boiler.

- In these boilers, the fuel is burnt in a firebox to produce hot combustion gases.

- The firebox is surrounded by a cooling jacket of water connected to the long, cylindrical boiler shell.

- The hot gases are directed along a series of fire tubes that penetrate the boiler and heat the water, in this way generating wet steam.

- Sometimes, this steam is passed through the superheater to dry the steam or superheat the steam.

Water tube package boiler

SPPU : May-10, 13

- A water tube boiler is a type of boiler in which water circulates in tubes heated externally by the fire.

- In these boilers, fuel is burned inside the furnace, creating hot gas which heats water in the steam generating tubes.

- In small boilers additional generating tubes are separate in the furnace while in large boilers water filled tubes make up the walls of the furnace to generate the steam. Refer Fig. 4.9.2.

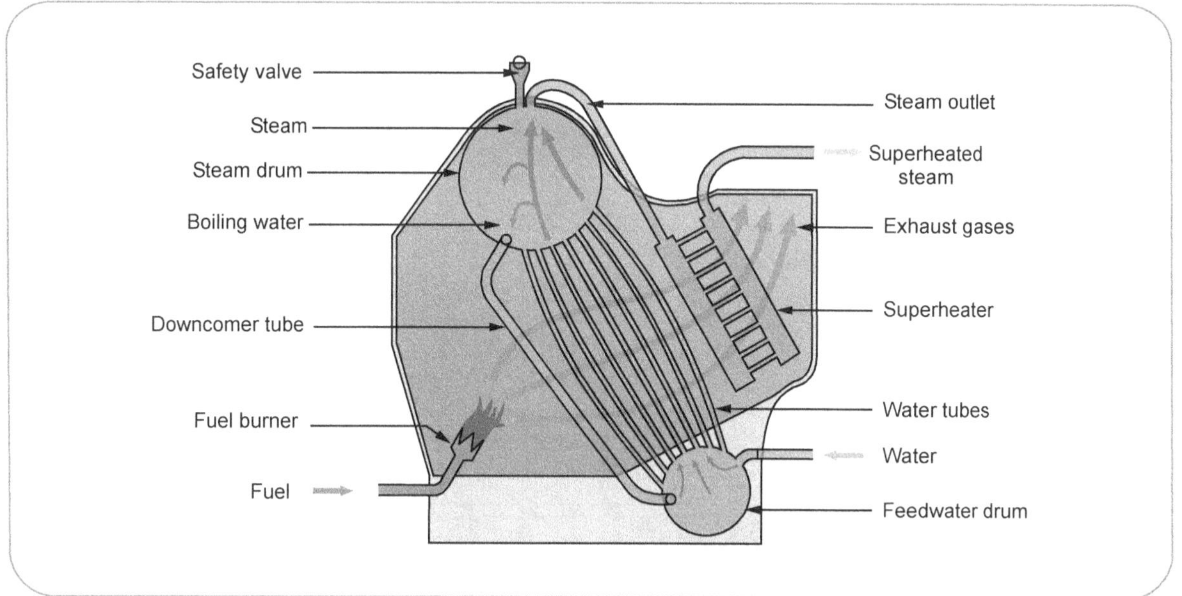

Fig. 4.9.2 Water tube package boiler

- The heated water then rises into the steam drum where saturated steam is drawn off the top of the drum.

- In some cases, the steam will pass through the superheater to become superheated.

- Cool water at the bottom of the steam drum returns to the feedwater drum through down-comer tubes, where it pre-heats the feedwater supply.

- These boilers are used for power generation, sugar industry, textile industry, food processing plant, marine applications, locomotives, etc.

- Babcock-Wilcox boiler, La-mont boiler, Benson boiler are the types of water tube boiler.

4.10.4 Comparison of Fire Tube and Water Tube Boiler   `SPPU : May-13`

Sr. No.	Parameter	Fire tube boiler	Water tube boiler
1.	Principle	In these boilers, the flue gases are passed through the tubes and the tubes are surrounded by water.	In these boilers, the water passes through the tubes and the flue gases pass through the external surface of tubes.
2.	Working pressure	These type of boilers works at low pressure upto 20 bar due to the large drum diameters.	These type of boilers can work at high pressure upto 200 bar due to the small drum diameters.
3.	Construction	Simple and rigid	Complex
4.	Initial cost	Low	High
5.	Operation and maintenance cost	High	Low
6.	Size	Small and compact	Bulky
7.	Transportation and Installation	Difficult	Easy
8.	Firing system	These are internally fired boilers.	These are externally fired boilers.
9.	Examples	Cochran, Lancashire and locomotive boilers	Babcock - Wilcox, La-mont, Benson boilers.

4.10.5 Boiler Mountings and Accessories   `SPPU : Dec.-03,04,09,12, May-07,09,12`

Boiler mountings

- For the operation and safety of the boiler different fittings and devices are necessary. These devices are called as **boiler mountings**.

- For example : Safery valve, feed check valve, water level indicator, steam stop valve, etc.

- Table 4.9.1 gives function and location of various boiler mountings :

Sr. No.	Boiler mounting	Location	Function
1.	Bourdon's pressure gauge	It is attached on the upper part of the front end plate.	It is used to indicate the steam pressure in the boiler.
2.	Safety valves	It is attached on the top of front end plate.	It is used to release the excess steam when the steam pressure inside the boiler exceeds.
3.	Water level indicator	It is attached to the lower part of front end plate.	It is used to indicate the water level inside the boiler.
4.	Fusible plug	It is fitted over the crown of the furnace or over the combustion chamber.	It is used to put off the fire in the furnace of boiler when the water level falls below unsafe level.
5.	Feed check valve	It is fitted to the shell below the water level of the boiler.	It is used to allow the supply of water at high pressure to the boiler and prevent the back flow of water.
6.	Blow-off cock	It is fitted to the lowest part of the boiler shell.	It is used to empty the boiler for cleaning, repair and inspection. It is also used to discharge the mud and sediments carried with the feed water.
7.	Steam stop valve	It is fitted to the highest part of the boiler shell.	It is used to regulate the flow of steam from the boiler to the engine and shut off the steam flow when not required.

Table 4.9.1 Location and function of various boiler mountings

Boiler accessories

- The auxillary parts which are used to increase the overall efficiency of the plant are called as **boiler accessories.**

- For example : Economiser, air-preheater, water feeding equipment, superheater, etc.

- Table 4.9.2 gives the function and location of various boiler accessories :

Sr. No.	Boiler accessory	Location	Function
1.	Economiser	It is fitted at the passage of flue gases from the boiler to chimney.	It extracts the waste heat of the chimney gases to preheat the water before feeding into the boiler. This reduces fuel consumption.
2.	Air-preheater	It is placed after the economiser and before the gases enter the chimney.	It extracts the waste heat of the flue gases and pre-heat the air supplied to the combustion chamber. It reduces fuel consumption.
3.	Superheater	It is fitted in the path of flue gases flowing to the chimney.	It is used to increase the temperature of the steam above its saturation temperature by passing the steam through a small set of tubes and hot gases over them.

Table 4.9.2 Location and function of various boiler accessories

4.10.6 Lancashire Boiler

- Lancashire boiler is a fire tube boiler as the hot flue gases flows inside the tube and the water flows around these tubes.

- It is stationary, internally fired and natural circulation type of boiler.

- Its shell consist of the brick work as shown in Fig. 4.9.3.

- This boiler consist of a cylindrical shell and two large tubes are passed through this shell.

- The brick work forms the channels for flow of flue gases known as bottom flue and side flue passage.

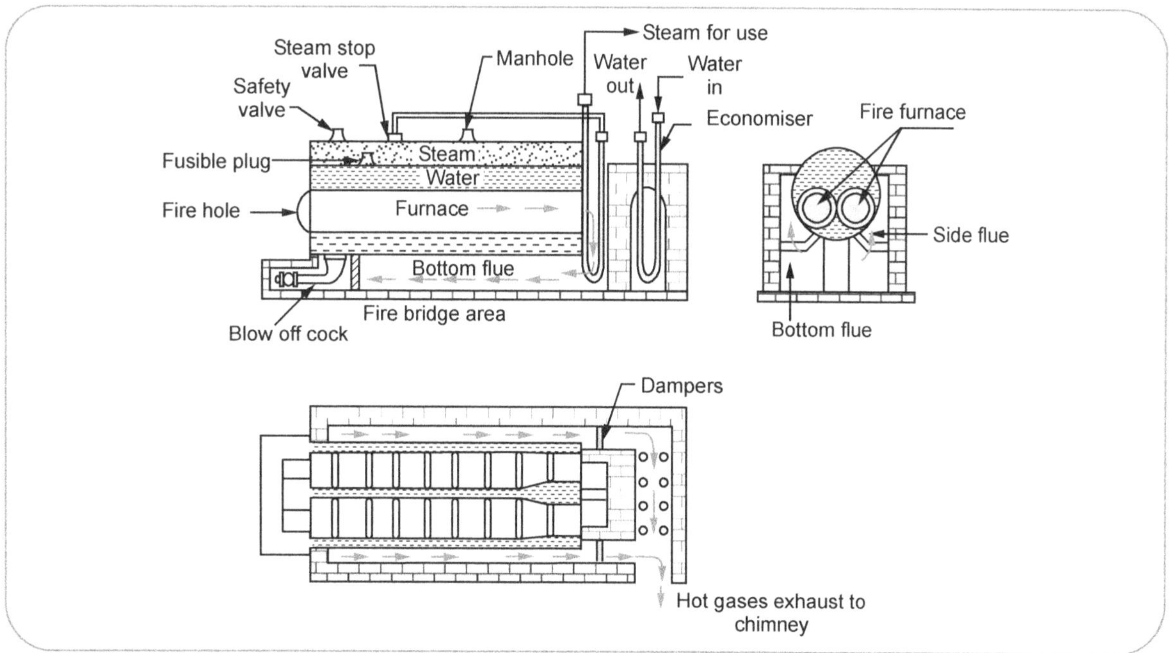

Fig. 4.9.3 : Lancashire boiler

- The grate is provided at the front end of the flue tube and the coal is fed to grate through the fire doors.

- A brick bridge arch is provided at the end of grate to avoid the coal and ash particle to pass through the fire tubes.

- This helps in preventing deposition of particles on the wall of fire tubes and enhance the heat transfer rate.

- The hot flue gases passed through the flow channels from the side and bottom of the boiler and transfers the heat to the water which converts it into the steam.

- Dampers are provided at the end of the fire tubes to control the flow of gases and to regulate the combustion.

- The superheater and the economiser are easily fitted in the lancashire boiler system.

- The superheater is placed at the end of main flue tube so that the flue gases can passed over the superheater before entering the bottom flow passage.

- The evaporative capacity of this boiler is upto 10000 kg/h and operating pressure is upto 15 bar.

Advantages of Lancashire Boiler

- Good evaporation quality.

- Inferior quality of coal can be used as a fuel.

- Heating surface area per unit volume is large.

- Easy maintenance.

- Efficiency is high (upto 80 - 85 %)

- Load fluctuations can be easily met.

Disadvantages of Lancashire Boiler

- It requires large space.

- Large amount of material and labour requires for boiler construction.

- Due to repeted heating and cooling, there is resultant expansion and contraction. This results in upsetting the brickwork and hence infiltration of air.

- Slow steam generation.

Applications of Lancashire Boiler

- Processing agent in textile, paper, sugar and chemical industries.

- To drive steam turbines, locomotives and marine applications.

4.9.7 Babcock and Wilcox Boiler

- Babcock and Wilcox boiler is a water tube boiler as the water is inside the tubes and hot flue gases flows over the tubes.

- Fig. 4.9.4 shows the schematic diagram of Babcock and Wilcox boiler with its different functional parts.

- The boiler shell consist of high quality of steel and is placed longitudinally. It is known as water and steam drum. The water level in the drum should be kept slightly above the center.

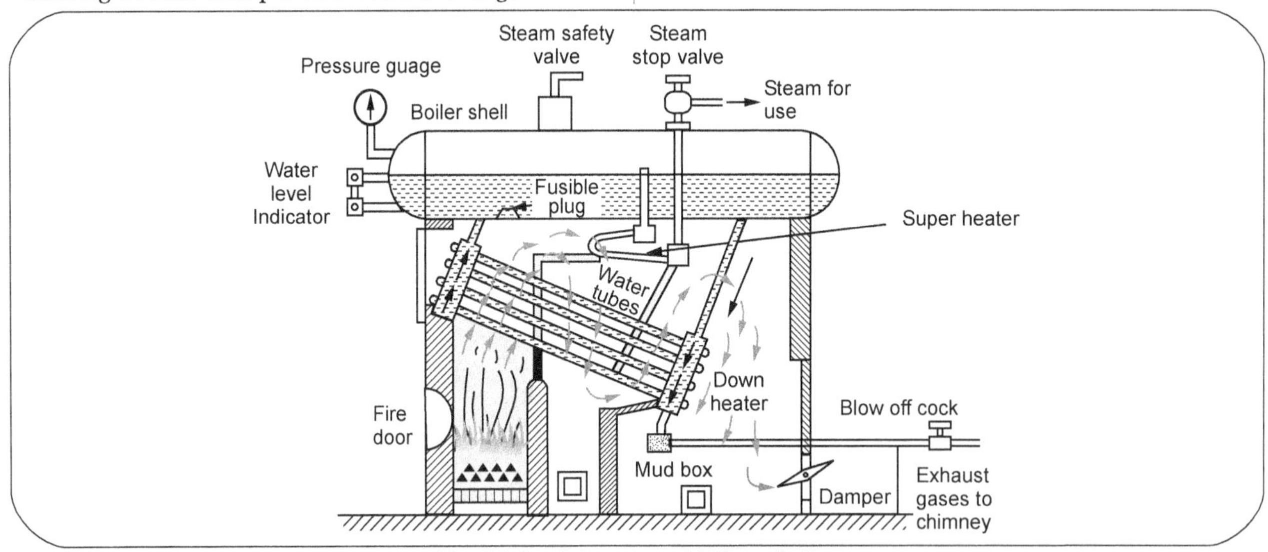

Fig. 4.9.4 : Babcock and Wilcox boiler

- The drum is connected by short tubes with uptake header and by long tubes with downtake header.

- A series of water tubes are connected to the uptake and downtake header at an angle of 15° to horizontal. This inclined position helps to make water flow.

- The grate is arranged below the uptake header. The fuel is supplied to grate through fire door. The fuel is burnt and forms hot flue gases.

- These flue gases are force to move in a specific path over the water tube due to baffle arrangement.

- The heat is transferred from hot flue gases to the water tubes and bottom cylindrical surface of the drum.

- The water flow is set due to the density difference in the water.

- The portion of water tubes at uptake header is subjected to the high temperature hot gases so the temperature of the water in this section rises due to decreased density and it flows to the drum via uptake header.

- Simultenously the water enters into tubes through downtake header and flow of water sets.

- The water and steam are seperated in the drum and as the steam is lighter it is collected in the upper portion of drum.

- To improve the quality of steam the superheater is placed between water drum and tubes.

- The steam formed in the drum is passed through the superheater and it becomes superheated.

- This superheated steam is then supplied to the turbine for electricity generation.

- The evaporative capacity for this boiler is upto 40000 kg/hr and operating pressure is 11 to 17 bar.

Review Questions

1. *What do you mean by heat? Explain the field of application of heat.*

2. *What are the different modes of heat transfer? Explain with suitable example.*

3. *State and explain Fourier's low of heat conduction. Also define each term in the equation.*

4. *Derive the relation for heat conduction through a thick wall.*

5. *Explain the electrical analogy of heat transfer.*

6. *Derive the relation for heat conduction through composite wall.*

7. *Explain the convection and radiation modes of heat transfer.*

8. *State and explain Newton's law of cooling.*

9. *Derive an expression for overall heat transfer coefficient.*

10. *Derive an expression for heat conduction through infinitely long cylinder and hollow sphere.*

11. *What do you mean by insulation? What are the different types of thermal insulating materials.*

12. *State and explain Stefan-Boltzmann law of radiation.*

13. *What is boiler ? Explain detail classification of boilers.*

14. *What are the requirements of good boiler ?*

15. *Explain the selection criterion of a boiler.*

16. *Write short note on following package boilers :*
    i) Fire tube package boiler
    ii) Water tube package boiler

17. *Explain any four boiler mountings and accessories.*

18. *Draw neat sketch of I.C. engine and explain its main components.*

19. *Explain four stroke S.I. engine with neat sketch.*

20. *Explain four stroke C.I. engine with neat sketch.*

21. *Compare S.I. engine and C.I. engine. (at least 8 points).*

22. *With neat sketch explain two stroke S.I. engine.*

23. *Compare two stroke engine and four stroke engine.*

❑❑❑

Notes

Unit - III
Vehicles and their Specifications

Chapter - 5 Vehicles and their Specifications   (5 - 1) to (5 - 34)

UNIT - III

5   Vehicles and their Specifications

Contents

Mind Map - Vehicles and their specifications

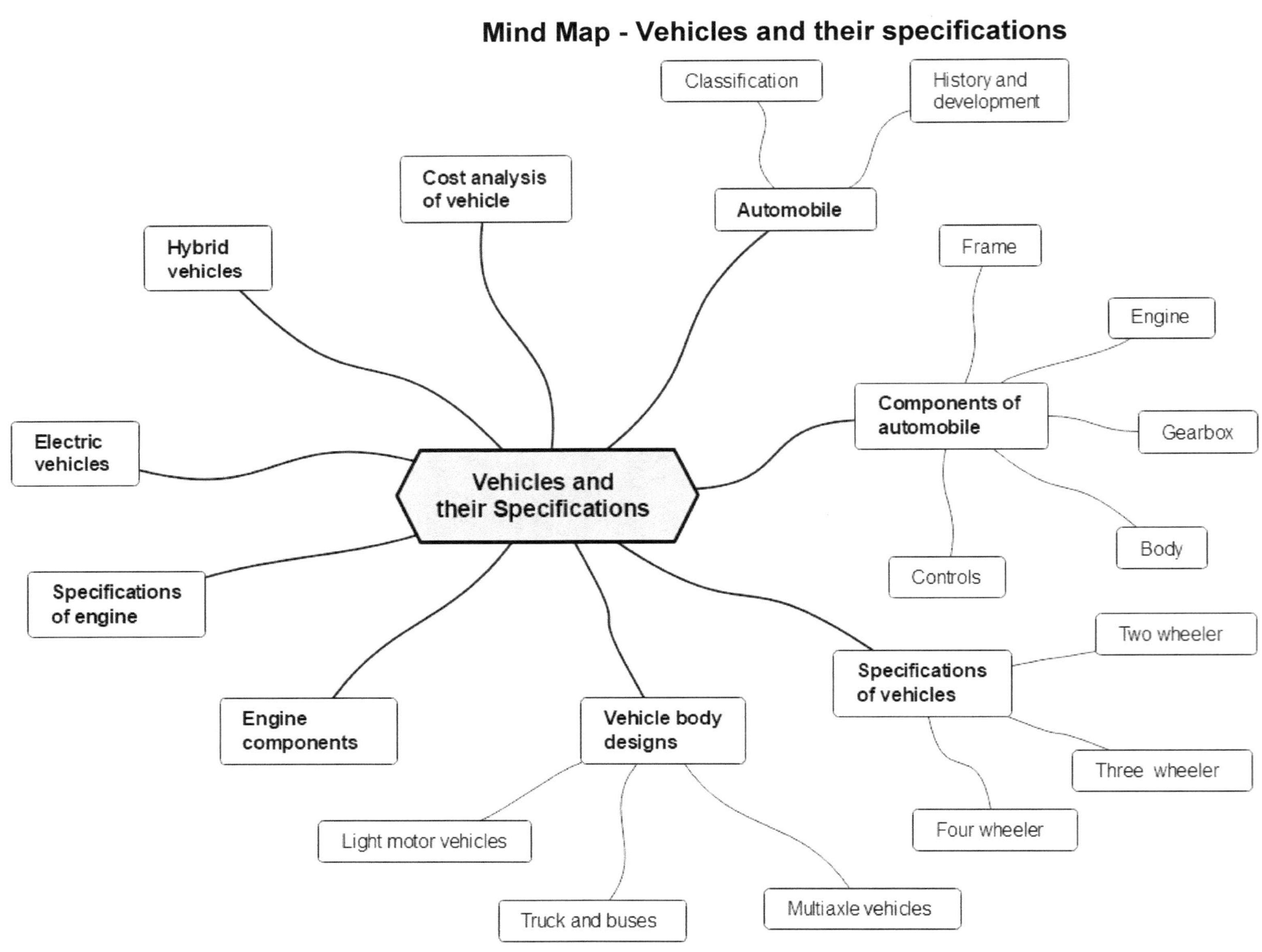

5.1 : Definition of Automobile

- An *automobile*, autocar or motor car is a self propelled wheeled motor vehicle used for transporting passengers and goods on ground.

- Automobiles are designed to run primarily on roads unlike other self propelled vehicles used for air transport (aeroplane, helicopter) and water transport (ship, motor boat).

5.2 : Automobile History and Development in India

- The history and development of automibile in India is summarized in Table 5.2.1.

Table 5.2.1 Automobile history and development in India

Time period	Highlights
1897 - 1930	• The first car was imported in India in 1897. • Until 1930s, cars were imported directly, but in small numbers.
1930 - 1940	• A few assembly plants were established in Bombay, Calcutta and Madras by leading foreign manufacturers.
1942 - 1944	• In order to develop and manufacture indigenous products, two Indian factories were started - Hindustan Motors Ltd. in Calcutta (1942) and Premier Automobiles Ltd. in Bombay (1944).
1947 - 1948	• Mahindra and Mahindra started assembly of jeep vehicles under license from Willys jeep.
1954	• Tata Motors ventured into manufacturing of commercial vehicles in joint venture with Daimler-Benz AG of Germany.
1979	• Maruti Udyog Ltd., the only public undertaking company came up in collaboration with Suzuki (Japan).
Present	• India is one of the fastest growing automobile market with sales of approximately (Year 2011) : • Passenger cars : 2.1 million • Commercial vehicles : 0.4 million • Three wheelers : 0.6 million • Two wheelers : 10 million • Almost all the global automotive manufacturers are now present in India.

5.2.1 Role of Auto Industry in National Growth

- The Automobile Industry is one of the fastest growing sectors in India. The increase in the demand for cars, and other vehicles, powered by the increase in the income is the main reason of growth of the automobile industry in India.

- Expanding population, low vehicle penetration (15 for every 1000 people), abundant availability of skilled talent, and a maturing automotive components segment play an important role globally to develop Indian automobile industry.

- The Indian Automobile industry includes two-wheelers, trucks, cars, buses and three-wheelers which play a crucial role in growth of the Indian economy. India has emerged as Asia's fourth largest exporter of automobiles, behind Japan, South Korea and Thailand.

- The Economic progress of this industry is indicated by the amount of goods and services produced which give the capacity for transportation and boost the sale of vehicles. There is a huge increase in automobile production with a catalyst effect by indirectly increasing the demand for a number of raw materials like steel, rubber, plastics, glass, paint, electronics and services.

- The well-developed Indian automotive industry fulfills this catalytic role by producing a wide variety of vehicles like passenger cars, light, medium and heavy commercial vehicles, multi-utility vehicles such as jeeps, scooters, motorcycles, mopeds, three wheelers, tractors etc.

- It contributes about 4% to India's Gross Domestic Product (GDP) and 5% to India's industrial production.

- Indian market before independence was seen as a market for imported vehicles while assembling of cars manufactured by General Motors and other brands was the order of the day. Indian automobile industry mainly focused on servicing, dealership, financing and maintenance of vehicles. Later only after a decade from independence manufacturing started.

- India has become one of the international players in the automobile market. In the year 2006-07, the Indian Automobile Industry produced 2.06 million four wheelers and 9 million two and three wheelers.

- As of 2019, the Indian automobile industry had contributed to almost 7 % of the country's GDP. During the time it provided 22 % of India's manufacturing GDP and provided around 18 % of excise duties to the state exchequer. The Indian automobile industry has also significantly increased the presence of the nation in international markets with a year-on-year increase in exports of approximately 18 %.

5.3 : Major Components / Systems of an Automobile

- The major components / systems used in an automobile are listed in Table 5.3.1.

Table 5.3.1 Major components / systems of an automobile

Sr. No.	Major components / Systems	Elements	
(1)	The Basic Structure (Frame Work)	(i)	Frame
		(ii)	Suspension system
		(iii)	Axles
		(iv)	Wheels
(2)	The Power Plant (Engine or Source of Power)	(i)	Air and fuel supply system
		(ii)	Engine
		(iii)	Exhaust system (Petrol, diesel or other)
(3)	The Power Train (Transmission system or Gearbox)	(i)	Clutch
		(ii)	Gearbox
		(iii)	Propeller shaft
		(iv)	Differential
(4)	The Body (Super - structure)		
(5)	The Controls	(i)	Steering
		(ii)	Accelerator pedal
		(iii)	Clutch pedal
		(iv)	Gear change knob, etc.
(6)	The Auxiliaries	(i)	Electrical supply system
		(ii)	Fuel system
		(iii)	Lubrication system
		(iv)	Cooling system

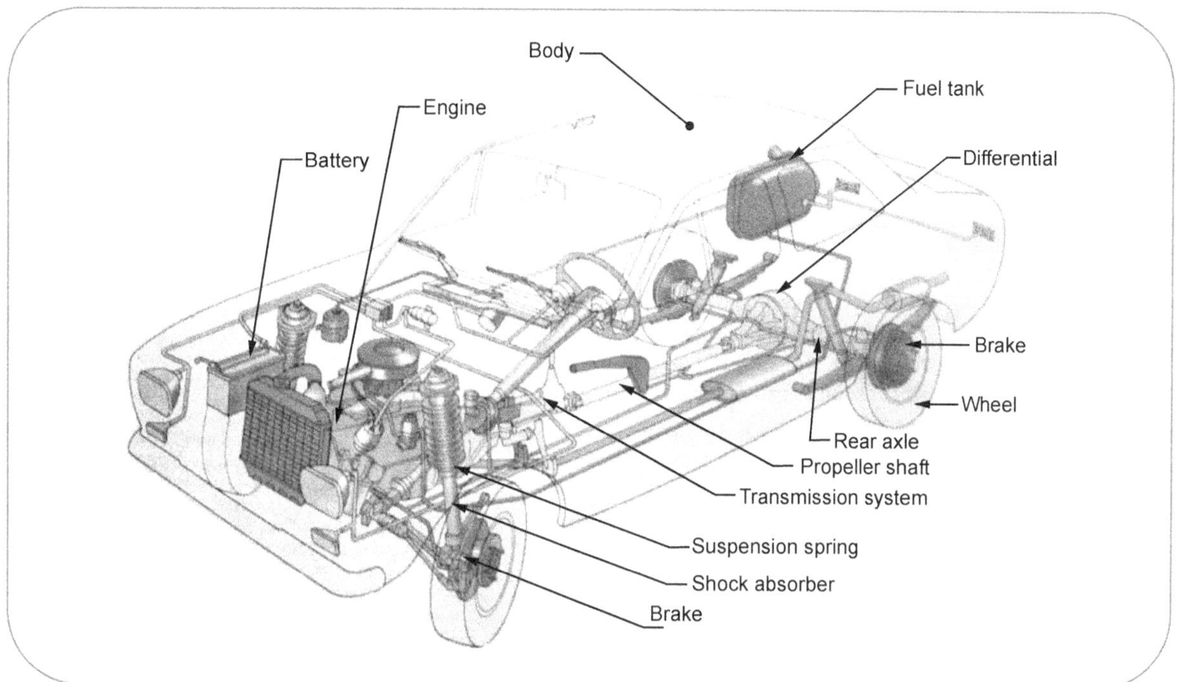

Fig. 5.3.1 Major components / systems of an automobile

- The major components / systems with their relative location in an automobile are shown in Fig. 5.3.1.

5.3.1 The Basic Structure (Framework)

- The basic structure or framework is the backbone of the vehicle components / assemblies.
- It is that important component of the vehicle on which the other units such as engine, car body, wheels, etc. are mounted.
- It consists of 4 major components :

(i) Frame

- It is the foundation member for carrying the engine, vehicle body, steering, etc.
- Frames may be made of box, tubular, channel section or riveted or welded.

(ii) Suspension system

- All major components are mounted on the frame by means of springing suspension system.
- The objective of suspension system is to prevent road shocks from being transmitted to vehicle and to preserve stability of vehicle in pitching or rolling.

(iii) Axles

- It can be considered as a beam carrying the vehicle load and supported at its two ends.
- The axles are subjected to vertical load of vehicle, side thrust while cornering, fore and aft load during braking and acceleration.

(iv) Wheels

- Wheels are the last components of the vehicle and are in direct contact with the road surface.
- Wheels can be wire spoked which are light in weight or made of pressed steel which are more common on private cars.

5.3.2 The Power Plant (Engine or Source of Power) and Exhaust System

- The power plant or engine provides the power for moving the car, wheels and other parts of the automobile.
- The power plant can be gas turbine engine or I.C. engine which may be either of spark - ignition (Petrol) or compression - ignition (Diesel) type.

- The power plant consists of three major elements or sub-systems as follows :

> i)   Air and fuel supply system
> ii) Engine
> iii) Exhaust system

(i) Air and fuel supply system

- The engine takes in fresh air from the atmosphere through an air filter.
- Fuel is pumped from the fuel tank and injected in the intake mainfold for a petrol engine, while injected in the cylinder directly for a diesel engine.

(ii) Engine

- The engine converts the heat energy generated by burning of fuel into mechanical energy.
- The burned fuel is forced out of the engine in the exhaust system.

(iii) Exhaust system

- The functions of the exhaust system are as follows :
  - Provide an easy passage to the flow of exhaust gases from the cylinder to the atmosphere.
    - To treat the exhaust gases (oxidize the unburnt fuel) before release to the atmosphere.
  - Damp the noise levels of the high speed exhaust gases.
- The main elements of the exhaust system are shown in Fig. 5.3.2 and its function is summarized in Table 5.3.2.

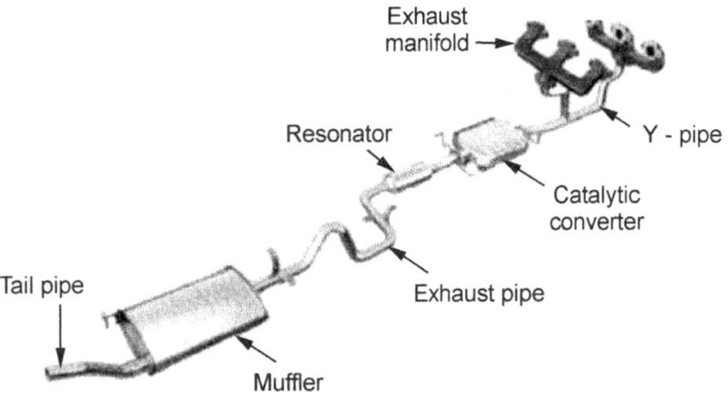

Fig. 5.3.2 Exhaust system of a car

Table 5.3.2 Elements of an exhaust system

Sr. No.	Element	Function
(i)	Exhaust manifold	It allows smooth out-flow of exhaust gases from the combustion chamber to the exhaust pipe.
(ii)	Exhaust pipe	It carries the exhaust gases from the exhaust manifold to the catalytic converter. The pipe should avoid heat transfer from exhaust gases to the surrounding system of the vehicle.
(iii)	Catalytic converter	It is used to oxidize the unburnt fuel in the exhaust gases. It reduces NO_x emission as well as CO and HC.
(iv)	Muffler	Its main function is to reduce the sound of the high speed out-going exhaust gases.

5.3.3 The Power Train (Transmission System or Gearbox)

- The function of power train system is to transfer power from engine to the wheels of the vehicle.
- It consist of a clutch, a gearbox, a propellor shaft, and a differential.

(i) Clutch

- It is usually a friction device.
- Its purpose is to enable the driver to gradually engage and disengage the drive from engine to road wheels when required.

(ii) Gearbox

- It consists of various types and sizes of gears.
- Gearbox provides the necessary leverage variation between the torque from engine to the road wheels.

(iii) Propeller shaft

- It consists of shaft having universal joint at its ends.

- Its function is to transmit power from output shaft of gearbox to input shaft of the differential.

(iv) Differential

- It consists of bevel pinion and crown wheel and a set of internal gears.

- Its purpose is to permit the two wheels on the axle to run at different speeds while cornerning a turn.

5.3.4 The Body (Super - Structure)

- The body or super - structure houses the vehicle passengers and protects them from atmospheric effects.

- It is conventionally a pressed steel frame with integral (frameless) design or separate member.

- The body contours should be aerodynamically designed to offer less resistance to the passage of air over it and at the same time should be pleasing and attractive at appearance.

5.3.5 The Controls

- In order to enable the driver to control the different parameters of engine and vehicle, a number of controls are provided around the driver seat, dash board and foot board.

- Important controls provided on all vehicles are steering, accelerator, clutch pedal, accelerator and brake pedal, gear change knob and various electrical switches and controls.

5.3.6 The Auxiliaries

- The auxiliaries are the important components of vehicle which assist in operation of other components.

- The major auxiliaries are :
  (i)   Electrical supply system - Consists of battery and alternator
  (ii) Fuel system - Consists of fuel pump or carburettors
  (iii) Lubrication system - Consists of oil pump, oil filter, etc.
  (iv) Cooling system - Consists of water pump, radiator, thermostat, etc.

5.4 : Classification of Automobiles

- There are numerous types of automobiles used on roads across the world.

- In general, automobiles can be classified on basis of following considerations :

Table 5.4.1 Classification of automobiles

Sr. No.	Considerations		Types
1.	On basis of its use	(i)	Auto cycles and mopeds
		(ii)	Motor cycles and scooters
		(iii)	Cars and jeeps
		(iv)	Trucks and buses
		(v)	Tractors

2.	On basis of capacity	(i)	Heavy Transport Vehicles (HTV) such as trucks, buses, tractors, etc.
		(ii)	Light Transport Vehicles (LTV) such as cars, jeeps, motor cycles, etc.
3.	On basis of make and model	(i)	Maruti Suzuki, Tata Motors, General Motors - cars and jeeps
		(ii)	Tata Motors, Ashok Leyland - buses and trucks
		(iii)	Hero Honda, Bajaj, TVS - motor cycles, scooters, etc.
4.	On basis of fuel used	(i)	Petrol vehicles - motor cycles, scooters, cars, etc.
		(ii)	Diesel vehicles
		(iii)	Electric vehicles - electric bikes, Reva car, etc.
5.	On basis of drive	(i)	With respect to vehicle being driven while sitting left or right : - Left hand drive - Right hand drive
		(ii)	With respect to driving axle : - Front wheel drive - Rear wheel drive - All wheel drive
6.	On basis of body style	(i)	Closed cars such as hatch back, sedan, saloons, etc.
		(ii)	Open cars such as convertible, sports car
		(iii)	Special cars such as station wagon, estate version
7.	On basis of wheels	(i)	Two wheelers such as scooters, motor cycles, etc.
		(ii)	Three wheelers such as autorickshaws, tempos, etc.
		(iii)	Four wheelers such as cars, jeeps etc.
		(iv)	Six wheelers such as trucks, buses, tractor-trailers, etc.
8.	On basis of transmission	(i)	Conventional - having ordinary crash type gear box
		(ii)	Semi automatic - having manual transmission with automatic clutch control
		(iii)	Fully automatic - having complete automatic control and uses epicyclic gearbox with torque converters
9.	On basis of suspension	(i)	Conventional - using leaf spring
		(ii)	Independent - using coil springs with torsion bar or air suspension

5.5 : Specifications of Vehicles

- Manufacturers produce automobiles of different types and purpose based on the customer needs.

- There are numerous models made by different manufacturers in a particular category of automobile.

- Hence to compare an automobile with another model, vehicle specifications are defined and provided with each automobile.

- Vehicle specification broadly specifies the following :

i) Vehical make and model	ii) Engine specification	iii) Fuel
iv) Transmission specification	v) Steering	vi) Brakes
vii) Wheels	viii) Overall dimensions of vehicle	

5.5.1 Specifications of Motorcycles

- The major specifications of popular motorcycles is provided in Table 5.5.1.

Table 5.5.1 Specifications of motorcycles

Make / Manufacturer	Honda Activa	Hero Splendor +
Model		
Variant	Activa Joy	Splendor +
Fuel type	Petrol	Petrol
Engine		
Engine type	Air cooled, 4-stroke, single cylinder.	Air cooled, 4-stroke, single cylinder.
Displacement, cc	109	97.2
Maximum Power, HP@rpm	8 HP @ 7500 rpm	7.5 HP @ 8000 rpm
Maximum Torque, Nm@rpm	8.8 Nm @ 5500 rpm	7.95 Nm @ 5000 rpm
Ignition Type	CDI	CDI
Transmission System		
Gearbox type	Vario-matic	4-speed constant mesh
Clutch	Centrifugal clutch	Wet, Multiplate
Chassis		
Chassis type	High rigidity underbone type	Tubular double cradle type
Suspension System		
Front	Spring loaded Hydraulic damper	Telescopic hydraulic shock absorbers
Rear	Spring loaded Hydraulic damper	Swing arm with hydraulic shock absorbers
Braking System		
Front	Internal expandable drum brakes	Internal expandable drum brakes
Rear	Internal expandable drum brakes	Internal expandable drum brakes
Wheels		
Type	Pressed steel	Spoke wheel/ Cast wheel
Tyres		
Front	90/100 - 10 53 J	2.75 X 18 - 42 P/ 4 PR
Rear	90/100 - 10 53 J	2.75 X 18 - 48 P/ 6 PR
Battery	12 V X 5 Ah	12 V X 2.5 Ah
Weight		
Kerb, kg	111	109
Fuel tank capacity, liters	5.3	11
Dimensions		
Overall length, mm	1761	1970
Overall width, mm	710	720
Overall height, mm	1147	1040
Wheel base, mm	1238	1230
Ground clearance, mm	145	159

Make / Manufacturer	Bajaj (Kawasaki) Caliber	Bajaj Discover
Model		
Variant	**Caliber** 115	**Discover** 100
Fuel type	**Petrol**	**Petrol**
Engine		
Engine type	Air cooled, 4-stroke, single cylinder	Air cooled, 4-stroke, single cylinder
Displacement, cc	111.6	94.36
Maximum Power, HP@rpm	9.5 HP @ 8000 rpm	7.7 HP @ 7500 rpm
Maximum Torque, Nm@rpm	9.15 Nm @ 6500 rpm	7.85 Nm @ 5000 rpm
Ignition Type	CDI	Digital ECU based
Transmission system		
Gearbox type	4-speed constant mesh	5-speed constant mesh
Clutch	Wet, Multiplate	Wet, Multiplate
Chassis		
Chassis type	Tubular semi-double cradle type	Single down tube
Suspension system		
Front	Telescopic hydraulic shock absorbers	Telescopic hydraulic shock absorbers
Rear	Trailing arm with hydraulic shock absorbers	Spring loaded nitrogen oxide assisted gas shock absorber
Braking system		
Front	Internal expandable drum brakes	Internal expandable drum brakes
Rear	Internal expandable drum brakes	Internal expandable drum brakes
Wheels		
Type	Spoke wheel	Alloy wheel
Tyres		
Front	2.75 X 18	2.75 X 17
Rear	3.00 X 18	3.00 X 17
Battery	12 V X 5 Ah	12 V X 5 Ah
Weight		
Kerb, kg	118	115
Fuel tank capacity, liters	14.8	8
Dimensions		
Overall length, mm	1995	2040
Overall width, mm	750	760
Overall height, mm	NA	1087
Wheel base, mm	1245	1305
Ground clearance, mm	150	165

Make / Manufacture	Hero CBZ X-Treme	Bajaj Pulsar	Honda CBR
Model			
Variant	CBZ X-Treme	Pulsar 150	CBR 250R
Fuel type	Petrol	Petrol	Petrol
Engine			
Engine type	Air cooled, 4-stroke, single cylinder	Air cooled, 4-stroke, single cylinder, DTS-i	Liquid cooled, 4-stroke, single cylinder
Displacement, cc	149.2	149.0	249.2
Maximum Power, HP@rpm	14.4 HP @ 8500 rpm	15.1 HP @ 9000 rpm	25 HP @ 8500 rpm
Maximum Torque, Nm@rpm	12.8 Nm @ 6500 rpm	12.5 Nm @ 6500 rpm	22.9 Nm @ 7000 rpm
Ignition type	Digital ECU based	Digital ECU based	Digital ECU based
Transmission system			
Gearbox type	5-speed constant mesh	5-speed constant mesh	6-speed constant mesh
Clutch	Wet, Multiplate	Wet, Multiplate	Wet, Multiplate
Chassis			
Chassis type	Tubular, diamond type	Tubular double cradle type	Twin spar type
Suspension system			
Front	Telescopic hydraulic shock absorbers	Telescopic hydraulic shock absorbers	Telescopic hydraulic shock absorbers
Rear	Swing arm with gas damper	Spring loaded nitrogen oxide assisted gas shock absorber	Pro link
Braking system			
Front	Disc brake	Disc brake	Disc brake
Rear	Disc brake	Internal expandable drum brakes	Disc brake
Wheels			
Type	Alloy wheel	Alloy wheel	Alloy wheel
Tyres			
Front	80/100 X 18 - 47 P	2.75 X 17	110 / 70 X 17 M/C
Rear	100/90 X 18 - 56 P	100/90 X 17	140/70 X 17 M/C
Battery	12 V X 4 Ah	12 V	12 V X 6 Ah
Weight			
Kerb, kg	148	143	167
Fuel tank capacity, liters	12.1	15	13
Dimensions			
Overall length, mm	2080	2055	2032
Overall width, mm	765	755	720
Overall height, mm	1145	1170	1127
Wheel base, mm	1325	1320	1367
Ground clearance, mm	145	165	145

5.5.2 Specifications of Cars

- The major specifications of popular cars is provided in Table 5.5.2.

Table 5.5.2 Specifications of cars

Make / Manufacturer	Maruti Suzuki M800	General Motors - Chevrolet
Model	M800	Beat
Variant	M800 AC	1.2 LT
Fuel type	Petrol	Petrol
Engine		
Engine type	8V SOHC 4-cyl MPFI	16V DOHC 4-cyl MPFI
Displacement, cc	796	1199
Maximum Power, HP@rpm	37 HP @ 5000 rpm	79 HP @ 6200 rpm
Clutch		
Type	Hydraulic	Hydraulic
Transmission system		
Gearbox type	4-speed manual with synchromesh on all gears	5-speed manual with synchromesh on all gears
Gear ratios		
1st	3.58	3.539
2nd	2.16	1.864
3rd	1.33	1.242
4th	0.9	0.974
5th	-	0.781
Reverse	NA	
Final drive ratio	4.35	4.19 : 1
Suspension system		
Front	MacPherson strut and coil spring	Independent, MacPherson strut
Rear	Coil spring with gas filled shock absorber	Non-independent, compound crank type
Steering System		
Type	Rack and pinion	Rack and pinion, power assisted
Turning radius, m	4.4 m	4.45 m
Braking system		
Front	Ventilated disc	Ventilated disc
Rear	Drum	Drum
Wheels		
Type	Pressed steel disc	Pressed steel disc
Size	4J X 12	4.5J X 14
Tyres	145/70 R12	155/70 R14, tubeless
Weight		
Kerb, kg	665	965
Dimensions		
Overall length, mm	3335	3640
Overall width, mm	1440	1595
Overall height, mm	1405	1520

Wheel base, mm	2175	2375
Front track, mm	1215	1408
Rear track, mm	1200	1403
Ground clearance, mm	170	165

Make / Manufacturer	Tata Indica	Hyundai Santro	Suzuki
Model	Indica	Santro	Swift
Variant	eV2	1L Xing	VDi
Fuel type	Diesel	Petrol	Diesel
Engine			
Engine Type	16V DOHC common rail	12V SOHC 4-cyl MPFI	DOHC, DDiS diesel common rail
Displacement, cc	1396	1086	1248
Maximum Power, HP@rpm	70 HP @ 4000 rpm	63 HP @ 5500 rpm	76 HP @ 4000 rpm
Clutch			
Type	Hydraulic	Hydraulic	Hydraulic
Transmission system			
Gearbox type	5-speed manual with synchromesh on all gears	5-speed manual with synchromesh on all gears	5-speed manual with synchromesh on all gears
Gear ratios			
1st	N.A.	3.833	3.545
2nd	N.A.	2.105	1.904
3rd	N.A.	1.31	1.233
4th	N.A.	0.919	0.911
5th	N.A.	0.784	0.725
Reverse	N.A.	N.A.	3.25
Final drive ratio	3.42:1	4.22 : 1	3.94 : 1
Suspension system			
Front	Independent, MacPherson strut	MacPherson strut and coil spring	MacPherson strut and coil spring
Rear	Independent, semi trailing	Torsion beam and coil spring	Torsion beam and coil spring
Steering system			
Type	Rack and pinion, power assisted	Rack and pinion, power assisted	Rack and pinion, power-assisted
Turning radius, m	4.9	4.4	4.7 m
Braking system			
Front	Ventilated disc	Ventilated disc	Ventilated disc
Rear	Drum	Drum	Drum
Wheels			
Type	Pressed steel disc	Pressed steel disc	Pressed steel disc
Size	14" inch	4.0J X 13	14" inch
Tyres	165/65 R14	155/70 R13, tubeless	165/80 R14 tubeless
Weight			
Kerb, kg	1080	886	1075

Dimensions			
Overall length, mm	3690	3565	3760
Overall width, mm	1685	1525	1690
Overall height, mm	1485	1590	1530
Wheel base, mm	2400	2380	2390
Front track, mm	1380	1315	1470
Rear track, mm	1360	1300	1480
Ground clearance, mm	165	N.A.	170

Make / Manufacturer	Ford	Honda	Toyota
Model	Fiesta	New City	Corolla Altis
Variant	1.4 TDCI Zxi Duratorq	V M/T	G M/T
Fuel Type	Diesel	Petrol	Petrol
Engine			
Engine type	8V SOHC common rail	16V SOHC 4-cyl in-line	16V DOHC 4-cyl VVT-i
Displacement, cc	1399	1497	1794
Maximum Power, HP@rpm	68 HP @ 4000 rpm	118 HP @ 6600 rpm	132 HP @ 6000 rpm
Clutch			
Type	Hydraulic	Hydraulic	Hydraulic
Transmission system			
Gearbox type	5-speed manual with synchromesh on all gears	5-speed manual with synchromesh on all gears	5-speed manual with synchromesh on all gears
Gear ratios			
1st	N.A.	N.A.	N.A.
2nd	N.A.	N.A.	N.A.
3rd	N.A.	N.A.	N.A.
4th	N.A.	N.A.	N.A.
5th	N.A.	N.A.	N.A.
Reverse	N.A.	N.A.	N.A.
Final drive ratio	3.82 : 1	N.A.	N.A.
Suspension system			
Front	Independent MacPherson strut with stabilizer bar	Independent MacPherson strut with stabilizer bar, coil spring	Independent MacPherson strut with stabilizer bar
Rear	Semi-independent twist-beam	Torsion beam axle with stabilizer, coil spring	Torsion beam type with stabilizer
Steering system			
Type	Rack and pinion, power-assisted	Rack and pinion, power-assisted	Rack and pinion, power-assisted
Turning radius, m	4.9 m	5.3 m	5.3 m
Braking system			
Front	Ventilated disc	Ventilated disc	Ventilated disc
Rear	Drum	Drum	Drum

Wheels			
Type	Pressed steel disc	Alloy wheel	Alloy wheel
Size	14" inch	15 x 5.5J	15" inch
Tyres	175/65R 14	175/65 R15	195 / 65 R15
Weight			
Kerb, kg	1150	1100	1200
Dimensions			
Overall length, mm	4282	4420	4540
Overall width, mm	1686	1695	1760
Overall height, mm	1468	1480	1480
Wheel base, mm	2486	2550	2600
Front track, mm	1474	N.A.	1530
Rear track, mm	1444	N.A.	1535
Ground clearance, mm	160	160	176

5.5.3 Specifications of Three Wheelers (Auto Rickshaw)

i) Bajaj RE Compact PETROL Auto Rickshaw Specifications. (Refer Fig. 5.5.1).

Power	6.6 kW@5000rpm
Torque	15.5 N.m@3300rpm
Cubic capacity	145.45 cc
Transmission	4 forward + 1 reverse gear
Clutch	Wet multidisc type
Kerb weight	307 kg
Wheel base	2000 mm
Overall width	1300 mm
Overall length	2635 mm
Overall height	1692 mm
Grade ability	18%

ii) Bajaj RE Compact LPG Auto Rickshaw Specifications. (Refer Fig. 5.5.2)

Power	6.0 kW@5200 rpm
Torque	14.10 N.m@3500 rpm
Cubic capacity	145.45 cc
Transmission	4 forward + 1 reverse gear
Clutch	Wet multidisc type

Kerb weight	344 kg
Wheel base	2000 mm
Overall width	1300 mm
Overall length	2635 mm
Overall height	1692 mm
Grade ability	16%

iii) Bajaj RE Compact CNG Auto Rickshaw Specifications. (Refer Fig. 5.5.3)

Power	5.3 kW@5200 rpm
Torque	12.9 N.m@3500 rpm
Cubic capacity	145.45 cc
Transmission	4 forward + 1 reverse gear
Clutch	Wet multidisc type
Kerb weight	368 kg
Wheel base	2000 mm
Overall width	1300 mm
Overall length	2635 mm
Overall Height	1692 mm
Grade ability	14%

Note : RE stands for Rear Engine, a term used commonly for **Bajaj's** three wheelers.

iv) TVS AutoRickshaw

(a) 4 Stroke LPG

India's first 200 cc 3 wheeler, the TVS King rules the road. Its 4 Stroke, electric start, high power engine with extra comfort and good mileage helps the driver earn more for a better life. Truly a "Car on 3 Wheels".

Technical Specifications

Engine Type	4 Stroke Single Cylinder Forced Air - Cooled Spark Ignition Engine
Cubic Capacity	199.26 cc
Max Power	6.0 kW @ 5000 rpm

Transmission	4 forwards and 1 reverse speed, constant mesh, fork-and-cam-type
Fuel Tank Capacity	20.6 litres (Water Equivalent)
Kerb weight	360 kg

(b) 4 Stroke DIESEL

TVS King Diesel DS comes with Proven Greaves GL 435, Single cylinder DI engine, gives More Miles (37Kmpl) & Smiles (Better Comfort, Safety, Style) to customers.

Technical Specifications

Engine Type	4 Stroke Single Cylinder Forced Air - Cooled Compression Ignition Engine
Cubic Capacity	436 cc
Max Power	5.5 kW @ 3600 +/- 50 rpm
Transmission	4 forwards and 1 reverse speed, constant mesh, fork-and-cam-type
Fuel Tank Capacity	9 litres
Kerb Weight	369 kg

Fig. 5.5.1 Bajaj-RE-Compact-Petrol-Auto Rickshaw

Fig. 5.5.2 Bajaj-RE-Compact-LPG-Auto Rickshaw

Fig. 5.5.3 Bajaj-RE-Compact-CNG-Auto Rickshaw

5.6 : Vehicle Chassis and Body

- A vehicle is made of two major assemblies - chassis and body.
- Chassis is the backbone of the vehicle.
- The chassis is an assembly of vehicle without body.
- The chassis includes frame, engine, clutch, gearbox, propellor shaft, universal joints, differential, axles, wheels, steering, brakes, fuel tank, suspension system, etc.
- Body is the top cover fitted on the chassis assembly.

- The body includes passenger compartment, bumpers, fenders, radiator grille, interior trim, glass, paint, etc.
- Refer Fig. 5.6.1.

Fig. 5.6.1 Vehicle chassis and body

5.7 : Vehicle Body

- The vehicle body is a major sub-assembly fitted on top of the chassis.
- In the earlier generation, vehicle body was a basic wooden cover body fitted on the horse carriage.
- The modern automobile consists of a metallic body providing more stylish appearance.

5.7.1 Functions of Body

- The main functions of a vehicle body are as follows :

(i) To provide protection to the passengers against weather and collision.

(ii) To provide comfort to the passengers.

(iii) To provide stylish (better look) appearance to the overall vehicle.

5.7.2 Vehicle Body Designs or Styles

- The vehicle body is designed to meet aesthetic and application requirements.
- Various body designs / styles have been developed over the generations and are summarised in Fig. 5.7.1.

(i) Saloon (Sedan)

- Saloon cars are three box / compartment cars.
- The front compartment consists of the engine.
- The second compartment accomodates a row of front and a row of rear seats without any partition between the driver and rear passenger seats.
- The passenger compartment is an enclosed compartment.
- The third compartment has space to accomodate luggage.

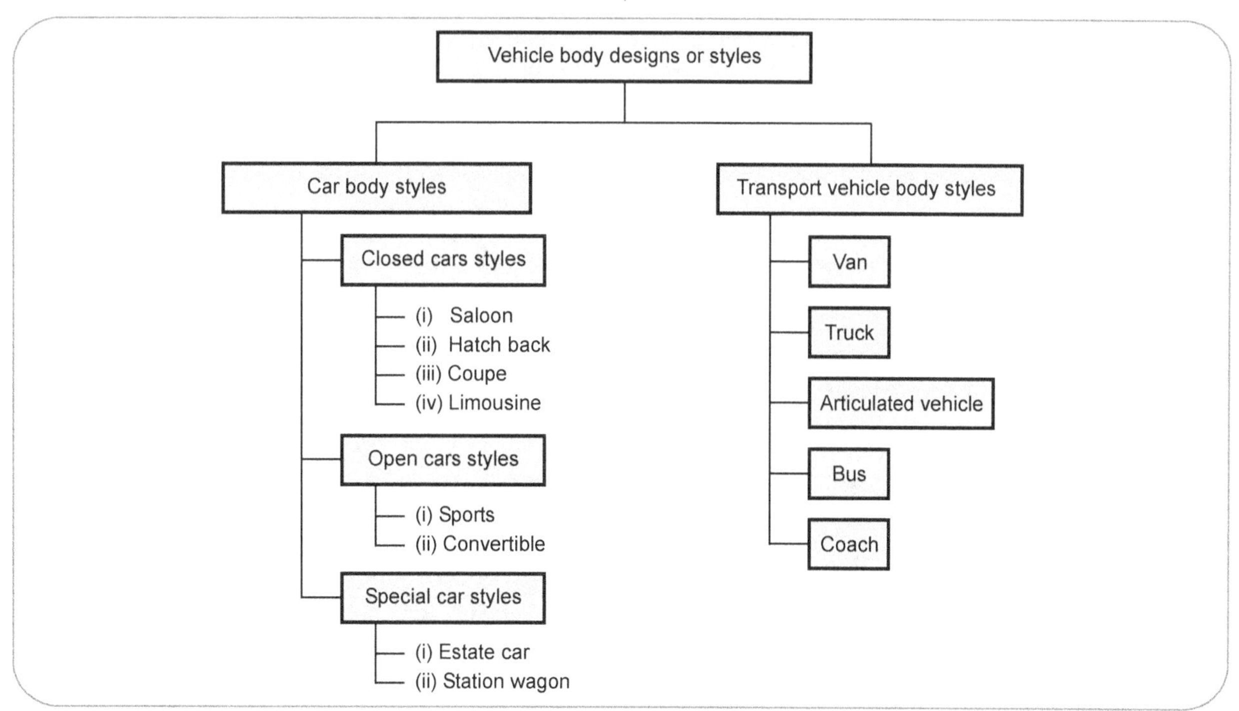

Fig. 5.7.1 Vehicle body designs or styles

- Saloon cars generally have two doors on each side of the car.
- Refer Fig. 5.7.2 for a typical saloon car.

Fig. 5.7.2 Saloon car

(ii) Hatchback

- Refer Fig. 5.7.3 for a typical hatchback car design.

Fig. 5.7.3 Hatchback car

- It is modification on the saloon car design.
- It consists of only two box / compartments without any dedicated luggage space.
- The two compartments of hatchback are used to accomodate the engine and two rows of passenger seats.

(iii) Coupe

- Refer Fig. 5.7.4 for a typical coupe car design.

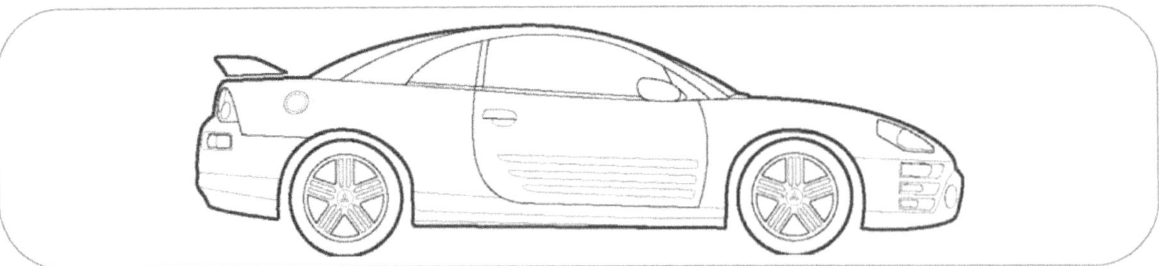

Fig. 5.7.4 Coupe car

- A coupe is similar to a saloon car having three compartments.
- A coupe has only one row of passenger seats unlike two rows in a saloon car.
- It has only one door on each side of the car.

(iv) Limousine

- Refer Fig. 5.7.5 for a limousine car body design.

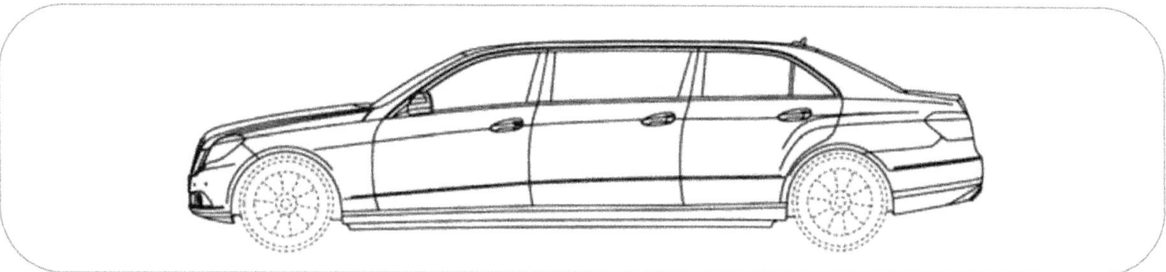

Fig. 5.7.5 Limousine car

- A limousine is also similar to a saloon car but has an extended passenger compartment.
- Within the passenger compartment, the driving compartment is separated from the rear compartment by a sliding glass division.

(v) Sports cars

- Refer Fig. 5.7.6 for a sports car body design.

Fig. 5.7.6 Sports car

- A sport car is a low height coupe car design.
- It has one door on each side and has aerodynamic body shape to offer least resistance to flowing air.
- It sometimes has a collapsible hood (bonnet), fold flat windscreen and also removable side screens.

(vi) Convertible car

- Refer Fig. 5.7.7 for a convertible car body design.
- It is also a three box / compartment saloon car but with a soft folding type roof.
- The roof of the convertible is a soft folding type and special wind-up designed windows.

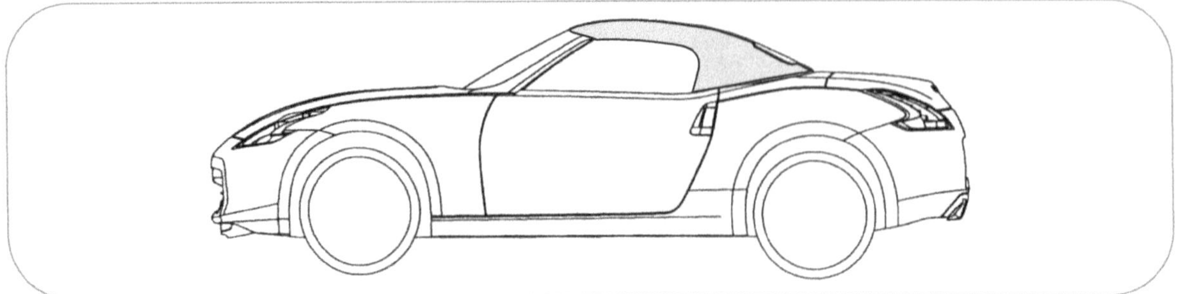

Fig. 5.7.7 Convertible car

- The convertible car can be driven as an enclosed passenger compartment or open passenger compartment.

(vii) Estate car

- An estate car is also similar to a saloon car design but with extended passenger roof.
- The passenger compartment roof is extended right up to the rear end.
- The rear seats are generally collapsible to provide access to the luggage area.
- This design provides maximum luggage space.

Fig. 5.7.8 Estate car

(viii) Stationwagon

- Refer Fig. 5.7.9 for a stationwagon car body design.

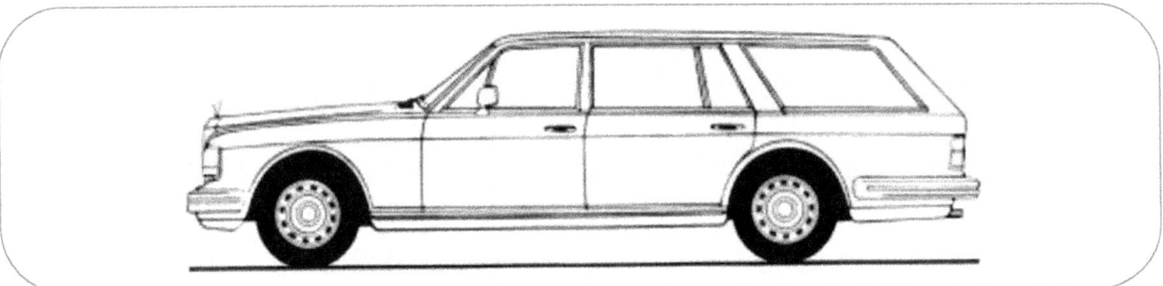

Fig. 5.7.9 Stationwagon

- A stationwagon is a tall body design of estate car.
- A stationwagon is generally desgined to carry heavy luggage.

(ix) Van

- Refer Fig. 5.7.10 for a van body design.

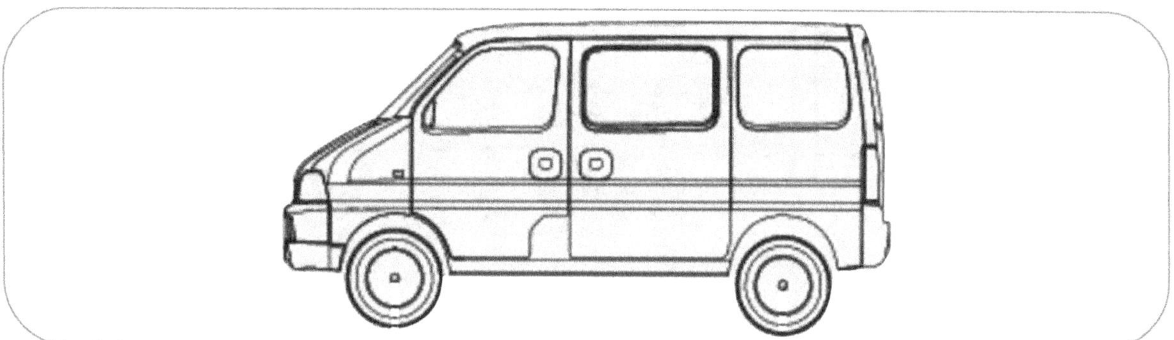

Fig. 5.7.10 Van

- Vans are light transport carrier vehicles used for short distance transport.
- The engine is generally located over or just in front of the driver compartment.
- Van body is designed to accomodate one row of seats in the front and only one additional row for other passengers.
- Sufficient space is provided to accomodate luggage.
- The side doors are generally of sliding type.
- The rear door is used for loading and unloading.

(x) Truck

- Refer Fig. 5.7.11 for a typical truck.

Fig. 5.7.11 Truck

- A truck is used for medium to heavy goods carriage.
- A truck has all its axles attached to a single frame.
- Generally twin wheels are fitted on the non-steered axles.
- Sometimes, a truck may have more than two axles depending on the load capacity.

(xi) Articulated vehicle

- Refer Fig. 5.7.12 for a typical articulated vehicle design. (Refer Fig. 5.7.12 on next page).
- It is a heavy goods vehicle consisting of a tractor and a semi-trailer.
- The tractor unit provides the propulsive power and is a short rigid chassis with two axles.

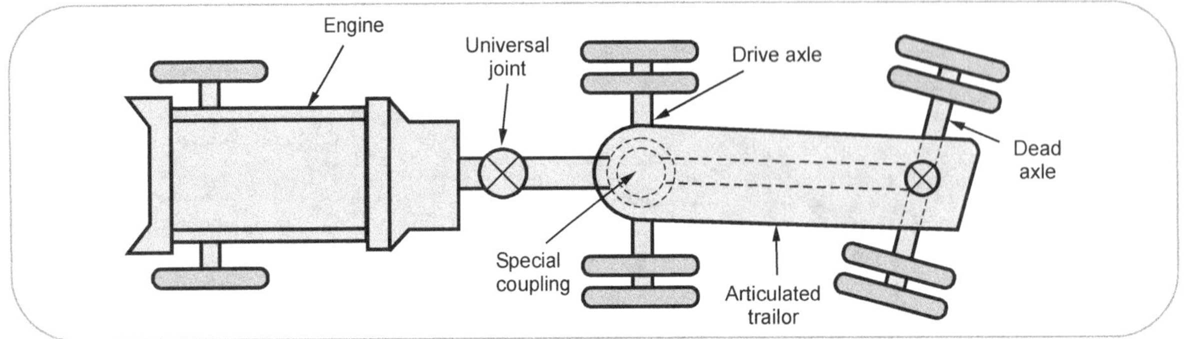

Fig. 5.7.12 Articulated vehicle

- The semi-trailer is used to carry heavy payload and is a long chassis with multiple axles.
- The two units are connected together by means of a fifth-wheel coupling.
- The trailer being detachable, the tractor can be directly moved off after reaching the destination with some other trailer unit without having to wait for unloading and re-loading.
- The front axle carries the steered road wheels.
- Since the trailer wheels do not follow the same path as the tractor wheels, sometimes steering becomes difficult.

Fig. 5.7.13

(xii) Bus

- Refer Fig. 5.7.14 for a bus body design.

Fig. 5.7.14 Bus

- Bus is used for carrying large number of people over short distance often in dense traffic.
- The interior of bus is designed such that adequate visibility is provided for the passengers to know where they are and where they have to get down.
- The bus has very limited space to accomodate luggage.
- It has two doors, the one at rear to get in and one at front to move out of bus.
- Sometimes, the buses have two floors called double-decker bus.

(xiii) Coach (Multi-axle vehicle)

- A multi-axle bus or coach has more than the conventional two axles (generally 3 or 4).

- Extra axles added for legal weight restriction reasons or to accomodate different vehicle designs like articulation.

- Extra axles may also be provided on shorter buses and coaches to accomodate extra equipment loads like passanger lifts, televisions, climate control etc.

- In some cases, the need is bus cargo transport when cargo compartments and heavy weight of cargo needs extra axles.

- Refer Fig. 5.7.15 for a typical coach body design.

Fig. 5.7.15 Coach

- A coach is similar to a bus, in general, meant for transporting passengers but over long distances.

- The interiors of a coach are designed to provide the best possible comfort and to minimize fatigue to passengers.

- For better visibility, large panelled windows are provided on either side extending the full length of vehicle.

- The engine can be mounted at the front or the rear.

> **Note : Light Motor Vehicles (LMV) or Light Transport Vehicle (LTV)** are nothing but cars, jeeps, motor cycles etc.

5.8 : Types of Driving License with their Full Forms

- The different types of driving license and their full forms are as follows :

- **MCWOG :** Motorcycle Without Gear

- **MCWG :** Motorcycle With Gear

- **LMV-NT :** Light Motor Vehicle Non Transport

- **LMV-CAB :** Light Motor Vehicle For Passenger Vehicle

- **LMV-TVA OR LMV-GV :** Light Motor Vehicle For Transportation Vehicle

- **3W-NT :** 3 (Three) Wheeler Vehicle For Non Transport Vehicle

- **3W-CAB :** 3 (Three) Wheeler Vehicle For Passenger Vehicle

- **3W-TVA :** 3 (Three) Wheeler Vehicle For Transportation Vehicle

- **HPV :** Heavy Passenger Vehicle

- **HGV :** Heavy Goods Vehicle

- **TVA :** Transport Vehicle Authority For Heavy Vehicle

- **PBUS :** Passenger Bus For Heavy Vehicle

5.9 : I.C. Engine and its Components

Fig. 5.9.1 shows reciprocating I.C. engine and its components. The main components of this 4-stroke petrol engine are as follows :

1. Cylinder	2. Piston
3. Crank and crankshaft	4. Connecting rod
5. Spark plug	6. Inlet and exhaust valve
7. Crank pin	8. Gudgeon pin
9. Cylinder head	10. Piston rings
11. Suction and exhaust manifold	
12. Carburettor	

1. **Cylinder :** It is the main body of an I.C. engine in which piston reciprocates to develop power. It has to withstand very high pressures (upto 75 bar) and temperatures (upto 2400 °C), because the the combustion takes place in the cylinder. It may be air cooled or water cooled. It is generally made of cast iron or alloy steel.

2. **Piston :** It is used to compress the charge (air-fuel mixture) during compression stroke and to transmit the gas force to the connecting rod and then crank during power stroke.

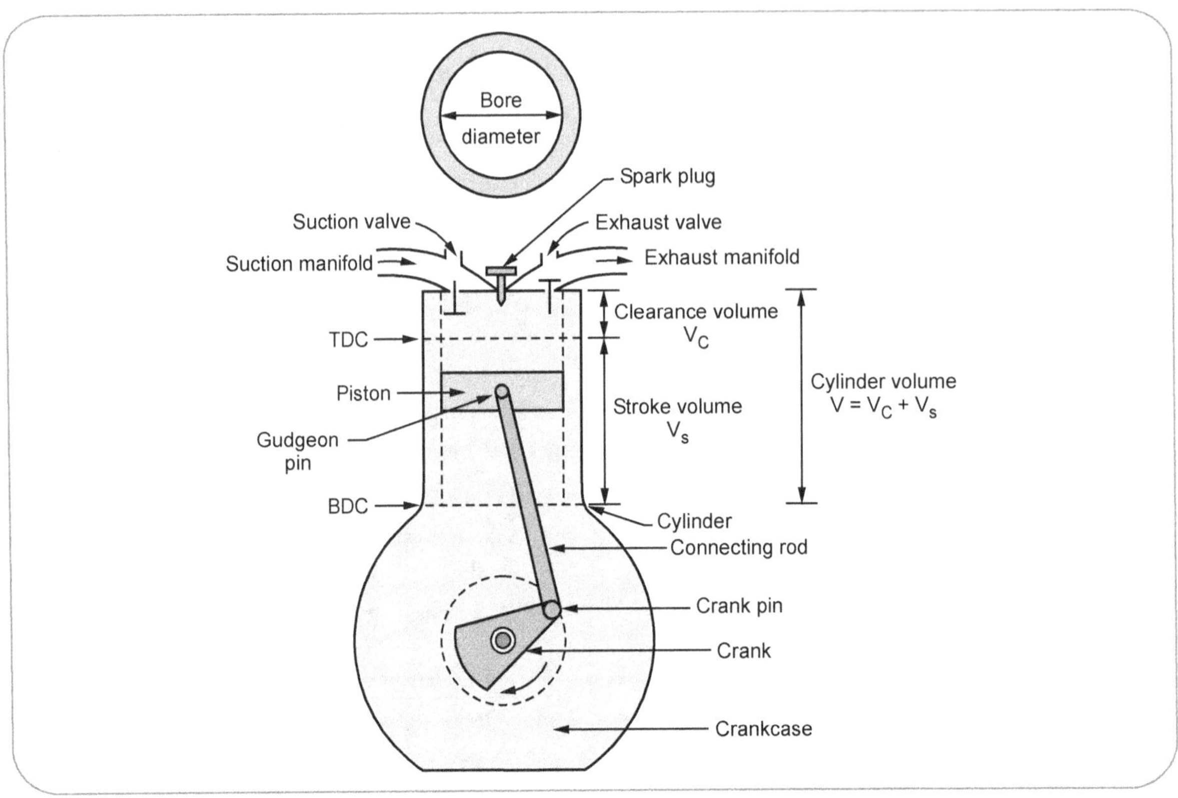

Fig. 5.9.1 Four stroke I.C. engine (petrol) and its components

3. **Crank and crankshaft :** Crank is the integral part of the crankshaft and crankshaft is the backbone of the engine, crankshaft is supported in main bearings and it has a heavy wheel called as **flywheel** to control the fluctuation of torque.

4. **Connecting rod :** It is a rod of circular or rectangular cross section. Its small end is connected to the piston and big end is connected to the crank. It converts reciprocating motion of pistion into rotary motion of crankshaft.

5. **Spark plug :** It is used to initiate the spark for the combustion of air-fuel mixture in petrol engines.

6. **Inlet and exhaust valve :** Inlet valve controls the admission of the charge during suction stroke and exhaust valve is used for the removal of exhaust gases after doing work on the pistion.

7. **Crank pin and Gudgeon pin :** Crank pin is used to connect big end of connecting rod to the crank and gudgeon pin is used to connect small end of connecting rod to the piston.

8. **Cylinder head :** It is used for closing the one end of the cylinder. It houses inlet and exhaust valves through which charge is taken inside the cylinder and burned gases are exhausted to the atmosphere from the cylinder.

9. **Piston rings :** Piston rings are housed in the circumferential grooves provided on the outer surface of the piston. It gives tight fit between the pistion and cylinder hence prevents the leakage of high pressure gases.

10. **Suction and exhaust manifold :** Suction manifold is the passage which carries the charge from carburettor to the engine where as exhaust manifold is the passage which carries the exhaust gases from the exhaust valve to the atmosphere.

11. **Carburettor :** It is used to supply the uniform air-fuel mixture to the cylinder of petrol engine through the intake manifold. It is controlled by a throttle valve.

5.9.1 Terminology used in I.C. Engines

In addition to the above components of I.C. engine, certain standard terminology used in I.C. engine is as follows :

1. Bore
2. Stroke
3. Top Dead Centre (T.D.C.)
4. Bottom Dead Centre (B.D.C.)
5. Clearance volume
6. Swept volume
7. Compression ratio

1. **Bore :** It is the inside diameter of the cylinder.

2. **Stroke :** It is the maximum distance travelled by the piston in the cylinder in one direction. It is generally twice the radius of crank.

3. **Top Dead Centre (T.D.C) :** It is the extreme position of the piston at the top of the cylinder. For horizontal engines it is called as inner dead centre (I.D.C).

4. **Bottom dead centre (B.D.C) :** It is the extreme position of the piston at the bottom of the cylinder. For horizontal engines it is called as outer dead centre (O.D.C).

5. **Clearance volume :** It is the volume contained in the cylinder above the top of the piston when it is at T.D.C. It is denoted by V_c.

6. **Swept volume :** It is the volume swept by the piston in moving between TDC and BDC. It is denoted by V_s.

7. **Compression ratio :** It is the ratio of the volume when the piston is at BDC to the volume when the piston is at TDC. It is denoted by r.

5.10 : Specifications of Engine

- Specifications of engine gives information about all parameters of engine.

- It may include the following parameters :

  - Power of engine

  - Cylinder capacity

  - Manufacturer of engine

  - Starting method of engine etc.

  - No. of cylinders

  - Revolution per minute (RPM)

- Piston (stroke) and bore (cylinder diameter) dimensions

- Type of transmission (gearbox)

- Some of the important parameters and their meanings are as follows :

1. Engine

- The engine is like the heart of the car. It is a machine that converts power into motion or produces the force that propels the vehicle into motion.

- Engines commonly are of three types - petrol (runs on gasoline), diesel (powered by diesel), electric (powered by batteries) and hybrid (using a combination of a fuel-driven engine and an electric one).

2. Number of cylinders

- A cylinder is a chamber where fuel is ignited to power the car into motion. So a 4-cylinder engine will have four of these compartments, a 3-cylinder, three, and so on.

- The cylinder houses a piston or large valve that creates compression that helps ignite the fuel.

3. Displacement (Eg. 1250 cc or 1.3-litre)

- The measurement of the volume of the engine cylinders or chambers is called displacement and is measured in litres or cubic centimeters (cc).

- So if you see "1250 cc", it means the volume of the cylinders together is 1250 cubic centimetres.

- "1.3-litre" on a specification sheet is just another way of saying the same thing, although not as precise, and means the volume of the cylinders is 1.3 litres in total.

4. Gearbox (5 speed transmission or 6 speed transmission, etc.)

- Much like we would choose different gears on a bicycle for varying inclines and pace, we need to choose gears depending on the speed of the car or the road situation.

- A 5-speed transmission has 5 gears or speeds that we can choose while driving or riding.

- A manual transmission is one that requires change speeds manually, while an automatic transmission picks gears automatically, requiring only operate the brakes and accelerator.

5. Power

- The easiest way to understand this is the unit of power-horsepower. One horsepower is informally defined as the amount of power one horse gives while pulling. So if your car has 75 horsepower, it has the pulling power of 75 thorough breds. To be a little more physicistical, it is the power required to raise 550 pounds a distance of one foot in one second.

6. Torque

- Torque is the ability (of the engine) to do work or a force that tends to cause rotation of the wheels in the case of a car or bike. Torque is measured in Newton Metres (Nm) and is usually higher in diesel engines when compared to similarly sized petrol ones.

7. Drivetrain

- This is the system in a vehicle that connects the transmission or gearbox to the drive axles and is broken down into three main categories - Front Wheel Drive (FWD), Rear Wheel Drive (RWD) or All-Wheel Drive (4x4/4WD/AWD).

- FWD means the power goes to the front wheels, therefore 'pulling the car', RWD means the 'push' comes from the rear wheels, and AWD means all four wheels transfer power.

8. Suspension

- The suspension is the system of springs and shock absorbers that connect the vehicle to its wheels, is designed to reduce the shock of bumps and potholes, and contribute to the way car handles or behaves on the road.

9. Fuel tank capacity

- The total amount of fuel your car or bike can store, which usually includes the volume of the reserve and measured in litres, is the fuel tank capacity of your vehicle.

- We can calculate the approximate range our vehicle is capable of by multiplying the capacity with the mileage of the car.

10. Wheelbase

- The distance between the front and rear axles (the rods connecting the centres of the wheels) of a vehicle is known as its wheelbase.

- Generally, the longer the wheelbase, the larger the cabin of the car can be, therefore leading to more interior room. Wheelbase is usually measured in millimetres.

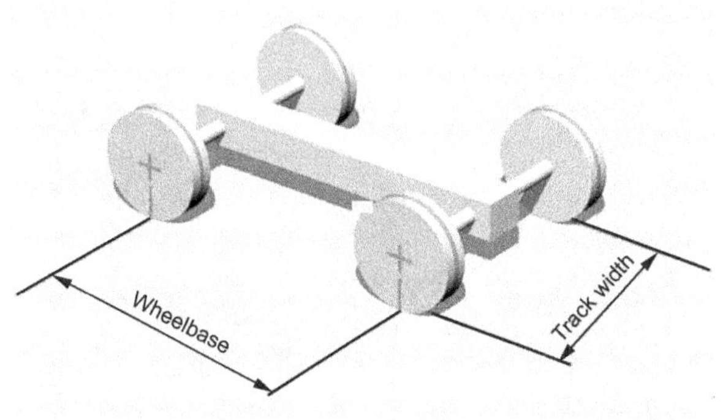

Fig. 5.10.1

11. Tread/track width

- The distance from tyre centre to tyre centre, measured width-of-the-car-wise, is called tread or track width. Usually, the wider the track width, the more stable and better around the corners a car is.

- The tread is usually denoted in millimetres.

12. Turning radius

- The radius of the smallest or tightest circular turn (or U-Turn) a car is capable of making is known as turning radius. It is measured in metres. Small cars generally, have a turning radius of around 5 metres.

13. Kerb weight

- Kerb weight is the weight of the car without any occupants or cargo and is usually denoted in kg.

- Small cars in India usually have a kerb weight hovering around the 1000-kg mark, while larger luxury vehicles can weigh more than double that.

5.10.1 Specifications of Different Vehicles

(a) Key Specifications of Maruti Swift

ARAI Mileage	28.4 kmpl
Fuel Type	Diesel
Engine Displacement (cc)	1248
Max Power (bhp@rpm)	74bhp@4000rpm
Max Torque (nm@rpm)	190Nm@2000rpm
Seating Capacity	5
TransmissionType	Automatic
Boot Space (Litres)	268
Fuel Tank Capacity	37
Body Type	Hatchback
Service Cost (Avg. of 5 years)	₹ 4,483

(b) Key Specifications of Hyundai Verna

ARAI Mileage	22.0 kmpl
City Mileage	18.0 kmpl
Fuel Type	Diesel
Engine Displacement (cc)	1582
Max Power (bhp@rpm)	126.2bhp@4000rpm
Max Torque (nm@rpm)	259.87nm@1500-3000rpm
Seating Capacity	5
TransmissionType	Automatic
Boot Space (Litres)	480 ers
Fuel Tank Capacity	45
Body Type	Sedan
Service Cost (Avg. of 5 years)	₹ 4,307

(c) Key Specifications of Toyota Innova Crysta

ARAI Mileage	11.36 kmpl
Fuel Type	Diesel
Engine Displacement (cc)	2755
Max Power (bhp@rpm)	171.5bhp@3400rpm
Max Torque (nm@rpm)	360Nm@1200-3400rpm
Seating Capacity	7
TransmissionType	Automatic
Fuel Tank Capacity	55
Body Type	MUV
(Service Cost (Avg. of 5 years)	₹ 4,589

5.11 : Introduction to Electric Vehicles (EVs)

- The fast depletion of crude oil reserves, frequent increase in crude oil prices, high atmospheric and noise pollution have created a need to develop automobile on alternate fuels.

- Electric vehicles (battery operated) can overcome most of these problems.

History of electric vehicles :

- Although it seems that electric vehicles have only recently taken the limelight, they have been around for a very long time.

- In the 1890s, a handful of new models of electrically-propelled cars were introduced by various inventors.

- In 1899, Belgian engineer Camille Jenatzy introduced one of the very first electric vehicles.

- Limitations in battery power, available horsepower from electric motors, and so forth were all holding the development of electric vehicles back.

- The recent technology development in electric power storage has been renewing the development and interest in electric vehicles.

5.11.1 Electric Vehicles (EVs)

Working of electric vehicles :

- Refer Fig. 5.11.1 for simplified layout of an electric vehicle. (Refer Fig. 5.11.1 on next page).

- A high capacity battery is the heart of the electric vehicle.

- Battery used can be of lead based, nickel based and most recently lithium ion battery.

- The battery gets the energy from the electric plug point outside the vehicle.

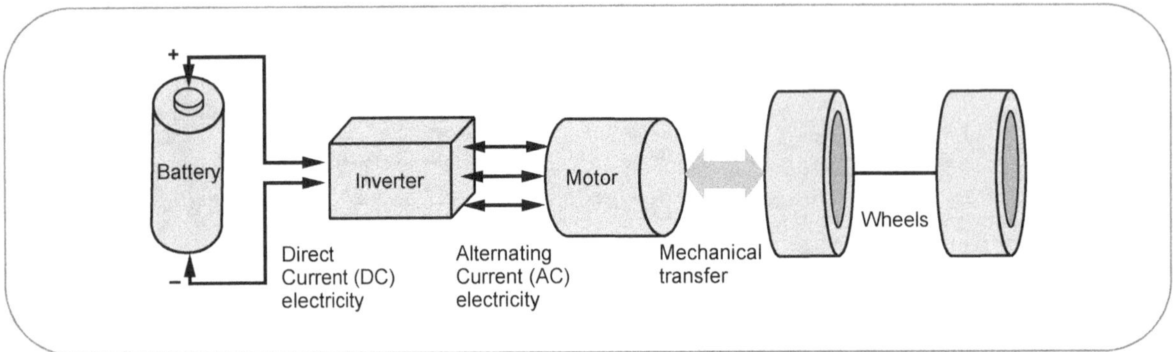

Fig. 5.11.1 Simplified electric vehicle layout

- The battery provides high current energy to the inverter.
- The inverter converts the high current DC signal into high current AC signal.
- The motor used has a DC series motor with shunt limiting winding.
- The motor generates the mechanical torque to power the wheels for propulsion.
- The speed of the motor is varied by varying the voltage across the motor smoothly and sleeplessly through the accelerator.
- The motor used is a special vehicle since it can reverse its function, i.e. it can also act as an electric generator and send back energy through the inverter to the battery.
- The practical EV vehicles have a microprocessor based controls.
- The microprocessor is used for speed control for optimal efficiency of drive system, for logic and sequencing, for indicating state of charge, self diagnostics, etc.
- Refer Fig. 5.11.2 for a practical electric vehicle.

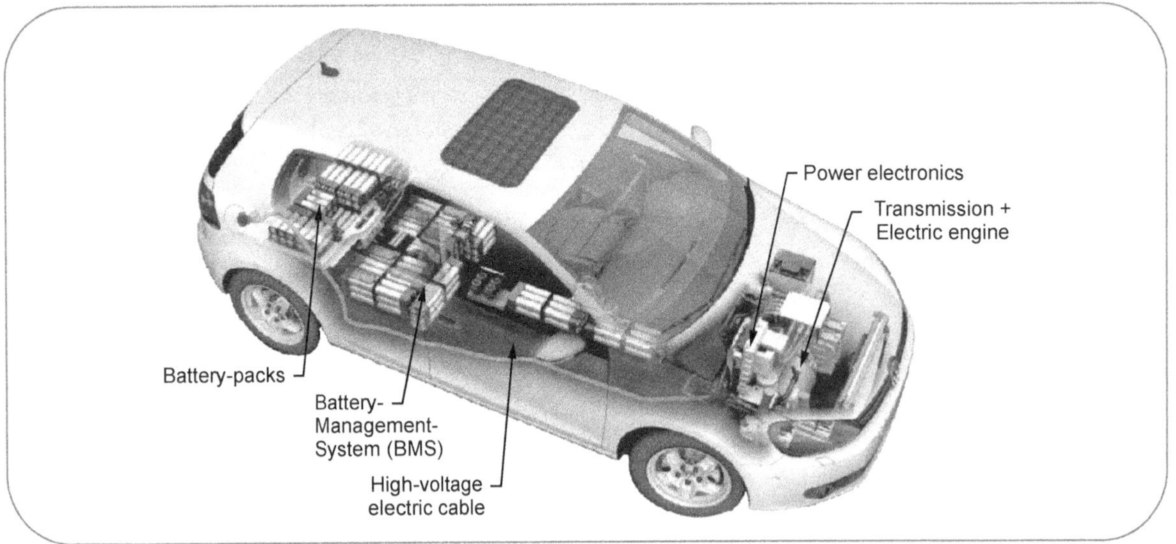

Fig. 5.11.2 Electric vehicle

- The characteristics of electric vehicles in relation with internal combustion engine is provided in Table 5.11.1.

Table 5.11.1 Characteristics of electric vehicle and internal combustion engine vehicle

Sr. No.	Characteristics	Internal combustion engine vehicle	Electric vehicle
1.	Initial purchase cost	1.0	1.5 to 1.8 times
2.	Running cost	1.0	0.2 to 0.4 times
3.	Maintenance cost	High	Negligible
4.	Life of vehicle	Low	High (except for battery)
5.	Atmospheric pollution	Very high	Moderate (pollution takes place mainly at the source of electric power generation plant)
6.	Noise pollution	Very high	Moderate (pollution takes place mainly at the source of electric power generation plant)

Advantages of EVs :

- It has limited number of components and are simple in design.

- EV vehicles require minimum or no tuning at regular intervals.

- They have high operating reliability.

- The running cost of EVs is much lower than conventional vehicles.

Limitations of EVs :

- EV's have a range of 80-100 miles using advanced battery technology.

- Batteries need frequent recharging.

- The initial cost of EVs is very high.

- The noise pollution near the vehicle surrounding is very low.

5.12 : Hybrid Electric Vehicles (HEVs)

- *The Hybrid Electric Vehicle (HEV) is progressive step in reducing the environmental impacts of automobile use without losing comfort, performance, storage room and extended driving range.*

- Hybrid electric vehicle is an intelligent combination of benefits of internal combustion engine with electric motor.

- Some of examples of recent HEVs are Toyota Camry Hybrid, Ford Escape Hybrid, Toyota High Lander Hybrid, Honda Insight, Honda Civic Hybrid, GM Volt and others.

- Refer Fig. 5.12.1 for hybrid vehicle layout.

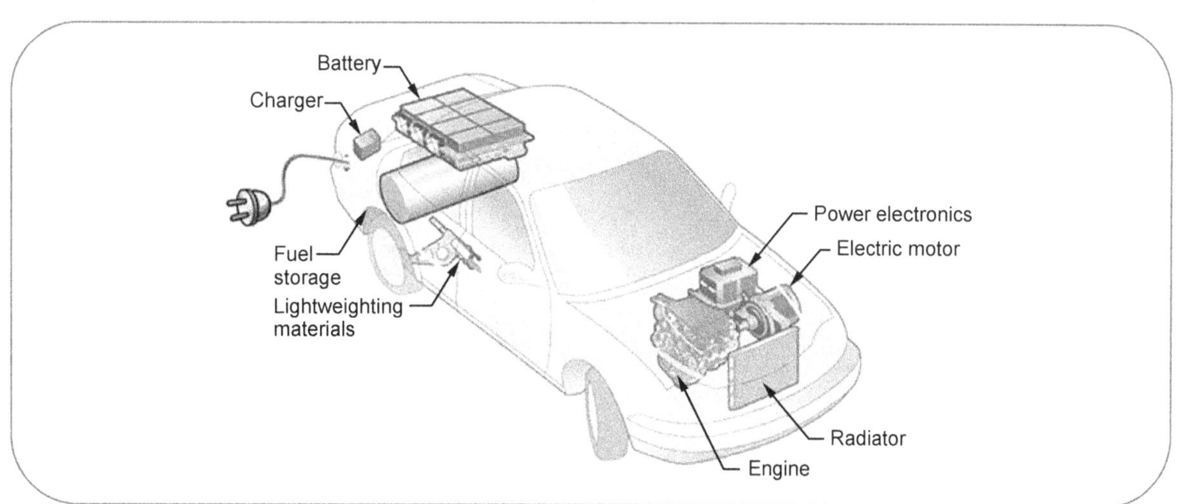

Fig. 5.12.1 Hybrid electric vehicle (HEV)

Components of the HEV System :

• The major components of the HEV system are as follows :

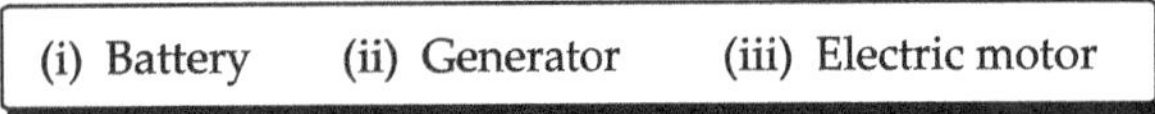

(i) Battery :

• The battery in an HEV system is the energy storage device.

• It consists of two or more electrochemical energy cells connected together to provide electrical energy.

(ii) Generator :

• The generator is similar to an electric motor.

• It converts rotary mechanical energy into electrical power.

(iii) Electric motor :

• The electric motor provides mechanical rotary energy to the shaft as per the electrical input provided.

• Advanced electronics allow it to act as a motor as well as a generator.

• For example, when it needs to, it can draw energy from the batteries to accelerate the car.

• But acting as a generator, it can slow the car down and return energy to the batteries.

Working of HEVs :

• Refer Fig. 5.12.2 for working of Hybrid Electric Vehicles.

• The HEV shown in Fig. 5.12.2 is of series / parallel type hybrid.

• At low vehicle speeds, the power and torque requirement of vehicle is low, hence the electric battery is used to supply the energy to motor connected to the wheels. Refer Fig. 5.12.2 (a).

• At high vehicle speeds, the power and torque requirement of vehicle is high, hence the internal combustion engine and electric battery are used in series to supply the energy to motor connected to the wheels. Refer Fig. 5.12.2 (b).

• During braking, the vehicle loses its kinetic energy.

• The inertia of vehicle drives the road wheels.

• The motor reverses its working and acts as a generator and charges the electric storage battery. Refer Fig. 5.12.2 (c).

(a) Scenario : Low speed - Power is provided by electric motor using energy supplied by the DC battery

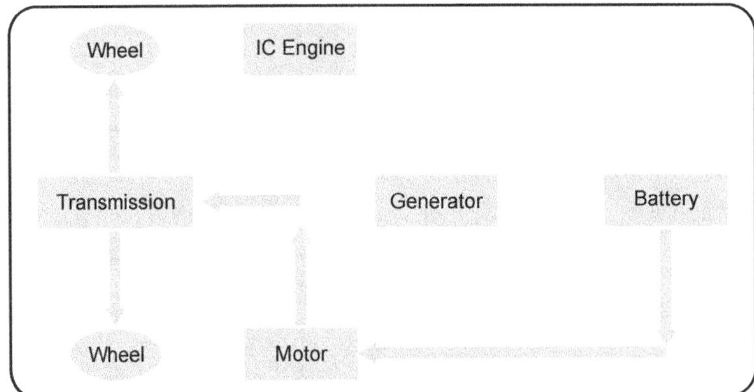

Fig. 5.12.2 (a)

(b) Scenario : High speed - Power is provided by IC engine and electric motor via generator. Generator also charges battery during high speed.

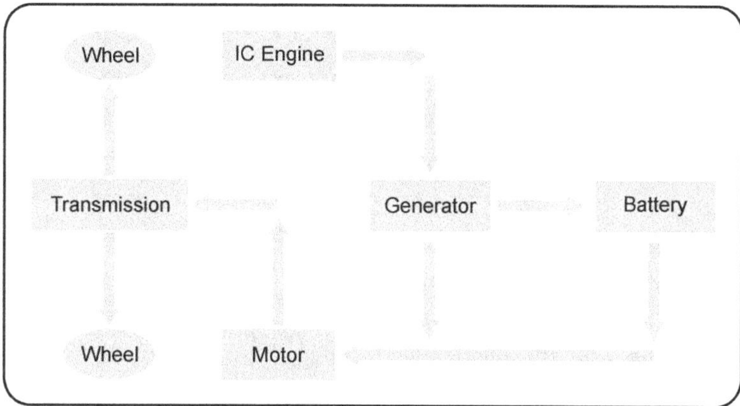

Fig. 5.12.2 (b)

(c) Scenario : Braking - Kinetic energy is converted to electric energy during regenerative braking by electric motor and supplied to battery.

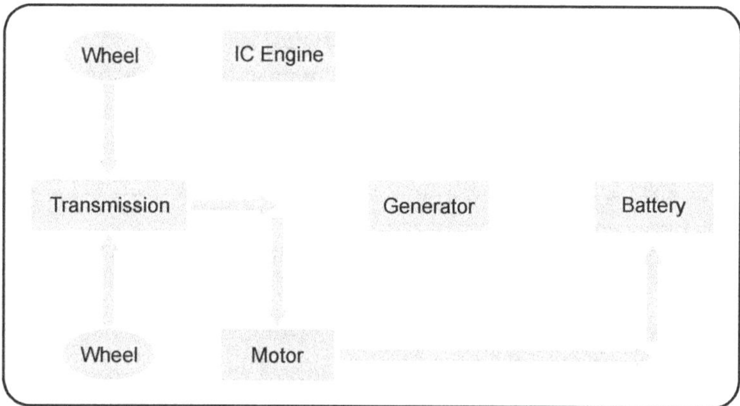

Fig. 5.12.2 (c)

Fig. 5.12.2 Operation of Hybrid Electric Vehicle (HEV)

Advantages of HEV's :

- The HEV system contains parts of both internal combustion engine and electric vehicles in an attempt to get the best of both worlds.

- It is able to operate nearly twice as efficiently as traditional internal combustion vehicles.

- It has equivalent power, range, cost and safety of a conventional vehicle while reducing fuel costs and harmful emissions.

- The battery is continuously recharged by a motor / generator driven by the internal combustion engine or by regenerative braking.

5.12.1 Types of Hybrid Electric Vehicles

- Hybrid electric vehicles can be classified according to the way in which the power is supplied to drive train :

> i) Parallel hybrids
> ii) Series hybrids
> iii) Power-split hybrids

i) Parallel hybrids

- In parallel hybrids, the internal combustion engine and the electric motor are connected to the mechanical transmission and transmits power to drive the wheels through a conventional transmission.

- The I.C. engine of many parallel hybrids can also act as a generator for supplemental recharging.

- Parallel hybrids are more efficient than comparable non-hybrid vehicles especially during urban stop and go conditions where the electric motor is permitted to contribute and during highway operation.

ii) Series hybrids

- In series hybrids only the electric motor drives the drive train and a smaller I.C. engine works as a generator to power electric motor or to recharge the batteries.

- Series hybrids have larger battery pack than the parallel hybrids, hence they are more expensive.

- Once the batteries are low, the small engine can generate power at its optimum settings at all times, making them more efficient in extensive city driving.

iii) Power-split hybrids

- Power-split hybrids have the advantages of a combination of series and parallel characteristics.

- Power-split hybrids are more efficient because series hybrids are efficient at lower speeds whereas parallel hybrids are more efficient at high speeds.

- The cost of power-split hybrids is more than the parallel hybrids.

> **Note :** In each of the above hybrids it is common to use regenerative braking to recharge the batteries.

5.12.2 Plug in Hybrid Electric Vehicles (PHEVs)

- Plug in hybrid electric vehicle is also called as plug in hybrid.

- It is a hybrid electric vehicle with rechargeable batteries that can be restored to full charge by connecting a plug to an external electric power source.

- PHEVs have the characteristics of both a conventional hybrid electric vehicle having an electric motor and an I.C. engine. Also, all electric vehicles have a plug to connect to the electrical grid.

- PHEVs have a much larger all electric range as compared to conventional gasoline electric hybrids.

- Also, it eliminate the fear that a vehicle has insufficient range to reach its destination, associated with all electric vehicles, because the combustion engine act as a backup when the batteries are discharged.

5.13 : Cost Analysis of Vehicle

- The cost of vehicle is one of the important factor for its sale in the market.

- The overall cost of vehicle in the showroom includes many parameters. Some of them are as follows :

i) **Type of engine :** According to the type of engine or its capacity the cost of vehicle changes. For high power engines, the cost of vehicle is very high.

ii) **Safety features :** If a vehicle includes safety features like air ABS, air bag etc. then the cost of vehicle increases. It also depends on the number of airbags in a vehicle.

iii) **Material of vehicle (Raw material) :** The material of vehicle is one of the important factor in cost analysis. For high grade material the cost of vehicle increases rapidly. Now-a-days some parts of vehicle are replaced by fiber parts to reduce the cost of vehicle.

iv) Production method : Some manufactures like BMW, Audi etc. use fully automatic machines for production of vehicle. This increases accuracy and finishing of vehicle. But this increases the cost of vehicle. Hence as per the production method the cost of vehicle changes.

v) Research and development (R and D) : It is a hidden cost of vehicle. It requires long time and number of technical people.

vi) Royality : The innovations liks ABS, Air bags, ESP etc. are all innovations by Bosch Company and not by car manufacturers. The manufacturers need to pay royalty to Bosch for using the technology.

vii) Dealer profit

viii) Insurance and taxtion

ix) Availability of spare parts

x) Advertisement (Posters, ads, campaigns etc.)

xi) Quantity of production

xii) Labour cost, etc.

Review Questions

1. What are the basic components / system of an automobile ?
2. State the purpose of transmission and suspension system.
3. State the function of brakes in an automobile.
4. How are automobiles classified ?
5. What is a automobile ? Explain automobile history and development in India.
6. Write short note on : Specifications of vehicles.
7. Explain spcifications of any two wheeler with suitable example.
8. Explain spcifications of any three wheeler with suitable example.
9. What is vehicle body ? State its functions.
10. Explain the following :
    i) Multi-axle vehicle ii) Light motor vehicles
11. What is meant by articulated vehicle ? Explain the neat sketch.
12. State the difference between saloon, hatchback and coupe car styles.
13. Explain any six components of I.C. Engine.
14. Explain engine specifications with suitable example.
15. Explain the meaning of : a) Engine b) Kerb weight c) Drive train d) Torque e) Horsepower
16. With neat sketch explain the construction and working of electric vehicles.
17. Explain construction and working of HEV.
18. What are the types of HEV ?

Notes

Unit - IV
Vehicle Systems

Syllabus : *Introduction of chassis layouts, steering system, suspension system, braking system, cooling system and fuel injection system and fuel supply system. Study of electric and hybrid vehicle systems. Study of power transmission system, clutch, gear box (Simple Numerical), propeller shaft, universal joint, differential gearbox and axles. Vehicle active and passive safety arrangements: seat, seat belts, airbags and antilock brake system.*

Chapter - 6 Vehicle Systems   (6 - 1) to (6 - 52)

Mind Map - Vehicle Systems

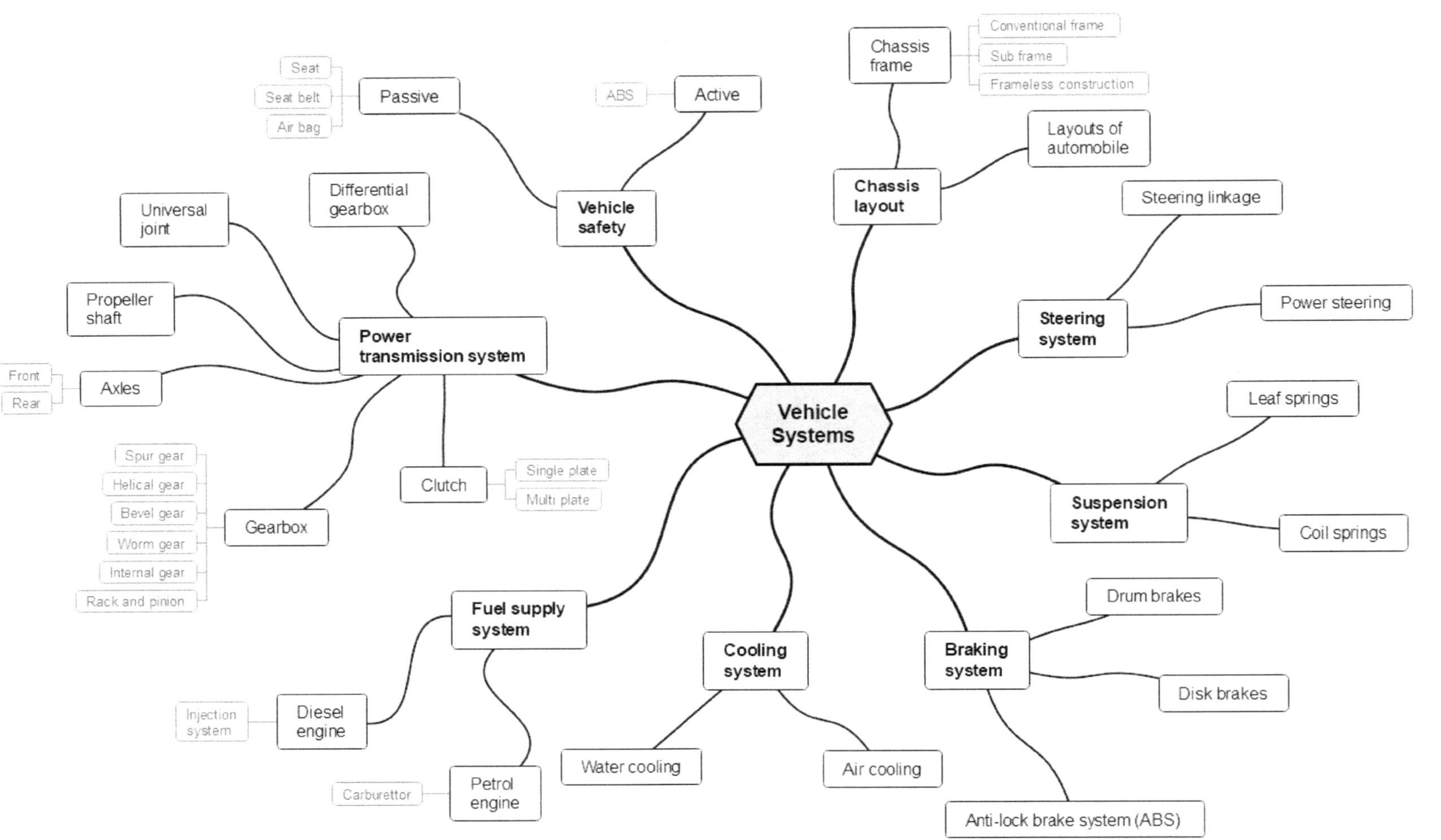

6.1 : Introduction of Chassis Layouts

- A chassis is basically the base frame of a car upon which the rest of the car is set.

- Vehicle chassis is a major sub-assembly in building an automobile.

- The chassis is also known as backbone of the automobile.

- The chassis includes frame, engine, clutch, gearbox, propellor shaft, universal joints, differential, axles, wheels, steering, brakes, fuel tank, suspension system, etc.

- The *chassis* is a french term which was originally used to denote the frame part but now has an extended list of components.

- All the components of the vehicle except the body is called a chassis.

- For commercial vehicles like truck a chassis consists of an assembly of all the required parts to be ready for operation on the road.

- A chassis of car will be different from chassis of truck or commercial vehicles because of the heavier loads and constant work use.

6.2 : Chassis Frame

- Frame is the most important component of a chassis.

- Depending on the arrangement of frame with the body, following are the types of chassis frame :

> i) Conventional frame   ii) Sub-frame
> iii) Frameless construction

6.2.1 Conventional Frame   `SPPU : Dec.-11`

- In this type of chassis construction, the frame is the basic unit on which various other chassis components are attached.

- The vehicle body is also bolted on the frame later on.

- Fig. 6.2.1 shows a typical conventional frame of a light duty truck.

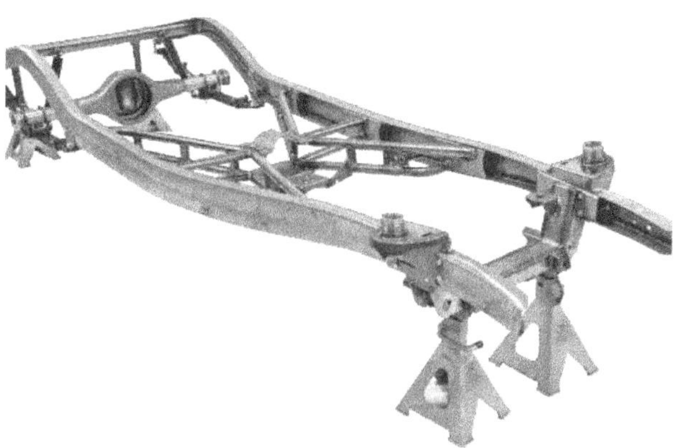

Fig. 6.2.1 Conventional frame

6.2.2 Sub - Frames

- In a conventional frame, the various chassis components are generally bolted or carried directly by the main frame members.

- But the engine, clutch and gearbox can be mounted on a separate frame called *sub-frame*.

- This sub-frame can be of simple construction and mounted on the main frame at three points through rubber blocks.

- Similarly, the suspension, wheels and final drive can also be mounted on a separate sub-frame.

- Fig. 6.2.2 shows a typical sub-frame for mounting of suspension system.

Fig. 6.2.2 Sub-frame

Advantages :

- The mass of sub-frame alone helps to damp vibrations.

- The provision of sub-frame isolates various chassis components from the effect of twisting and flexing of the main frame.

- The sub-frame is relatively simple in construction and cheaper to repair in case of accidental damages.

- The provision of sub-frame simplifies production on vehicle assembly line.

Applications :

- The sub-frame construction is used in most light commercial vehicles and passenger cars.

6.2.3 Frameless Construction   SPPU : May-11

- The frameless chassis construction was first introduced in 1934.

- In this type of construction, the heavy side members used in conventional frame construction are eliminated.

- The floor is strengthened by cross-members and the body, all welded together.

- It is also known as *chassisless, unitary, monocoque or integral construction.*

- This type of construction is now-a-days widely used for passenger cars.

- Fig. 6.2.3 shows a typical frameless construction.

- Basically, the structure includes an underframe or floor structure having side members, cross-members, floor and other components.

- They are all welded together as one assembly.

- The pressed steel body shell is also welded or bolted directly to the structure.

- Sometimes, the engine, clutch and gearbox are mounted on a sub-frame which is also attached to the front of the body shell.

- The welding of wings, bonnet sides and steel facia at the front and similarly welding of luggage compartment and other structures at rear, additional strengthening of the structure can be achieved.

- To improve stiffness of floor, grooves are pressed into the steel floor and side panels.

- To improve visibility, thinner pillars are used.

Advantages :

- This construction reduces the weight and cost of the vehicle.

- Stresses are evenly distrubuted throughout the frameless structure.

- The welding of various members and sub-assemblies to the structure provides good torsional rigidity and resistance to bending.

Disadvantages :

- Cost of repairs to body is quite high.

- This type of construction can be used only for small passenger cars.

Applications :

- This type of frameless construction is used for most passenger cars and some light utility vehicles.

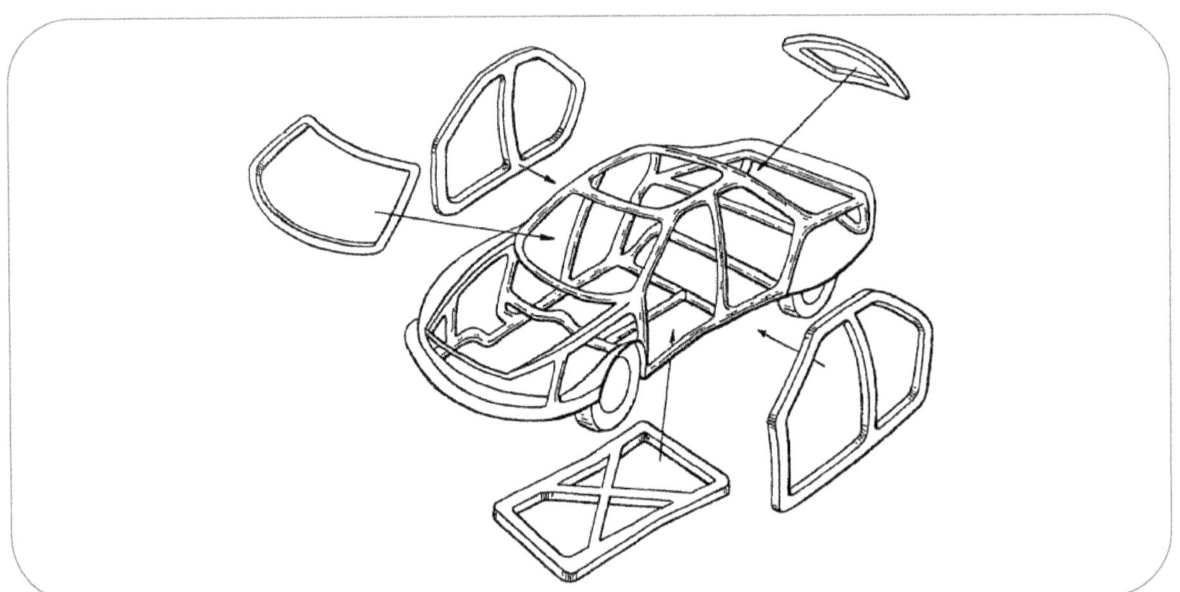

Fig. 6.2.3 Frameless construction

6.3 : Layout of an Automobile

SPPU : Dec.-07,13, May-09,12, April-17

- Layout of an automobile is the basic design of the vehicle.
- The layout indicates the various components / assemblies that transfers the drive from the engine to the wheel.
- The layout also indicates the relative positioning of all drive related components.
- A conventional layout of an automobile is as shown in the Fig. 6.3.1.

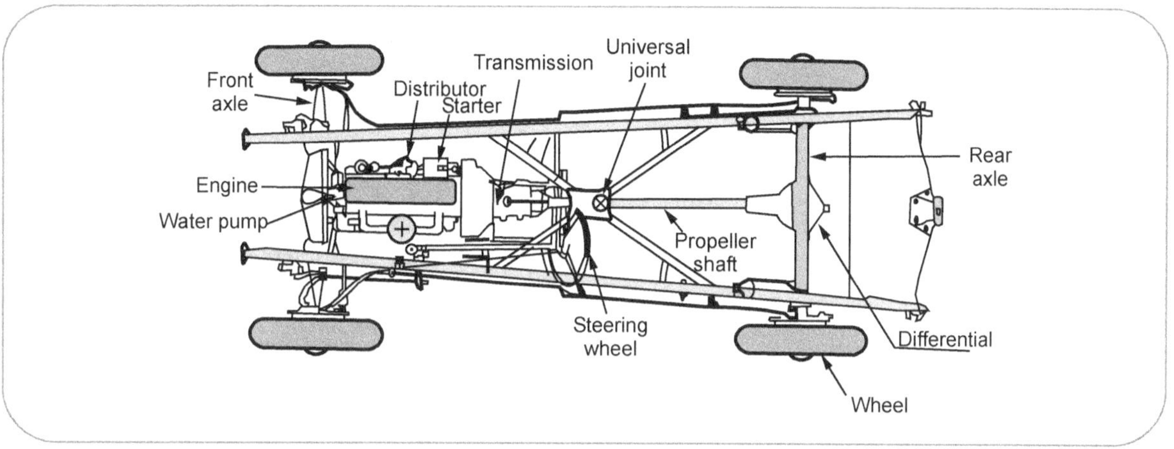

Fig. 6.3.1 Layout of an automobile

- The various components, their relative location and purpose is as mentioned below :

i) Frame	ii) Engine	iii) Clutch	iv) Gear box
v) Propeller shaft	vi) Differential	vii) Wheels	viii) Electrical Equipments

(i) Frame

- Frame is the basic metallic structure on which various other components / assemblies are mounted.

(ii) Engine

- It consists of an internal combustion engine which converts thermal energy of fuel into mechanical rotary energy at the engine flywheel.

(iii) Clutch

- It is connected after the engine and transfers the drive from engine to gearbox at the will of driver.
- The clutch and hence the drive can be easily engaged and disengaged by a pedal provided at the foot region in the driver's cabin.

(iv) Gearbox

- The gearbox is the next element is the drive chain.
- It provides the necessary variation of torque available from the engine to the propeller shaft.

(v) Propeller shaft

- It is the drive element in between the gearbox output shaft and the differential on the rear axle to counter for any shaft misalignment due to uneven road surface.

(vi) Differential

- The differential transfers the drive from the propeller shaft to the rear wheel axle.

- It ensures equal distribution of torque between the two wheels irrespective of their relative speeds.

(vii) Rear axle

- The rear axle consists of driving shaft enclosed in a tubular structure.

- On the extreme end of the axle are mounted the wheels.

- In this case, the rear axle is the drive axle.

(viii) Front axle

- The front axle consists of the steering mechanism, the stub axle and wheels mounted on the extreme ends.

(ix) Wheels

- The wheels are the last link in the drive chain.

- The wheels support the load of vehicle and passengers.

(x) Electrical equipment

- The layout also indicates various electrical equipments such as starter, alternator, etc.

6.4 : Types of Layout in an Automobile

- The automobile layout constitutes the drive from engine (starting element) to the drive axle (last element).

- Based on relative positioning of engine and drive axle, layout of an automobile can be grouped as :

> i)   Front engine rear wheel drive
> ii)   Rear engine rear wheel drive
> iii)   Front engine front wheel drive
> iv)   Four wheel drive
> v)   Articulated vehicle layout

6.4.1 Front Engine Rear Wheel Drive

SPPU : May-08,10,12,13, Dec.-09,12,13

- Refer Fig. 6.4.1 for a front engine rear wheel drive. (See Fig. 6.4.1 on next page)

- It is the most conventional type of layout and as the name suggests, the engine is mounted in the front part of vehicle and the drive is transmitted to the rear axle.

- The drive chain in this layout is :

Engine $\rightarrow$ clutch $\rightarrow$ gearbox $\rightarrow$ universal joint $\rightarrow$ propeller shaft $\rightarrow$ differential $\rightarrow$ rear axle $\rightarrow$ wheels

Advantages :

- Balanced weight distribution in a vehicle (front axle carries the engine weight and rear axle carries the passenger weight).

- Simple front axle design with steering mechanism.

- Increased luggage carrying capacity at rear.

- Linkages controlling the engine are short and simple such as clutch, accelerator linkages, etc.

- Accessibility to various engine components is easier.

- Better engine cooling by taking full benefits of natural air stream flowing across the radiator.

Disadvantages :

- It requires long propeller shaft to transmit the drive from gearbox to differential.

- The rear floor houses the propeller shaft and hence rear leg space is limited.

- Higher noise transmitted from front engine to driver cabin.

- It requires larger brake pads at front wheels because of higher weight being transferred on front wheels while deceleration.

6.4.2 Rear Engine Rear Wheel Drive

SPPU : May-09, 11

- This type of layout eliminates the necessity of a propeller shaft.

- The engine is mounted at the rear and drive is also transmitted to the rear axle.

- The drive chain for this layout is.

Engine $\rightarrow$ clutch $\rightarrow$ gearbox $\rightarrow$ universal joints $\rightarrow$ differential $\rightarrow$ rear axle

- Refer Fig. 6.4.2 for layout of rear engine rear wheel drive. (See Fig. 6.4.2 on next page)

Advantages :

- The front axle consists of a very simple design and houses the steering mechanism only.

- Because of high weight on the driving axle, it provides excellent traction and grip on steep hills.

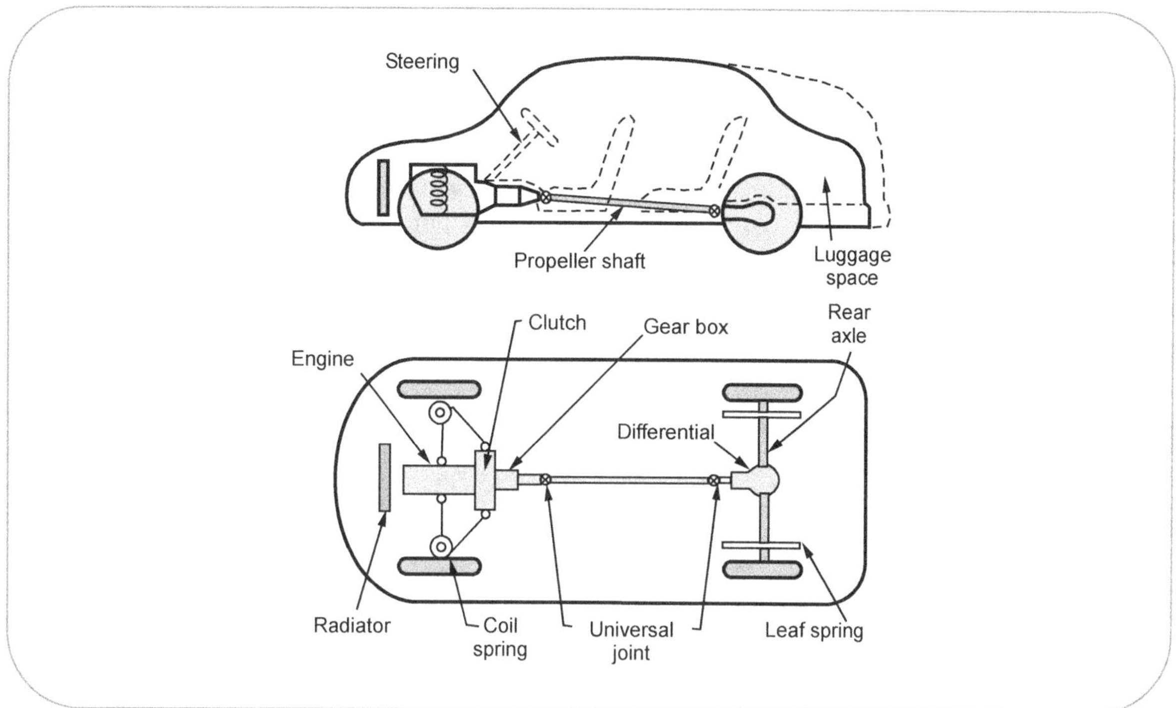

Fig. 6.4.1 Layout of front engine rear wheel drive

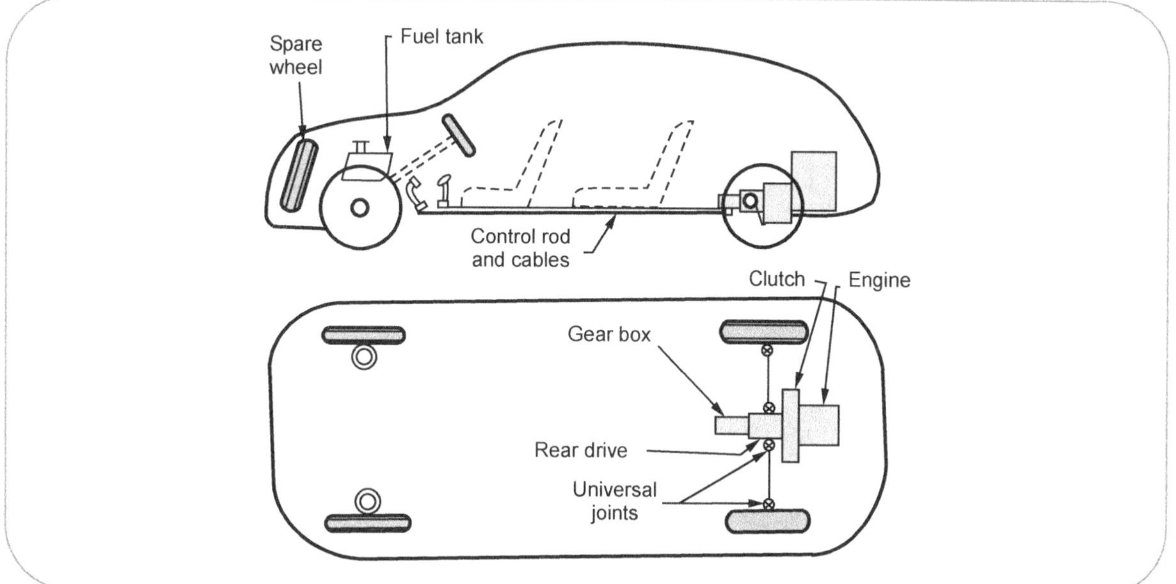

Fig. 6.4.2 Layout of rear engine rear wheel drive

- The rear floor can be made flat due to absence of propeller shaft.
- Driver cabin is well isolated from engine noise.
- Because of elimination of front engine packaging constraints, the front body can be designed as per styling.

Disadvantages :
- Natural air cooling of engine is not possible, hence it requires a powerful radiator fan.
- The clutch and gear shifting mechanism is long and complex.

- Because of higher weight concentration at rear, the vehicle has a tendency to oversteer while taking a sharp turn.
- Luggage space at front is restricted due to small compartment that houses the fuel tank and spare wheel.

6.4.3 Front Engine Front Wheel Drive
SPPU : May-08,09, Dec.-11, April-17

- This layout is the most compact layout.
- It is being popularly used on most hatchback cars in India.
- Refer Fig. 6.4.3 for layout for front engine front wheel drive.

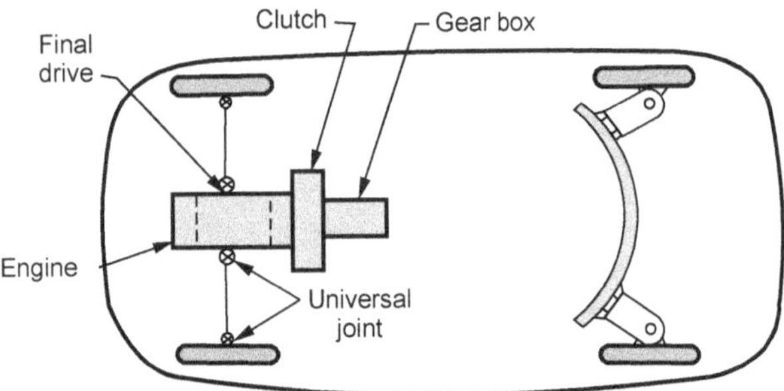

Fig. 6.4.3 Layout for front engine front wheel drive

- The drive chain for this layout is :

Engine → clutch → gearbox → universal joints → differential → front axle → wheels

Advantages :
- Compact design of vehicle.
- Due to higher weight of the steering wheels, it provides stable steering during turns.
- Reduced skidding on slippery roads.
- No propeller shaft is required.
- It has a flat rear floor due to absence of propeller shaft.
- Better engine cooling takes place due to natural air stream flowing across the engine radiator.

Disadvantages :
- The front wheel design is complicated because it supports the engine, differential and steering mechanism.
- High noise transmission from engine compartment to driver cabin.

- Poor ability of the vehicle to climb a steep hill because of lower weight on rear wheel.

6.4.4 Four Wheel Drive
SPPU : May-08,11,13, Dec.-08,09,11

- Sometimes while driving on rural or uneven or rough unconstructed roads, automobile with conventional layout (2 - wheel drive) gets stuck in the pothole.
- It becomes difficult for driver to get the automobile out of the pothole or ditch.
- The case is more severe if the drive wheel itself gets stuck in such a pothole or ditch.
- Hence to provide better maneuverability i.e. drive and traction on all the four wheels of the vehicle, a four wheel drive train is used.
- Refer Fig. 6.4.4 for a four wheel drive layout. (See Fig. 6.4.4 on next page)

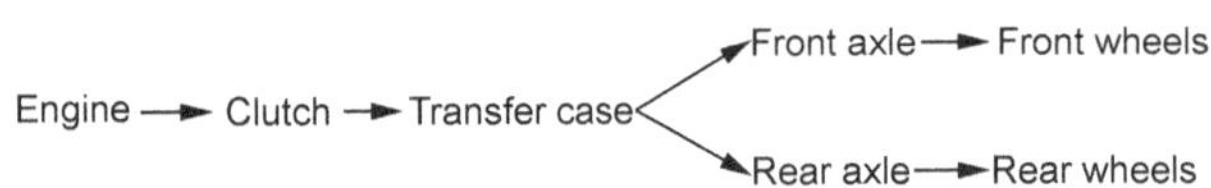

- The drive chain for this layout is :
- A vehicle with four wheel drive is denoted as 4 WD or all - wheel drive.
- An additional lever is provided in the driver cabin to engage or disengage the 4 WD.

Advantages :
- Better traction on all the four wheels.
- Excellent road handling characteristics.
- Vehicle can be used on uneven or rough or unconstructed road.

Disadvantages :
- Vehicle weight is high.
- Vehicle cost is higher.
- Accessibility to various components is difficult due to complicated design.
- It requires special transfer case.

6.4.5 Articulated Vehicle Layout
SPPU : Dec.-08

- This layout is typically used for very high load carrying vehicles such trucks with more then 25 T (tonnes) load carrying capacity.

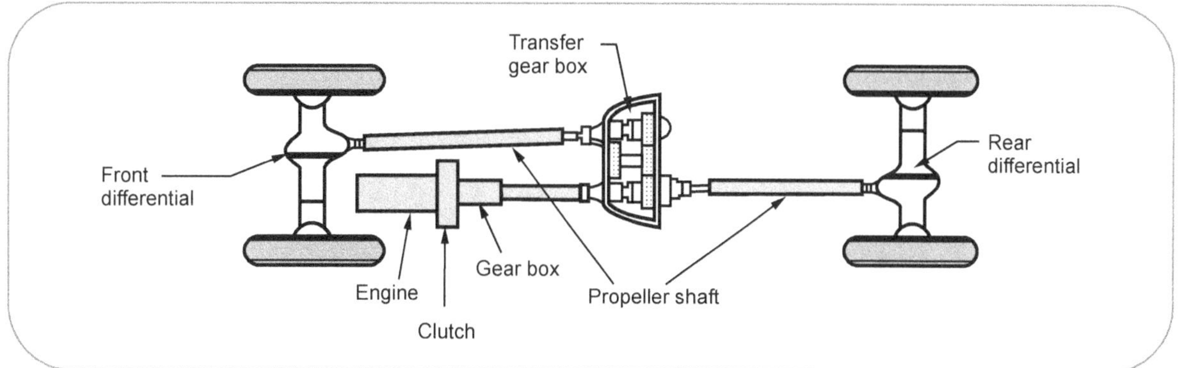

Fig. 6.4.4 Layout of a four wheel drive

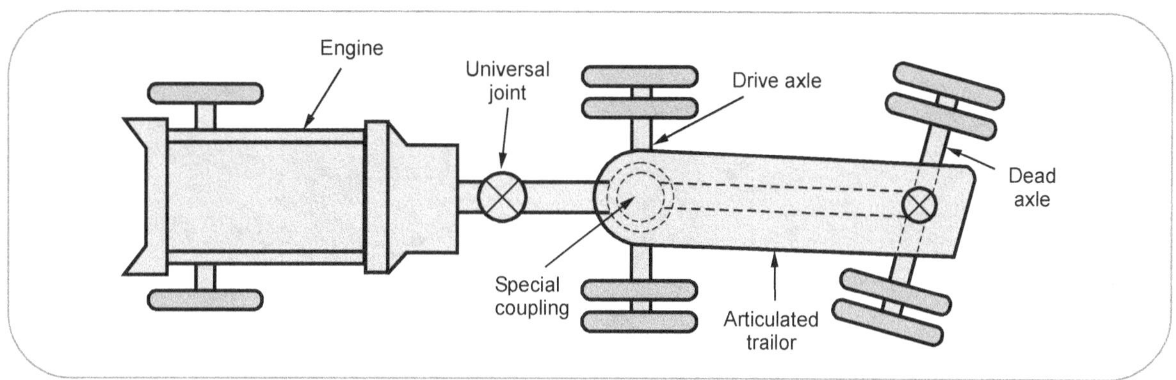

Fig. 6.4.5 Layout of articulated vehicle

- It is also referred to as tractor - trailor type of layout.

- Refer Fig. 6.4.5 for articulated vehicle layout.

- It consists of a very heavy and powerful engine at the front (tractor compartment).

- The tractor part of vehicle can be a four wheeled or six wheeled and consists of the drive axle.

- The trailor carries a dead axle at the rear.

- By means of a special coupling, the trailor is attached to the tractor.

Advantages :

- It provides large luggage / load carrying space.

- It greater maneuverability.

- Tractor compartment can be easily disengaged and connected to another pre-loaded trailor, hence reduced loading / unloading time.

Disadvantages :

- Difficult to drive on sharp turns.

- Heavy vehicle weight is uneconomical for fuel efficiency.

6.5 : Steering System

- Steering is the collection of components, linkages, etc. which allows any vehicle to follow the desired course.

- The primary purpose of the steering system is to allow the driver to guide the vehicle.

- Steering linkages are the links that transmit the rotary motion of the steering wheel to the turning of the front wheels.

- Steering linkages depend on the type of front axle, i.e. whether it is independent front suspension or a rigid type front axle suspension.

Steering Linkage for Vehicle with Rigid Axle Front Suspension :

- Refer Fig. 6.5.1 for steering linkages for vehicle with rigid front axle suspension.

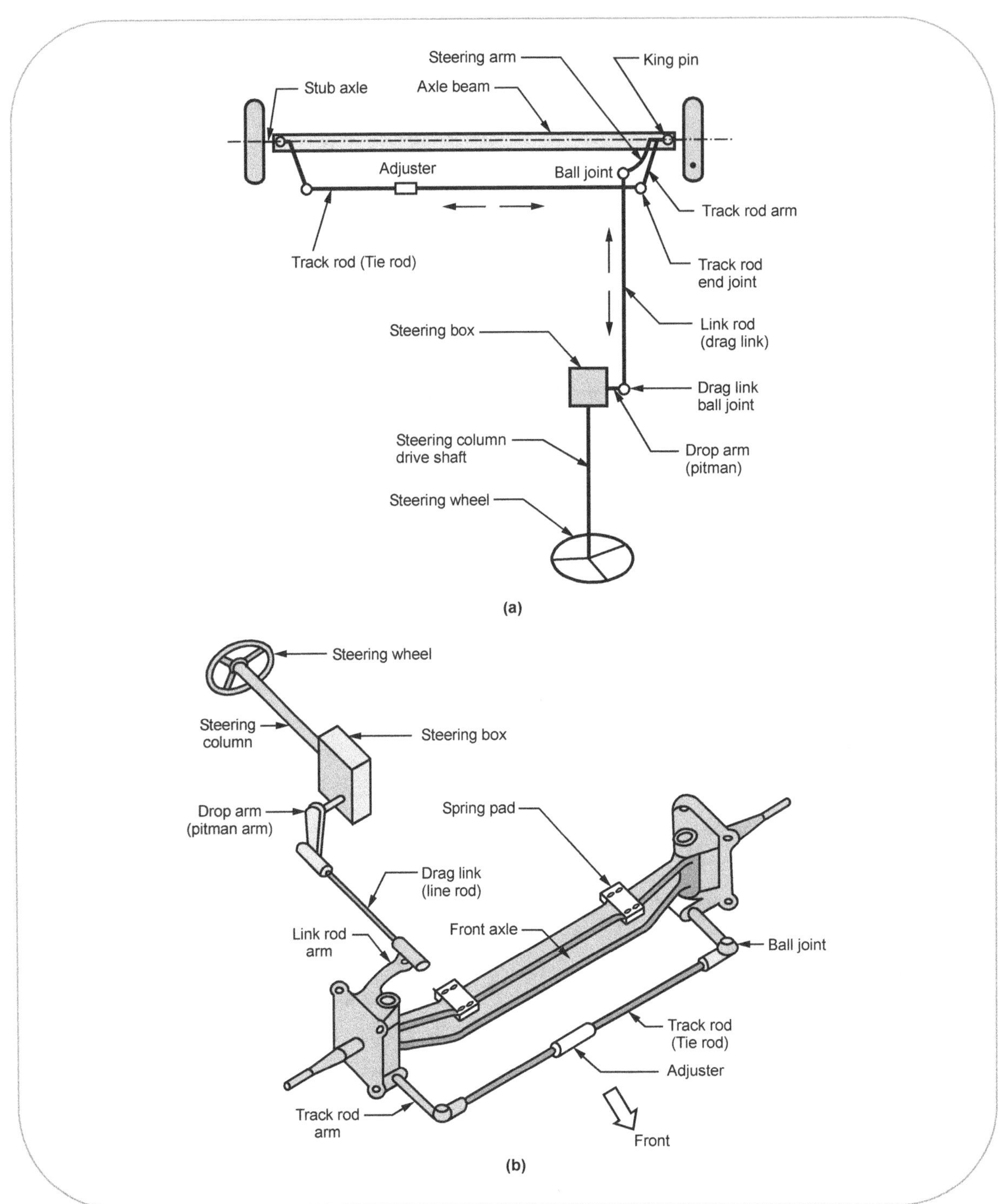

Fig. 6.5.1 Steering linkage for rigid axle front suspension

The major elements of steering linkages are :

> (i) Steering wheel   (ii) Steering column
> (iii) Steering gear box   (iv) Pitman or drop arm
> (v) Drag link (link rod)   (vi) Stub axle
> (vii) Track rod arm and tie rod

(i) Steering wheel

- It is provided in the driver's cabin.

- The driver can rotate the steering wheel in the direction as required to steer the vehicle.

(ii) Steering column

- It transmits the rotary motion of steering wheel to the input of steering gearbox.

(iii) Steering gearbox

- It provides the mechanical advantage so that only a small effort is required at the steering wheel to apply much larger force at the wheels.

(iv) Pitman or drop arm

- Pitman is rigidly connected to the output of steering gearbox.

- The other end of pitman is connected to the drag link by a ball joint.

- It converts the rotary motion of gearbox output shaft to linear movement of drag link.

(v) Drag link (Link rod)

- The drag link is connected to the stub axle.

(vi) Stub axle

- The stub axle is swiveled about the king pin.

- One end of stub axle is connected to the drag link by a link rod arm.

- The two stub axles are connected to each other by track rod arm and tie rod.

(vii) Track rod arm and tie rod

- Track rod arm and tie rod form a 4-bar link mechanism.

- They connect both the stub axles.

- A slider joint is provided in the tie rod for adjustment of steering linkages for wheel alignment.

6.5.1 Steering Column Mounted Controls

SPPU : Dec.-07

- The steering column can be used as a foundation column to mount various driver controls.

- Steering column is easily accessible to driver for operating the controls.

- Following are some of the controls mounted on the steering column : (Refer Fig. 6.5.2)
  (See Fig. 6.5.2 on next page)

> (i) Anti theft lock   (ii) Direction indicators
> (iii) Head lamp switch   (iv) Windshield wiper
> (v) Horn   (vi) Audio controls
> (vii) Steering tilt mechanism   (viii) Gear selector lever

(i) Anti theft lock

- It is a combined steering column lock and ignition lock.

- When the ignition key is removed and steering wheel turned to one of its extreme, the column gets locked.

(ii) Direction indicators

- It is an ON/OFF switch provided perpendicular to the steering column.

- Usually on Indian vehicles, the direction indicators are provided on the right side of driver on the steering column.

(iii) Head lamp switch

- The head lamp switch is integrated with the direction indicator.

- Rotating the direction indicator along its center axis, turns the head lamp bulb ON / OFF.

(iv) Windshield wiper

- It is a lever similar to that of direction indicator but mounted on the opposite side.

- The windshield lever has an integrated control to operate the wiper blades and the washer system.

(v) Horn

- The horn switch is mounted on the top of the steering column at its center.

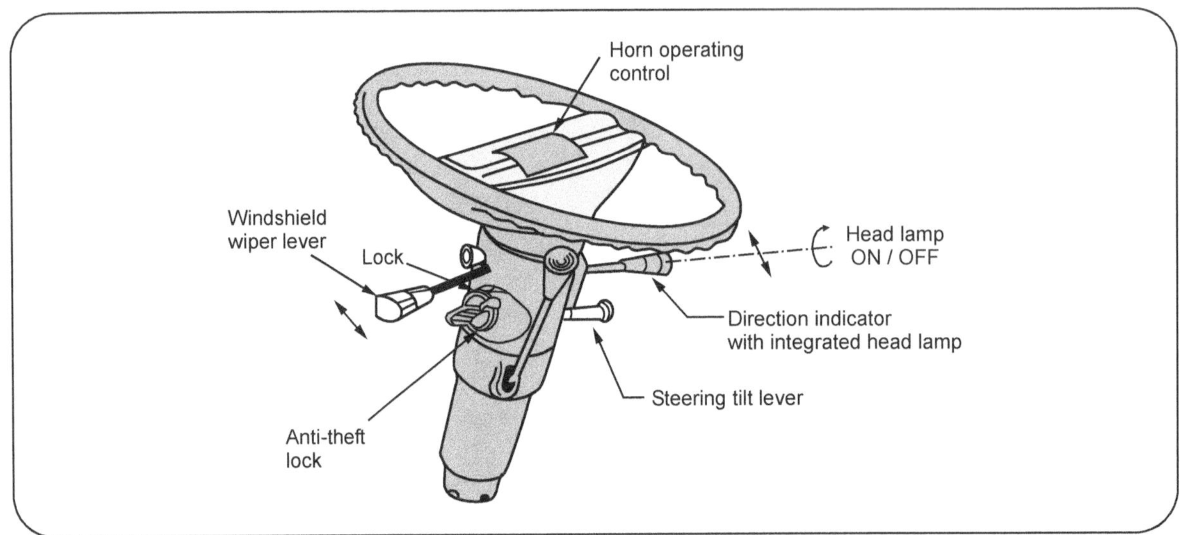

Fig. 6.5.2 Steering column mounted controls

- Pressing the horn switch closes the electric circuit of horn, thus producing the required sound.

- On releasing the horn switch, the electric circuit closes.

(vi) Audio controls

- Now-a-days, modern cars have integrated audio controls along the circumference of steering wheel.

(vii) Steering tilt mechanism

- This is an option provided on certain high end cars.

- This provision allows the driver to adjust the steering wheel position (proximity to driver seat) according to his requirement.

- It is useful for easy ingress and egress of the driver from the vehicle.

(viii) Gear selector lever

- In earlier generation passenger cars, the gear selector lever was provided on the steering column.

- But now-a-days, the selector lever is provided on top of the gearbox.

6.5.2 Power Steering

> **SPPU : May-08,10, Dec.-08,11,13, April-17**

- Bigger vehicles offer high resistance for steering of wheels.

- Thus, large amount of torque is required to be applied by driver for steering the wheels.

- Hence, to reduce the effort of driver, some power assistance is provided to steer the wheels.

- Power steering provides hydraulic or electric assistance to the turning effort applied to the manual steering.

- When the manual effort at the steering wheel exceeds a pre-determined value, the power steering become operative.

- In case of any failure of power steering, the driver is still able to steer the wheels but with much larger effort.

- The arrangement of hydraulic power steering is shown in Fig. 6.5.3 for a recirculating ball type worm and wheel steering gear.

- The major components of a hydraulic power steering are - pump, filter, steering gears, rotary valve and fluid lines.

- Almost all the existing vehicles have power steering as standard or as optional feature, e.g. Maruti Suzuki Alto, Swift, Hyundai ilo, i20, Tata Indica Vista etc.

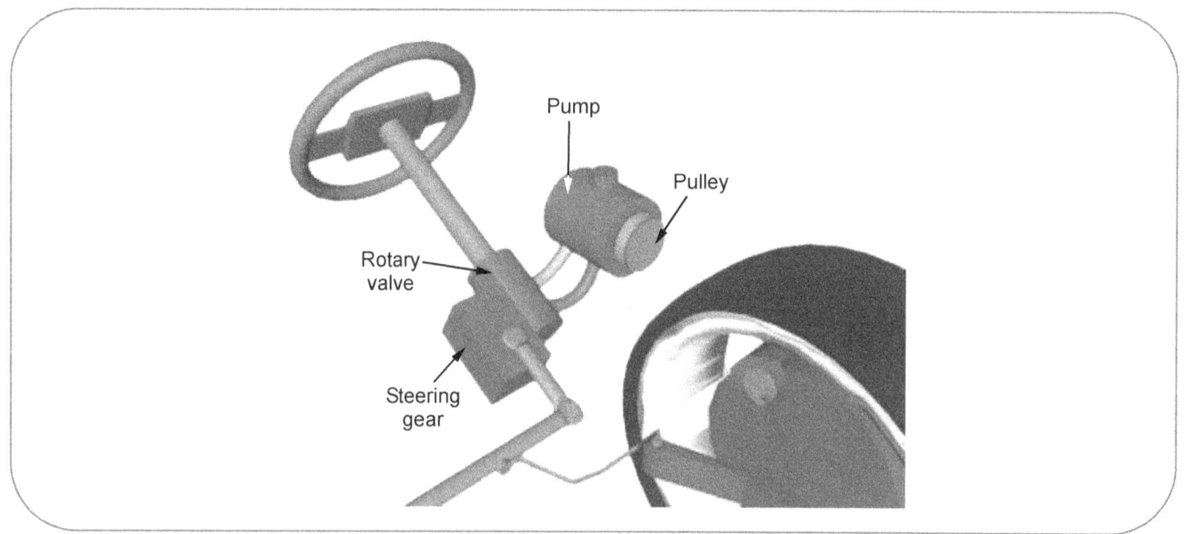

Fig. 6.5.3 Hydraulic power steering

6.6 : Suspension System

- A vehicle is run on different types of road conditions such as even, uneven, rough, etc.

- To isolate the passengers and other vehicle components from the irregularities of road surface, the vehicle chassis is not directly mounted on the axles.

- Springs or suspension system mounted in between the passenger compartment and axle, isolates the vehicle chassis from the road conditions.

- Also during cornering, the vehicle is subjected to undesirable effects such as pitch, roll or sway.

- These effects are absorbed by the suspension system thereby avoiding stress development in the vehicle frame and body.

- The suspension system consists of a spring and damper assembly.

6.6.1 Objectives of Suspension System

SPPU : May-10, Dec.-10

The major objectives of suspension system are as under :

- To safeguard the passengers against road shocks and provide riding comfort.

- To prevent the road shocks from being transmitted to vehicle frame and other components.

- To keep the vehicle on even keel (inclination) while traveling on rough or inclined roads or while

cornering and thereby minimize the effects of rolling, pitching and swaying.

- To maintain traction between the tyres and road surface.

- To support the vehicle weight.

- To hold the vehicle wheels in alignment.

6.6.2 Sprung Weight and Unsprung Weight

SPPU : May-08,10, Dec.-10

Sprung Weight :

- Sprung weight is the weight supported by the suspension system.

- It includes the weight of passengers, engine, gearbox, etc.

Unsprung Weight :

- Unsprung weight is the vehicle weight that is not supported by the suspension system.

- It includes the weight of drive axles, axles shafts, wheels and tyres.

- To have a smooth drive, the unsprung weight should be as low as possible.

6.6.3 Leaf Springs / Conventional Springs

SPPU : Dec.-05,08,09,10, May-09,10,11,12

- Leaf springs are the most traditional used types of suspension system.

- They are used for light, medium and heavy commercial vehicles.

• Refer Fig. 6.6.1, 6.6.2 and 6.6.3 for construction and mounting of leaf spring on the rear axle.

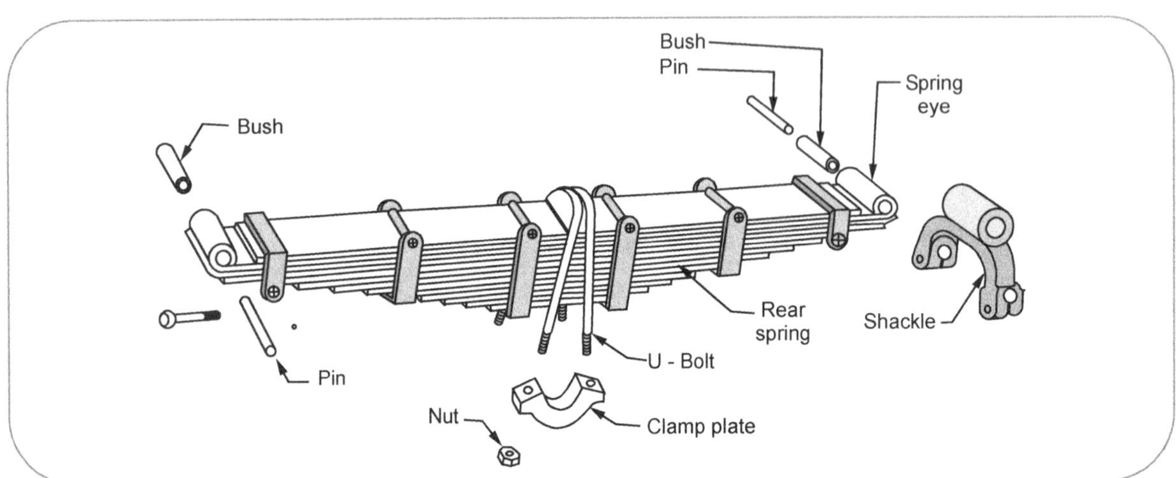

Fig. 6.6.1 Exploded view of leaf spring for rear axle

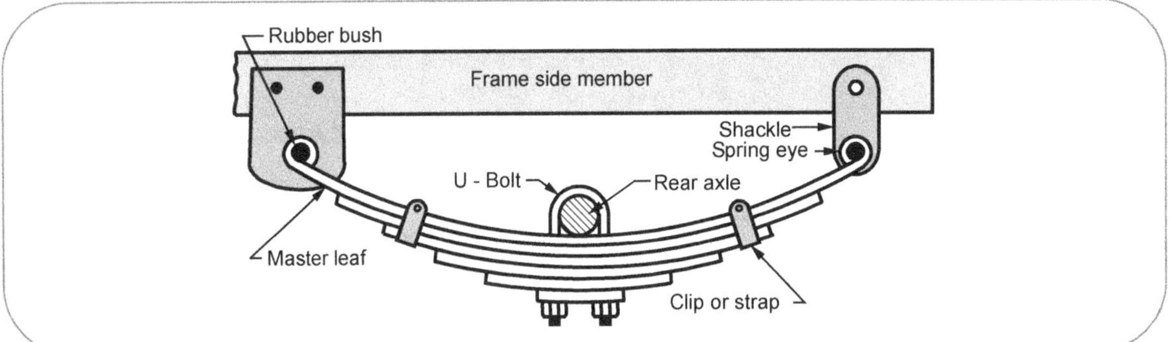

Fig.6.6.2 Assembly of leaf spring for rear axle

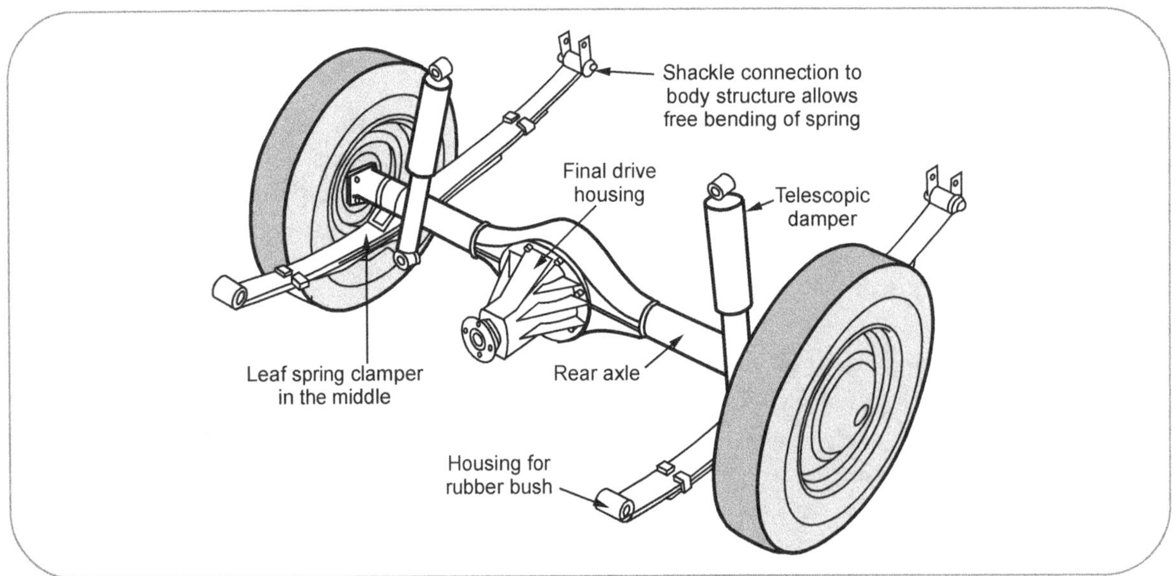

Fig. 6.6.3 Rear axle mounted leaf spring

Construction :

The leaf spring consists of following major components :

(i) Leafs / blades

- The leaves of the suspension system are also called blades.
- These blades are made of semi-elliptical cross-section to ensure uniform stress distribution across the cross section area.
- The longest leaf is mounted on the top and is called as master leaf.
- The blades of smaller length are clamped below the master leaf in descending order of their lengths.
- The master leaf has eyes at its ends.
- They are made of carbon steel or chrome-vanadium steel or silico-manganese steel.

(ii) Clamps

- These are made of steel and are used to hold all the leaves together.

(iii) Shackles

- Shackles are special couplings which allow small rotary motion about its longitudinal axis.
- Refer Fig. 6.6.1 for construction of shackle.
- For a rear suspension, the shackle is provided on the rear end of master leaf.
- Shackle connects the leaf spring with the frame side member.
- For a front suspension, the shackle is provided on the front eye of master leaf.

(iv) U-bolt

- The U-bolt is used to mount the leaf spring on the axle.
- These are made of steel.

(v) Bushes

- Bushes are placed at the eyes of the master leaf.
- They provide bearing surface for small rotary movement.
- They are made of phosphor bronze or rubber.

- Rubber bushes are less noisy and have reduced wear.

Working :

- Consider Fig. 6.6.2 for working of leaf spring mounted on rear axle.
- The leaves are initially cambered to allow its deflection during operation.
- When the vehicle moves over a bump on road surface, the wheel moves up.
- This causes the leaves to deflect, hence changing the length of leaf between the spring eyes.
- If both the ends of the leaf are fixed, then it will not be able to accomodate this change in length. Hence a shackle is provided.
- Since the front end is fixed the leaf spring has a center of rotation at the front fixed end.
- Also the propellor shaft is connected to the universal joint near the front end.
- Thus, the leaf spring and propellor shaft (with axle) swivel about the front end while the shackle permits this swivelling of the rear eye of leaf.

6.6.4 Coil Springs

- Coil springs are most commonly used with independent suspension.
- Coil springs can be manufactured for varying spring rates.
- They can be accommodated in confined space.

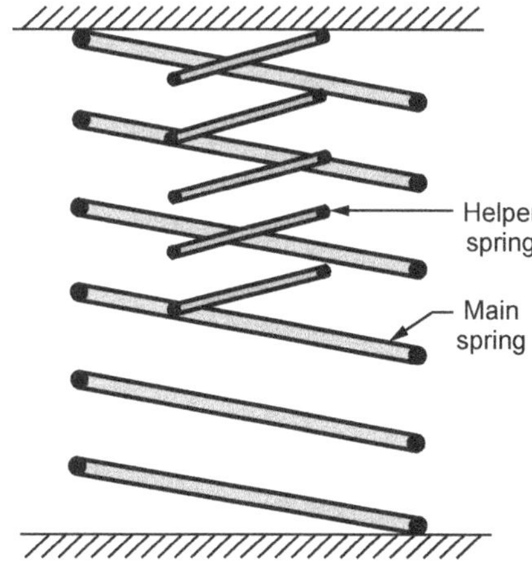

Fig. 6.6.4 Coil spring with helper spring

- The energy stored per unit volume is almost double in the case of coil springs as compared to that in leaf springs.

- As compared to leaf springs, they weigh only half to withstand the same vertical load.

- Coil springs take the vertical shear as well as bending loads of the vehicle.

- But coil springs cannot take torque and side thrust load during acceleration, braking and cornering.

- Hence, sometimes a helper spring is provided to withstand increased vertical and other loads.

6.7 : Braking System

- Braking system is the most efficient controlling mechanism of the vehicle on run.

- The braking system provides the driver run time control of the speed of the vehicle.

- Braking system also provides service to a parked vehicle.

- Brakes are used to slow the vehicle and stop as per requirement.

- Hence, braking system is required to be one of the most efficient and critical system of the vehicle.

6.7.1 Purpose of Braking System

The main purpose / functions of the braking system are as follows :

(i) To provide a mean to slow the vehicle without affecting the engine speed

(ii) To provide a mean to stop the vehicle while the engine is still running

(iii) To provide emergency stop of the vehicle within lowest time and distance

6.7.2 Drum Brakes

- Drum brakes are the cheapest and most widely used brakes.

- It consists of a rotating brake drum mounted on the wheel and two semi-circular brake shoes attached on a stationary back plate.

- The pressing of brake shoes on the rotating drum causes friction resulting in braking.

Construction :

- Refer Fig. 6.7.1 and 6.7.2 for a drum brake system and its components.

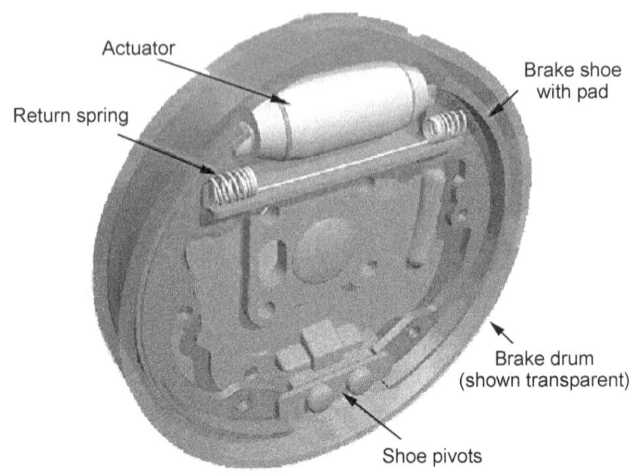

Fig. 6.7.1 Drum brake system assembly

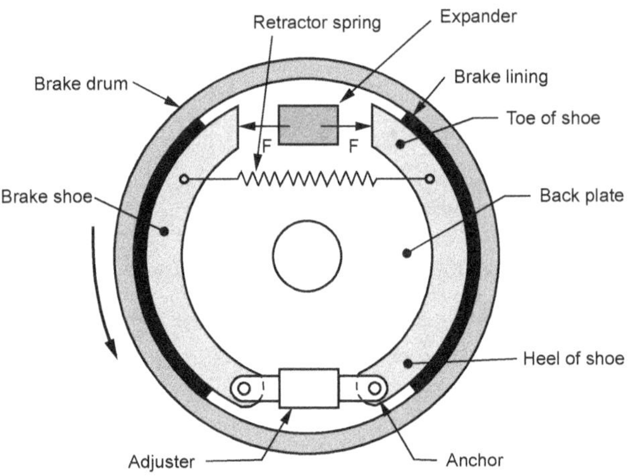

Fig. 6.7.2 Simplified drum brakes assembly

- The brake drum is a hollow cylinder type construction made of cast iron.

- It is mounted concentric to the axle hub and rotates alongwith the wheel.

- A separate back plate is mounted on the stationary axle casing and behind the brake drum.

- The back plate houses the assembly of semi-circular brake shoes mounted on an anchor.

- The back plate alongwith the brake shoe assembly remains stationary.

- The back plate is made of pressed steel sheet and absorbs the torsional loads due to friction between brake shoes and drum.

- It also covers the entire brake assembly, thereby protecting it from dirt, dust and mud.

- The two brake shoes are pivoted to the anchor at one of its ends.

- The other end of brake shoe is held by an expander.

- Expander can be mechanical cam type or a hydraulic piston type.

- The compression type retractor spring connected between the brake shoes always try to pull them inwards.

Working :

- The drum mounted on the axle hub continues to spin along with the wheel.

- When the brake pedal is pressed, the braking effort is transferred to the expander through an activation mechanism.

- This results in a force 'F' at the expander, pushing the brake shoes on the drum, against the spring force.

- The friction between the stationary brake shoes and the revolving brake drum provides the braking action.

- When the brake pedal is released, the retractor spring pulls the brake shoes inward to disengage the brakes.

- An adjuster is also provided to compensate for the wear and tear of the brake shoes.

Advantages :

- It consists of large friction shoes providing better braking.

- The overall system design and construction is simple.

- The drum brakes are most economical at cost.

Limitations :

- The wear and tear on brake shoes is not uniform.

- The overall size of the system is more.

- Replacement of brake shoe requires dismantling of entire brake drum and back plate assembly.

6.7.3 Disc Brakes   SPPU : May-12, 17

- Disc brakes are more efficient and now-a-days being adopted on large scale in the automotive segment.

- It consists of a rotating brake disc mounted on the wheel and two friction pads positioned on either side of the disc.

- The pressing of the stationary brake pads on the revolving disc causes friction, resulting in braking.

Construction

- Refer Fig. 6.7.3 and 6.7.4 for the construction and assembly of disc brake.

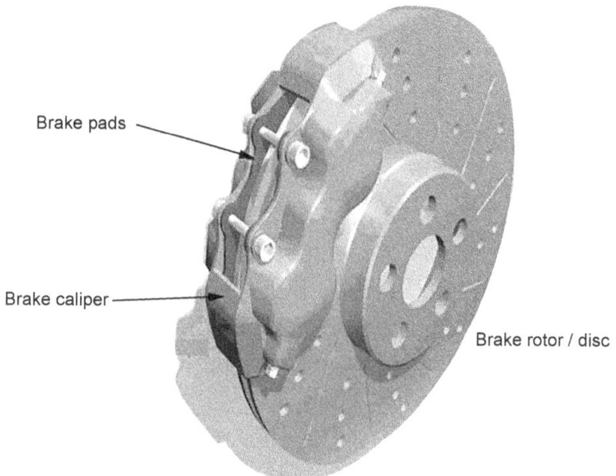

Fig. 6.7.3 Disc brake assembly

- It consists of a cast iron or steel pressed disc bolted on the wheel hub.

- The brake disc revolves along with the wheel.

- The outer circumference of brake disc is housed in the hydraulic caliper.

- The hydraulic caliper consists of two sliding pistons.

- The outer surface of sliding pistons is provided with friction pad riveted on it.

- Fluid lines connect the caliper to the brake lever or pedal.

- Retractor springs (not shown in Fig. 6.7.4) are mounted in between the piston and caliper housing.

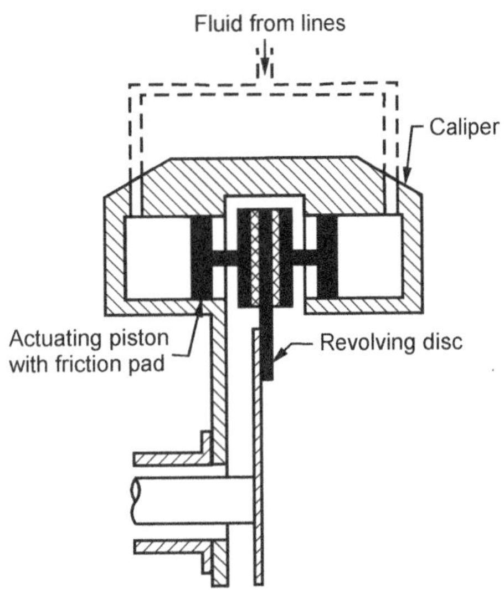

Fig. 6.7.4 Simplified disc brake assembly

Working :

- When the vehicle is running, the brake disc revolves along with the wheel.

- When the brake lever or pedal is operated, the braking effort is transmitted to the hydraulic caliper through pressurized fluid.

- The fluid pressure pushes the two piston towards the brake disc.

- The friction between the brake pads on the stationary pistons and the rotating brake disc causes the braking of the vehicle.

- When the brake lever is released, the two pistons are pushed back by the retractor springs.

Advantages :

- The operation and assembly of disc brake is much simpler.

- As the friction pads are flat, the wear and tear is uniform.

- Heat dissipation is faster.

Limitations :

- The overall system cost is higher due to hydraulic caliper and fluid lines.

- The frictional area of pads is less, thereby requiring high pressure intensity fluid.

6.7.4 Anti-Skid Brake System or Anti-Lock Brake System (A.B.S)

- One of the important requirement from a brake system is that the wheels should not lock or skid while the brakes are slowing down the vehicle.

- A wheel that is locked or skids on the road, looses its traction and the vehicle becomes non-controllable.

- An Anti-lock Brake System (A.B.S.) is a safety system which prevents the wheels from locking during braking operation.

- The anti-lock braking system ensures :

  (i) Maintains vehicle control

  (ii) Directional stability

  (iii) Optimum deceleration while braking

Principle of anti-lock brake system :

- The system monitors the rotational speed of each wheel continuously and controls the brake line pressure to each wheel during braking.

- This prevents the wheel from locking up.

Layout and construction :

- Fig. 6.7.5 shows the layout of various components of an anti-lock brake system. (See Fig. 6.7.5 on next page)

The major components of anti-lock brake system and functions of each are as mentioned below :

(i) Speed sensor with toothed ring

- A circular toothed ring is mounted at each of the wheels.

- A magnetic speed sensor is located perpendicular to the toothed ring.

- As per the speed of the ring, it cuts the magnetic flux of the sensor.

- The speed sensor transmits this speed data of each wheel to the E.C.U. of A.B.S.

(ii) A.B.S. E.C.U. (Electronic Control Unit)

- The E.C.U. is the heart of the A.B.S.

- The E.C.U. receives speed input of each wheel and computes the state of rotation of each wheel.

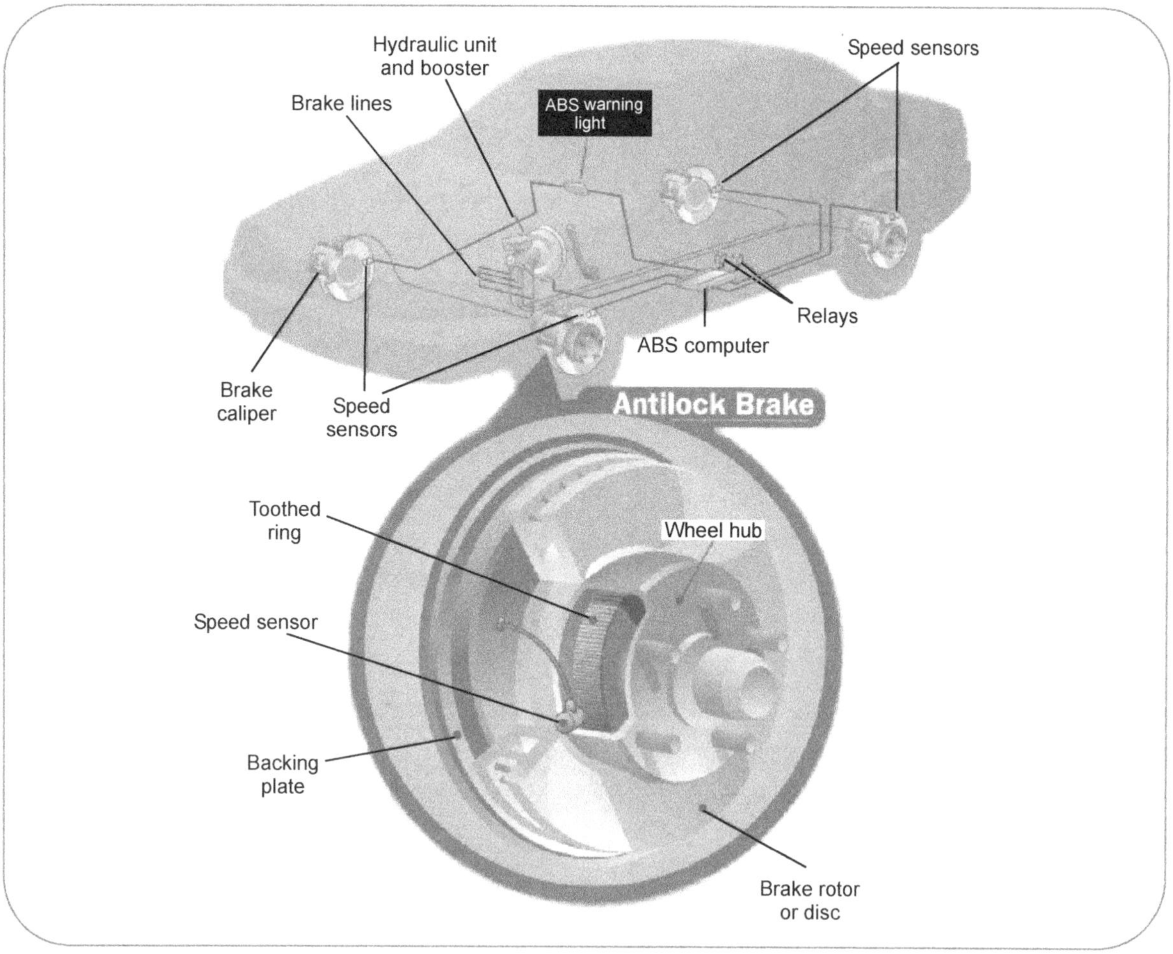

Fig. 6.7.5 Anti-lock brake system

- E.C.U. then directs the hydraulic booster to supply varying pressure fluid to each brake wheel line.

(iii) Hydraulic booster

- It is an electrically operated booster.

- It has four outlet brake lines connecting to each wheel.

Working of the A.B.S. system :

- The sensor at each wheel provides the varying voltage signal to the E.C.U. of the brake system.

- The E.C.U. computes this voltage signal and compares it with programmed information and determines whether a wheel is about to lock or skid.

- When the wheel is about to lock the E.C.U. signals the hydraulic unit to reduce hydraulic pressure at that wheel's brake caliper.

- Thus, the particular wheel starts to spin again and avoids locking state.

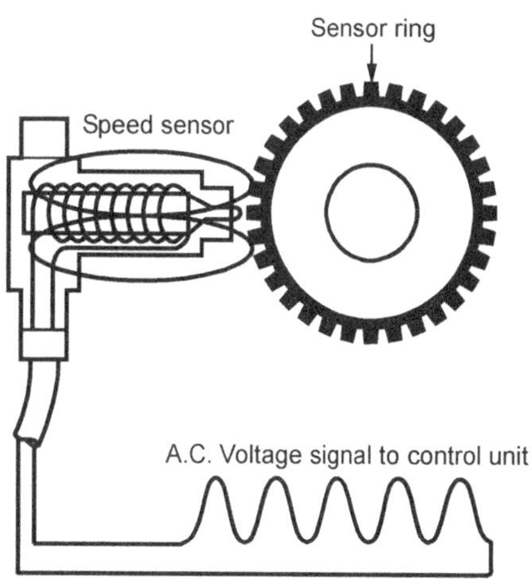

Fig. 6.7.6 Speed sensor

- The E.C.U. then instructs the hydraulic booster to re-apply full pressure and again measures the rotation of wheel.

- This on-off-measure cycle happens around 15 to 30 times in a second.

Advantages of A.B.S. :

- A.B.S. provides safety to the passengers.

- It provides better traction during brake application.

- It ensures directional stability of vehicle on turns.

Limitations of A.B.S. :

- Under slippery surface, the stopping distance is high.

- During braking, a vibratory feeling is produced at the brake pedal, thereby creating a false sense of security concern among drivers who do not understand the operation.

- The system cost is very high.

6.8 : Necessity of Cooling

SPPU : Dec.-15

- In case of I.C. engines, the energy input to the engine cylinder is by burning the fuel with air. The utilisation of this heat energy in terms of percentage is as follows :

  - 30-35 % of energy is used for conversion into useful work.

  - 30-35 % of energy is carried away by exhaust gases.

  - 10-15 % of energy is lost by radiation, convection and conduction.

  - 20-25 % of energy flows from gases to cylinder walls and raises its temperature.

- The temperature of gases inside the engine cylinder may vary from 30 °C to 2500 °C.

- If no external cooling is provided and the engine is allowed to run, the average temperature inside the engine cylinder would be 1200 °C to 1500 °C.

- Therefore, the cylinder walls, piston and cylinder head will be exposed to high temperature.

- At higher temperature, strength of material reduces. Hence, it is required to keep the engine temperature within certain limits. The cylinder wall temperature should be limited to 65-70 °C as lubricating oil will start evaporating beyond this limit.

- In SI engines, cooling must be satisfactory to prevent pre-ignition of air-fuel mixture and knocking. In CI engines, cooling must be adequate to allow the components of the engine work properly.

- Hence, cooling is a process of equalization of temperatures to prevent overheating of components and to remove excess heat energy to maintain the practical overall operating temperature.

- Therefore, it becomes necessary to provide cooling system to maintain the temperatures within certain limits to obtain the maximum performance from the engine.

6.8.1 Effects of Overheating

The effects of overheating of engine are as follows :

- High temperature of cylinder walls will increase the tendency of pre-ignition which in turn can lead to undesirable effects like detonation and knocking.

- Reduction in mechanical strength of engine parts, thereby reducing life of the engine.

- Evaporation of lubricating oil, thereby further increasing friction losses.

- Uneven expansion of piston in cylinder may result in seizure of piston.

6.8.2 Effects of Overcooling

The effects of overcooling of engine are as follows :

- Reduction in thermal efficiency as more heat is carried away by the coolant.

- Lower temperature will increase viscosity of lubricating oil, thereby increasing friction losses.

- Starting problems may arise due to overcooling of the engine.

- It also affects the loss of power and fuel economy.

6.8.3 Functions of Engine Cooling System

The functions of the engine cooling system are as follows :

- To absorb and dissipate heat to maintain average temperatures in the range of 160 °C to 250 °C for maximum power, smooth running and operation of the engine.

- To prevent the damage of vital engine components.
- To ensure complete combustion of charge.
- To promote higher volumetric efficiency.

6.8.4 Types of Cooling System

- The cooling system for I.C. engine can be broadly classified as follows :

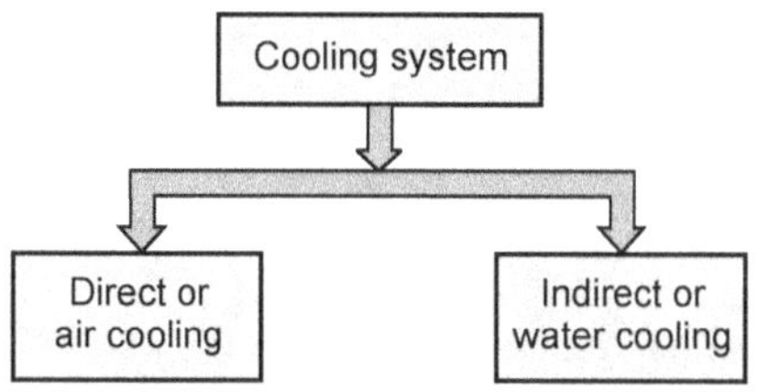

Fig. 6.8.1 : Classification of cooling system

6.9 : Air Cooling

- In air cooling, the object to be cooled have a larger surface area (surface area is increased by providing cooling fins) or have an increased flow of air over its surface or both.
- In small capacity I.C. engines, the cylinder block and cylinder head are provided with cooling fins on its outside surface and air is allowed to flow past the cylinder, allowing it to convey heat by convection process.

- The method of convection employed can be natural convection or forced convection (using a circulating fan) or combination of both.
- The air cooling is mostly applied to small capacity engines like motorcycles, scooters etc. Sometimes, it is also employed in stationary engines having smaller capacity.
- The cooling rate between material and air is lower due to lower heat transfer coefficient. Due to lower heat transfer coefficients the cylinder wall temperatures of air cooled engines are higher than that of water cooled engines.
- In order to increase heat transfer rates, it is necessary to either increase the rate of air flow or increase surface area of engine which is exposed to the atmosphere or a combination of both.
- Fig. 6.9.1 (a) shows the air cooling system.
- The amount of heat dissipated depends upon :
  - Cooling surface area in contact with passsing air.
  - Rate of mass flow of air over cooling area.
  - Effective temperature difference between the cylinder and air.
  - Coefficient of heat transfer of metal and air.

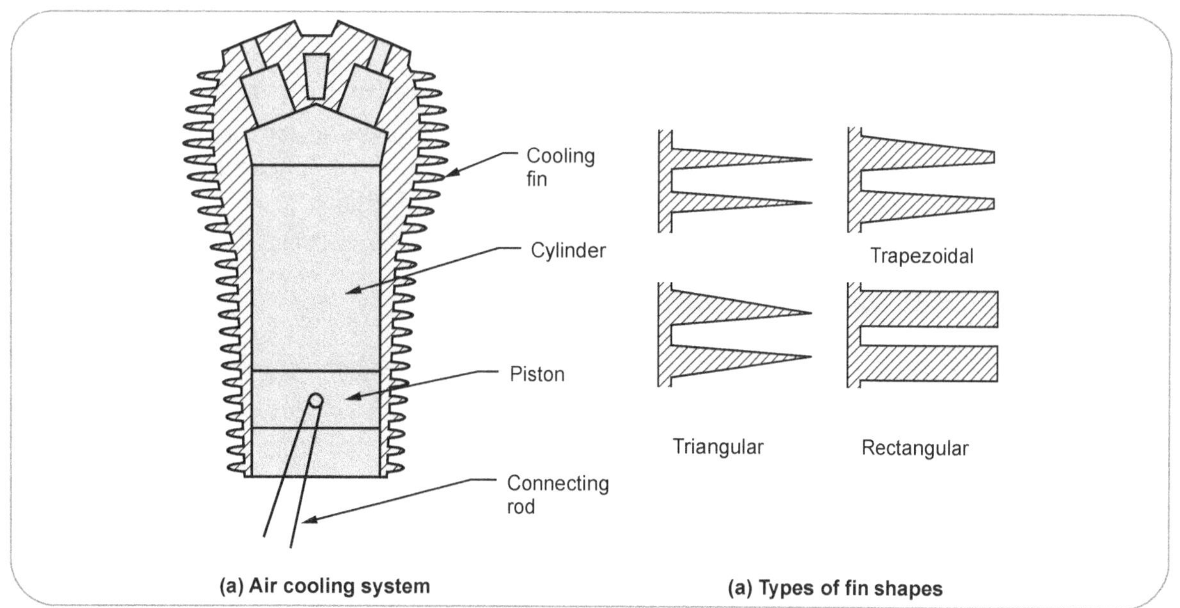

Fig. 6.9.1 : Air cooling system

6.9.1 Cooling Fins

- Cooling fins are used to increase heat dissipation of the engine by increasing the cooling surface area.

- The cooling fins are either cast integrally with the engine or attached separately to the external surface of the engine.

- The heat transfer capacity of fin depends on the type of cross-section and length of fin.

- Fig. 6.9.1 (b) shows the various types of cooling fins cross-section.

- The rectangular cross section is found to have lowest temperature drop and less efficient, also these fins increase the weight of the engine.

- The trapezoidal or triangular fins have high heat dissipation rate and low weight. But the manufacturing of these fins is difficult.

- The commonly used dimension of fins are :

   1) Tip thickness = 0.5 to 1.5 mm

   2) Length of fin = 25 to 50 mm

   3) Clearance between fins = 2.5 to 5 mm (at root).

6.10 : Water Cooling

- Water cooling is mainly preferred in medium and large size engines.

- Water is circulated through the water jackets provided around the cylinder walls. The water extracts the heat energy from the cylinder walls and releases it to the atmosphere.

- The water jacket is connected to a radiator.

- Radiator is a kind of heat exchanger which cools the hot water and it recirculates the cooled water to engine for cooling purpose.

6.10.1 Types of Water Cooling Systems

Water cooling systems can be classified as follows :

> 1) Non-return cooling
>
> 2) Thermo-syphon cooling.
>
> 3) Forced circulation system
>
> 4) Pressurized water cooling
>
> 5) Evaporative cooling

6.10.1.1 Non-Return Cooling System

- This system is mainly used for large capacity engines where ample amount of water is available.

- The water is stored in storage tanks. The water from these tanks is directly circulated through the cooling water jackets.

- The cooling water extracts the heat energy from the cylinder walls and then these water is directly discharged. This discharged water cannot be reused.

- The system is more efficient but consumes more amount of water.

6.10.1.2 Thermo-Syphon Cooling System

- Fig. 6.10.1 shows the schematic arrangement of an engine cooled on Thermo-syphon principle.

- The system is so designed that the water may circulate naturally (without any external power source) because of the density difference of hot water and cold water.

- The system consists of a radiator having upper and lower tanks connected to upper and lower water jackets of the cylinder respectively through the pipes.

- The hot water in the jacket rises and flows into the upper tank due to lower density compared to cold water and the cold water from radiator flows to lower water jacket to replace the hot water.

- From upper tank the water travels down the radiator tubes across which the cool air passes drawn by the fan driven by the engine crankshaft.

- In order to increase the rate of heat transfer, the surface area of the radiator exposed to the air blast is provided with fins.

- This system is suitable for low capacity engines only.

Advantages

- There is no need of circulation pump.

- Cost of thermo-syphon cooling system is lower than other systems.

Disadvantages

- It circulates water only when engine becomes hot.

- This system is not suitable for heavy duty engines where high heat transfer rates are required.

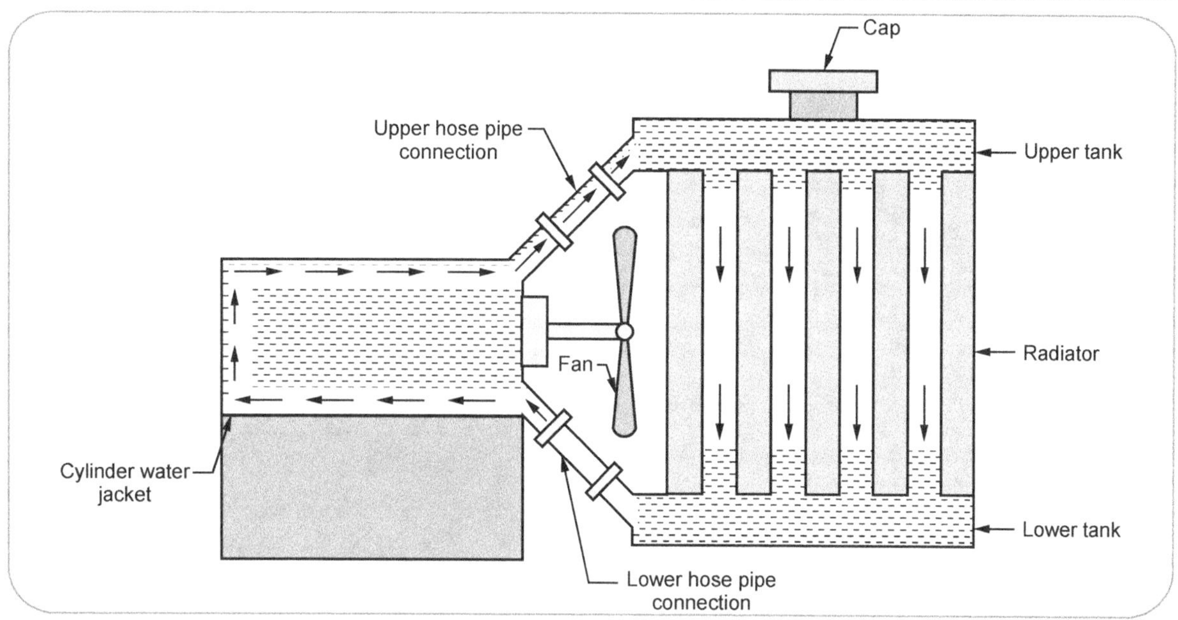

Fig. 6.10.1 : Thermo-syphon cooling system

- Temperature of cooling water should not be allowed to exceed beyond 80 °C in order to avoid evaporation of water.

6.10.1.3 Forced Circulation System

SPPU : May-14, Dec.-15,16

- It is also known as **Pump Assisted Thermo-syphon Cooling System.**

- The drawbacks of thermo-syphon system are eliminated by introducing a water circulation pump to assist the water circulation in the water jackets.

- Introduction of pump ensures the positive circulation of water under all operating conditions (Refer Fig. 6.10.2).

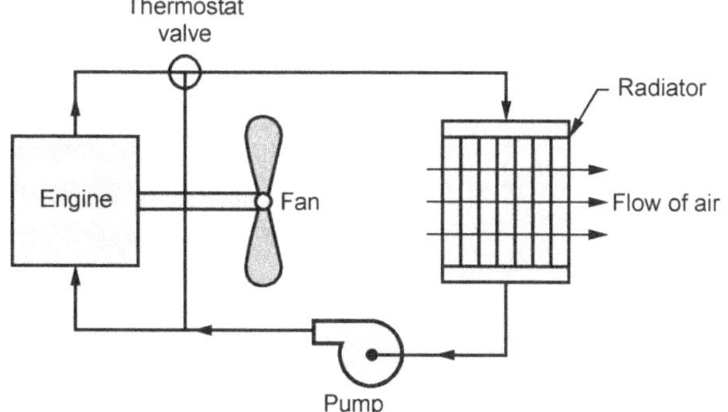

Fig. 6.10.2 : Forced circulation system

- As the cooling of the engine is independent of temperature difference of hot and cold water, it may result into over cooling of the engine which affects the thermal efficiency of the engine and its working.

- To avoid overcooling of the engine, **thermostat** is used. The thermostat maintains the constant temperature of cooling water which in turn maintains the temperature of engine within desirable limits.

6.10.2 Radiators

- The main function of radiator is to extract the heat from the hot water which circulated around the block. This extracted heat is dissipated to the surrounding.

- The size of radiator depends upon the following factors :

  ○ Rate of heat dissipation needed.

  ○ Velocity of air and water.

  ○ Design of radiator.

- The rate of heat transfer from hot water carried in tubes is increased by providing fins over the surface of tubes.

- The fins are arranged in such a manner that it creates some turbulence in air passage to enhance heat transfer coefficient to the air side.

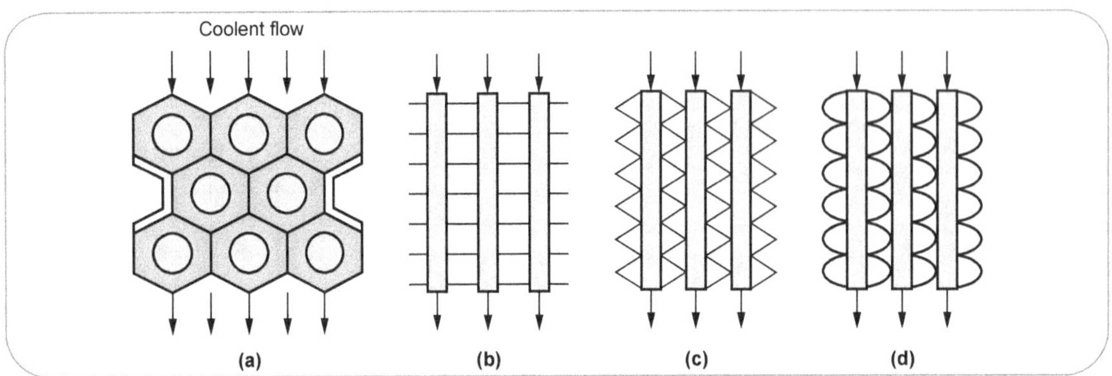

Fig. 6.10.3 : Types of radiators

- The radiator can be classified into following types :

> 1) Honeycomb core radiators
>
> 2) Tube and fin type radiators
>
> 3) Ribbon-cellular or fin type radiators
>
> 4) Corrugated fin type radiators

- In **honeycomb core radiators** a large number of hexagonal shaped tubes are joined together by brazing. It provides passage for hot water to flow and the cooling air is passed through circular tubes. Refer Fig. 6.10.3 (a).

- In **tube and fin type radiators,** the cross-section used can be of elliptical or circular type. It consists of long tubes with straight metallic fins extending from top to bottom of radiator. Refer Fig. 6.10.3 (b).

- In **ribbon cellular or fin type radiators,** a set of thin metal ribbons are brazed to the tubes carrying hot water. The ribbons are brazed in zig-zag manner which increase the turbulence of air flow. Due to the air turbulence, the heat dissipation rate is improved. Refer Fig. 6.10.3 (c).

- In **corrugated fin type radiators,** water tubes are made of oval shape section with zig-zag copper ribbons to enhance the surface area and the air turbulence. It improves the heat transfer rate to atmosphere.

6.11 : Fuel Supply System for Petrol Engines

- The basic function of this fuel system, is to supply vaporized air fuel (A/F) mixture to engine cylinder.

- The basic fuel supply system in an automobile with petrol engine consists of following elements :

> a) Fuel tank   b) Fuel lines
>
> c) Fuel pump   d) Fuel filter
>
> e) Air cleaner   f) Carburettor
>
> g) Inlet manifold
>
> h) Supply and return pipelines

- The types of system which are used for the supply of fuel from the fuel tank to engine cylinder are as follows :

> a) Gravity system   b) Pressure system
>
> c) Vacuum pump   d) Pump system
>
> e) Fuel injection system

a) Gravity system :

- In this system fuel tank is mounted at the highest position from where the fuel drops into the carburettor float chamber by gravity.

- This is very simple and low cost system. It is mainly suitable for two wheelers.

> **Note :**
> - Carburetion is the process of breaking up and mixing of air with fuel in correct proportions to meet the engine requirements.
> - It is a system which consists of all the mechanisms that are required to get clean air from the atmosphere as well as clean fuel from the fuel tank.
> - The process of formation of a combustible fuel-air mixture by mixing the proper amount of fuel with air before admission to the cylinder is called as **Carburetion** and the device which does this job is called a **carburetor.**

b) Pressure system

- In this system pressure is created in the tank by means of engine exhaust or separate air pump.

- For starting, the priming of pump is by hand.

- In such system there are chances of pressure leak hence now-a-days these systems are not in use.

c) Vacuum system

- It is based on the fact that the engine suction can be used for sucking fuel from the main tank to the auxiliary fuel tank from it flows by gravity to carburettor float chamber.

- These systems are also not in use now-a-days.

d) Pump system

- This system is commonly used in modern cars.

- In this system a steel pipe carries petrol to the fuel pump which pumps it into the float chamber of carburettor through a flexible pipe.

- Mechanical type of fuel pumps are operated by camshaft hence placed near the engine. Electrically operated type of fuel pump can be placed anywhere. Refer Fig. 6.11.1.

e) Fuel injection system

- This is the most accurate fuel supply system. Hence, modern cars are mainly equipped with this system.

- In this system carburettor is dispensed with altogether.

- For each cylinder separate injectors are used while the mixture under different load conditions and speed is controlled electronically.

- The fuel is atomised by using injector nozzle and then delivered into an air stream.

6.11.1 Simple Carburettor

- We know that, the purpose of carburetion process is to provide a homogeneous air fuel mixture.

- The simple carburettor consists of the following elements :

  1) Float chamber

  2) Fuel discharge nozzle

  3) A metering orifice

  4) A venturi

  5) Throttle valve

  6) Choke

- The float and needle valve maintain the fuel level in the tank.

- If the fuel level goes down, the float goes down admitting fuel opening the float needle valve. When the fuel level is optimum, the flow is stopped.

- The fuel strainer is used to trap the impurities from the fuel and prevent choking of the fuel nozzle. The strainer can be removed periodically for cleaning.

- The fuel level in the float chamber is maintained just below the tip of nozzle to avoid fuel overflow from the nozzle.

- The difference between top of nozzle and fuel level in float chamber is known as **jet tip.** Jet tip is denoted by 'h'.

- The nozzle acts as a discharging jet, which sprays fuel into the air stream.

- The cross sectional area of venturi is smaller at throat which provides constriction to air flow.

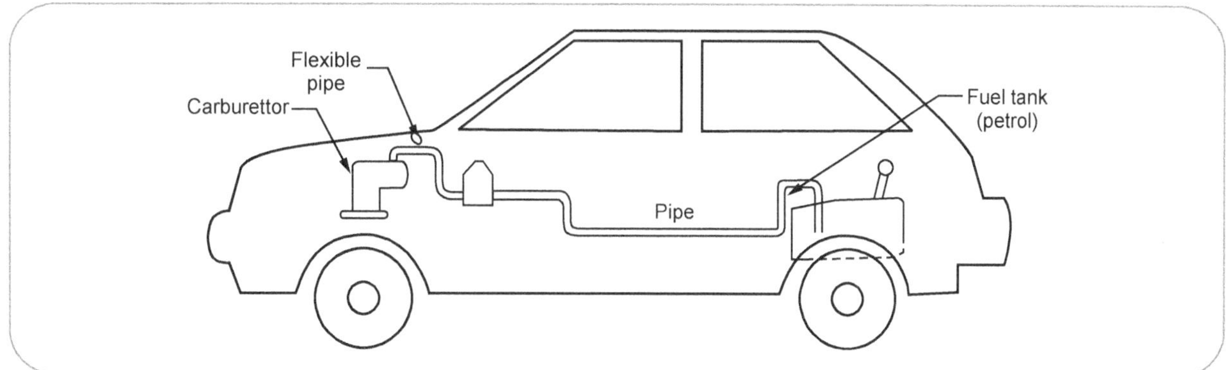

Fig. 6.11.1 Pump system of fuel supply

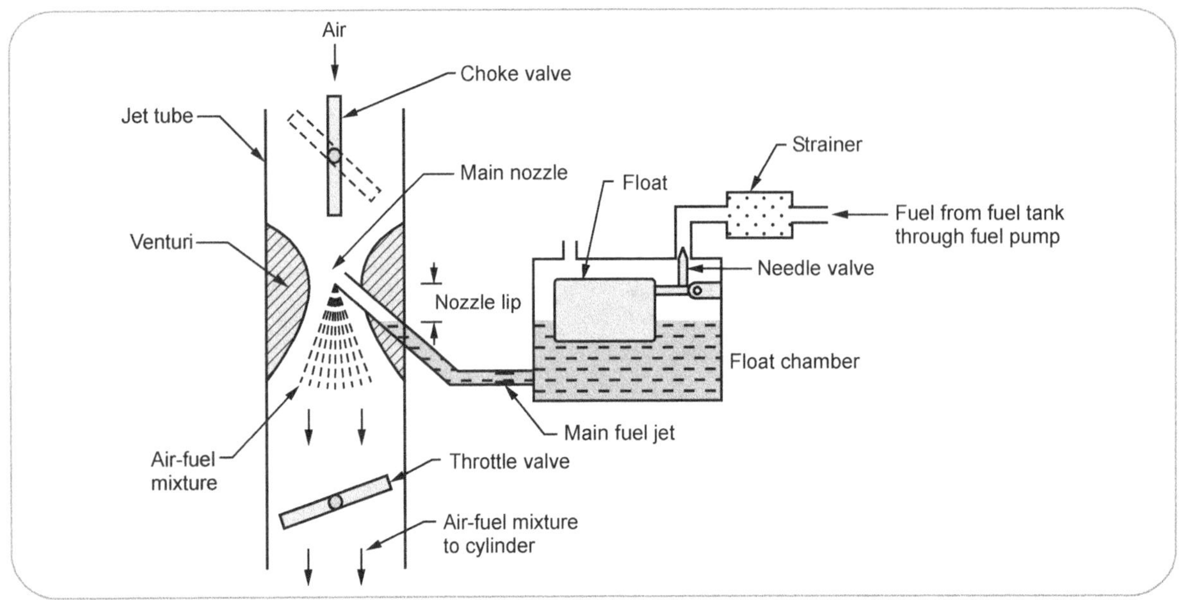

Fig. 6.11.2 : Simple carburetor

- When the piston moves from TDC and BDC it creates vacuum in the engine cylinder. Due to suction, the air flows from the intake manifold and carburetor to the cylinder.

- As it passes through the venturi, the velocity of air increases and pressure falls below atmospheric pressure.

- Due to decreased pressured at venturi, a pressure differential is created between nozzle and float chamber.

- Due to this pressure difference the fuel is sprayed into the air stream which mixes with air. This air fuel mixture is supplied to the engine cylinder.

- The quantity of air-fuel (A/F) mixture delivered is controlled by the throttle valve.

- The simple carburetor is used in small capacity stationary engines which runs at constant speed.

6.12 : Fuel Injection System (Fuel Supply System for Diesel Engines)

- The fuel injection system is one of the most important part of the diesel engines.

- It supplies, meters, injects and atomizes the fuel. It is a direct method of introducing fuel into the engine without the use of carburetor for mixing air and fuel.

- It is considered to be the heart of the diesel engine as it plays a very important role.

- The ignition of fuel is dependent on the atomisation of the fuel and pressure of fuel injection. The performance of engine is dependent on the performance of fuel injection system.

- The pressure of fuel injection in commercial vehicles has reached to around 1800 bar. The use of high pressure fuel injection system depends on the type of vehicle, manufacturer of vehicle, etc.

- The fuel injection system consists of a fuel tank, fuel filter, fuel pump, fuel supply line and fuel injectors. The simple layout of fuel injection system is given in Fig. 6.12.1. (Refer Fig. 6.12.1 on next page)

- The fuel from the fuel tank is drawn into the fuel supply pump through the coarse filter. In this, the pressure of fuel is increased slightly.

- Then the fuel is supplied to the fine filter in which all dust and dirt particles are removed. Then the fuel is supplied to the high pressure fuel injection pump where the pressure of fuel is raised to around 250 bar and then it is injected into the cylinder by injector.

- The return line collects the extra fuel and returns it to the fuel tank.

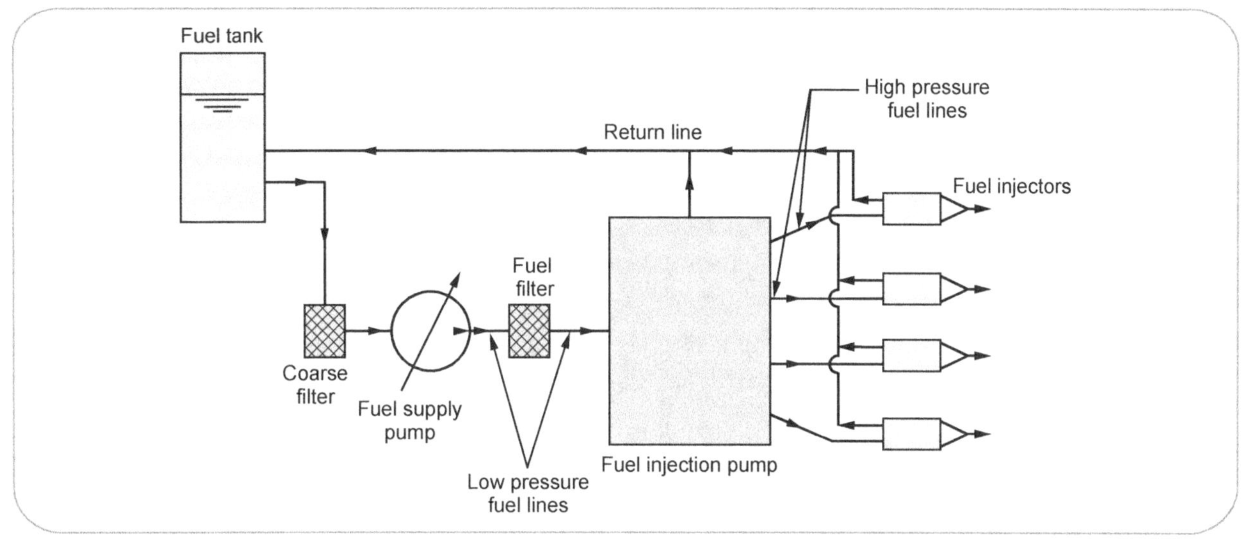

Fig. 6.12.1 : Layout of fuel injection system

6.12.1 Requirements of Fuel Injection System

The main requirements of fuel injection system for good performance of engine are as follows :

- To meter the exact quantity of fuel per cycle as per the load variations of the engine.

- The timing of fuel injection should be controlled such that maximum power output will be obtained with least amount of fuel consumption and ensuring the clean burning.

- It should atomise the fuel in fine particles.

- It should ensure proper spray pattern of fuel such that it results into rapid mixing of fuel with air.

- It should ensure uniform distribution of fuel throughout the combustion chamber.

- It should ensure uniform distribution of metered fuel into each cylinder of a multi-cylinder engine.

- The injection of fuel should start and terminate (end) instantaneously.

6.12.2 Air Injection System

- In case of air injection system, the fuel is injected by means of high pressure air [about 70 bar] into the combustion chamber.

- This system needs a compressor to supply compressed air and the fuel pump to draw the desired fuel from the fuel tank.

- Since it requires a multistage air compressor, resulting in increase of weight and reduction in

power, therefore this system is not used very much universally.

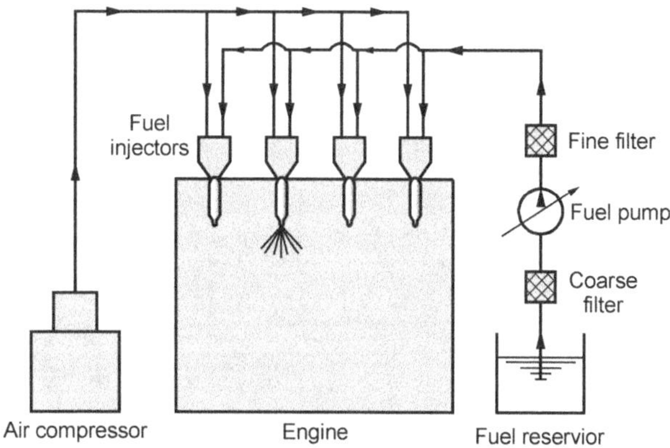

Fig. 6.12.2 : Air injection system

Advantages :

- It provides better atomisation of fuel.

- Heavy viscous fuels which have low cost can be used.

- It gives higher brake mean effective pressure.

Disadvantages :

- System is bulky and expensive.

- Increased weight of engine because of the use of compressor.

- Air compressor needs extra maintenance.

Note : Electric and hybrid vehicle systems are already discussed in previous chapter.

6.13 : Power Transmission System

- The function of power train system is to transfer power from engine to the wheels of the vehicle.

- It consist of a clutch, a gearbox, a propellor shaft, and a differential.

(i) Clutch

- It is usually a friction device.

- Its purpose is to enable the driver to gradually engage and disengage the drive from engine to road wheels when required.

(ii) Gearbox

- It consists of various types and sizes of gears.

- Gearbox provides the necessary leverage variation between the torque from engine to the road wheels.

(iii) Propeller shaft

- It consists of shaft having universal joint at its ends.

- Its function is to transmit power from output shaft of gearbox to input shaft of the differential.

(iv) Differential

- It consists of bevel pinion and crown wheel and a set of internal gears.

- Its purpose is to permit the two wheels on the axle to run at different speeds while cornerning a turn.

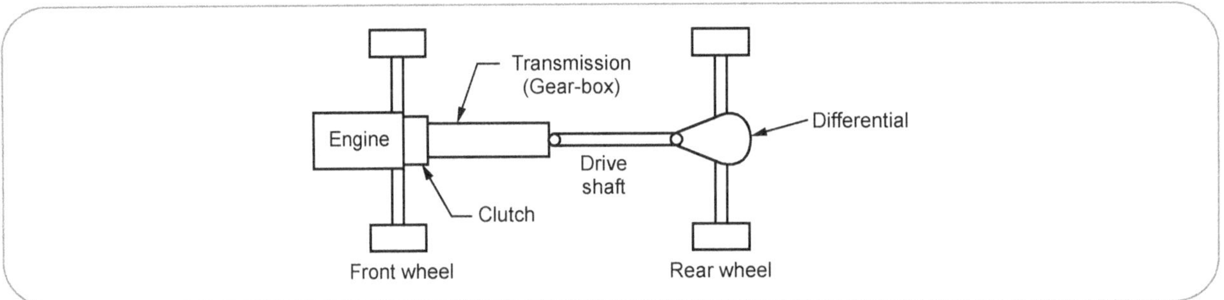

Fig. 6.13.1 Block diagram of power transmission system

6.14 : Clutch

- *Clutch* is a mechanism to transmit rotary motion from one shaft (driving shaft) to another coincident shaft, (driven shaft), as and when required, without stopping the driving shaft.

Purpose of Clutches   SPPU : Dec.-10, May-17

- During initial pickup of vehicle from stationary to moving, the engine is running but the vehicle is stationary.

- To over come the vehicle's self inertia, increasing amount of power is required to be transmitted to the gearbox and hence to the wheels.

- This increasing power transmission should be gradual and should not result in any jerks to the passengers.

- Hence, for gradual engagement of rotary motion from engine to gearbox shaft, a special friction coupling called "clutch" is used.

- Clutches are also required to disengage the drive from engine to gearbox for changing the gears.

- During slowing of vehicle or stopping, the clutch is used to disengage engine from drive wheels and enable smooth stopping of vehicle.

- Since clutch is of friction material, it also takes care of speed and torque variation for engine crankshaft to gearbox input shaft.

Functions of the Clutch   **SPPU : May-08, Dec.-08**

The clutch has following four major functions :

(i) When clutch is engaged (clutch pedal position-up), the clutch transmits maximum power from engine crankshaft to gearbox input shaft.

(ii) When clutch is disengaged (clutch pedal position-down), the clutch allows driver to shift the transission in various gear positions (first, second, third, etc.)

(iii) When clutch is disengaged (clutch pedal position-down), the engine can be cranked freely without transmiting the drive to wheels.

(iv) When clutch is engaging (clutch pedal position-moving up), the clutch accomodates for minor slippages and hence provides smooth drive transmission without jerks.

6.14.1 Single Plate Clutch   **SPPU : May-08,17**

- Single plate clutch is the most commonly used type of clutch on automobiles.

- It provides quicker disengagement.

- It consists of clutch disc, pressure plate and a cover assembly which are bolted to the engine flywheel.

- Refer Fig. 6.14.1 for a simplified assembly of a single plate clutch.

Construction :

The construction features of major components of single plate clutch is as explained below :

(i)	Flywheel
(ii)	Clutch plate (Friction plate)
(iii)	Pressure plate
(iv)	Thrust spring
(v)	Release lever
(vi)	Clutch cover
(vii)	Clutch shaft
(viii)	Thrust bearing

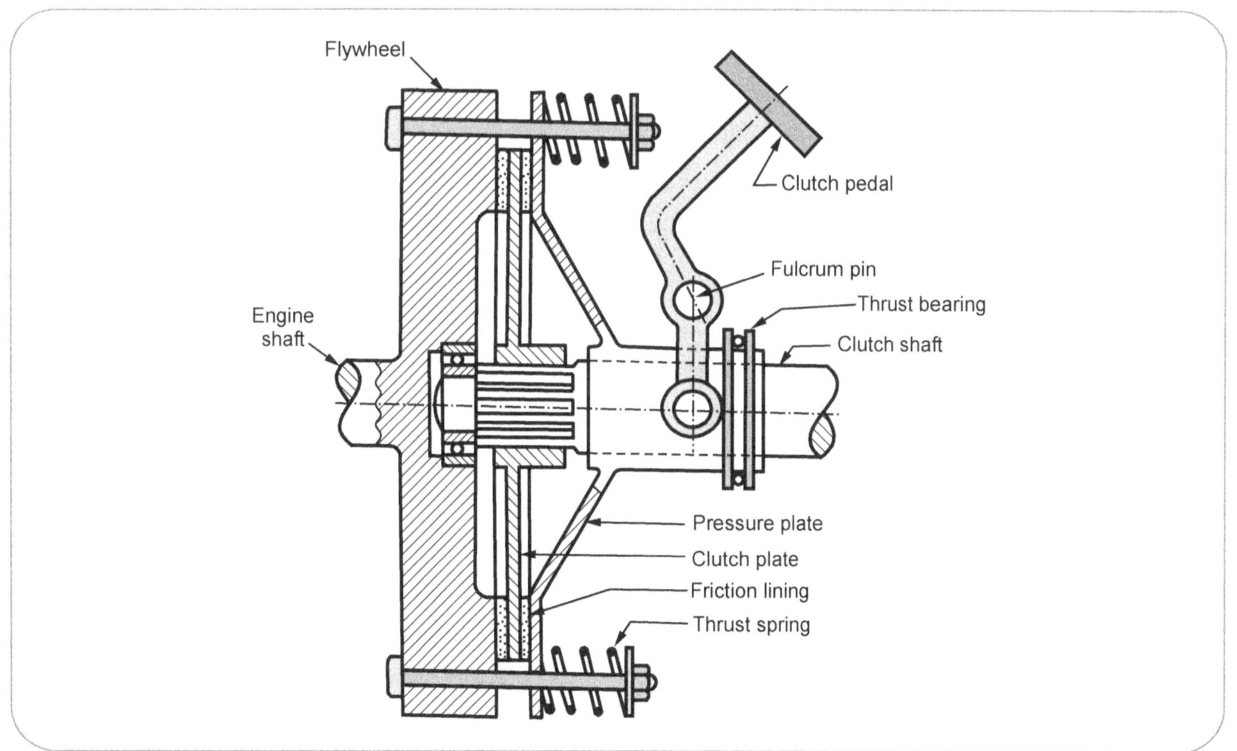

Fig. 6.14.1 Single plate clutch

(i) Flywheel

- Flywheel is a heavy mass disc bolted on the engine crankshaft.

- It has a flat surface that engages with the clutch / friction plate.

(ii) Clutch plate (Friction plate)

- It is a mechanical disc with friction linings on both its sides.

- It consists of a central hub machined internally so as to slide on a splined shaft.

- To take care of the torsional vibrations and damping actions, coil springs are mounted radially on the clutch plate.

(iii) Pressure plate

- It is a heavy mass plate with a flat surface on one of its side that engages with the clutch plate.

(iv) Thrust springs

- Thrust springs are axially mounted in between the clutch plate and pressure plate.

- These springs keep the clutch in engaged position when the pedal position is up.

(v) Release lever

- They are pivoted on fulcrum pin and are levers that are used to disengage the clutch plate.

(vi) Clutch cover

- It is a sheet metal or forged cover which surrounds the overall clutch asembly and protects it from dirt and dust.

(vii)Clutch shaft

- It is the shaft that is connected to the gearbox input shaft.

- It is splined near the central hub on which the clutch plate can slide.

(viii)Thrust bearing

- The thrust bearing takes care of the axial load that comes during sudden release of thrust springs.

Working :

- Fig. 6.14.1 shows the clutch assembly when in engaged position (pedal position-up).

- When the pedal position is up, the axial force of thrust springs ensures the pressure plate is pressed against the flywheel with clutch plate being sandwiched between the two of them.

- The drive is hence transmitted from the flywheel to clutch plate through friction and from clutch plate to clutch shaft through mechanical splines.

- When the clutch pedal is pressed down, the release lever pulls the pressure plate against the thrust springs.

- This loosens the contact between clutch plate and flywheel and hence the drive gets disengaged to the gearbox.

Advantages :

- Simple design of construction and working.

- Better heat dissipation from single plate.

- Gear changing with single plate clutch is easier.

- It has better torsional vibration absorbing capacity.

Disadvantages :

- For higher power transmission, the surface area of clutch plate increases and thereby increasing the overall size of clutch.

- Clutch pedal force required is higher.

Applications :

- It is used on vehicles wherein space limitations are not severe such as trucks, buses, etc.

6.14.2 Multi - Plate Clutch `SPPU : May-09,10,11,12`

- Multi-plate clutch is an extension of single plate clutch and works on the same principle as that of single plate clutch.

- As the name indicates, it has multiple number of frictional and metal plates.

- This provides increased friction surface area and hence higher torque transmitting capacity of the clutch.

- It is used on applications wherein space available is constrained.

- Refer Fig. 6.14.2 for a simplified assembly of a multi-plate clutch.

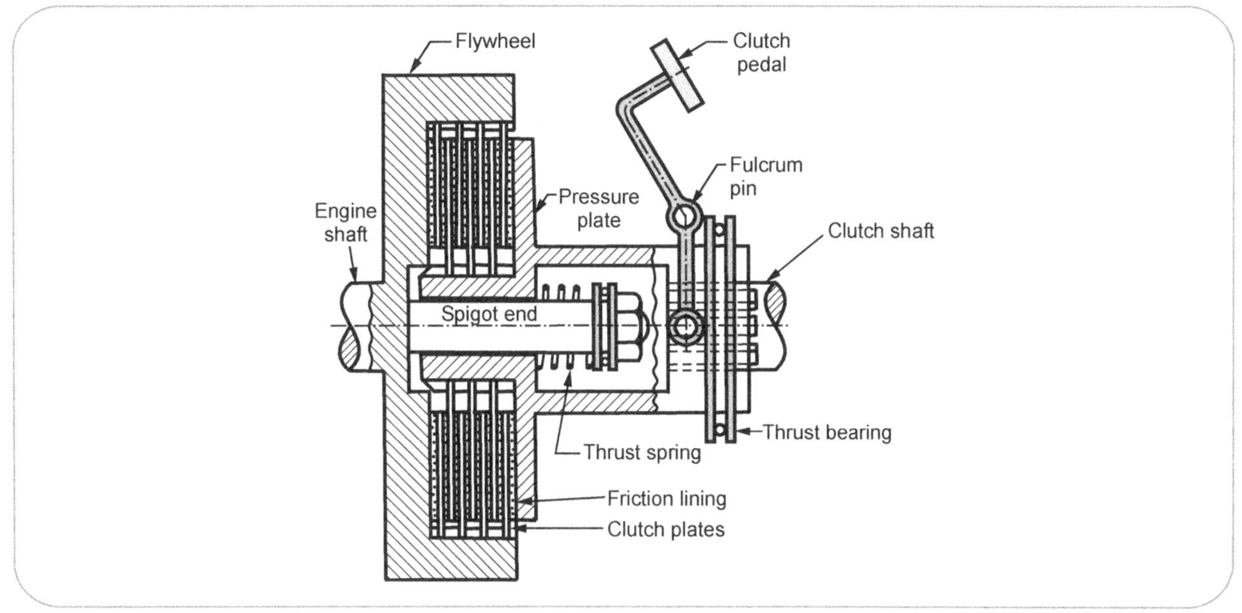

Fig. 6.14.2 Multi-plate clutch

Construction :

> (i) Flywheel
> (ii) Clutch plate (Friction plate)
> (iii) Pressure plate
> (iv) Thrust spring
> (v) Release lever
> (vii) Clutch shaft
> (viii) Thrust bearing

(i) Flywheel

- It consists of a cylindrical flywheel bolted on the engine crankshaft.

- The friction surface is provided on the vertical side of the flywheel.

- The cylindrical portion of flywheel contains internal teeth for engaging with outer clutch plates.

(ii) Clutch plate

- The clutch plates used are in two sets, i.e. one with outer teeth and other with inner teeth.

- Refer Fig. 6.14.3 for clutch plate set.

- The outer teeth type plate engages with the inner teeth of flywheel.

- The inner teeth type plate engages with the splined clutch shaft.

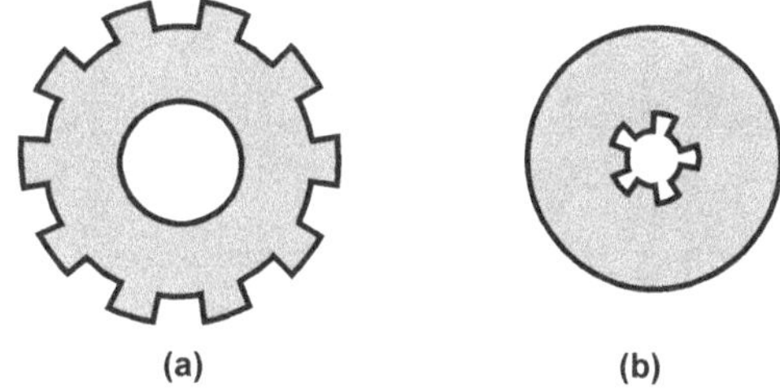

Fig. 6.14.3 Clutch plate set

(iii) Pressure plate

- It is a heavy mass plate with a flat surface on one of its side that engages with the clutch plate.

(iv) Thrust springs

- Thrust springs are axially mounted in between the clutch plate and pressure plate.

- These springs keep the clutch in engaged position when the pedal position is up.

(v) Release lever

- They are pivoted on fulcrum pin and are levers that are used to disengage the clutch plate.

(vi) Clutch shaft

- It is the shaft that is connected to the gearbox input shaft.

- It is splined near the central hub on which the clutch plate can slide.

(vii) Thrust bearing

- The thrust bearing takes care of the axial load that comes during sudden release of thrust springs.

Working :

- When the clutch is fully engaged (as shown in Fig. 6.14.2), i.e. clutch pedal position-up, the thrust springs ensure pushing the pressure plate over the clutch plate and flywheel.

- The drive is transmitted from flywheel to outer teeth type plate to inner teeth type clutch plate and hence to the clutch shaft.

- When the clutch pedal is pressed down, the release lever pulls the pressure plate away against the thrust springs.

- It thus loosens the frictional contact between the plates and hence disengages the drive to clutch shaft.

- If 'n' is the number of friction plates then the number of frictional contact surfaces are (n – 1).

Advantages :

- The overall size of the clutch is smaller.
- It has higher torque transmitting capacity.
- Drive transmission is smoother.
- Wear and tear of clutch plate is lower.

Disadvantages :

- The design of clutch plate set is complicated.
- It is difficult to service.
- The cost of multi-plate clutch is higher.

Applications :

- It is used where high torque transmission is required such as racing cars, etc.

- It is also used where overall space is constrained such as scooters, motorbikes, etc.

6.15 : Gearbox

6.15.1 Introduction to Transmission System

- Transmission system or gearbox is an important element of the power train system.

- *Transmission system* is referred to the whole mechanism responsible for transmitting power from engine to road wheels.

- Gearbox is that part of power train system that provides a mean to have suitable variation of the engine torque to the road wheels.

6.15.2 Functions of Gearbox

SPPU : May-08, Dec.-08

(a) The engine produces torque or tractive effort over a limited engine speed. But an automobile is driven under different conditions requiring large variation in torque at road wheels.

The main function of gearbox is to provide a mean to vary the torque ratio between the engine and the road wheels as and when required.

(b) Gearbox also provides a mean to move the automobile backward with a reverse gear.

(c) It also enables a neutral position for starting the engine and keep it running without transmitting drive to the road wheels.

6.16 Gear Drive

SPPU : May-06,08,11,12, Dec.-06,08

- In precision machines or mechanisms like watch gear mechanism, definite speed ratio is required which is obtained by **gear drive** or **gear wheels.**

- Gear drive is a positive drive and it is provided when the distance between the driver and follower is very small.

- **Gears** are defined as toothed wheels which can transmit power and motion from one shaft to another shaft by means of successive engagement of teeth.

- It is important to note that, both the gears which are engaged, always rotate in opposite direction.

- The gear drive consists of two wheels. The smaller wheel is called as **pinion** and larger wheel is called as **gear**. Refer Fig. 6.16.1.

- Both pinion and gear are mounted on separate shafts.

6.16.1 Speed Ratio or Gear Ratio   **SPPU : May-14**

- Speed ratio or gear ratio is defined as the ratio of pinion speed to the gear speed.

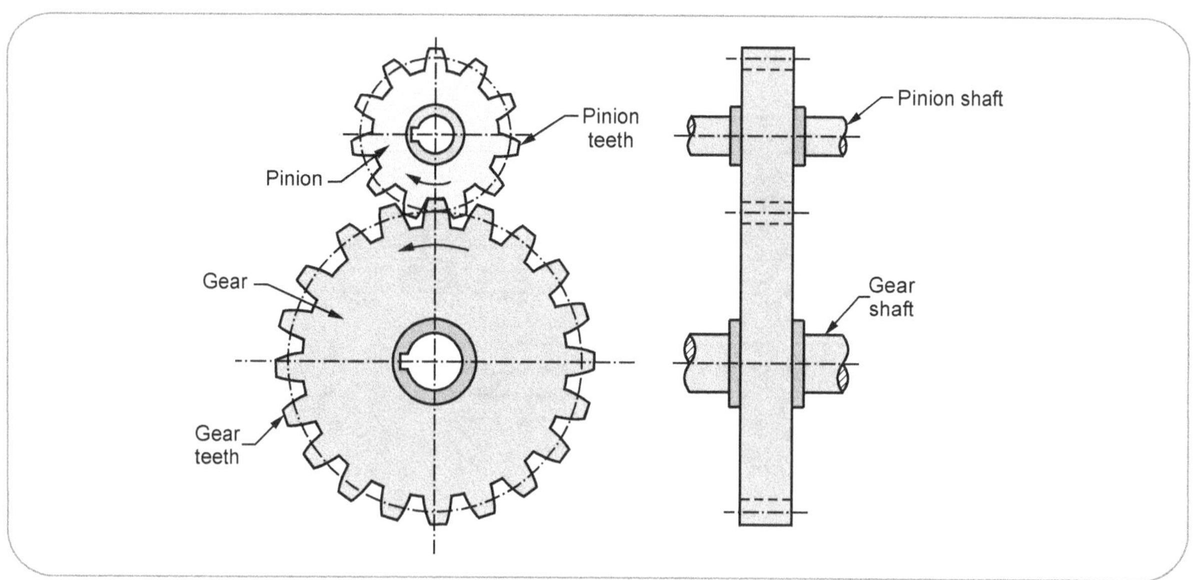

Fig. 6.16.1 Gear drive

- It is also defined as the ratio of number of teeth on gear to the number of teeth on pinion or it is the ratio of diameter of gear to the diameter of pinion.

Let, n_G and n_P = Gear and pinion speeds in r.p.m.

d_G and d_P = Diameter of gear and pinion in mm

Z_G and Z_P = Number of teeths on gear and pinion

$$\text{Gear ratio or speed ratio (G)} = \frac{n_P}{n_G} = \frac{d_G}{d_P} = \frac{Z_G}{Z_P}$$

6.17 : Types of Gears

- The gears or toothed wheel are generally classified as per the position of axes of the two shafts between which the motion is to be transmitted. They are grouped as follows :

1. Parallel shaft axes gears
2. Intersecting shaft axes gears
3. Non-intersecting and perpendicular shaft axes gears
4. Non-intersecting and non-perpendicular shaft axes gears

Refer Fig. 6.17.1 on next page.

6.17.1 Parallel Shaft Axes Gears

SPPU : May-14,16,17

For the transmitting the motion, two parallel shafts are connected by following types of gears :

a) Spur gears   b) Helical gears   c) Herringbone gears
d) Rack and pinion   e) Internal gears

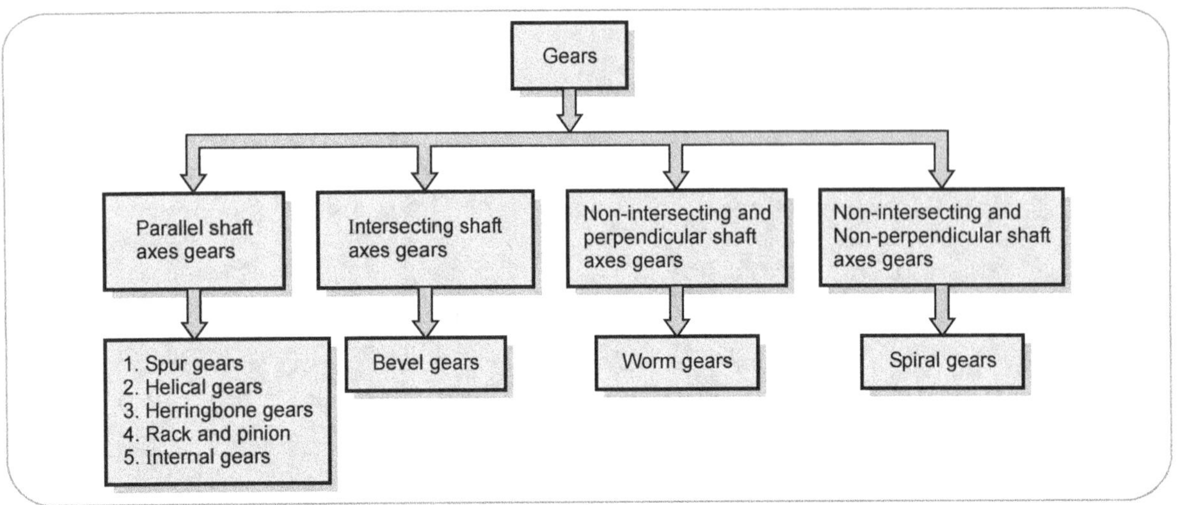

Fig. 6.17.1 Types of gears

a) Spur gears

- Spur gears are used to transmit motion between two parallel axes shafts. Refer Fig. 6.17.2.

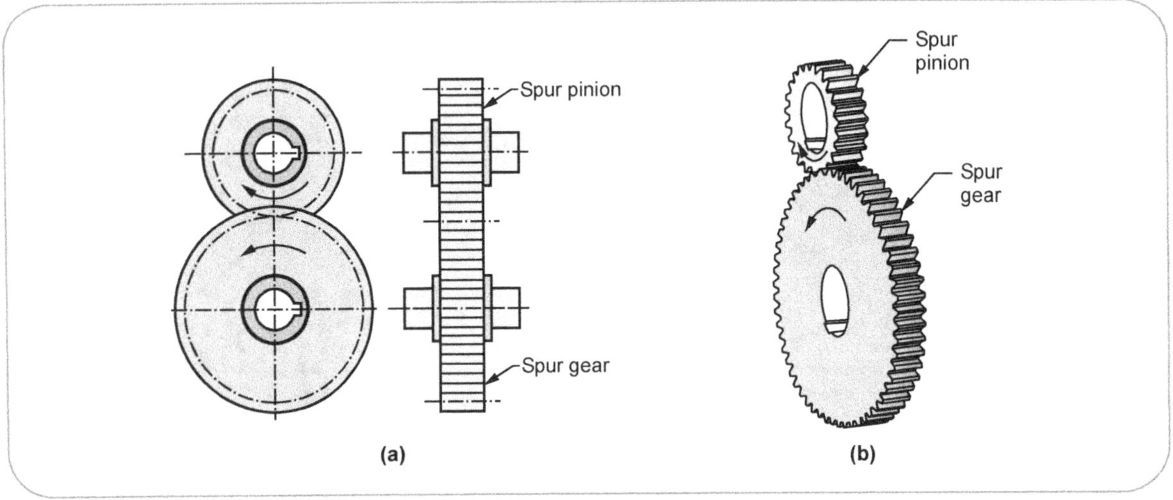

Fig. 6.17.2 Spur gears

- They are the simplest of all the gears and easiest in production.
- The teeth of spur gears are cut along the periphery and parallel to the axis of gear.
- Spur gears can be made of steel, brass, other metals and plastics.

Advantages

- Spur gears are easy to manufacture.
- They are made in variety of sizes from less than 25 mm to several cm in diameter.
- These gears are less expensive.
- Efficiency of spur gear is high (upto 98 %).

Disadvantages/Limitations

- Spur gears are not used for high speed applications.
- The operation of spur gear is noisy.

Applications

- Spur gears are commonly used in machine tool gear box, automobile gear box, watches, etc.

b) Helical gears :

SPPU : May-04, Dec.-05

- Helical gears are similar to spur gears but its teeth are cut at an angle with the axis of rotation of the gear. Refer Fig. 6.17.3.

- Helical gears can transmit motion from one shaft another shaft which are parallel to each other.

- The helix angles of gear and pinion are same in magnitude but of different hands. For example, right hand pinion meshes with left hand gear.

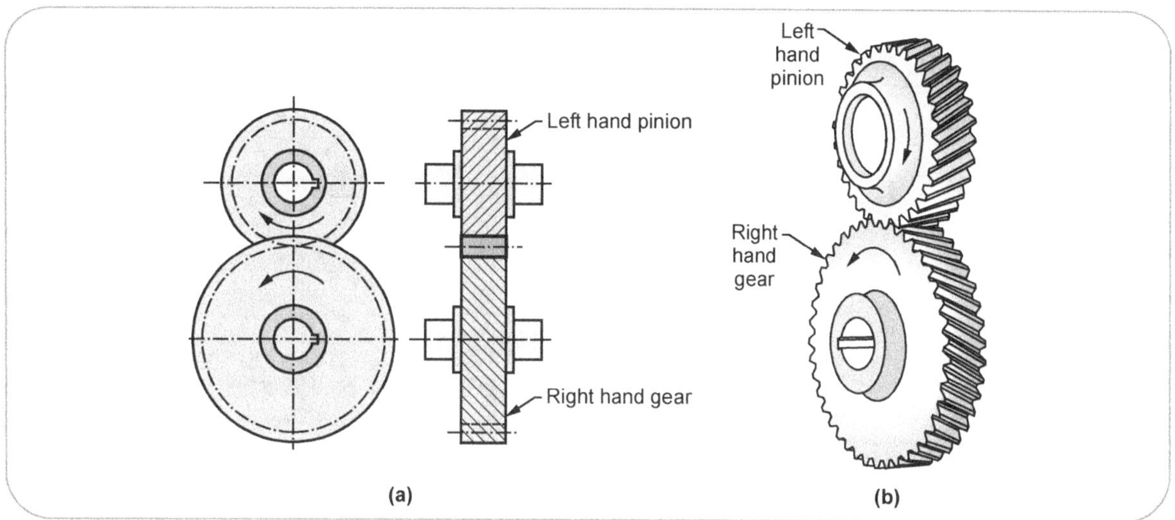

Fig. 6.17.3 Helical gears

Advantages

- Helical gears operate more quitely and smoothly than the spur gears.
- Teeth of helical gears are stronger than the spur gear teeth.
- Efficiency of helical gears is high (upto 99 %).
- These gears are suitable for high speed applications.

Disadvantages / Limitations

- Helical gears are difficult to manufacture as compared to spur gears.
- Helical gears exert an end thrust which is absorbed by thrust bearings.
- In helical gears, the friction, heat generation and wear is high.
- These gears are more expensive than the spur gears.

Applications

- Helical gears are commonly used in high speed automobile gear box, machine tool gear box, rolling mills, steam and gas turbines and other high speed applications.

c) Herringbone gears

- Herringbone gears looks like two single helical gears, one right hand and other left hand, placed side by side. Refer Fig. 6.17.4. (See Fig. 6.17.4 on next page)
- These gears reduces the end thrust acting on the thrust bearings.

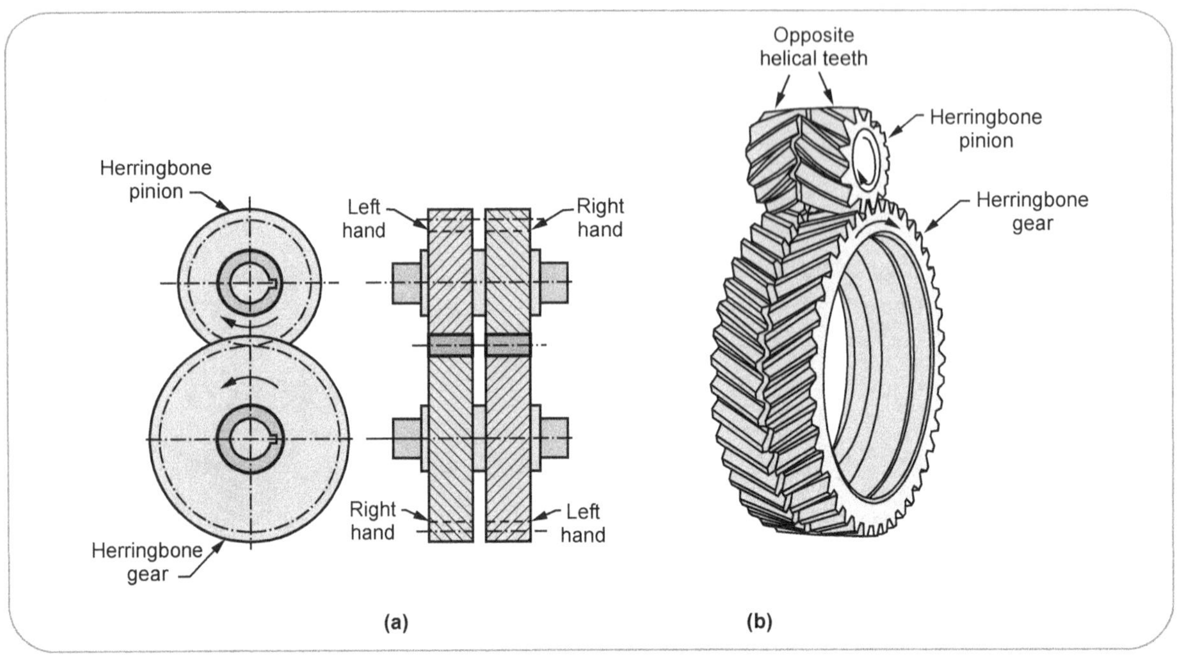

Fig. 6.17.4 Herringbone gears

- These gears are used to transmit motion from one shaft to another shaft which are parallel to each other.
- Herringbone gears are also called as **double helical gears.**

Advantages

- Herringbone gears operate more quitely and smoothly.
- Load carrying capacity of these gears is high.
- These gears reduces the end thrust on the bearings.
- Efficiency of these gears is high (between 98 % and 99 %).

Disadvantages / Limitations

- Herringbone gears are difficult to manufacture.
- These gears are more expensive than the spur and helical gears.
- Machining of teeth of these gears is very difficult.

Applications

- Herringbone gears are used in high speed applications and especially in hearvy duty machinery.

d) Rack and pinion

SPPU : May-04

- Rack is a straight gear which has no curvature and it represents a gear of infinite radius. Refer Fig. 6.17.5.
- Rack gears may have either straight or helical teeth.

- During the operation, a rack gear meshes with a pinion and converts rotary motion of pinion into reciprocating motion of rack or reverse.

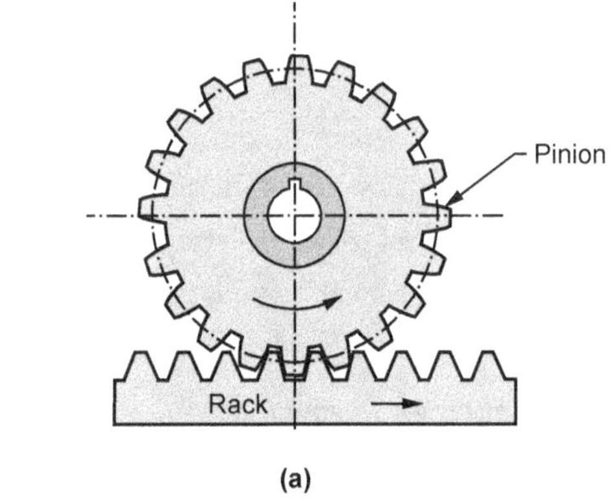

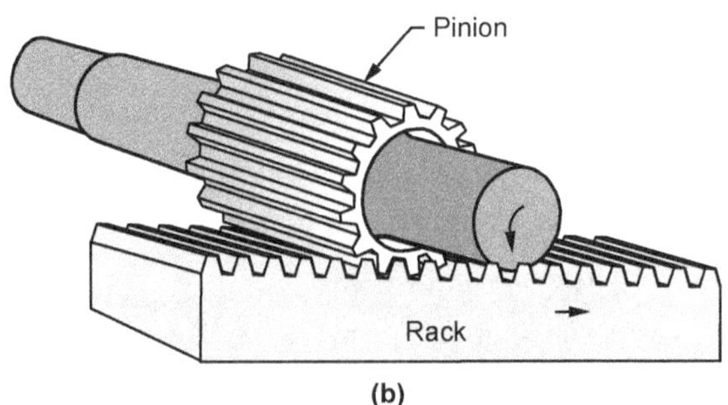

Fig. 6.17.5 Rack and pinion

Applications

- Rack and pinion arrangement is used in feeding mechanisms, reciprocating drives, etc.

- They are also used in gauges and small equipments and lathe, milling, drilling and other machine tools.

e) Internal gears

- All the above gears carry teeth on their peripheries or external surfaces.

- But, internal gears have teeth on their inner surface which may be spur, helical, etc. Refer Fig. 6.17.6.

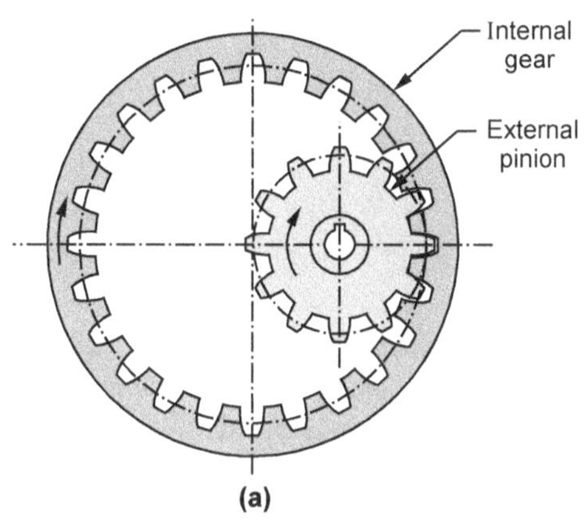

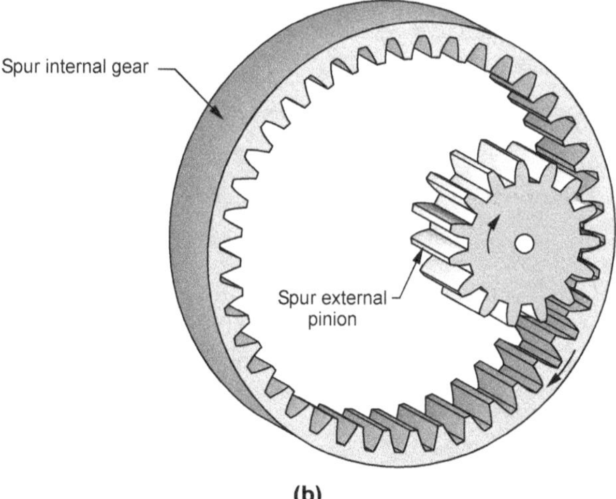

Fig. 6.17.6 Internal gears

- It is important to note that, in internal gears both the wheels rotate in the same direction.

Advantages

- As the centres of the gears are closer together, less space is required.

- During the motion, there is less friction between the teeth.

- Efficiency of these gears is high.

Applications

- Internal gears are commonly used in internal clutches and heavy vehicle drives.

6.17.2 Intersecting Shaft Axes Gears

SPPU : Dec.-05, 10, May-07, 09

- For transmitting the motion between two intersecting shafts or shafts at desired angle, **bevel gears** are commonly used.

- The surface of the bevel gear is like a frustrum of a cone and its size decreases towards the apex of cone. Refer Fig. 6.17.7. (See Fig. 6.17.7 on next page)

- The teeth of bevel gears are cut so that they radiate from the apex of a cone and lie on the conical surface.

- Generally bevel gears are used when two shafts are at right angle to each other; but it can be used for any desired shaft angle.

- Bevel gears are available with straight, helical and spiral teeth.

Advantages

- Bevel gears can transmit the motion between two intersecting shafts.

- Bevel gears are available in various sizes.

- While running at high speeds, there is less noise.

- As compared to spur and helical gears, less space is required.

- Efficiency of these gears is high.

Disadvantages / Limitations

- Manufacturing of bevel gears is difficult.

- These gears are more expensive than the spur and helical gears.

Applications

- Bevel gears are commonly used in differential gear box of automobiles and for connecting two intersecting shafts.

6.17.3 Non-Intersecting and Perpendicular Shaft Axes Gears

SPPU : May-04, Dec.-05,09

- For transmitting the motion between two non-intersecting and perpendicular shafts, **worm and worm gears** are used.

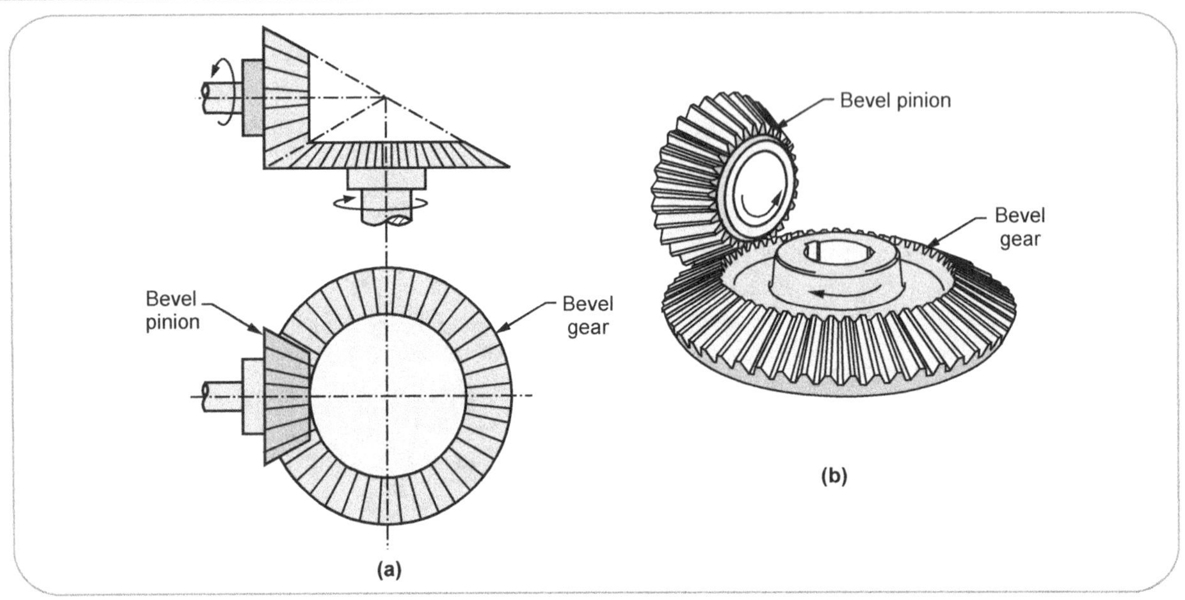

Fig. 6.17.7 Straight bevel gears

- In worm and worm gears, worm is more or less similar to screw having single or multiple start threads which forms the teeth of the worm. Refer Fig. 6.17.8.
- Worm gear is a helical gear with concave face to accommodate a portion of worm periphery.
- This worm drives a worm gear or worm wheel while transmitting the motion.
- In this type of gear pair, worm and worm gear have same hand.
- Worm and worm gears are used when large speed reduction ratio is required from one shaft to other shaft.

- The important property of this pair is that, it is *non-reversible* it means the worm is always driver and worm gear is driven or worm gear cannot drive the worm.

Advantages
- Worm gears are compact than the spur, helical and bevel gears.
- They can be used for high reduction ratio.
- These gears are self-locking or non-reversible.
- The operation of worm gears is silent and smooth.

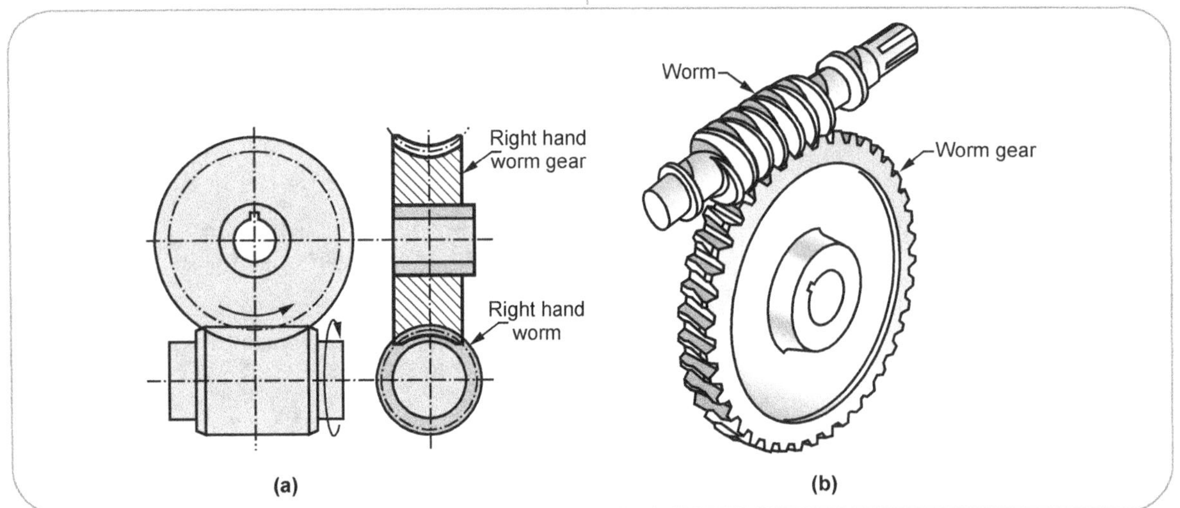

Fig. 6.17.8 Worm gears

Disadvantages / Limitations

- The efficiency of worm gears is less.

- During the operation, more heat is generated hence effective cooling is required.

- Worm gears are more expensive than the all gears.

- Worm gears are difficult to manufacture.

Applications

- Because of self-locking or non-reversible property, worm gears are commonly used in steering mechanisms, hoisting devices, cranes, dam gates, lifts, etc.

- They are also used in musical instruments like guitar, for locking the strings.

6.18 : Solved Examples on Gear Drive

Ex. 6.18.1 : *A pinion with 120 mm pitch circle diameter meshes with a gear of 400 mm pitch circle diameter. The number of teeths on pinion is 18 and it rotates at 1440 r.p.m. Determine : i) Gear ratio ii) Number of teeth on gear iii) Speed of the gear.*

Sol. : Given data : d_P = 120 mm, d_G = 400 mm, z_P = 18, n_P = 1440 r.p.m.

To find : (i) G   (ii) z_G (iii) n_G

Step - 1 : Calculate the gear ratio.

$$G = \frac{d_G}{d_P} = \frac{400}{120} = 3.3333 \qquad \text{... Ans.}$$

Step - 2 : Calculate the number of teeth of gear

We know that, $G = \dfrac{z_G}{z_P}$ $\therefore 3.3333 = \dfrac{z_G}{18}$

$\therefore \qquad z_G = 60 \qquad\qquad$ ... Ans.

Step - 3 : Calculate the speed of the gear

$$G = \frac{n_P}{n_G} \qquad \therefore 3.3333 = \frac{1440}{n_G}$$

$\therefore \qquad n_G = 432$ r.p.m. $\qquad$ ... Ans.

Ex. 6.18.2 : *A pinion of pitch circle diameter 150 mm meshes with a gear having 80 teeth. Gear ratio is 4 and speed of gear is 500 r.p.m. Determine :*
(i) Diameter of gear (ii) Speed of pinion (iii) Number of teeth of pinion

Sol. : Given data : d_P = 150 mm, G = 4, z_G = 80, n_G = 500 r.p.m.

Step - 1 : Calculate the diameter of gear

We know that, $\quad G = \dfrac{d_G}{d_P}$ $\therefore 4 = \dfrac{d_G}{150}$

$\therefore \qquad\qquad d_G = 600$ mm $\qquad$ ... Ans.

Step - 2 : Calculate the speed of pinion

We know that, $\quad G = \dfrac{n_P}{n_G}$ $\therefore 4 = \dfrac{n_P}{500}$

$\therefore \qquad\qquad n_P = 2000$ r.p.m. $\qquad$ ... Ans.

Step - 3 : Calculate the number of teeth on pinion

We know that, $\quad G = \dfrac{z_G}{z_P}$ $\therefore 4 = \dfrac{80}{z_P}$

$\therefore \qquad\qquad z_P = 20 \qquad\qquad$ ... Ans.

Ex. 6.18.3 : *The pitch circle diameter of pinion is 200 mm. Gear ratio is 3. The pinion is attached to motor having 1440 r.p.m. and 20 kW power. The number of teeth on pinion is 20. Determine :*
(i) Torque required to transmit the power at pinion
(ii) Number of teeth of gear

Sol. : Given data : d_P = 200 mm, G = 3, n_P = 1440 r.p.m., P = 20 kW = 20×10^3 W z_P = 20.

To find : (i) T_P (ii) z_G

Step - 1 : Calculate the torque required to transmit the power

We know that, $\qquad P = \dfrac{2\,\pi\,n_P\,T_P}{60}$

$\therefore \qquad\qquad 20 \times 10^3 = \dfrac{2\,\pi \times 1440 \times T_P}{60}$

$\therefore \qquad\qquad T_P = 132.6291$ N.m $\qquad$ ... Ans.

Step - 2 Calculate the number of teeth on gear

The gear ratio is, $\quad G = \dfrac{z_G}{z_P} \qquad \therefore\ 3 = \dfrac{z_G}{20}$

$$\therefore \qquad z_P = 60 \qquad \text{... Ans.}$$

6.19 : Propeller Shaft

- The *propeller shaft* is the driving shaft connected between the gearbox output shaft and the differential input shaft.
- The main functions of a propeller shaft are :
  - (i) To transmit rotary motion from gearbox to the differential
  - (ii) To adjust for the variation in the angle between the gearbox output shaft and differential
- Fig. 6.19.1 (a) shows a propeller shaft.
- A propeller shaft assembly consists of three major components as mentioned below :

(i) Shaft
- It is a solid or hollow tubular cross-section.
- It is subjected to varying torsional loads in between the gearbox output shaft and differential input shaft.
- Vehicle with large wheel base have long propeller shaft.
- A long shaft has a tendency to sag and whirl.
- Whirl can be imagined as the movement of rope revolving when held at its ends.
- At a certain speed, the whirling intensity becomes critical and the shaft vibrates violently and can break.
- Hence to reduce whirl, the diameter of shaft is increased or length of shaft is reduced by splitting it.

(ii) Universal joint
- Depending on the type of the rear axle arrangement, one or two universal joints can be used.
- The universal joints account for the variation in the angle between the gearbox output shaft and differential input shaft.

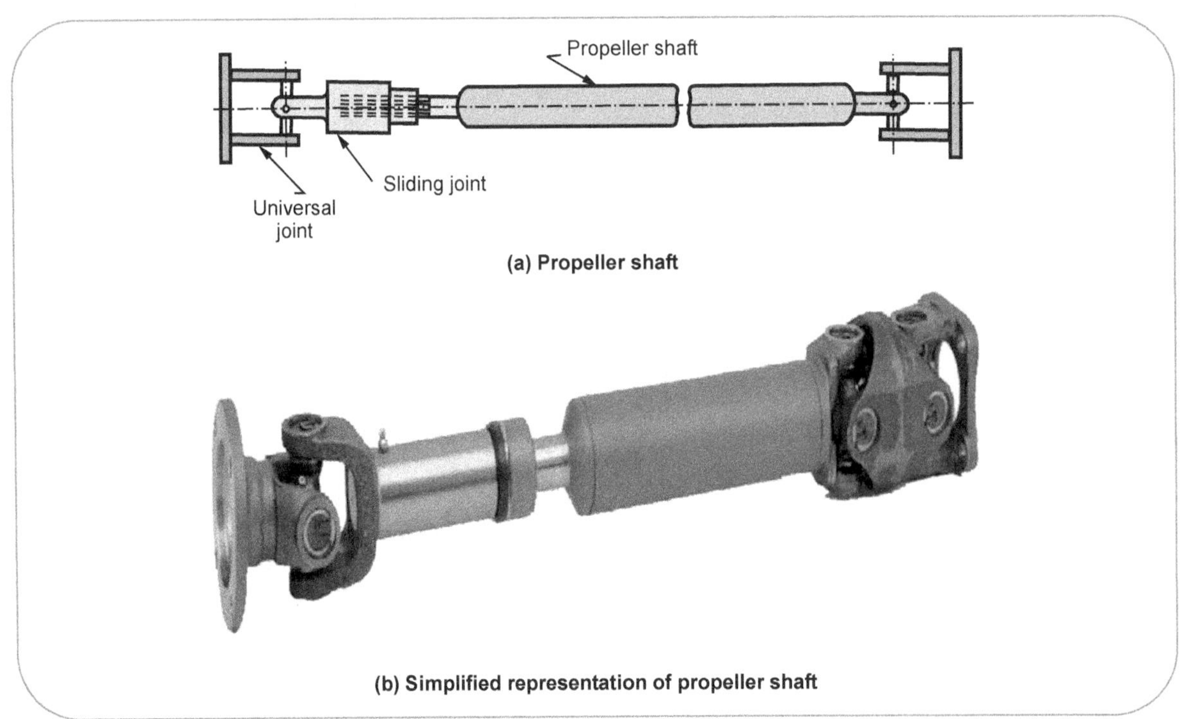

Fig. 6.19.1 Simplified representation of propeller shaft

- It thereby accommodates for the up and down movement of the drive axle w.r.t. the gearbox output shaft when the vehicle is running.

(iii) Slip joint

- Slip joint is an arrangement of splined shaft rotating within a splined hub.
- It accounts for the variation in the length of propeller shaft when the drive axle moves up or down during running of the vehicle.

6.20 : Universal Joint

- A universal joint is a special type of joint used to transmit power between two shafts that are inclined to each other.
- It also accounts for the continuous variation in the angle between the two connecting shafts.
- In an automobile, the gearbox is rigidly mounted on the frame while the drive axle (usually rear axle) continuously varies with respect to gearbox output shaft due to the action of suspension system.
- Therefore, a universal joint is used to transmit power from the gearbox output shaft to the differential input shaft.
- There are different types of universal joints, the simplest being the yoke type as shown in Fig. 6.20.1.
- The yoke type is the most simple, compact and widely used universal joint.

Construction :

- A yoke arrangement is provided at the end of the two shafts to be connected.
- A 4-arm cross spider connects the two yokes.
- The opposite arms of the cross are supported in the bushes provided in the yoke.
- As seen in the Fig. 6.20.1, the two connected shafts (A and B) are inclined at an angle.

Working :

- Shaft with yoke 'A' can have angular rotation about axis X-X while shaft with yoke 'B' can have angular rotation about axis Y-Y.
- The rotation of yoke 'A' is transmitted to yoke 'B' through the cross.
- As the connection between the two connecting shafts is mechanical, a positive drive transmission takes place and still allowing some angular movement between the shafts.

Types of universal joints :

- Different types of universal joints are used in automobile applications.
- The major types are as follows :

> (i)   Hooke's joints
> (ii)   Flexible ring coupling
> (iii) Yoke type universal joint
> (iv) Hardy spicer needle bearing universal joint
> (v)   Layrub coupling
> (vi) Constant velocity universal joint

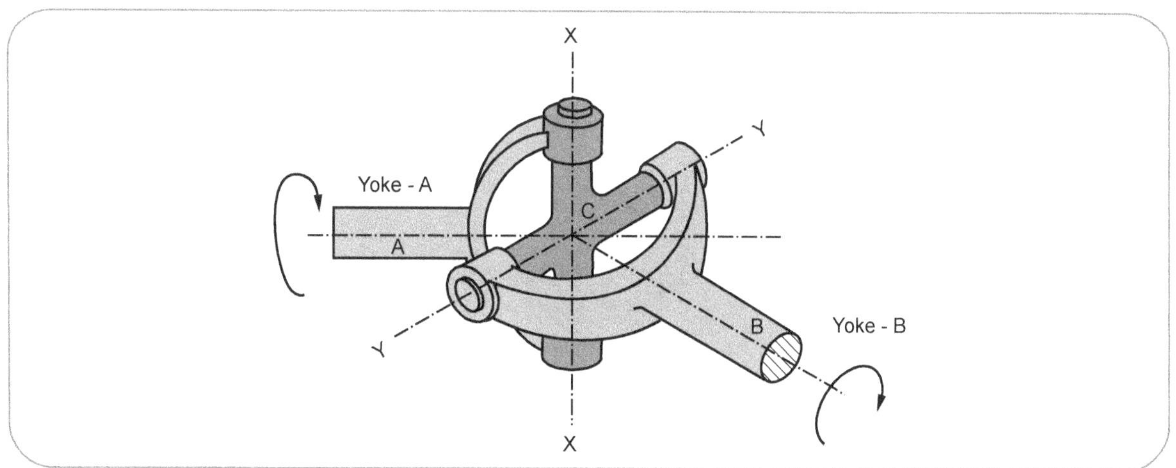

Fig. 6.20.1 Yoke type universal joint

6.20.1 Hooke's Joint   **SPPU : May-11**

- It is similar to a yoke type universal joint.

- The two connecting shafts are splined at the ends.

- An internal splined fork is fitted on each of the splined shaft.

- Fig. 6.20.2 shows a Hooke's joint.

- A cross spider having four pins is used to connect the two forks.

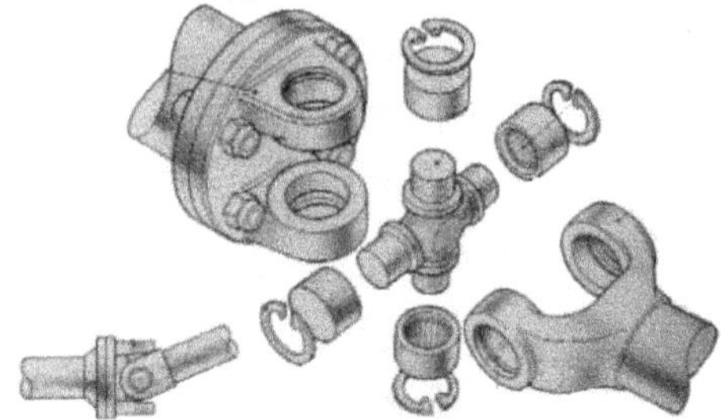

Fig. 6.20.2 Hooke's joint

- A phosphor bronze bearing is provided in between the pins of spider and holes of the fork.

- The cross spider ensures keeping the forks at right angle to each other.

6.20.2 Flexible Ring Universal Joint

- It consists of a flexible ring instead of the cross spider.

- Fig. 6.20.3 shows a flexible coupling.

- The flexible ring is made of number of layers of fabric impregnated with rubber.

- Six holes are provided on the surface of the flexible ring.

- A three arm spider is mounted on the splines of the connecting shafts.

- The spider is fitted on each side of the ring by means of bolts and nuts.

- This type of universal joint is used only for small angular deflection.

- It does not require any lubrication.

6.20.3 Yoke Type Universal Joint

- It consists of simple yokes with internal splines connected with the shaft.

- A 4-arm cross spider is used to connect the two yokes.

- Yoke type universal joint is as explained in Fig. 6.20.1.

- The cross spider is provided in a phosphor bronze bearing bushes at its ends.

6.20.4 Hardy Spicer Needle Bearing Universal Joint

- It is similar to a Hooke's joint, the difference being in the bearing contact between the cross spider and the yoke.

- It uses needle roller bearing to support the cross in the yokes.

- The needle roller bearing results in increased joint efficiency.

- Fig. 6.20.4 shows the exploded view of cross spider with needle roller bearing.

- The needle bearing ensures uniform loading on the arms of the spider.

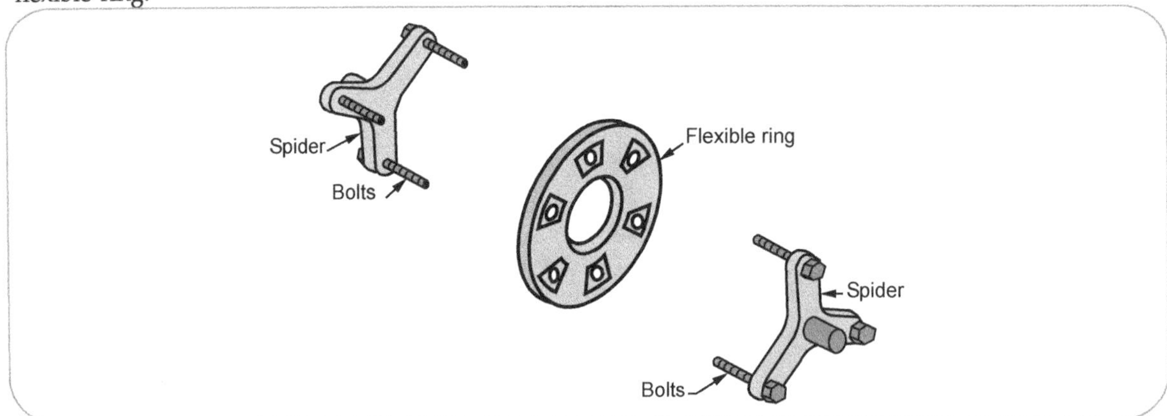

Fig. 6.20.3 Flexible ring universal joint

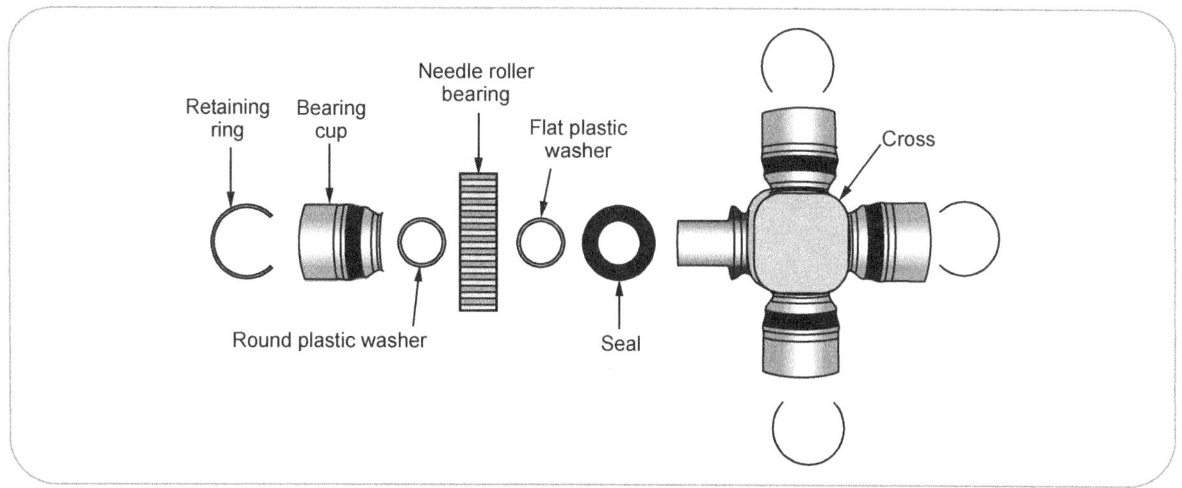

Fig. 6.20.4 Exploded view of hardy spider needle roller bearing universal joint

6.21 : Differential Gearbox

- Differential is a mechanical geared arrangement that distributes the power amongst each half axle based on the resistance offered to motion of the half axle.

Need for differential :

- When the vehicle is running on a straight road, both the wheels should rotate at the same speed.

- This condition can be fulfilled by a solid rear axle.

- But, when the vehicle is taking a turn, the outer wheel will have to travel a greater distance as compared to the inner wheel.

- If the vehicle is fitted with a solid rear axle, then both the wheels will rotate at the same speed, causing the inner wheel to skid thus, loosening the control on steering of vehicle.

- Hence to avoid skidding, some mechanism or unit must be installed which should reduce the speed of inner wheels and increase the speed of the outer wheels while taking turns.

- This requirement is fulfilled by the differential unit.

Functions :

The main functions of a differential unit are as follows :

(i) To compensate for the difference in distance that the outer wheel travels while the vehicle is taking a turn.

(ii) To ensure equal speeds of the drive wheels when the vehicle is moving on a straight road.

(iii) To transmit power to the drive wheels.

Construction :

- Refer Fig. 6.21.1 for the assembly of differential unit without its casing. (See Fig. 6.21.1 on next page)

- The bevel pinion and the crown wheel are part of the final drive unit attached to the differential.

- A cage is mounted on the crown wheel and houses the cross pin or spider.

- Two planet pinions are mounted on the cross pin or spider.

- The planet pinions are free to rotate about the cross pin.

- The two sun gears are in mesh with the planet pinions forming a mechanical chain.

- The sun gear is in fact, an extension of the half axle.

Working :

- The working of the differential can be discussed during the following two modes of operation :

  (i) When the vehicle is running on a straight road.

  (ii) When the vehicle is taking a turn.

- Refer Fig. 6.21.2 for working of a differential unit.

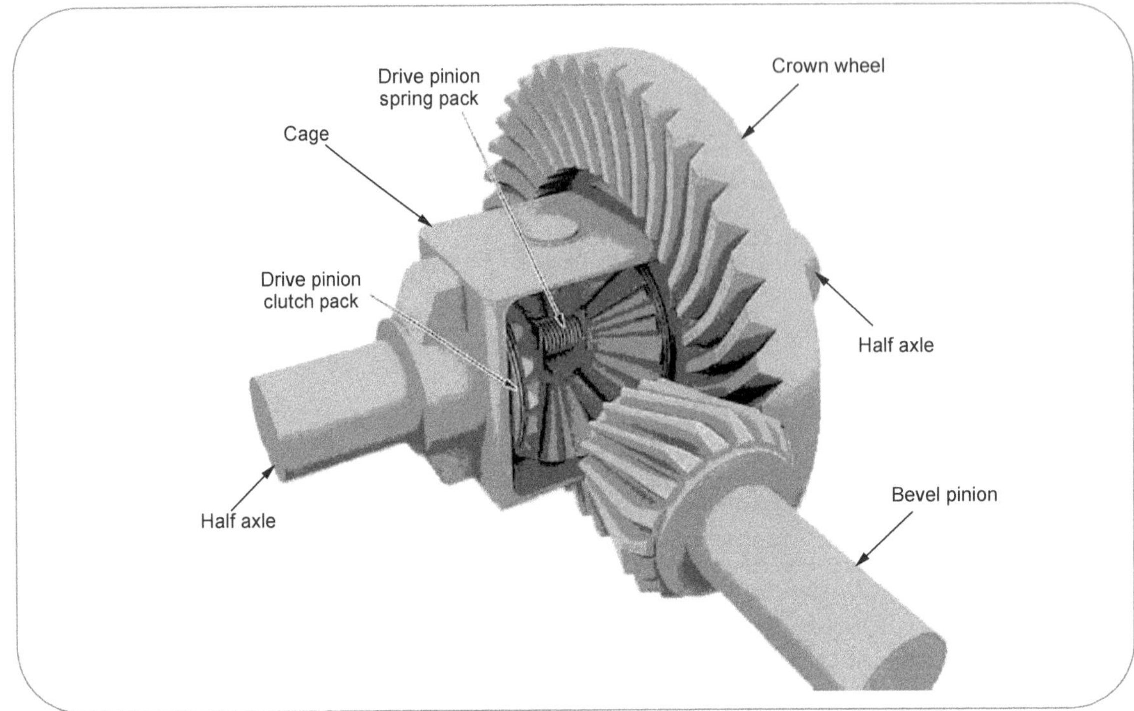

Fig. 6.21.1 Differential gear assembly

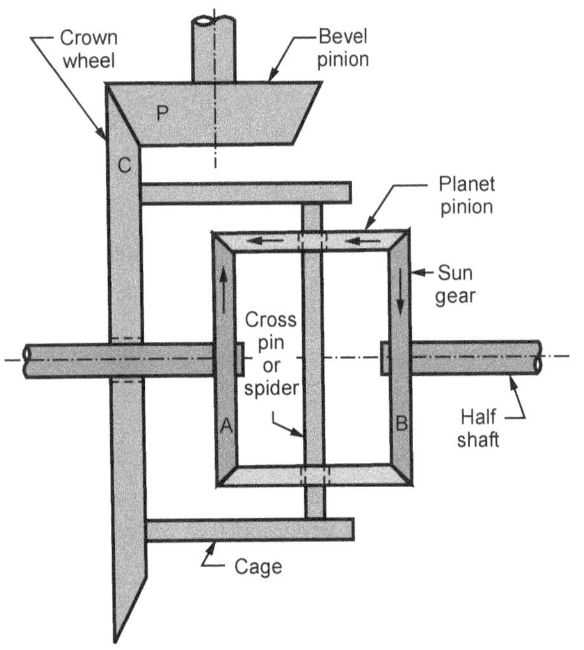

Fig. 6.21.2 Working principle of differential unit

(i) When the vehicle is running on a straight road:

- In this case, both the drive wheels have to overcome equal amount of rolling resistance.

- The crown wheel causes rotation of the cage.

- The two planet pinions and the sun gears form a closed mechanical chain and rotate as a single unit.

- There is no relative motion among the various elements of the differential gear unit.

(ii) When the vehicle is taking a turn :

- Let us consider, the vehicle is taking a left turn.

- This causes increased rolling resistance on the half axle with sun gear 'A'.

- Thus, the sun gear 'A' is not able to keep pace with the speed of the cage.

- This causes the planet pinion to roll on the circumference of sun gear 'B'.

- The sun gear 'B' thus rotates faster than sun gear 'A'.

- If 'N' is the normal running speed of the sun gears while vehicle is running on a straight road, then due to differential action, the inner wheel will rotate at (N - n) r.p.m. and outer wheel will rotate at (N + n) r.p.m.

- The differential transmits torque to both drive axles depending on the speed ratio across it.

6.22 : Axles

- An axle is a central shaft for rotating a wheel or gear.

- It may be fixed to the wheels, rotating with them or fixed to the vehicle with the wheels rotating around the axle.

6.22.1 Front Axle

- Front axle is usually a forged member carrying the road wheels at its ends.

- It carries the front weight of the vehicle.

- It also carries the steering mechanism which is used to direct the vehicle in the desired direction.

- Modern cars have live front axle while heavy vehicles use conventional dead front axle.

6.22.1.1 Purpose / Functions of Front Axle

- The front axle performs the following functions :

    (i) It carries the front weight of the vehicle.

    (ii) It also carries the steering mechanism which is used to direct the vehicle in the desired direction.

    (iii) It also houses the front suspension system which absorbs the shocks produced due to road surface variations.

    (iv) In case of front wheel drive, it also carries the differential unit.

6.22.1.2 Front Axle   **SPPU : Dec.-07**

- Front axle is used for steering the front wheels carried on the stub axle.

- It is the major component of a vehicle designed to transmit the vehicle weight to the front steering wheels through the front suspension system.

- Fig. 6.22.1 shows a conventional front axle.

- The front axle has to take vertical vehicle load, bending loads due to weight, and torque loads due to braking and acceleration.

- To withstand the above loads, the axle is made of 'I' section in the central portion while the ends are made elliptical.

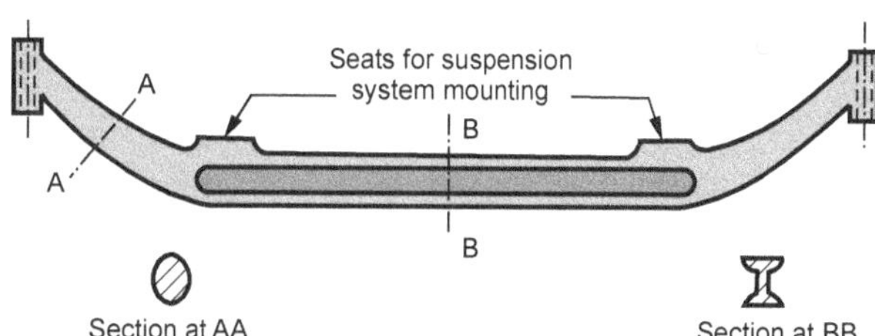

Fig. 6.22.1 Conventional dead front axle

- Now-a-days with independent suspension, the front axle is split and hence the front wheels are independent to rise and fall vertically relative to each other.

- The front axle assembly includes stub axles, brake assembly and tie rod units.

- The central portion of front axle is given a downward sweep to keep a low chassis height.

6.22.2 Rear Axle

- Rear axle drives are employed on front engine rear wheel drive vehicles.

- The rear axle drive system consists of important sub-systems such as universal joints, slip joint, propeller shaft, suspension leaf springs, rear axle, rear axle casing, final drive and differential.

- The rear axle drive system is subjected to the following forces and torques.

(i) Weight of the vehicle

- It acts vertically downward and causes bending moment in the axle shaft.

(ii) Driving thrust

- It acts torsionally in the longitudinal direction of the vehicle.

(iii) Torque reaction

- It also acts torsionally in the longitudinal direction of the vehicle and is caused due to acceleration or braking of the road wheels.

(iv) Side thrust

- It acts from the sides of vehicle due to cornering force, etc.

6.23 : Introduction to Vehicle Safety

SPPU : Dec.-12

- Now-a-days, vehicles are driven for long distances, at high speeds, with heavy loads and other adverse conditions.

- It is hence important to ensure safety of passengers inside the vehicle, pedestrians on the road and the luggage carried by the vehicle.

- *Vehicle safety* is the study and practice of design, construction, equipment and regulation to minimize the occurrence and consequences of automobile accidents.

- There are basically two types of safety systems :

> (i) Active safety system
> (ii) Passive safety system

6.24 : Vehicle Safety System

SPPU : Dec.-11,13, May-12,13

6.24.1 Active Safety System

- *Active safety systems* are those that improve driving safety through technology designed to prevent accidents before they happen.

- These are technology systems assisting in the prevention of an accident.

- They are also referred as *primary safety system.*

- Active safety systems use an understanding of the state of the vehicle to both avoid and minimize the effects of a crash.

- The modern vehicle has various electronic sensors and microprocessor controlled actuators that assist driver to have better control on the vehicle.

- These forward-looking technologies are expected to play an increasing role in accident avoidance and mitigation.

- Examples of active safety systems include :
  - ABS - Antilock Braking System
  - TCS - Traction Control System
  - ESP - Electronic Stability Program
  - Cruise Control

6.24.2 Passive Safety System

- *Passive safety systems* are those that protect the passengers once an accident has occurred, by reducing the risk and severity of injury.

- These are technology systems for reducing the impact of crash on passengers.

- They are also referred as *secondary safety system.*

- These systems are only deployed or effective in response to a vehicle accident.

- Examples of passive safety systems include :

> (i) Seatbelt
> (ii) Pretensioners
> (iii) Airbag - front and side airbags

6.25 : Seat Belts

SPPU : Dec.-11, May-12, 17

- A *seat belt* is a safety harness designed to secure the occupant of a vehicle against harmful effects resulting during a collision or a sudden stop.

- The seat belt is sometimes also referred as *safety belt.*

- A seat belt reduces the likelihood and severity of injury in a traffic collision by stopping the vehicle occupant from hitting hard against interior elements of the vehicle or other passengers.

- The seat belt keeps the occupant positioned correctly on its seat for maximum benefit from the impact.

Need for seat belt :

- Any moving or stationary element has inertia, i.e. a tendency to continue in its state of motion or rest unless some external forces disturbs its state.

- If a car is speeding along at 50 miles per hour, then the components and passengers inside the car also keep moving along the car at 50 miles per hour.

- But if the car were to crash into a stationary pole, the force of the pole would bring the car to an abrupt stop.

- Without a seatbelt, the occupant would either slam into the steering wheel at 50 miles per hour or go flying through the windshield at 50 miles per hour.

- If the occupant hits the windshield with its head, the stopping power is concentrated on one of the most vulnerable parts of the human body causing severe injury or fatal death.

- A seatbelt applies the stopping force to more durable parts of the body over a longer period of time.

- The basic idea of a seatbelt is to keep the occupant from flying through the windshield or hurdling toward the dashboard when the car comes to an abrupt stop.

Working of seat belts :

- Car seatbelts have the ability to extend and retract.

- The occupant can lean forward easily while the belt stays fairly loose.

- But in a collision, the belt will hold the occupant in place.

- Refer Fig. 6.25.1 for construction and working of seat belt.

Fig. 6.25.1 Construction and working of seat belt

- The seat belt consists of a strap (webbing) connected on one of its end to spring loaded reel (retractor).

- The coil spring applies a rotation force or torque to the reel to rotate the reel so that it winds up any loose strap.

- When the occupant pulls the strap out, the reel rotates counter-clockwise, which turns the attached spring in the same direction.

- Effectively, the rotating reel works to untwist the spring.

- The spring wants to return to its original shape, so it resists this twisting motion.

- If the occupant releases the strap, the spring will tighten up, rotating the reel clockwise until there is no more slack in the belt.

- The reel spool (retractor) has a locking mechanism that stops the reel from rotating rapidly when the car is involved in a collision.

- There are two sorts of locking systems in common use today :

  (i) Systems triggered by the car's movement

  (ii) Systems triggered by the belt's movement

- The locking mechanism triggered by the car's movement is shown in Fig. 6.25.2.

- This locking system locks the reel when the car decelerates rapidly (when it hits something).

- The central operating element in this mechanism is a weighted pendulum.

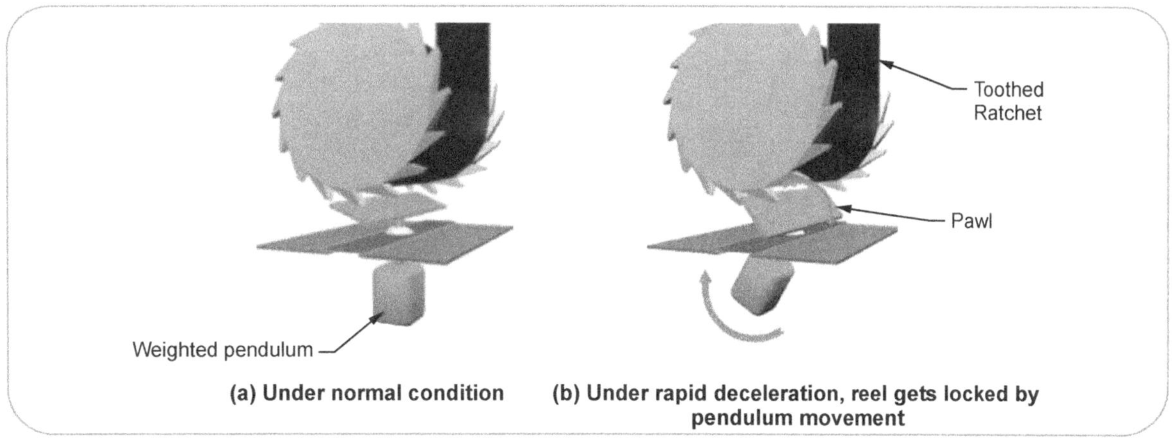

Fig. 6.25.2 Seat Belt locking mechanism triggered by car's movement

- When the car comes to a sudden stop, the inertia causes the pendulum to swing forward.

- The pawl on the other end of the pendulum catches hold of a toothed ratchet gear attached to the reel.

- Since the pawl grips one of its teeth, the gear can't rotate counter-clockwise, and neither can the connected reel.

- When the strap loosens again after the crash, the gear rotates clockwise and releases the toothed ratchet.

6.26 : Air Bag

SPPU : May-12, 17

- The airbag was invented to perform as a safety cushion between the passengers and the hard surfaces of the car.

- It is also referred as *Supplemental Restraint System (SRS), Air Cushion Restraint System (ACRS)* and *Supplemental Inflatable Restraint (SIR).*

- It consists of a flexible envelope designed to inflate rapidly during an automobile collision.

- Its purpose is to cushion occupants during a crash and provide protection to their bodies when they strike interior objects such as the steering wheel or a window.

- The airbag is designed to only inflate in moderate to severe frontal crashes.

- Airbags are normally designed with the intention of supplementing the protection to an occupant who is correctly restrained with a seatbelt.

- Most designs are inflated through pyrotechnic means and can only be operated once.

- This airbag inflation is controlled by the Airbag Electronic Control Unit (AECU).

- After deployment, the inflated airbags need to be replaced and the ECU usually needs to be reset.

Components of Airbag system :

The major components of airbag system are :

(i)   Crash sensors

(ii) Airbag Electronic Control Unit (AECU)

(iii) Detonator

(iv) Inflatable airbag

Triggering conditions :

- Airbags are designed to deploy in frontal and near-frontal collisions with severity beyond a certain threshold.

- This threshold is defined by the regulations governing vehicle construction as per regional bodies, such ARAI in India.

- Real-world crashes typically occur at angles other than directly into the front of the vehicle, and the crash forces usually are not evenly distributed across the front of the vehicle.

- The airbag system uses a MEMS (Micro Electro Mechanical System) accelerometer sensor, which is a small circuit with integrated micro mechanical elements.

- The microscopic mechanical element moves in response to rapid deceleration, and this motion causes a change in capacitance, which is detected by the electronics on the chip that then sends a signal to fire the airbag.

- In case of vehicle fire, wherein temperatures reach 150-200 °C (300-400 °F), the airbag is programmed to inflate.

- This safety feature helps to ensure that such temperatures do not cause an explosion of the entire airbag module.

- The airbag triggering algorithms are very complex and considers inputs from various sensors and vehicle controls.

- They try to reduce unnecessary deployments and to adapt the deployment speed to the crash conditions.

Working of airbag :

- Refer Fig. 6.26.1 for construction and working of airbag system.

- The airbag system employs various MEMS based accelerometer, impact, side pressure, wheel speed, gyroscope and other sensors.

- The signals from the various sensors are fed into the airbag control unit, which determines from them the angle of impact, the severity or force of the crash, along with other variables.

- Depending on the result of these calculations, the AECU may also deploy various additional restraint devices, such as seat belt pre-tensioners and / or

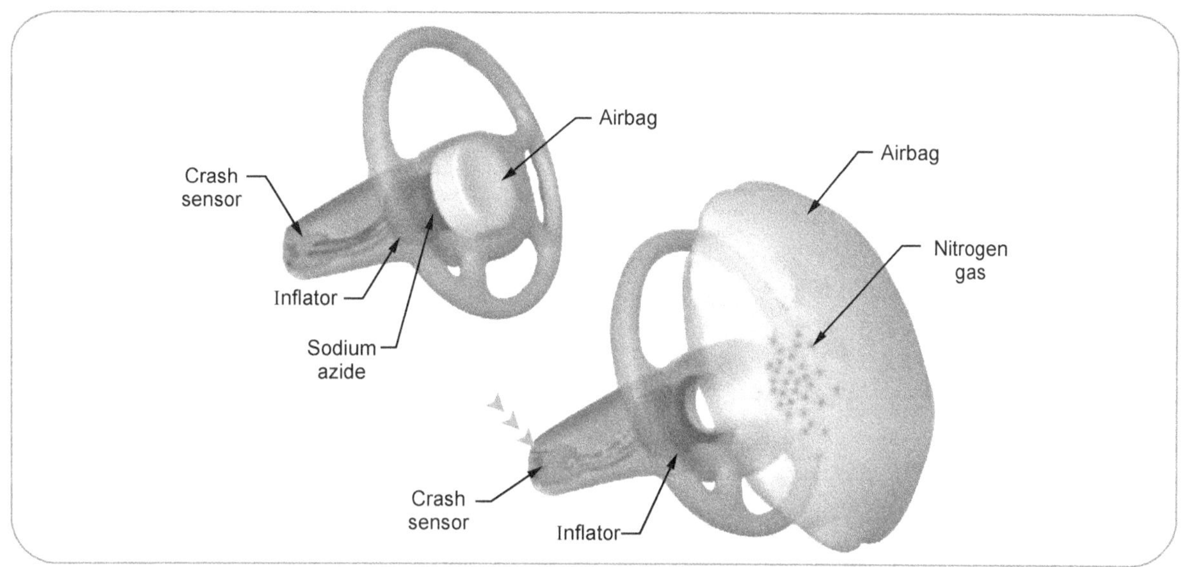

Fig. 6.26.1 Construction and working of airbag system

airbags (including frontal bags for driver and front passenger).

- When the requisite threshold has been reached or exceeded, the airbag control unit will trigger the ignition of a gas generator propellant to rapidly inflate a fabric bag.

- The decision to deploy an airbag in a frontal crash is made within 15 to 30 milliseconds after the onset of the crash, and both the driver and passenger airbags are fully inflated within approximately 60-80 milliseconds after the first moment of vehicle contact.

- As the vehicle occupant collides with and squeezes the bag, the gas escapes in a controlled manner through small vent holes.

- The airbag's volume and the size of the vents in the bag are tailored to each vehicle type, to spread out the deceleration of (and thus force experienced by) the occupant over time and over the occupant's body, compared to a seat belt alone.

- An air bag contains a mixture of sodium azide (NaN_3), KNO_3, and SiO_2.

- The production of nitrogen gas in an airbag sytem takes place in 3 stages as follows :

   (i) $2\ NaN_3 \rightarrow 2\ Na + 3\ N_2\ (g)$

   (ii) $10\ Na + 2\ KNO_3 \rightarrow K_2O + 5\ Na_2O + N_2\ (g)$

   (iii) $K_2O + Na_2O + 2\ SiO_2 \rightarrow K_2O_3Si + Na_2O_3Si$
        (silicate glass)

- The nitrogen gas produced in second reaction is used to inflate the airbag.

- The final reaction is used to eliminate the K_2O and Na_2O produced in the previous reactions because the first-period metal oxides are highly reactive.

- These products react with SiO_2 to produce a silicate glass which is a harmless and stable compound.

Types of airbag :

- There are several types of airbags designed to maximize protection and used in new high end cars.

- These includes :

   (i)   Frontal driver airbag

   (ii)   Frontal passenger airbag

   (iii) Side airbag

   (iv) Curtain airbag

   (v)   Backseat airbag

   (vi) Knee airbag

6.27 : Seat

- A car seat is the seat used in automobiles. Most of the car seats are made from inexpensive but durable material to withstand prolonged use.

- The commonly used material for seats is polyester.

- In latest cars the arrangement of seat is **bucket type**.

- It is a separate seat with encountered platform designed to accomodate one person distinct from a bench seat i.e. a platform designed to seat upto three people.

- Individual bucket seats typically have rounded backs and may offer a variety of adjustments to fit different passangers.

- Some sedan models offer fold-down rear seats to gain cargo space when they are occupied by passangers.

- According to motor vehicle safety act the seats should be properly anchored and the construction of automobile seats should be proper.

- In addition to this, the act also includes that a child to sit up front is should be 5 feet and they must weight upto 40 kg.

- Side airbags are often built right into the side of the seat.

- Some modern cars like BMW, Audi, Benz etc. are equipped with a battery powered automatic control to adjust how the seat sits into a car.

- Some vehicle let the driver save the adjustments in memory (memory seat) for later recall with the push of a button.

- Car seat covers are accessories that protect the original seat from wear and tear. It also add a custom look to a vehicle's interior. This can maintain the resale value of the car and maximize the comfort of the driver and passangers.

6.28 : Sliding Mesh Gearbox   `SPPU : Dec.-13`

- Refer Fig. 6.28.1 for construction and working of a sliding mesh gearbox. (see Fig. 6.28.1 on next page)

- It shows a typical 3-forward and 1-reverse gear configuration.

- It is referred to as sliding mesh type because herein meshing of gears take place by axial sliding of gears.

Construction :

The arrangement and function of various components of a sliding mesh gearbox is as mentioned below :

(i)	Clutch shaft	(ii)	Lay shaft
(iii)	Main shaft	(iv)	Bearings
(v)	Reverse idler shaft	(vi)	Transmission gears
(vii)	Selector mechanism	(viii)	Transmission case

(i) Clutch shaft

- It is the input shaft to the gerbox.

- Its outer end is connected to the clutch disc.

- It has a gear machined at its inner end (gear 'A') that meshes with respective gear on the lay shaft (in this case, it is gear 'B').

(ii) Lay shaft

- The lay shaft is freely suspended in bearing mounted on the transmission case.

- It has gears rigidly mounted / machined on it (gears 'B', 'E', 'C' and 'G').

(iii) Main shaft

- It is the output shaft of the gearbox.

- It has splines cut across its length to accomodate axial movement of gears on it (gears 'F' and 'D').

- Its outer end is connected to the propeller shaft through a universal joint.

(iv) Bearings

- Generally, taper roller bearings are used.

- These bearings are required to take radial and thrust loads during gear engagement.

(v) Reverse idler shaft

- It is a short shaft that supports the reverse idler gear.

(vi) Transmission gears

- In sliding mesh gearboxes, generally spur gears are used.

- These gears can be grouped into :

  Gears on clutch shaft - rigidly attached / machined

  Gears on lay shaft - rigidly attached / machined

  Gears on main shaft - free to slide on main shaft

  Gear on reverse idler shaft - rigidly attached / machined

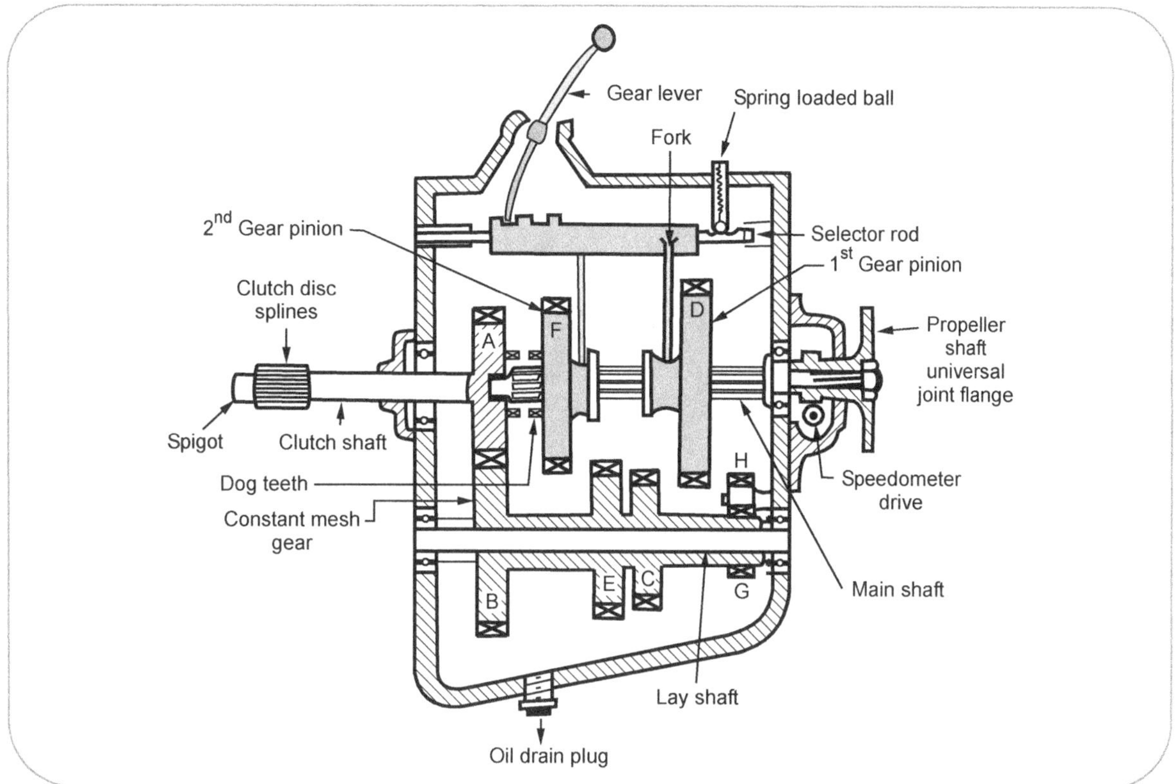

Fig. 6.28.1 Sliding mesh gearbox

(vii) Selector mechanism

- Fig. 6.28.1 shows a mechanical selector mechanism for shifting of gear pairs.

(viii) Transmission case

- It provides support for the bearings and shafts.
- It also provides an enclosure for lubricating oil.
- It is generally made up of aluminium to reduce weight.
- It has a vent on its top surface to ensure atmospheric pressure inside the gearbox.

(i) Ist Gear

- To engage Ist gear, gear 'D' is shifted towards left such that it meshes with gear 'C'.
- Refer Fig. 6.28.2 (a) for power flow in Ist gear.

$$\text{Gear ratio } (G_1) = \frac{N_A}{N_D} = \frac{T_D}{T_C} \times \frac{T_B}{T_A}$$

(ii) IInd Gear

- To engage the IInd gear, Ist gear is disengaged.

- Then gear 'F' is shifted to mesh with gear 'E'.
- Refer Fig. 6.28.2 (b) for power flow in IInd gear.

$$\text{Gear ratio } (G_2) = \frac{N_A}{N_F} = \frac{T_F}{T_E} \times \frac{T_B}{T_A}$$

(iii) IIIrd Gear

- To engage the IIIrd gear, the sliding gear 'F' is shifted towards left such that its dog teeth meshes with dog teeth on gear 'A'.
- This is also called as direct drive.
- Refer Fig. 6.28.2 (c) for power flow in IIIrd gear.

$$\text{Gear ratio } (G_3) = N_A = N_F$$

(iv) Reverse Gear

- To engage the reverse gear all other gears are disengaged.
- The sliding gear 'D' is shifted towards right such that it meshes with reverse idler gear 'H'.
- Refer Fig. 6.28.2 (d) for power flow in reverse gear.

Working :

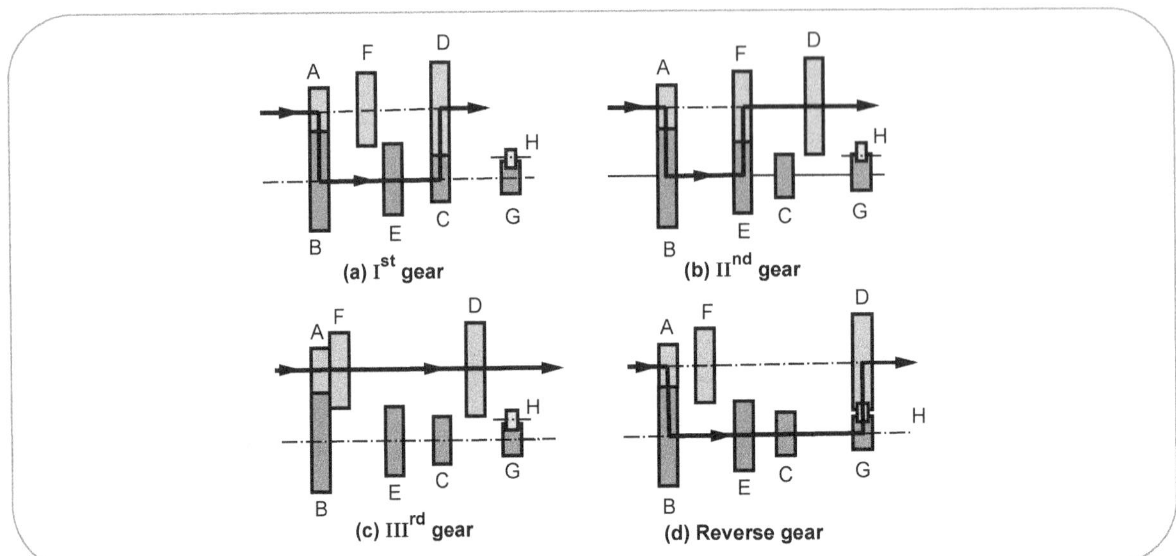

Fig. 2.9 Various gear positions in a sliding mesh gearbox

$$\text{Reverse Gear Ratio } (G_R) = \frac{N_A}{N_D} = \frac{T_D}{T_H} \times \frac{T_H}{T_G} \times \frac{T_B}{T_A}$$

Advantages of sliding mesh gearbox :

- It is the most simplest in construction and working.
- Since it uses simple spur gear, it is cheaper than other types of gears.

Limitations of sliding mesh gearbox :

- Spur gears used are noisy in operation.
- Teeth wear out is more as only 2-3 teeth are in mesh during operation.
- Spur gears cannot take any thrust loads.

Review Questions

1. *What is chassis ?*
2. *What are the different layouts of vehicle ?*
3. *Explain the types of chassis frame.*
4. *Explain steering system with neat sketch.*
5. *With neat sketch explain suspension system (leaf spring arrangement).*
6. *What is the need of suspension system ?*
7. *State the functions of braking system.*
8. *Explain drum brake with neat sketch.*
9. *What is the necessity of cooling system ? What are the different methods of it ?*
10. *Explain fuel supply system of petrol engine.*
11. *Explain with neat sketch air injection system for diesel engine.*
12. *With block diagram explain power transmission system.*
13. *State the functions of clutch. Explain single plate clutch with neat sketch.*
14. *Draw a neat sketch of multiplate clutch.*
15. *What is gear drive ? What are the types of gear ?*
16. *What is gear ratio ? State advantages, limitations and applications of gear drive.*
17. *Explain propeller shaft with neat sketch.*
18. *Write short note : a) Universal joint*
    b) Differential gear box
19. *What is axle ? Explain front and rear axle.*
20. *What is meant by active and passive safety ?*
21. *Write short note on :*
    a) Seat b) Seat belt c) Air bag d) ABS

□□□

Unit - V
Introduction to Manufacturing

Syllabus : ***Conventional Manufacturing Processes*** *: Casting, Forging, Metal forming (Drawing, Extrusion, etc.), Sheet metal working, Metal joining, etc. Metal cutting processes and machining operations- Turning, Milling and Drilling, etc. Micromachining. Additive manufacturing and 3D Printing. Reconfigurable manufacturing system and IOT, Basic CNC programming: Concept of Computer Numerical Controlled machines.*

Chapter - 7 Conventional Manufacturing
Processes   (7 - 1) to (7 - 74)

UNIT - V

7   Conventional Manufacturing Processes

Syllabus

Casting, Forging, Metal forming (Drawing, Extrusion, etc.), Sheet metal working, Metal joining, etc. Metal cutting processes and machining operations- Turning, Milling and Drilling, etc. Micromachining. Additive manufacturing and 3D Printing. Reconfigurable manufacturing system and IOT, Basic CNC programming: Concept of Computer Numerical Controlled machines.

Contents

Mind Map – Conventional Manufacturing Processes

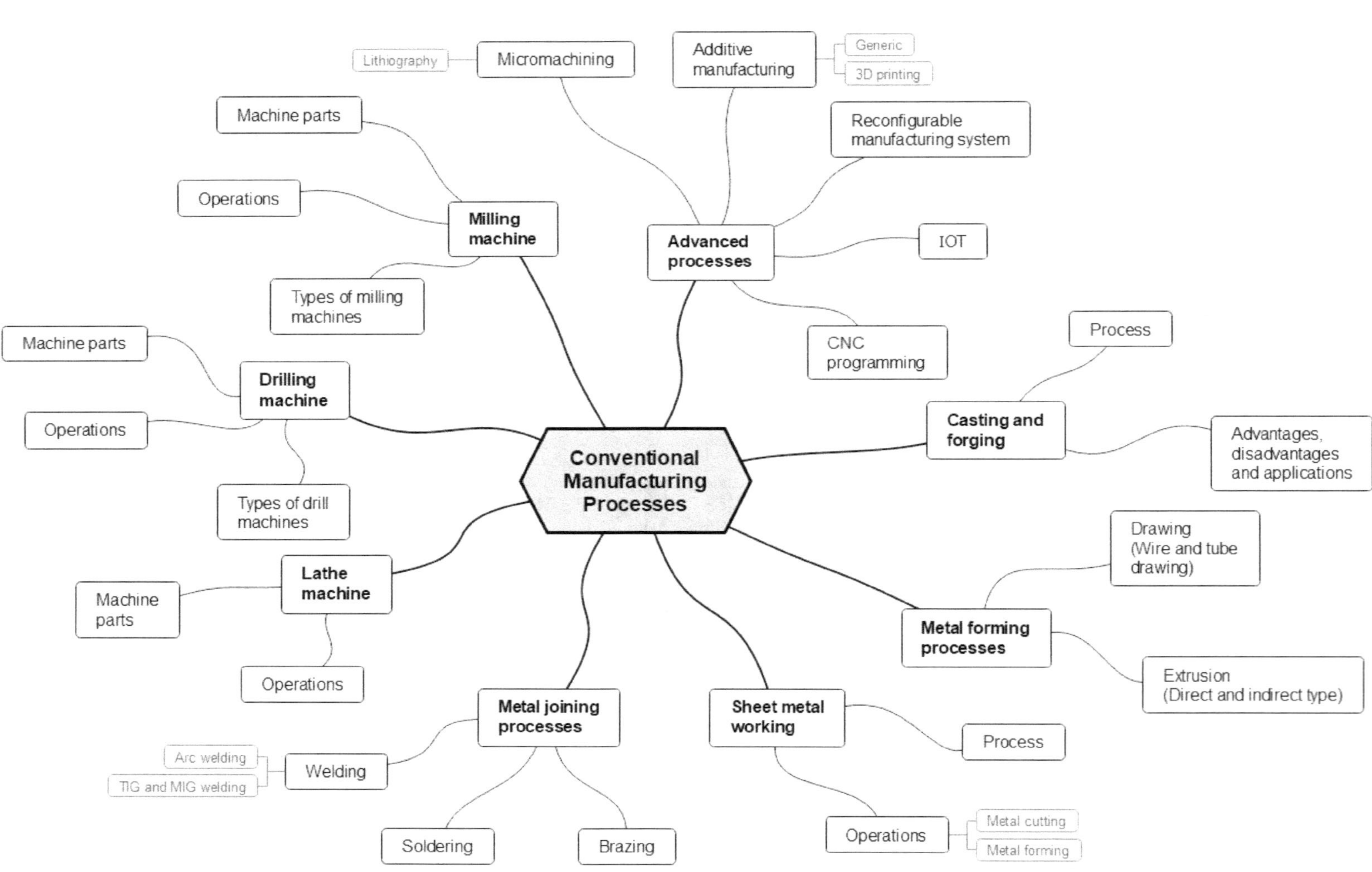

7.1 : Introduction

- **Manufacturing process** of a workpiece involves transforming a raw material from its original state to a finished state by changing its shape or the properties of the material in a series of steps.

- The design engineer should have complete knowledge of manufacturing processes.

- Actually, manufacturing process is the part of production process directly concerned with the changes in dimensions, shapes and properties of raw material. It is accomplished in definite sequence.

7.1.1 Classification of Manufacturing Process

Manufacturing processes can be divided into the following groups (Refer Fig. 7.1.1) :

> 1. Primary shaping processes
> 2. Deforming processes
> 3. Machining processes
> 4. Joining processes
> 5. Surface finishing processes
> 6. Material properties modification processes

1. Primary shaping processes

- Primary shaping is the manufacturing of a solid body from a molten state or gaseous state or amorphous material (gases, liquids, powders, fibres, chips, etc.).

- A primary shaping process contains a molten metal like cast iron which poured into the hollow space (mould of desired shape). After solidification, it attains the shape of hollow space.

- Some of the important primary shaping processes are as follows :

> 1. Casting processes
> 2. Powder metallurgy processes
> 3. Processing of plastics

2. Deforming processes

- In deforming processes, a metal is in cold or hot condition, which is deformed plastically into the desired shapes without changing its mass or metal composition.

- In deforming processes, no metal is removed; only it is deformed and displaced.

- This process makes use of suitable stresses like tension, compression, etc. to cause plastic deformation.

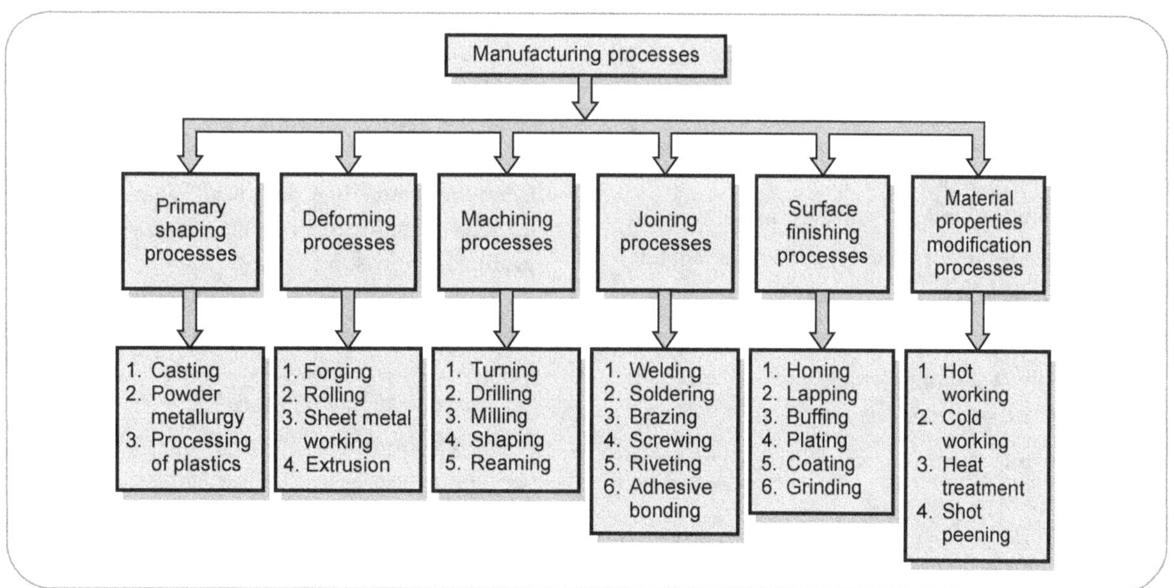

Fig. 7.1.1 Classification of manufacturing processes

- Some of the important deforming processes are as follows :

1. Forging	2. Rolling
3. Sheet metal working	4. Extrusion

3. Machining processes

- In machining processes, the material is removed by providing suitable relative motions between the workpiece and tool, so as to generate the required size and shape on the component.

- As the material is removed, these processes are also called as **removing processes.**

- Some of the important machining processes are as follows :

1. Turning	2. Drilling	3. Milling
4. Shaping	5. Reaming	

4. Joining processes

- In these processes, two or more pieces of metal parts are joined together to make a final component.

- The joining process can be carried out by fusing, pressing, rubbing, etc.

- Most of the processes require heat and pressure for joining of metal pieces.

- Some of the important joining processes are as follows :

1. Welding	2. Soldring	3. Brazing
4. Screwing	5. Riveting	6. Adhesive bonding

5. Surface finishing processes

- These processes are used only to provide good surface finish or decorative or protective coating on the metal surface of a workpiece.

- During the process, dimensions of the part are not changed, only a negligible amount of metal is removed from the workpiece.

- Some of the important surface finishing processes are as follows :

1. Honing	2. Lapping	3. Buffing
4. Plating	5. Coating	6. Grinding

6. Material properties modification processes

- These processes are used to provide certain specific properties to the metal parts so as to make them suitable for particular operations.

- In these processes, shape of the workpiece remains same.

- Some of the important material properties modification processes are as follows :

1. Hot working	2. Cold working
3. Heat treatment	4. Shot peening

7.2 : Casting Process

SPPU : Dec.-11,12,13, May-13,17

- **Casting** or **founding** is the process of producing metal or alloy parts.

- The parts of desired shapes are produced by pouring the molten metal or alloy into a prepared mould and then allowing the metal or alloy to cool and solidify.

- This solidified piece of metal or alloy is called as **casting**. Refer Fig. 7.2.1. (See Fig. 7.2.1 on next page)

7.2.1 Advantages, Disadvantages and Applications of Casting Process

SPPU : Dec.-13

Advantages

- Casting is one of the most versatile manufacturing processes.

- It provides the greatest freedom of design in terms of shape, size and quality of product.

- Casting provides uniform directional properties and better vibration damping capacity to the cast components.

- Complex and uneconomical shapes which are difficult to produce by other processes can be easily produced by casting process.

- A product obtained by casting is one piece; hence there is no need of metal joining processes.

- Very heavy and bulky parts which are difficult to get fabricated, may be cast.

- It also produces machinable parts.

- Casting process can be mechanised and generally used for mass production of components.

Disadvantages

- Cast components require more machine finish.

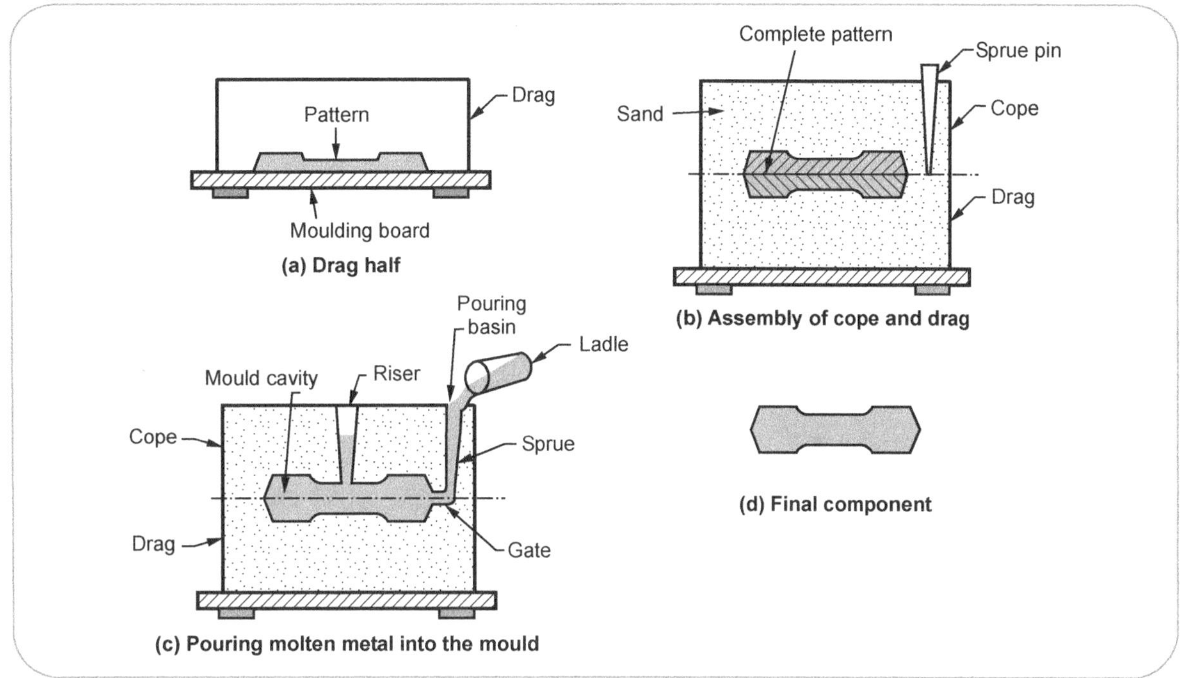

Fig. 7.2.1 Casting process

- Cast components are generally brittle i.e. weak in tension.
- Welding of cast components is difficult.
- Defects like cracks, blow holes, etc. make the casting weak and unsuitable for use.
- Accuracy of cast components is less.
- During the casting process, metal is melted for which large amount of heat and space is required. This step in casting process pollutes the atmosphere.

Applications

Casting or cast components are widely used in engineering field. Few applications of cast components are as follows :

- Automobile parts (pistons, cylinders, clutch and gear housings, gear blanks, etc.)
- Machine parts (pulleys, gear blanks, beds, frames, etc.).
- Aircraft parts (Engine blades, motor housings, etc.).
- Turbine vanes, power generators, pump parts, filters, valves, etc.
- Agricultural parts, railway crossings, sanitary fittings, etc.
- Construction, communication and atomic energy applications.

7.2.2 Basic Terminology in Casting Process

SPPU : Dec.-13

1. **Mould :** It is a container having a cavity of the shape which is to be manufactured. The molten metal is poured in the mould. The process of making the mould is called as **mould making.**

2. **Pattern :** It is a model or replica of the object to be cast. For effective casting, various types of allowances are provided on the pattern. If only one object has to be cast, then also pattern is required. Pattern is surrounded with sand to give rise to a mould cavity, in which molten metal is poured and casting is produced.

3. **Core :** Core is an obstruction which when positioned in the mould, does not permit the molten poured metal to fill the space occupied by the core, hence produces hollow casting.

4. **Mould Box :** Mould box is combination of two halves; upper half and lower half. Upper half is called as **cope** and lower half is called as **drag.** Refer Fig. 7.2.1 (a).

   It is important to note that, *drag half is inverted (turn-over) and then kept on the board.* Refer Fig. 7.2.1 (b).

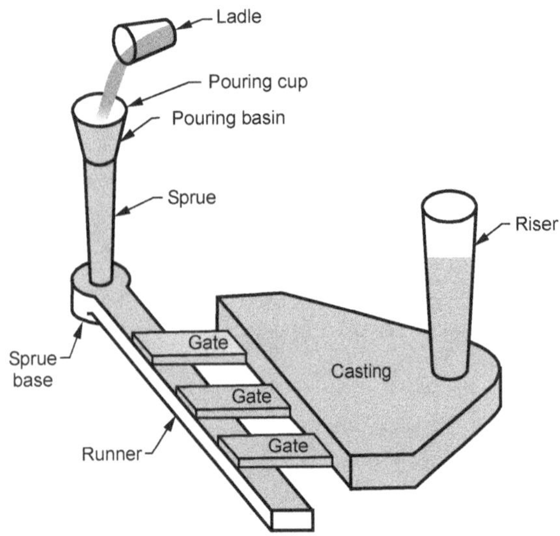

Fig. 7.2.2 Components of gating system

5. **Mould material :** Mould material is the one out of which mould is made. Mould material should be such that, the cavity of the mould retains its shape till the metal has solidified. Moulds can be made up of wax, plaster of paris, ceramics, etc.

6. **Pouring cup :** It is a funnel shaped cup which forms the top portion of the sprue. It makes easier to the operator to direct the flow of metal from crucible to sprue. Refer Fig. 7.2.2.

7. **Pouring basin :** Molten metal is initially poured into a pouring basin which acts as a reservoir from which it moves smoothly into the sprue. It holds back the slag and dirt which floats on the top and allows only the clean metal to enter into the sprue.

8. **Sprue :** It is the channel through which the molten metal is brought into the parting plane where it enters the runner and gate. It may be square or round in shape.

9. **Runner :** In case of large components, molten metal is carried from the sprue base to the several gates through a passage called as runner.

10. **Gate :** A gate is a channel which connects the runner with the mould cavity, through which molten metal flows to fill the mould cavity. For large components, more than one gates are used.

11. **Riser :** It is a passage of sand, made in the cope to permit the molten metal to rise above the highest point in the casting after the mould cavity is filled up. It permits the escape of air and mould gases. It indicates that, the mould cavity has been completely filled or not.

12. **Ladle :** Ladles are used to carry the molten metal from the furnace to the moulding boxes.

7.2.3 Steps Involved in Casting Process

Following are the steps to be followed while making a **sand casting :**

1. **Pattern making :** Make the pattern of wood, metal or plastic.

2. **Sand mixing and preparation :** Select a particular sand, test it and prepare the necessary sand mixtures for mould and core making.

3. **Core making :** With the help of patterns prepare the mould and required cores.

4. **Melting :** Melt the metal or alloy to be cast.

5. **Pouring :** Pour the molten metal or alloy into the mould and remove the casting from the mould after solidification of metal.

6. **Finishing :** Clean and finish the casting.

7. **Testing :** Test and inspect the casting and remove the defects, if any.

8. **Heat treatment :** Relieve the casting stresses by using various heat treatments.

9. **Re-testing :** Again inspect the casting and deliver it.

7.3 : Forging Process

- Forging is the process of shaping heated metal by the application of sudden blows (hammer forging) or steady pressure (press forging) and makes use of the characteristic plasticity of the material.

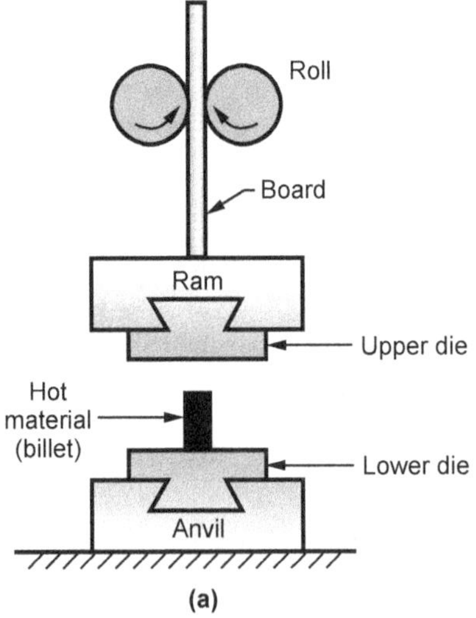

Fig. 7.3.1 Forging process

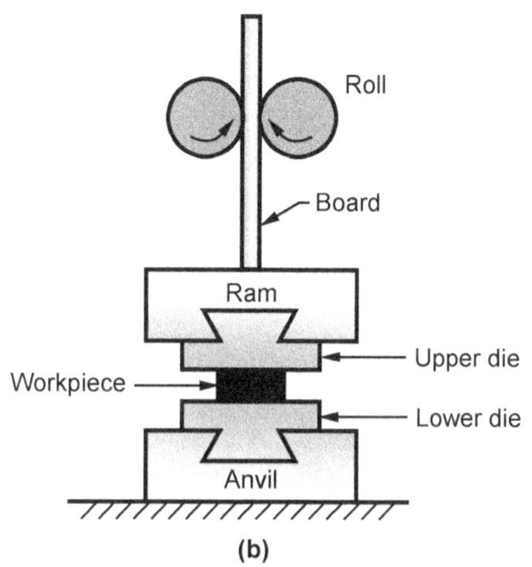

Fig. 7.3.1 Forging process

- Forging is metal forming process which may be done by hand or by machine.

- In case of hand forging, hammering is done by hand; whereas forging by machine involves the use of dies and it is mostly used in mass production. Refer Fig. 7.3.1.

- Whatever may be the method of applying pressure for shapping the metal, the primary requirement is to heat the metal to a definite temperature to bring it into the plastic state.

- This may done in an open hearth, called as **Smith's forge** for small jobs or in closed furnaces for large jobs.

- The shop in which the work is carried out is called as **Smithy or Smith's shop.**

- The metals which are used in forging process must possess the required ductility.

- We know that ductility refers to the capacity of a material to undergo deformation, under tension without failure.

- The commonly used forging materials are : Aluminium alloys, copper alloys, low carbon steels, alloy steels, nickel alloys, tungsten alloys, magnesium alloys, titanium alloys, beryllium, etc.

7.3.1 Advantages, Disadvantages and Applications of Forging Process

Advantages

- In forging process, grain flow is continuous and uninterrupted. It gives greater strength and toughness to the forged components. Fig 7.3.2

shows three spanners, produced through three different methods casting, machining and forging. The cast spanner is the weakest of all, the machined spanner is relatively stronger and the strongest will be the forged spanner.

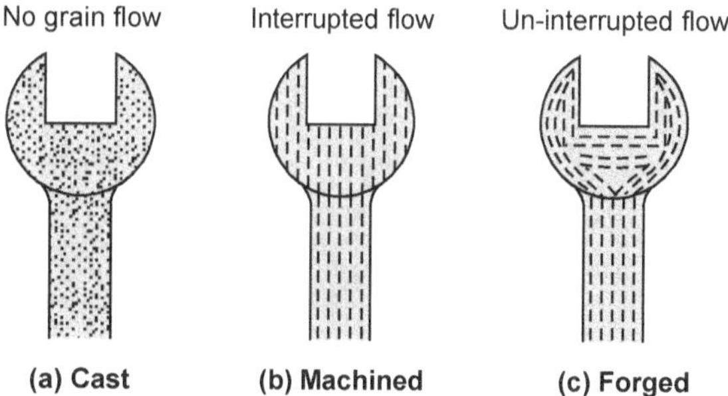

Fig. 7.3.2 Spanners produced through three different methods

- Forged components requires minimum surface finish.

- The forging process gives the high dimensional accuracy.

- Forged components have better mechanical properties like strength, toughness, etc.

- Forged components have better resistance to shock and vibrations.

- Welding of forged parts is easy.

Disadvantages

- Complicated shapes cannot be forged easily.
- Forging process is mostly suitable for large parts.
- Forging of brittle materials is difficult.
- Due to high cost of forging dies, forging process is costly.
- More noise and vibrations are produced during the process.

Applications

Forging process is used in the manufacturing of following components :

- Car axles, crankshafts, connecting rods, leaf springs, crane hooks, jet engine turbine dies and blades.
- Levers, flanges, propellers, hollow bodies, railway wheel disks, tank bottoms.
- Air-craft and rocket parts, knife blades, bolts, nuts, washers, collars, gear blanks, etc.

7.3.2 Types of Forging Processes

Forging processes are classified as follows :

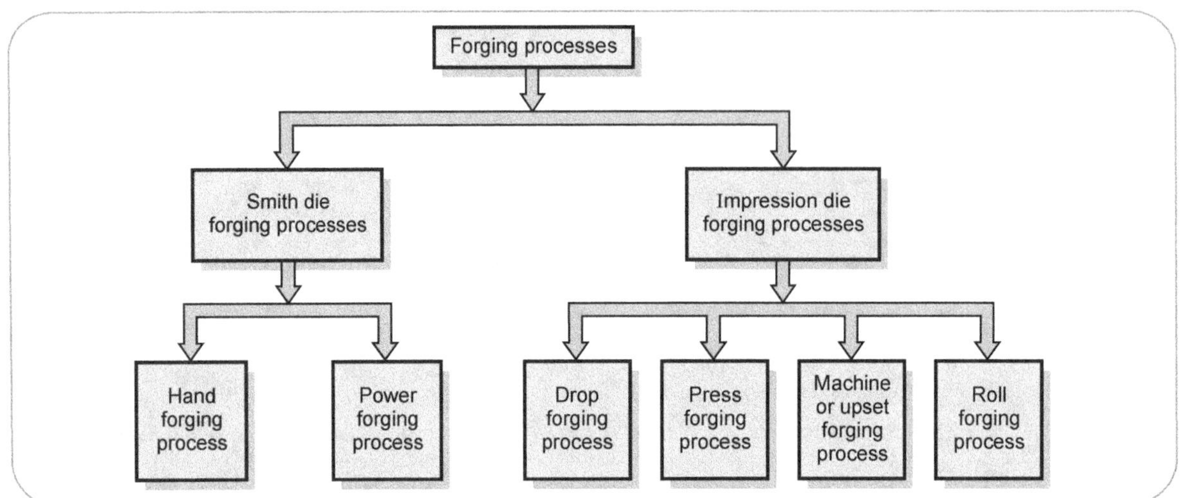

Fig. 7.3.3 Types of forging processes

1. Hand forging process

- It is the process of deforming the hot material/workpiece into the required shape by applying the repeated blows of hammer held in hand. Refer Fig. 7.3.4.

- The hot material is held in one hand by using suitable clamping device and kept on anvil block whereas, hammer blows are given by hammer held in another hand.

- Initial cost of this hand forging equipments is low hence cost of components forged from this process is also low.

- This process is only used for making simple components of small size.

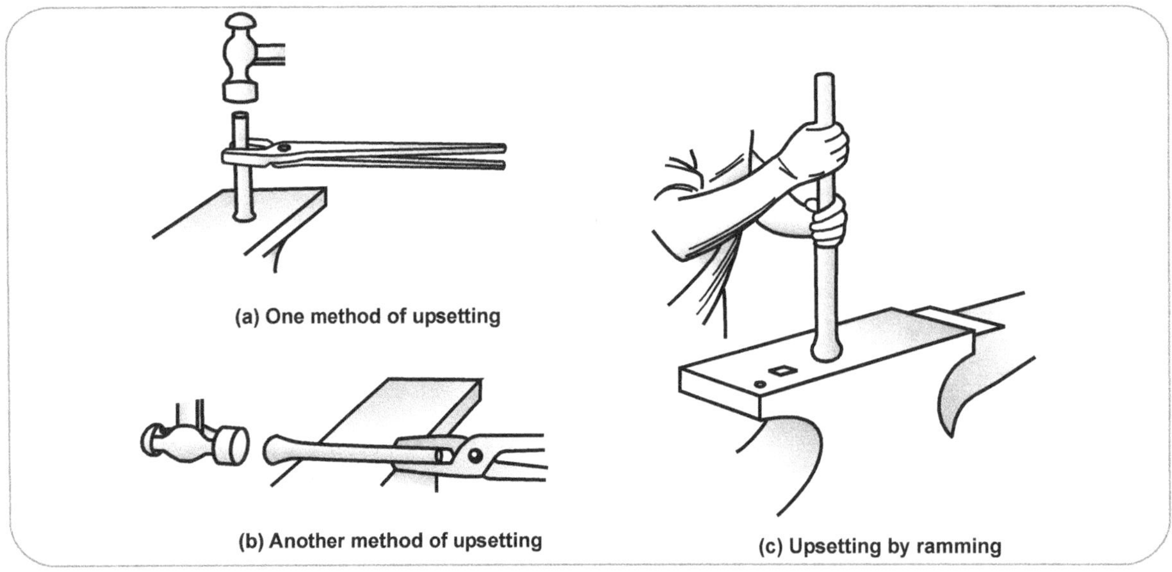

Fig. 7.3.4 Hand forging process

- This process is not used for high production rate.

2. Drop forging process

- This process utilises closed impression die to obtain the required shape of the component.
- The dies are matched and separately attached to the movable ram and the fixed anvil. Refer Fig. 7.3.5.

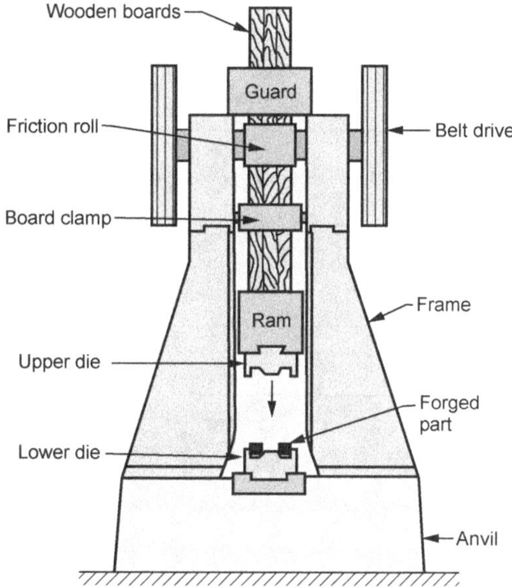

Fig. 7.3.5 Drop forging process

- The forging is produced by the impact or pressure which forces the hot metal (billet) to form the shape of the dies.
- During the operation, there is a drastic flow of metal in the dies caused by repeated blows of hammers on the metal.
- To ensure proper flow of the metal, the operation is divided into number of steps. Each step changes the form gradually, until the final shape is obtained.
- The equipment used for applying the blows is called as **drop hammer.**
- During the process, the ram is raised to a definite height and then it is allowed to drop or fall freely under its own weight.
- The workpiece is kept on the lower die while the ram delivers four to five blows on the metal in quick succession, hence the metal spreads and fills the die cavity.

- Drop forging process is less expensive as compared to other forging processes.

Applications

- This process is used for manufacturing of car axles, connecting rods, crankshafts, leaf springs, crane hooks, jet engine blades, etc.

3. Press forging process

- In press forging process, the hammering action is relatively slow squeezing instead of delivering heavy blows. Refer Fig. 7.3.6.
- Press forging process is more accurate than the drop forging because in press forging metal penetrates deeply as it gives the time to flow the metal.
- The life of the press and dies is longer than that of the hammer.

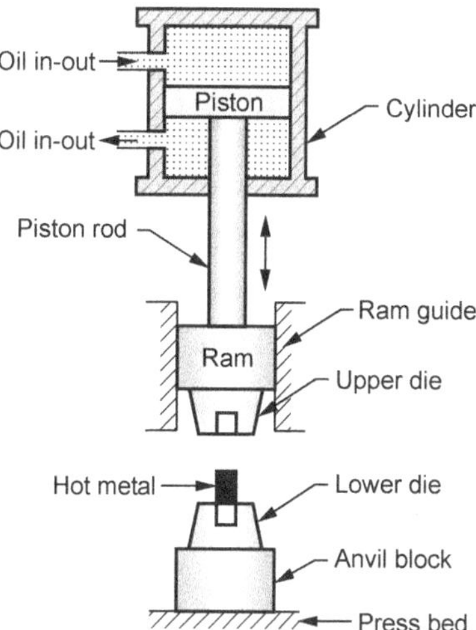

Fig. 7.3.6 Press forging process

- The process does not require highly skilled operator because the speed, pressure, etc. are automatically controlled.
- In press forging process, ram is connected to the piston rod which is operated by piston.
- For motion of piston, two ports are provided in the cylinder from which hydraulic fluid (oil) flows.
- During the downward motion, oil is supplied from the upper port similarly, during the upward motion oil is supplied from the lower port.
- In this process, there are less vibrations and noise as compared to drop forging.

Applications

- This process is used for manufacturing of large levers flanges, propellers, hollow bodies, railway wheel disks, panels, etc.

4. Machine or Upset forging process

- It consists of applying pressure longitudinally on a hot bar, which is gripped firmly between grooved dies, to upset a required portion of its length.

- All forgeable metals can be upset through this process.

- They may have any shape of cross-section, but round shape is most commonly used. The equipment used for this type of forging is known as forging machine or upsetter.

- The machine provides forging pressure in a horizontal direction.

- The dies are so designed that, the complete operation is performed in several stages and the final shape is attained gradually.

- The operation is performed by using die and punch which is called as **heading tool**, as shown in Fig. 7.3.7.

- The die is either made hollow to receive the round bar through it or in two parts to open out and receive the bar.

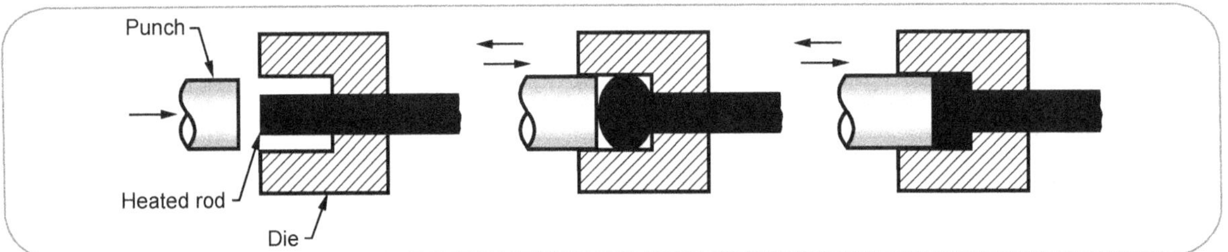

Fig. 7.3.7 Upset forging

Applications :

- This process is used in the manufacturing of all types of bolts, nuts washers, collars, gear blanks, etc.

7.3.3 Comparison between Forging and Casting Processes

Sr. No.	Forging	Casting
1.	In forging process, grain flow is continuous and uninterrupted. Refer Fig.	In casting process, there is no grain flow. Refer Fig.
2.	Due to improved grain size and true grain flow, forging give greater strength and toughness.	Due to no grain flow and weak crystalline structure, casting is weak in withstanding working stresses.
3.	Requires minimum machine finish.	Requires more machine finish.
4.	Forged components have better mechanical properties like strength, toughness, resistance to shock and vibrations.	Cast components are brittle i.e. weak in tension. Also they have poor resistance to shock and vibrations.

5.	Welding of forged parts is easy.	Welding of cast parts is difficult.
6.	During the operation, cracks and blow holes are welded up.	Defects like cracks and blow holes make the casting weak and unsuitable for use.
7.	Accuracy is more.	Accuracy is less.
8.	Complicated shapes cannot be produced.	Complicated shapes can be produced.
9.	Generally used for large parts.	Generally used for small parts.
10.	Because of cost of dies, process is costly.	As there are no dies, casting is less expensive.

7.4 : Mechanical Working of Metals SPPU : May-14

- **Mechanical working** of a metal is a simply plastic deformation performed to change the dimensions, properties and surface conditions with the help of mechanical pressure.
- Depending upon the temperature and strain rate, mechanical working may be either **hot working** or **cold working,** such that recovery process takes place simultaneously with the deformation.
- During deformation the metal is said to flow, which is called as **plastic flow** of the metal and grain shapes are changed.
- If the deformation is carried out at higher temperatures, then the new grains start growing at the locations of internal stresses.
- When the temperature is sufficiently high, the grain growth is accelerated and continues till the metal comprises fully of new grains only.
- This process of formation of new grains is called as **recrystallisation** and the corresponding temperature is the **recrystallisation temperature** of the metal.
- Recrystallisation temperature is the point which differentiates hot working and cold working.
- Mechanical working of metals above the recrystallisation temperature, but below the melting or burning point is known as **hot working** whereas; below the recrystallisation temperature, is known as **cold working.**

7.4.1 Hot Working

- Hot working is accomplished at a temperature above the recrystallisation temperature but below the melting or the burning point of the metal, because above the melting or the burning point, the metal will burn and become unsuitable for use.
- Every metal has a characteristic hot working temperature range over which hot working may be performed.
- The upper limit of working temperature depends on composition of metal, prior deformation and impurities within the metal.
- The changes in structure from hot working improves mechanical properties such as ductility, toughness, resistance to shock and vibration, % elongation, % reduction in area, etc.
- Hot working of metals have following advantages and disadvantages :

Advantages

- Due to hot working, no residual stresses are introduced in the metal.
- Hot working refines grain structure and improves physical properties of the metal.
- Any impurities in the metal are disintegrated and distributed throughout the metal.
- Porosity of the metal is minimised by the hot working.

- During hot working, as the metal is in plastic state, larger deformation can be accomplished and more rapidly.
- Hot working produces raw material which is to be used for subsequent cold working operations.

Disadvantages

- As hot working is carried out at high temperatures, a rapid oxidation or scale formation takes place on the metal surface which leads to poor surface finish and loss of metal.
- Due to the loss of carbon from the surface of the steel piece being worked, the surface layer loses its strength.
- This weakening of the surface layer may give rise to fatigue crack which results in failure of the part.
- Close tolerances cannot be obtained.
- Hot working involves excessive expenditure on account of high tooling cost.

7.4.2 Cold Working

- The working of metals at temperatures below their recrystallisation temperature is called as **cold working**.
- Most of the cold working processes are performed at room temperature.
- Unlike hot working, it distorts the grain structure and does not provide an appreciable reduction in size.
- Cold working requires much higher pressure than hot working.
- If the material is more ductile, it can be more cold worked.
- Residual stresses are setup during the process, hence to neutralise these stresses a suitable heat treatment is required.
- Cold working of metals have following advantages and disadvantages :

Advantages

- Better dimensional control is possible because there is not much reduction in size.
- Surface finish of the component is better because no oxidation takes place during the process.
- Strength (tensile strength and yield strength) and hardness of metal are increased.
- It is an ideal method for increasing hardness of those metals which do not respond to the heat treatment.

Disadvantages

- Ductility of the metal is decreased during the process.
- Only ductile metals can be shaped through the cold working.
- Over-working of metal results in brittleness and it has to be annealed to remove this brittleness.
- To remove the residual stresses setup during the process, subsequent heat treatment is mostly required.

7.4.3 Comparison between Hot Working and Cold Working
SPPU : May-14, Dec.-18

Sr. No.	Hot working	Cold working
1.	Hot working is carried out above the recrystallisation temperature but below the melting point, hence deformation of metal and recovery takes place simultaneously.	Cold working is carried out below the recrystallisation temperature and as such there is not appreciable recovery of metal.
2.	During the process, residual stresses are not developed in the metal.	During the process, residual stresses are developed in the metal.
3.	Because of higher deformation temperature used, the stress required for deformation is less.	The stress required to cause deformation is much higher.
4.	Hot working refines metal grains, resulting in improved mechanical properties.	Cold working leads to distortion of grains.
5.	No hardening of metal takes place.	Metal gets work hardened.
6.	If the process is properly performed, it does not affect ultimate tensile strength, hardness, corrosion and fatigue resistance of the metal.	It improves ultimate tensile strength, yield and fatigue strength but reduces corrosion resistance of the metal.
7.	It also improves some mechanical properties like impact strength and elongation.	During the process, impact strength and elongation are reduced.
8.	Due to oxidation and scaling, poor surface finish is obtained.	Cold worked parts carry better surface finish.
9.	Close dimensional tolerances cannot be maintained.	Superior dimensional accuracy can be obtained.
10.	Hot working is most preferred where heavy deformation is required.	Cold working is preferred where work hardening is required.

7.5 : Metal Forming

- Metal forming includes a large number of manufacturing processes in which plastic deformation property is used to change the shape and size of metal workpieces.

- During the process, for deformation purpose, a tool is used which is called as die. It applies stresses to the material to exceed the yield strength of the metal.

- Due to this the metal deforms into the shape of the die. Generally, the stresses applied to deform the metal plastically are compressive.

- But, in some forming processes metal stretches, bends or shear stresses are also applied to the metal.

- For better forming of metal, the desirable properties of metal are low yield strength and high ductility.

- These properties are highly affected by the temperature. When the temperature of the metal is increased, its ductility increases and yield strength decreases.

- The other factors which affect the performance of metal forming process are, strain rate, friction, lubrication, etc.

- Metal forming processes can be classified as follows :

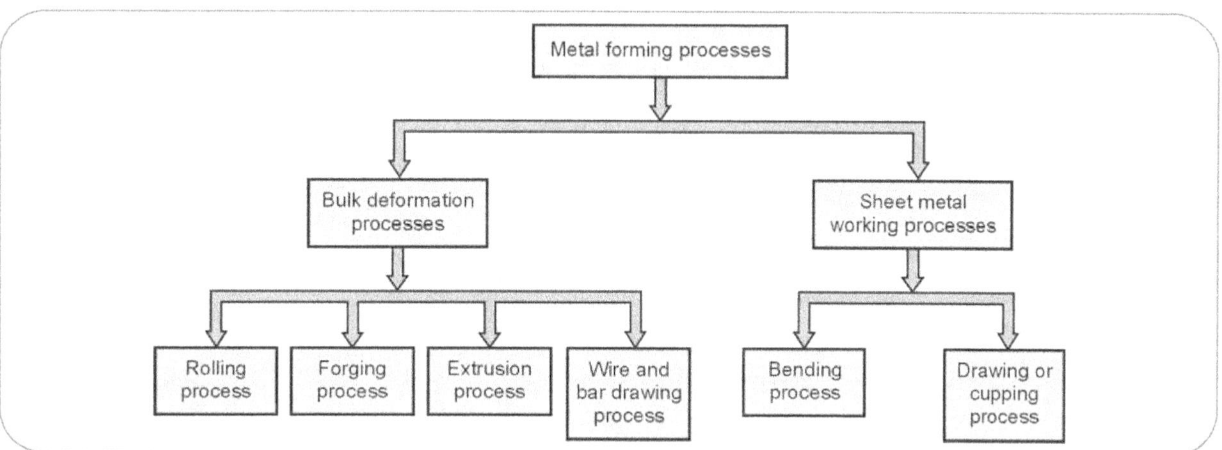

Fig. 7.5.1 : Classification of metal forming processes

7.5.1 Bulk Deformation Processes

- These processes are characterised by significant deformations and massive shape changes but the surface area to the volume of the work is relatively small.

- The workpieces which have this low area to volume ratio is called as **bulk**.

- Initial workpiece shapes for bulk deformation processes include cylindrical billet (hot material) and rectangular bars.

- Fig. 7.5.2 shows the basic operations in bulk deformation process.

1. Rolling

- It is a compressive deformation process in which the thickness of a plate or slab (hot) is reduced by two opposing cylindrical rolls.

- The rolls rotate in order to draw the workpiece into the gap between them and squeeze the workpiece. Refer Fig. 7.5.2 (a).

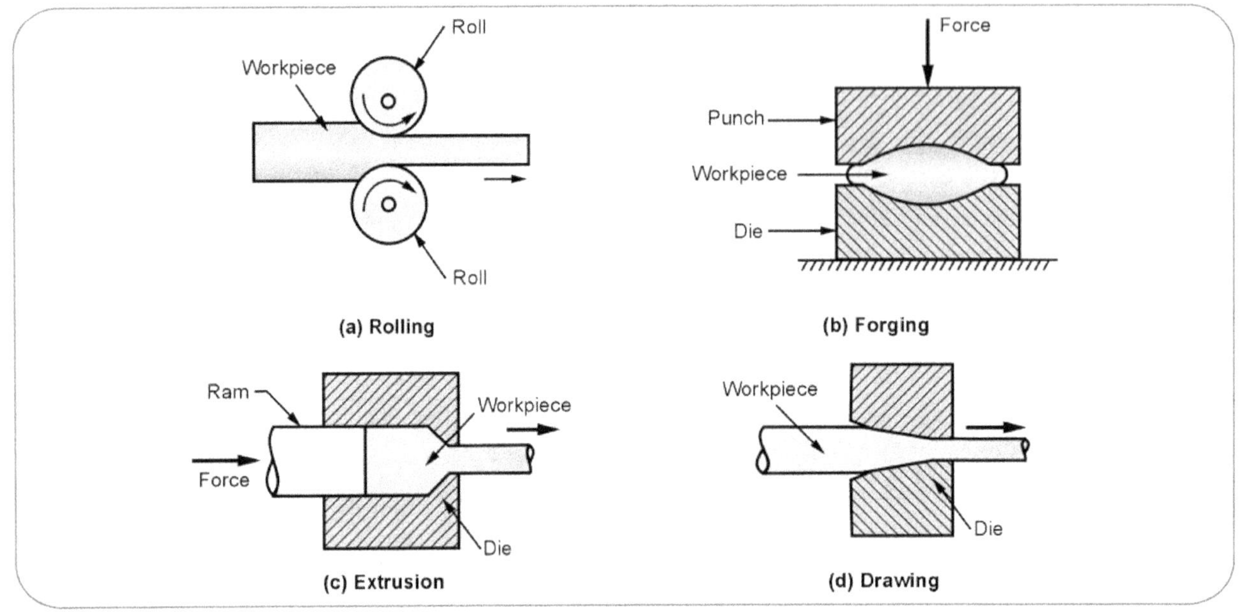

Fig. 7.5.2 : Basic bulk deformation processes

2. Forging

- In this process, the workpiece is compressed between the two opposing dies in order to produce the die shapes on the workpiece. Refer Fig. 7.5.2 (b).

- It is generally a hot working process but sometimes it can be included in cold working also.

3. Extrusion

- It is a compressive deformation process in which the work metal is forced to flow through a die opening as shown in Fig. 7.5.2 (c).

- During the flow through a die, the work metal takes the shape of the opening as its cross-section.

4. Wire drawing

- In this type of forming process, the diameter of a round bar (billet) is reduced by pulling it through a die opening.

- Fig. 7.5.2 (d) shows the drawing process.

7.5.2 Sheet Metal Working Processes

- In this type of metal forming processes, the operations are performed on metal sheets, strips and coils.

- In these processes, the surface area to *volume ratio* is high.

- Generally, the sheet metal working processes are carried out on **punching press machine**, hence sheet metal working is also called as **pressworking**.

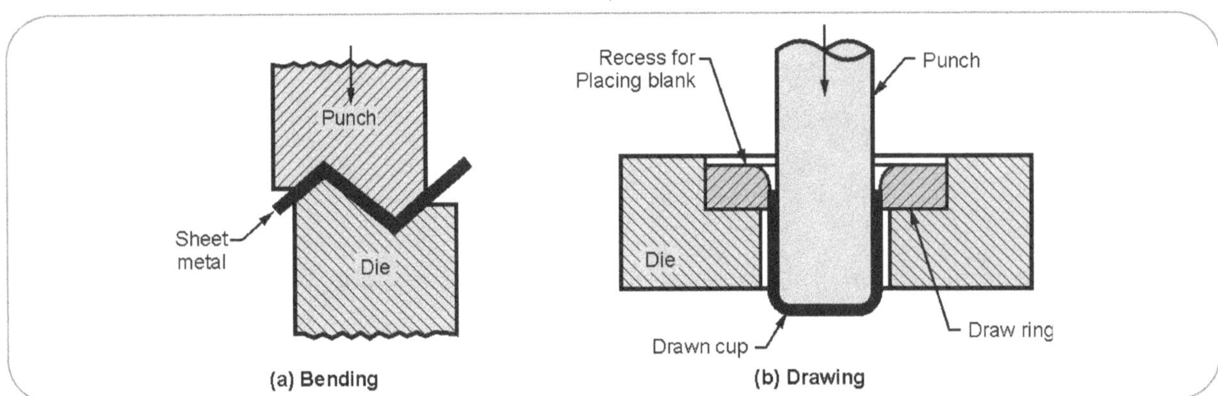

Fig. 7.5.3 : Basic sheet metal working operations

- A component produced by sheet metal working process is called as stamping.
- These operations are performed as cold working processes. The tools used for the operations is called as **punch** and **die**.
- The punch is a positive portion whereas the die is a negative portion of the tool set.
- Fig. 7.5.3 shows the basic operations in sheet metal working process. (Refer Fig. 7.5.3 on previous page)

1. Bending

- In this process, there is straining of metal sheet or plate to take an angle along a straight axis. Refer Fig. 7.5.3 (a).
- The bending may be of V shape, U shape or any other shape.

2. Drawing or Cupping

- It refers to the forming of a flat metal sheet into a hollow or concave shape like a cup by stretching the metal.
- During the process, a blank holder is used to hold the blank and the punch pushes into the sheet metal. Refer Fig. 7.5.3 (b).

7.6 : Extrusion

- Extrusion is a compression process in which the work metal is forced to flow through a small opening which is called as **die** to produce a required cross-sectional shape.
- The extrusion process is similar to squeezing toothpaste or cream from a tube.
- Almost any solid or hollow cross-section may be produced by extrusion process.

- As the geometry of the die remains same during the operation, extruded parts have the same cross-section.
- During the process, a heated cylindrical billet is placed in the container and it is forced out through a steel die with the help of a ram or plunger.
- The products made by extrusion process are tubes, rods, railings for sliding doors, structural and architectural shapes, door and window frames, etc.
- Extrusion process is suitable for the non-ferrous alloys, steel alloys, non-ferrous metals, stainless steel, etc.
- Extrusion process is carried out on horizontal hydraulic press machines which are rated from 250 to 5500 tonnes in capacity.
- Extrusion process is classified as follows :
1. *According to physical configuration*
    a) Direct (Forward) extrusion
    b) Indirect (Backward) extrusion
2. *According to working temperature*
    a) Hot extrusion
    b) Cold extrusion

7.6.1 Direct Extrusion (Forward Extrusion)

- **Direct** or **forward hot extrusion** is most widely used and the maximum numbers of extruded parts are produced by this method.
- Fig. 7.6.1 shows the direct extrusion process in which the raw material is a billet.
- A billet is heated to its forging temperature and fed into the machine chamber.

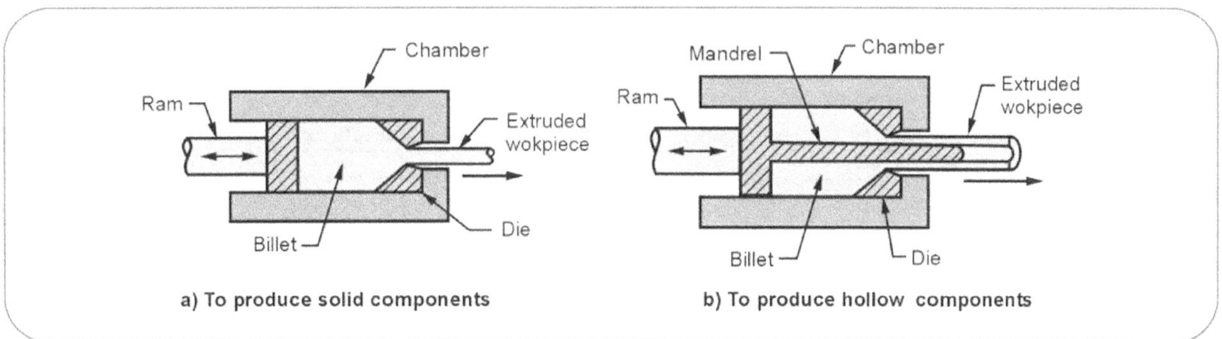

Fig. 7.6.1 : Direct Extrusion

- Pressure is applied to the billet with the help of ram or plunger which forces the material through the die.

- The length of extruded part will depend on the billet size and cross-section of the die.

- The extruded part is then cut to the required length.

- As the ram approaches the die, a small portion of billet remains which cannot be forced through the die opening. This extra portion is known as **butt** which is separated from the product at the end.

- When the billet is forced to flow through the die opening, there is friction between the workpiece and chamber walls. This friction is overcome by providing additional ram force. This is the major problem with this process.

- To overcome this problem oxide layer is provided on the billet or dummy block is used between the ram and billet.

- Direct extrusion process is also used to produce hollow or semi-hollow sections.

- To produce hollow sections, by direct extrusion process, a *mandrel* is used. Refer Fig. 7.6.1 (b).

- When the billet is compressed, the material is forced to flow through the gap between the mandrel and die opening. This results in tubular cross-section.

7.6.2 Indirect Extrusion (Backward Extrusion)

- Indirect extrusion is also called as **backward extrusion**.

- In this type, the ram or plunger used is hollow and as it presses the billet against the backwall of the closed chamber, the metal is extruded back into the plunger. Refer Fig. 7.6.2.

- It involves no friction between the metal billet and the chamber because the billet does not move inside the chamber.

- As compared to direct extrusion, less total force is required in this method.

- But the equipment used is mechanically complicated in order to support the passage of the extruded shape through the centre of the hollow ram.

- Indirect extrusion is also used to solid as well as hollow components. For producing solid parts, ram is hollow whereas for producing hollow parts ram is solid. Refer Fig. 7.6.2 (b).

7.6.3 Comparison between Direct and Indirect Extrusion

Sr. No.	Direct / Forward extrusion	Indirect / Backward extrusion
1.	During the process a solid ram is used.	A hollow ram is used for the process.
2.	Flow of metal is in the same direction as the movement of the ram.	Flow of metal is in the opposite direction as the movement of the ram.
3.	A dummy block is used during the operation.	Dummy block may or may not be used. Die plays a part of the dummy block.

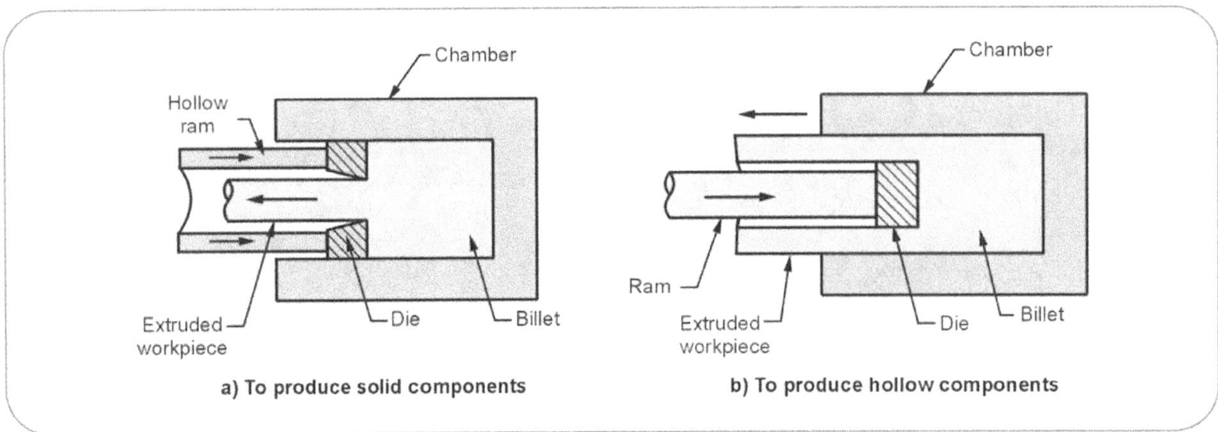

Fig. 7.6.2 : Indirect or backward or reverse extrusion

4.	Die is mounted on the cylinder or container.	Die is mounted over the bore of the ram.
5.	Because of relative motion between the heated metal billet and the cylinder walls, friction problem arises.	As the billet in the container remains stationary, there is no friction.
6.	Large amount of force is required to move the billet in the cylinder.	As the billet is stationary, process does not require large amount of force.
7.	Handling of extruded metal is very easy.	Handling of extruded metal is difficult.

7.7 : Wire Drawing

- **Drawing** is an operation in which the cross-section of a bar, rod or wire is reduced by pulling it through a die opening.

- The general features of the drawing process are similar to extrusion. But the difference is that, *in drawing the workpiece is pulled through the die whereas in extrusion workpiece is pushed* through the die.

- During the process, tensile as well as compressive stresses are produced in the material.

- The main difference beween the bar drawing and wire drawing is the stock size (workpiece size). *Bar drawing is used for large diameter (bar and rod) stock whereas wire drawing is used for small diameter stock.*

- Wire sizes upto 0.03 mm can be drawn by wire drawing process.

- The process consists of pulling the hot drawn bar or rod through a die of which the bore size is similar to the finished product size.

- Depending upon the material to be drawn and the amount of reduction required, total drawing can be accomplished in a single die or in a series of successive dies.

- One end of the rod to be drawn into wire is made pointed, entered through the die and gripped at the other end by using tongs.

- After pulling a certain length, this end is wound to a reel or draw pulley.

- When the pulley or reel is rotated, the rod is pulled through the die and its diameter reduces.

Refer Fig. 7.7.1

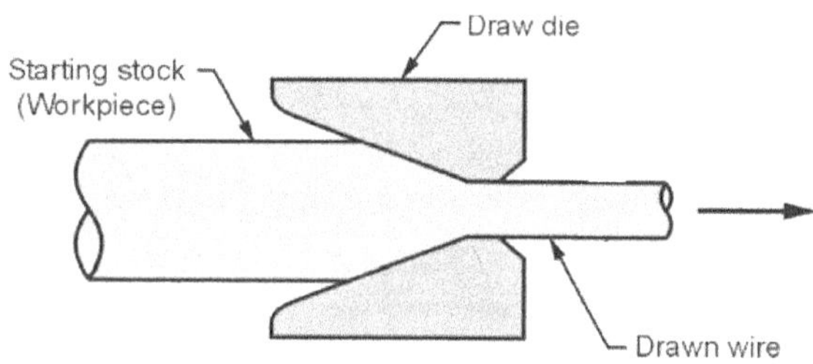

Fig. 7.7.1 : Wire drawing

- The die is made of highly wear resistant material.

- Generally, tungsten carbide is used for die making.

- The die made of tungsten carbide is suitably supported in a die holder which is made of mild steel or brass.

7.7.1 Tube Drawing

- As the initial tubing has been produced by other processes like extrusion, drawing can be used to reduce the diameter or wall thickness of seamless tubes and pipes.

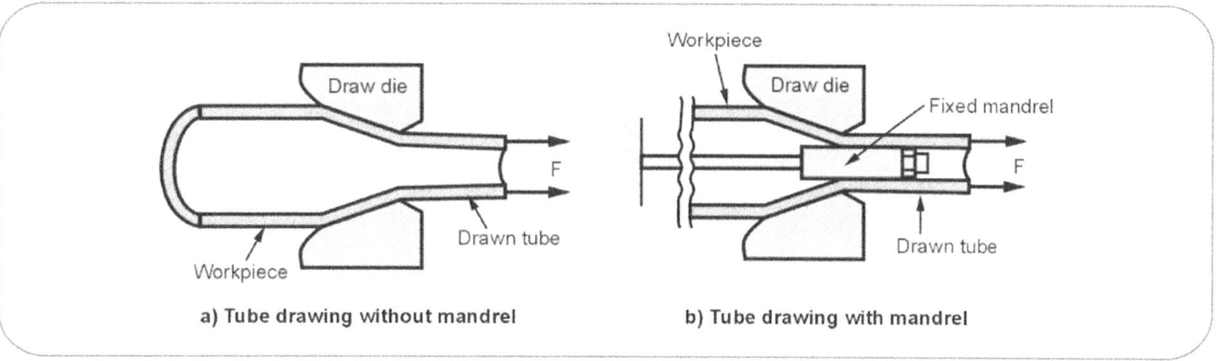

Fig. 7.7.2 : Tube drawing

- Tube drawing can be carried out either with mandrel or without mandrel.

- The simplest method of producing tubes and pipes is shown in Fig. 7.7.2 (a) in which mandrel is not used. This method is also called as **tube sinking**.

- In tube sinking method there is no control over the inner diameter and wall thickness of tube.

- To overcome this drawback, mandrels are used in the process.

- Fig. 7.7.2 (b) shows tube drawing with mandrel. In this method, mandrel is fixed and attached to a long support bar to produce inside diameter and wall thickness during the process.

7.8 : Sheet Metal Working   SPPU : Dec.-03, 04, 11

- Sheet metal working is generally associated with **press machines** and **press working.**

- Press working is a chipless manufacturing process by which various components are produced form sheet metal.

- The thickness of metal varies from 0.1 mm to 10 mm.

- Press machine consists of a frame which supports a ram and bed and a mechanism for operating the ram. Refer Fig. 7.8.1.

- The ram is equipped with **punch** whereas **die block** is attached to the bed.

- The punch and die block assembly is called as **die** or **die-set.**

- Die block is a stationary part which contains die cavity and punch is a moving part which enters in the die cavity.

- During the operation, metal sheet is kept on the die block and punch moves downward. The punch forces the metal sheet into the die cavity, hence metal sheet will form the shape of the die cavity.

- There is always some clearance between the punch and die block.

- On the press machine, various operations can be performed and all the operations are done at the room temperature.

7.8.1 Advantages, Disadvantages and Applications of Sheet Metal Working

Advantages

- Sheet metal working is associated with press machine, on which number of operations can be performed.

- Metal sheets of less thickness can be formed into various shapes.

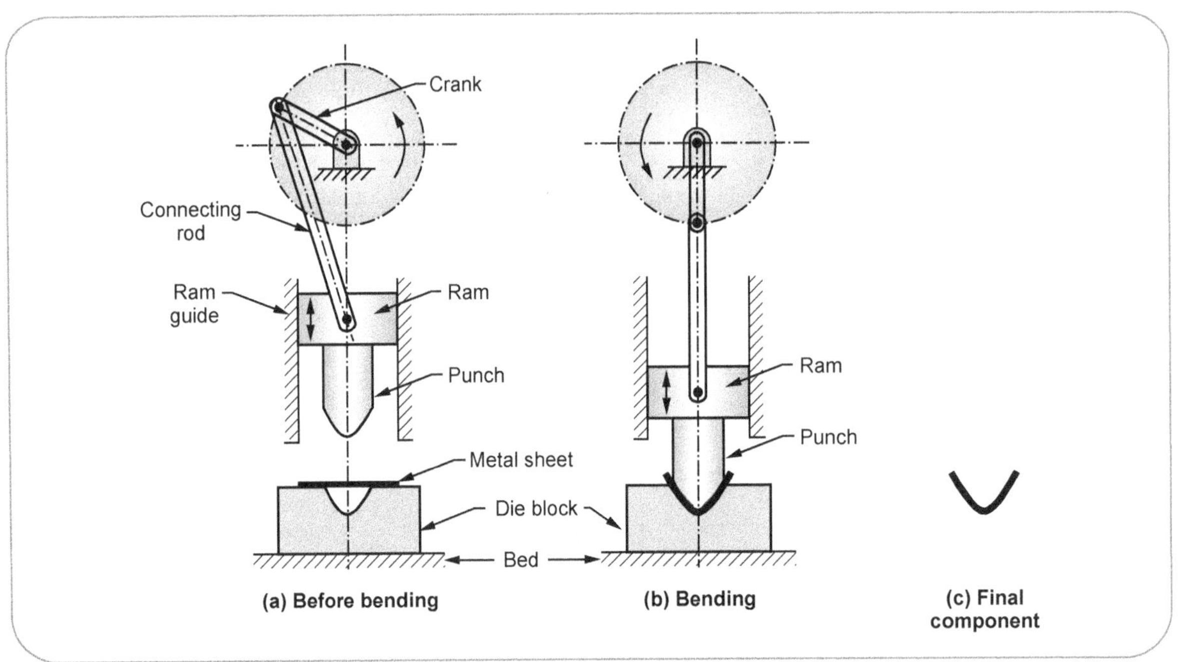

Fig. 7.8.1 Sheet metal working

- The components produced by sheet metal working are of low cost.

- Production rate of press machine is very high.

- The process does not require skilled labour.

Disadvantages

- Sheet metal working is only used for mass production.

- The cost of die is very high.

- Initial cost of press machine is also high.

- Metals of thickness more than 10 mm are difficult to form.

- The operation produces more noise and vibrations.

Applications

The components produced by sheet metal working are as follows :

- Press parts are widely used in automobile (bikes, cars, trucks, buses, etc.) industry. Vehicle parts like doors, roofs, fuel tanks, front guards, etc. can be produced.

- Aircraft industry, radio and telephone industry, electrical parts, etc.

7.9 : Types of Sheet Metal Working

SPPU : May-04,05,06,07,08,09,10,11,12,13,14, Dec.-03,04,05,06,07,08,10,11,12,13,16

Press operations may be grouped into two categories i.e. *cutting operations* and *forming operations*. In cutting operations, the sheet metal is stressed beyond its ultimate strength whereas, in forming operations the stresses are below the ultimate strength of the metal. Refer Fig. 7.9.1.

7.9.1 Metal Cutting Operations

SPPU : May-11,13,19 Dec.-13,18

- In sheet metal cutting operations, the metal gets sheared hence these operations are also called as **shearing operations.**

- In these operations, the metal sheet is stressed beyond its ultimate strength.

- Metal cutting operations include following operations :

1. Blanking	2. Punching (Piercing)
3. Notching	4. Perforating
5. Slitting	6. Lancing
7. Shaving	8. Shearing

1. Blanking

SPPU : May-04, 05, 06, 07, 12, 14, Dec.-05, 06, 08, 16

- Blanking is the cutting operation of a flat metal sheet and the article punched out is known as **blank.**

- Blank is the required product of the operation and the metal left behind is considered as a waste. Refer Fig. 7.9.2 and Fig. 7.9.3 (a).

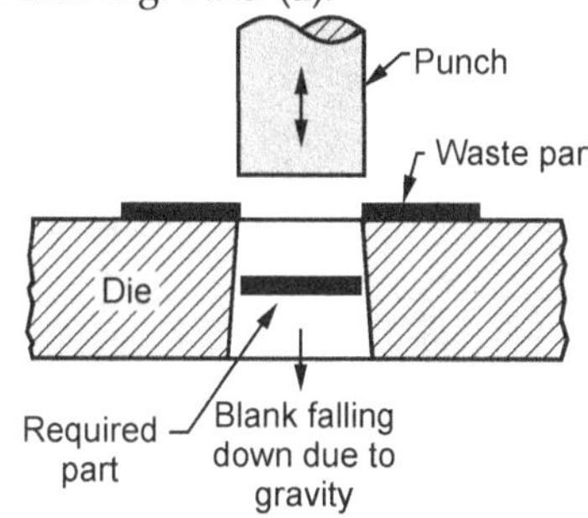

Fig. 7.9.2 Blanking

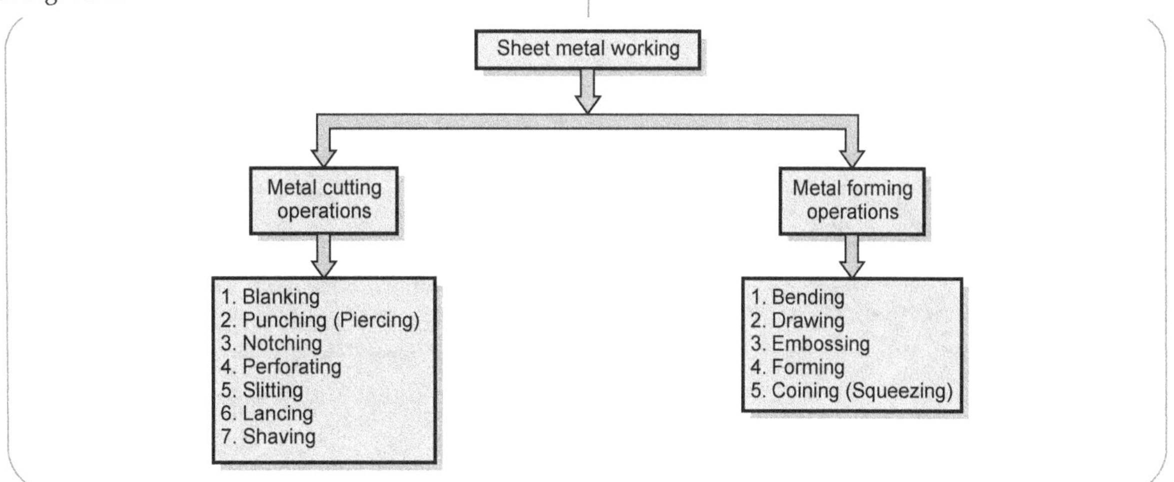

Fig. 7.9.1 Types of sheet metal working

2. Punching (Piercing)

SPPU : May-04, 05, 06, 07, 12, Dec.-05, 06, 10

- It is the cutting operation with the help of which holes of various shapes are produced in the sheet metal.

- It is similar to blanking; only the main difference is that, the hole is the required product and the material punched out to form a hole is considered as a waste. Refer Fig. 7.9.3 (b).

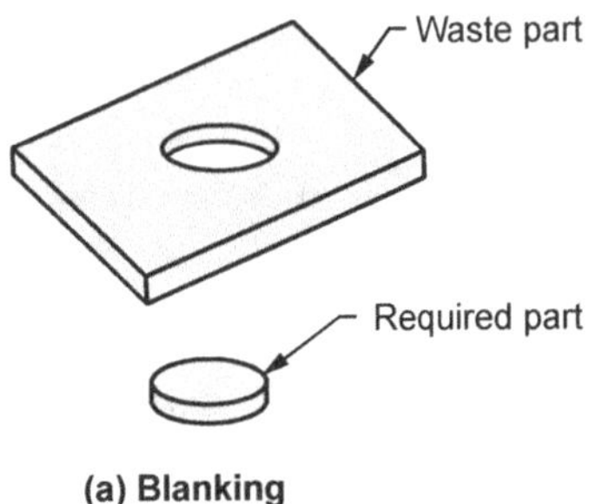

(a) Blanking

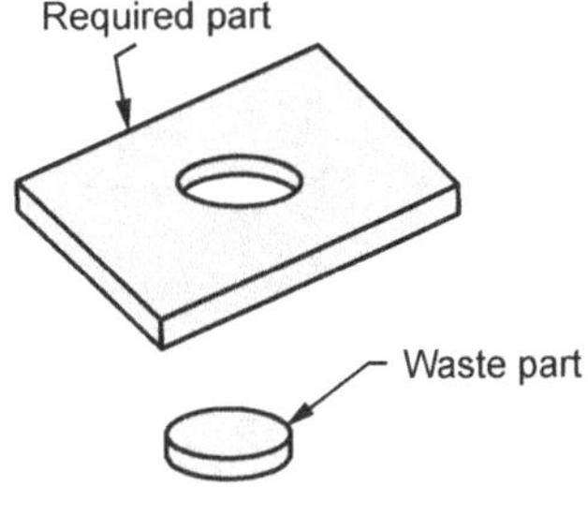

(b) Punching (Pirecing)

Fig. 7.9.3 Blanking and punching (piercing)

3. Notching   **SPPU : Dec.-03, 06, 08, 11, May-05, 08**

- It is similar to blanking operation, but in this full surface of punch does not cut the metal.

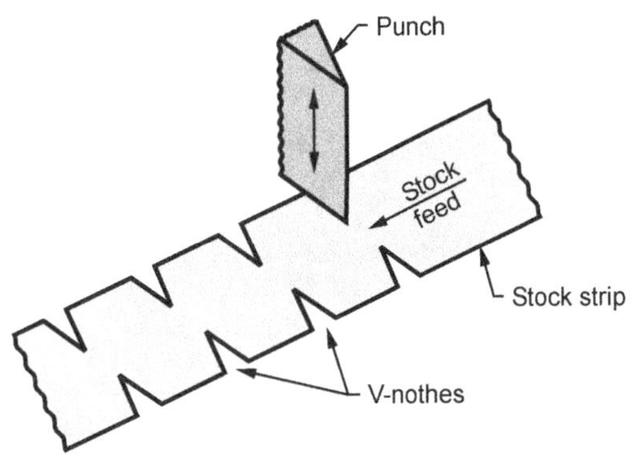

Fig. 7.9.4 Notching

- In this operation, metal pieces are cut from the edges of a sheet. Refer Fig. 7.9.4.

4. Perforating   **Dec.-07, 10, 11, May-12**

- It is similar to piercing but the difference is that, to produce holes the punch is not of round shape.

- In this process, multiple holes which are very small and close together are cut in the sheet metal. Refer Fig. 7.9.5.

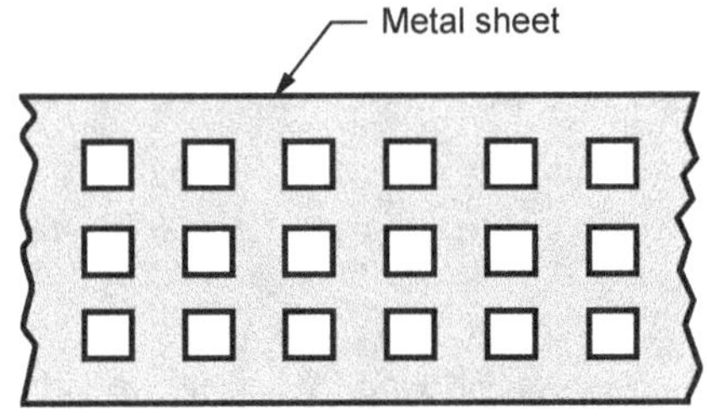

Fig. 7.9.5 Perforating

5. Slitting

- It is the operation of making an unfinished cut through a limited length only. Refer Fig. 7.9.6.

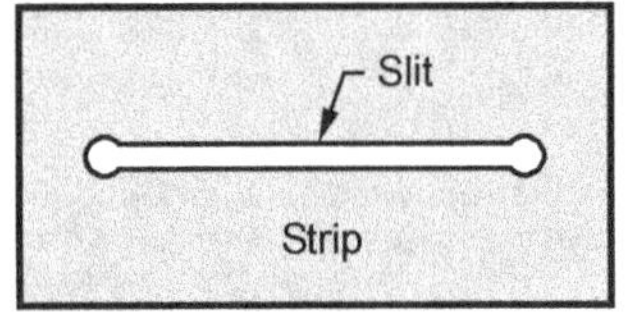

Fig. 7.9.6 Slitting

6. Lancing   **May-12**

- In this operation, there is a cutting of sheet metal through a small length and bending this small cut portion downwards. Refer Fig. 7.9.7.

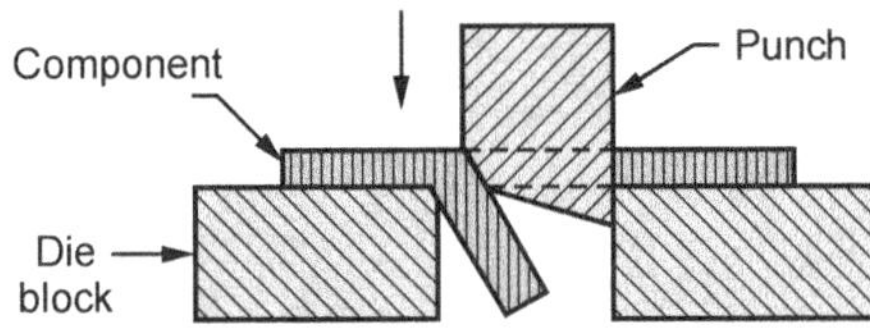

Fig. 7.9.7 Lancing

7. Shaving

- This operation is used for cutting unwanted excess material from the periphery of a previously formed workpiece. Refer Fig. 7.9.8.

- In this process very small amount of material is removed.

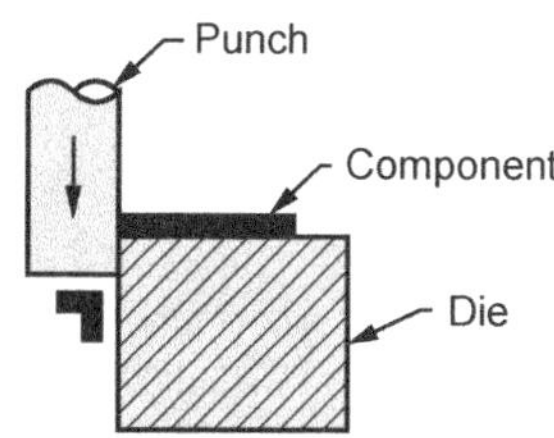

Fig. 7.9.8 Shaving

8. Shearing   `SPPU : Dec.-07,12, May-09, 10`

- It is an operation through which a metal is cut along a single line which is generally a straigh line. Refer Fig. 7.9.9.

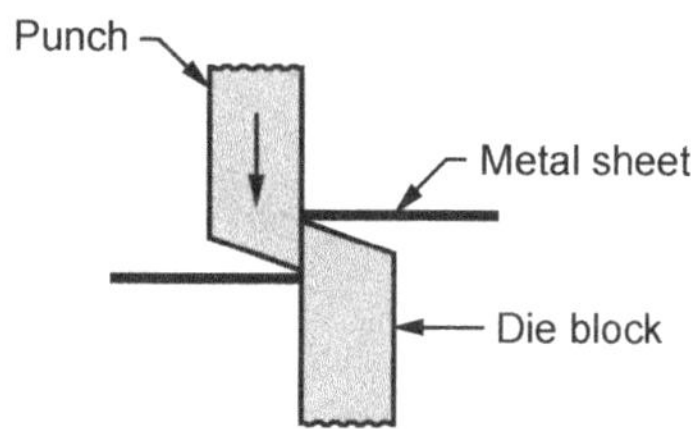

Fig. 7.9.9 Shearing

7.9.2 Metal Forming Operations   `SPPU : Dec.-13`

- In metal forming operations, the sheet metal is stressed below the ultimate strength of the metal.

- In these operations, no material is removed hence there is no wastage.

- Metal forming operations include following operations :

1. Bending	2. Drawing	3. Embossing
4. Forming	5. Coining (Squeezing)	

1. Bending   `SPPU : Dec.-06, 10, 12, May-09, 10`

- It is a metal forming operation in which the straight metal sheet is transformed into a curved form.

- In bending operations, the sheet metal is subjected to both tensile and compressive stresses.

- During the operation, plastic deformation of material takes place beyond its elastic limit but below its ultimate strength.

- The bending methods which are commoly used are as follows :

a) U-bending	b) V-bending
c) Angle bending	d) Curling

a) U-bending   `SPPU : Dec.-08`

- Fig. 7.9.10 shows U-bending operation which is also called as **channel bending.**

- In this operation, the die cavity is in the form of U, due to which component forms the shape of U.

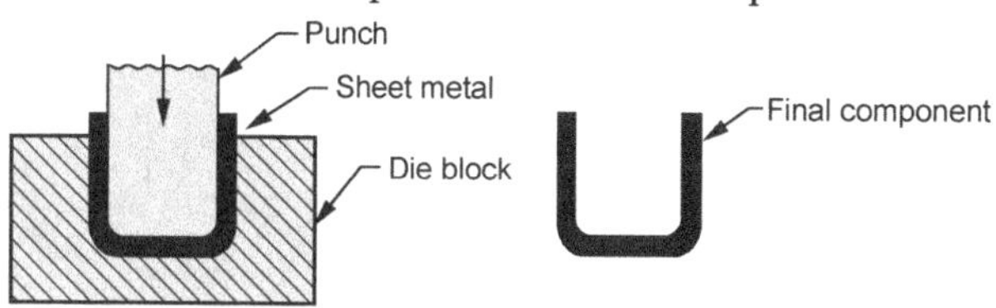

Fig. 7.9.10 U-bending

b) V-bending   `SPPU : Dec.-08`

- Fig. 7.9.11 shows V-bending operation in which wedge shape punch is used.

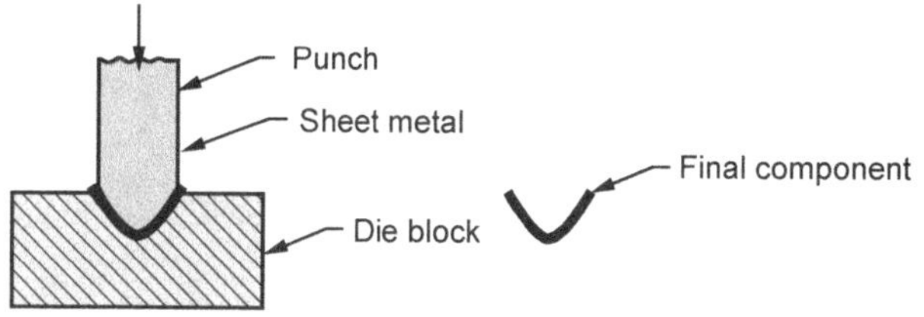

Fig. 7.9.11 V-bending

- The angle of V may be acute, 90° or obtuse.

c) Angle bending   `SPPU : Dec.-03, 04, 06, 11, May-06`

- In this operation, there is a bending of a sheet metal at a sharp angle. Refer Fig. 7.9.12.

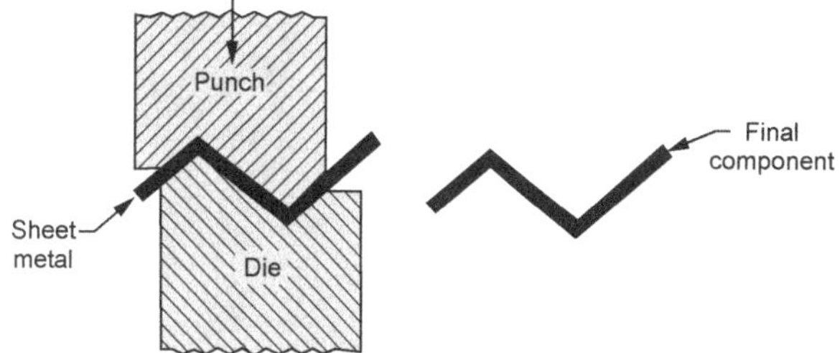

Fig. 7.9.12 Angle bending

d) Curling   `SPPU : May-04, Dec.-05, 06, 07, 12`

- In this operation, the edge of a sheet metal is curled around.

- The punch and die both are made to contain the cavity for cutting partially.

- After the operation, punch moves up and workpiece is ejected out with the help of plunger as shown in Fig. 7.9.13. (See Fig. 7.9.13 on next page)

- This process is used in the manufacturing of drums, pots, vessels, pans, etc.

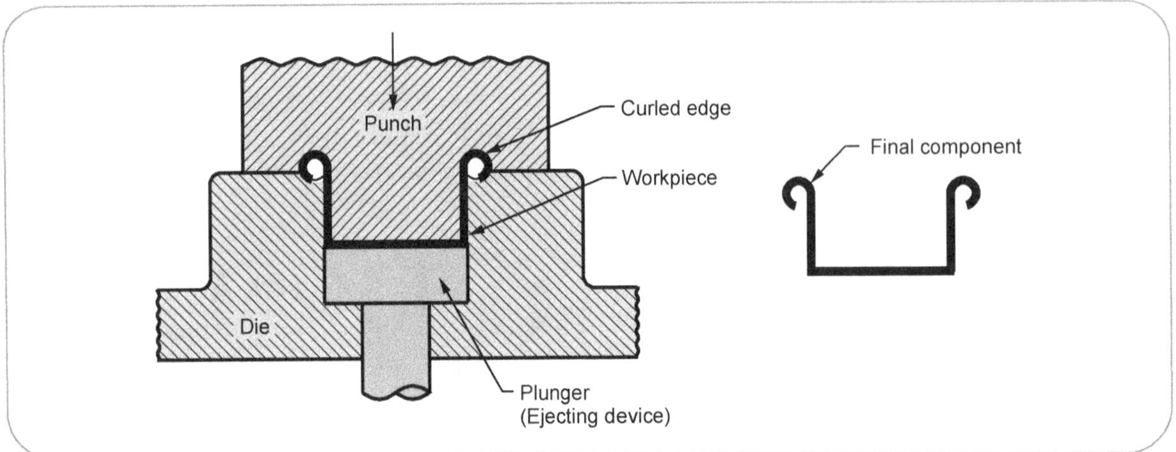

Fig. 7.9.13 Curling

2. Drawing

SPPU : May-04, 09, 10, 14, Dec.-12, 16

- In this operation, punch forces a sheet metal blank to flow plastically into the clearance between the punch and die.
- Finally, the blank takes a shape of cup. Refer Fig. 7.9.14.

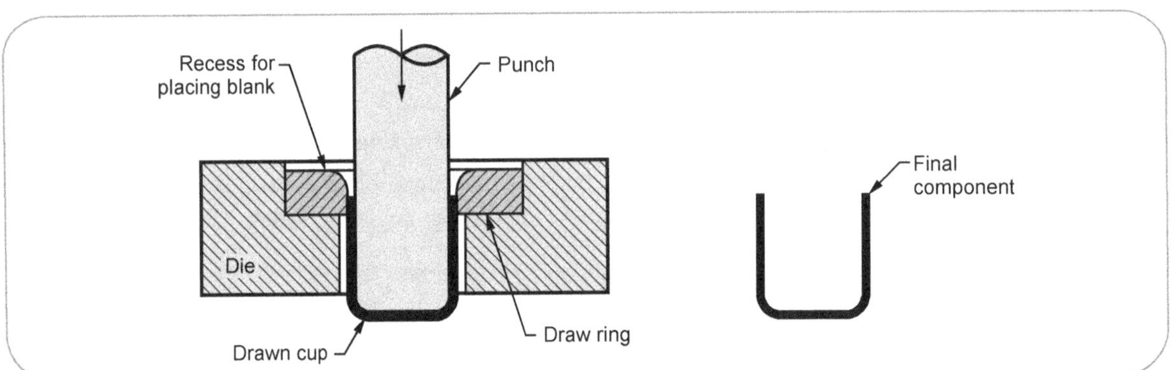

Fig. 7.9.14 Drawing

3. Embossing

SPPU : Dec.-03, 10, May-08

- With the help of this operation, specific shapes or figures are produced on the sheet metal.
- It is used for decorative purposes or giving details like names, trade marks, specifications, etc. on the sheet metal. Refer Fig. 7.9.15.

4. Forming

SPPU : Dec.-04, 05, 12, 16

- In forming operation, sheet metal is stressed beyond its yield point so that it takes a permanent set and retains the new shape.

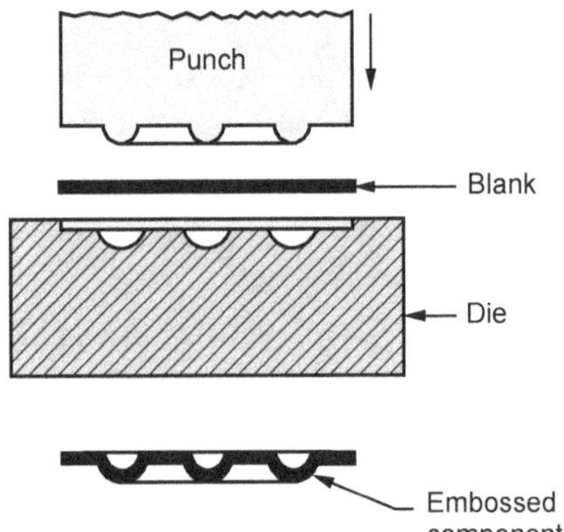

Fig. 7.9.15 Embossing

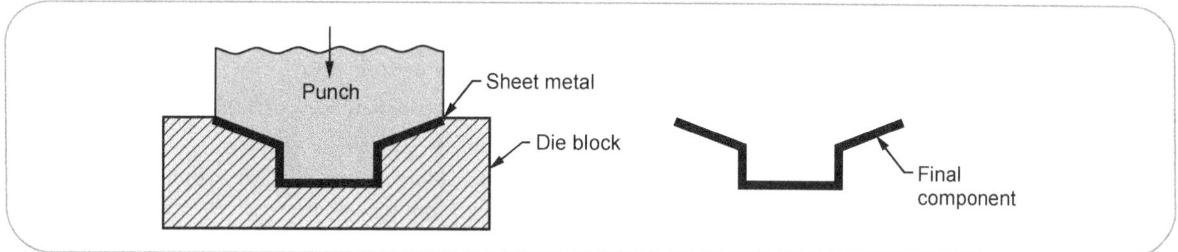

Fig. 7.9.16 Forming

- In this process, the shape of punch and die surface is directly reproduced without any metal flow. Refer Fig. 7.9.16.

- This operation is used in the manufacturing of door ponels, steel furnitures, air-craft bodies, etc.

5. Coining (Squeezing)　　May-09, 10, Dec.-12

- In coining operation, the metal having good plasticity and of proper size is placed within the punch and die and a tremendous pressure is applied on the blank from both ends. Refer Fig. 7.9.17.

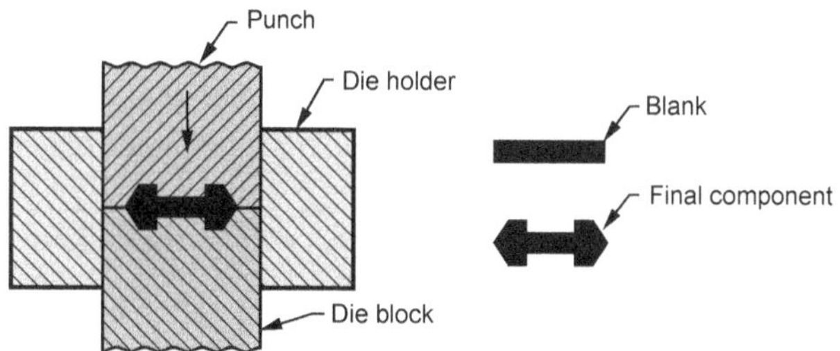

Fig. 7.9.17 Coining

- Under severe compressive loads, the metal flows in the cold state and fills up the cavity of the punch and die.

- This operation is used in the manufacturing of coins, medals, ornamental parts, etc.

7.9.3 Deep Drawing　　SPPU : May-05, Dec.-06

- It is a process of making the cup-shaped parts from a flat sheet-metal blank.

- To provide necessary plasticity for working, the blank is first heated and then placed in position over the die or cavity. Refer Fig. 7.9.19.

- The punch descends and pushes the metal through the die to form a cup; hence this process is also called as **cupping**.

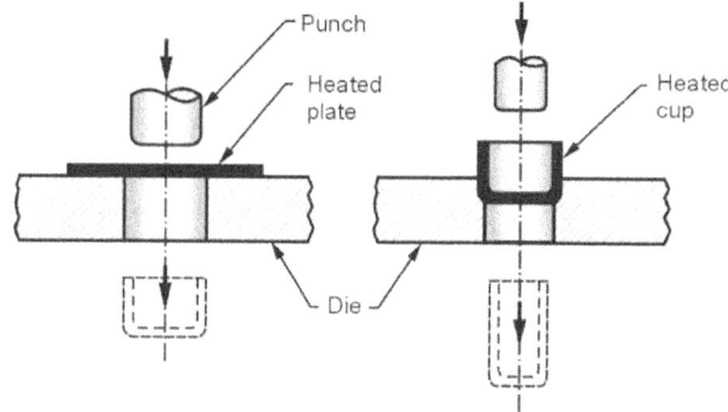

Fig. 7.9.19 Deep drawing

- To obtain cup shaped pieces of the desired size and wall thickness, the process may be continued through a series of successively smaller dies and punches.

7.10 : Metal Joining Processes

- Joining processes are used for joining metal parts and in general fabrication work.

- Such requirements generally occur when large/lengthy sections are required or when several pieces are to be joined together to fabricate a required structure.

- These processes are also applied when steam or water-tight joints are needed.

- Temporary or permanent type of fastening is also enabled by these processes.

- The commonly used joining processes are as follows :

1. Welding	2. Soldering
3. Brazing	4. Adhesive bonding

7.11 : Welding Process

SPPU : Dec.-04,11,17, May-05,18

- It is a joining process used for various materials.

- The large varieties of materials that can be welded are metals and their alloys, although the term welding is also applied to the joining of other materials such as thermoplastics.

- Welding joins various metals with the help of a number of processes in which heat is supplied either electrically or by using a gas torch.

- For joining two or more pieces of metal together by one of the welding processes, the most essential requirement is **heat**.

- **Pressure** is also required, but not in all the processes.

7.11.1 Advantages, Disadvantages and Applications of Welding Process

Advantages

- A large number of metals or alloys, both similar and dissimilar can be joined by welding.

- Welding can be mechanized.

- A good weld is as strong as the base metal.

- Welding can join workpieces through spots, end to end and in a number of other configurations.

- General welding equipments are not very costly.

- Portable welding equipments are also available.

Disadvantages

- Welding gives out harmful radiations, fumes and spatter.

- Welding results in residual stresses and distortion of the workpieces.

- To hold and position the parts to be welded, generally jigs and fixtures are required.

- To produce a good welding job, a skilled welder is required.

- Heat generated due to welding produces metallurgical changes hence, the structure of welded joint is different than that of parent metal.

- Before welding, generally edge preparation of the workpieces is required.

Applications

Some of the important applications of welding process are as follows :

- Aircraft construction (Welding of engine parts, turbine frame, ducts, etc.)

- Rail-road equipments (Air receiver, engine, front and rear hoods, etc.)

- Pipings and pipelines (Open pipe joints, oil and gas pipelines, etc.).

- Pressure vessels (boilers) and tanks (Joining of nozzles, ends, tanks, etc.).

- Buildings and bridges (Columns base plates, erection of structures, etc.).

- Automobile parts (Trucks, buses, cars, cranes, bikes parts, etc.).

- Machine parts (Frames, beds, tools, dies, etc.).

7.11.2 Classification of Welding Process

SPPU : May-05

- There are different ways of classifying welding and allied processes.

- Modern methods of welding may be classified under two broad categories i.e. plastic welding and fusion welding.

- In **plastic (pressure) welding,** the metal pieces to be joined are heated to a plastic state and then forced together by external pressure. In this method filler metal is not required.
  For example, forge welding, electric resistance welding, etc.

- In **fusion (non-pressure) welding,** the material at the joint is heated to molten state and allowed to solidify. In this method, filler material is required during the process.
  For example, arc welding, gas welding, etc.

- In general, welding and allied processes are classified as follows :

1. **Arc welding :**
   a) Gas tungsten arc welding (TIG)
   b) Gas metal arc welding (MIG)
   c) Electro-slag welding
   d) Stud arc welding
   e) Shielded metal arc welding
   f) Submerged arc welding
   g) Plasma arc welding

2. Gas welding :

a)   Oxy-acetylene welding

b)   Oxy-hydrocarbon welding

c)   Air acetylene welding

d)   Pressure gas welding

3. Resistance welding :

a)   Spot welding   d) Percussion welding

b)   Seam welding   e) Resistance butt welding

c)   Projection welding

7.12 Arc Welding Process

SPPU : Dec.-04,05,06,07,17, May-05,07,08,09,14,18

- Electric arc welding is a fusion welding process in which welding heat is obtained from an electric arc between an electrode and the workpiece.

- The electrode is first allowed to touch the workpiece to form an electric circuit and then separated continuously to flow through the gaseous medium.

- The temperature produced at the centre of the arc is 6000 ºC to 7000 ºC.

- In this, the base metal is melted by the temperature of the arc, forming a pool of molten metal which is forced out of the pool by blast from the arc.

- Electrode metal also gets melted and deposited at the weld.

- Either A.C. or D.C. supply is used for arc welding process.

- The electrodes used in the process are of two types i.e. bare electrodes and coated electrodes.

- Bare electrodes are cheaper but welds produced through these are of poor quality whereas, coated electrodes are used in modern welding machines as they carry a core of bare metallic wire produced with coating on the outer surface.

- The length of electrodes varies from 250 mm to 450 mm whereas, diameter varies from 1.6 mm to 9 mm.

Arc Welding Equipments

The most commonly used equipments for arc welding are as follows :

a)   A.C. or D.C. machine

b)   Wire brush

c)   Cables and connectors

d)   Earthing clamps

e)   Chipping hammer

f)   Helmet

g)   Safety goggles

h)   Cable plug

i)   Hand gloves, apron, etc.

7.12.1 Advantages, Disadvantages and Applications of Arc Welding Process

SPPU : Dec.-04, 07

Advantages

- It is the most versatile process which can be applied for thin and thick setions.

- Welding of complicated shapes can also be done.

- Welding can be done in any position with high weld quality.

- Very neat appearance and smooth weld shapes can be obtained.

- Generally, edge preparation is not required.

- The equipments are portable and less expensive.

Disadvantages

- As the electrodes are coated, the chances of slag entrapment and their related defects are more.

- Welding control is difficult.

- The process needed filler material.

- It is a slow process.

Applications

Arc welding is commonly used in the manufacturing of following parts :

- Air receiver, boilers, pressure vessels fabrication.

- Automobile, chemical and aircraft industry.

- Ship buiding and bridge construction.

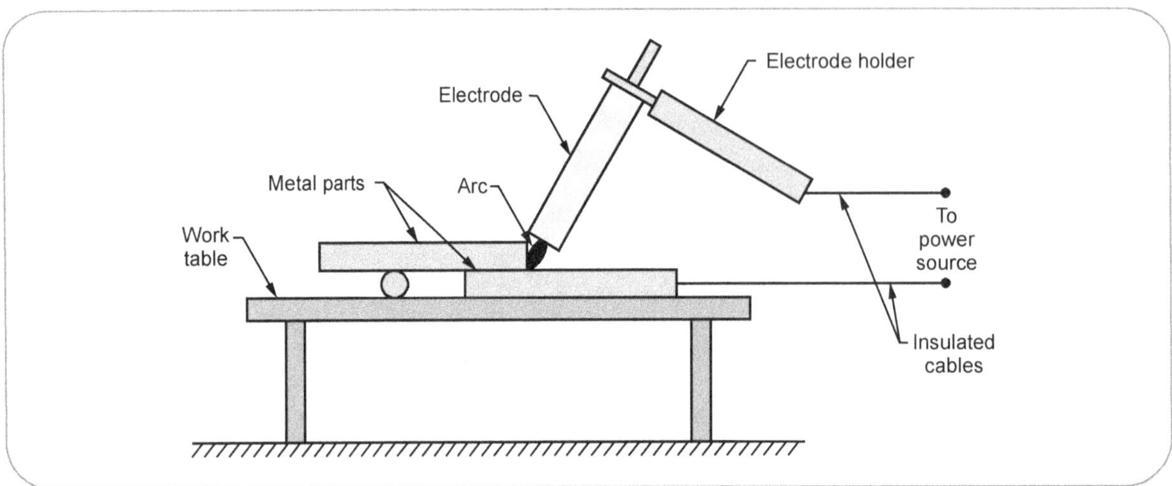

Fig. 7.12.1 Arc welding principle

7.12.2 Types of Arc Welding Process

The main types of arc welding process are as follows :

1. Shielded metal arc welding
2. Gas tungsten arc welding (TIG)
3. Gas metal arc welding (MIG)
4. Electroslag welding
5. Plasma arc welding
6. Stud arc welding

7.12.3 Shielded Metal Arc Welding (SMAW) Process   `SPPU : May-09`

- It is also called as **Flux shielded metal arc welding.**

- It is an arc welding process where coalescence is produced by heating the workpiece with an electric arc set up between the flux coated electrode and the workpiece.

- Due to arc heat, the flux covering decomposes and performs functions of arc stability, weld metal protection, etc.

- The electrode itself melts and supplies the necessary filler metal.

Principle :

- Required heat for welding is obtained from the arc struck between the coated electrode and the workpiece.

- By employing higher or lower currents, the arc temperatures and thus arc heat can be increased or decreased.

- Generally, the arc temperature is about 2400 °C to 2600 °C.

- A high current with a smaller arc length produces very intense heat.

- The arc melts the electrode and the workpiece.

- Material droplets are transferred from the electrode to workpiece through the arc and are deposited along the welded joint.

- The coating of flux melts and produces a gaseous shield and slag to prevent atmospheric contamination of the molten weld metal.

- Fig. 7.12.2 shows the working of flux shielded metal arc welding.

Advantages

- It is the simplest of all arc welding processes.

- The equipment is portable and less expensive.

- Various metals and their alloys can be welded.

- Welding can be done in any position with high weld quality.

Limitations

- Due to limited length of electrode and brittle flux coating on it, mechanisation is difficult.

- It is a slow process.

- Because of flux coated electrodes, the chances of slag entrapment and other related defects are more.

- Welding control in this process is difficult.

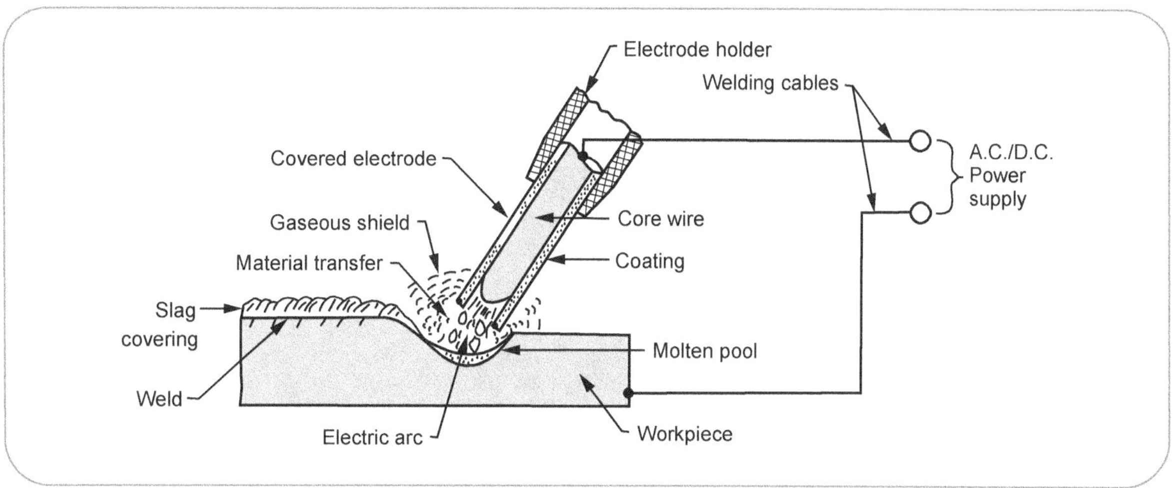

Fig. 7.12.2 Shielded metal arc welding (SMAW)

Applications

- Used for all commonly used metals and their alloys.
- It is used as a fabrication process and for maintenance and repair jobs.
- Also used in,
  - Air receiver, boiler and pressure vessel fabrications
  - Automotive and aircraft industry
  - Ship building and bridge construction.

7.12.4 Gas Tungsten Arc Welding (GTAW) Process

SPPU : May-05, 08, Dec.-06

- It is also called as **Tungsten Inert Gas (TIG) Welding.**
- It is an arc welding process where coalescence is produced by heating the workpiece with an electric arc struck between tungsten electrode and workpiece.
- To avoid atmospheric contamination of the molten weld pool, a shielding gas is used.
- If required, a filler metal may be added.

Principle :

- To start the process, welding current, water and inert gas supply are turned on.
- The arc is struck by touching the electrode with a scrap metal tungsten piece.
- Then the welding torch is brought near to the workpiece.

- When electrode tip reaches within a distance of 2 to 3 mm from the workpiece, a spark jumps across the air gap between the electrode and the workpiece.
- Then the air path gets ionised and arc is established.
- Fig. 7.12.3 shows the manual welding torch.
- The welding continues by moving the torch along the joint and at the far end of the workpiece arc is broken by increasing the air gap.
- To avoid atmospheric contamination, a shielding gas is impinged on the solidifying part for a few seconds.
- The welding torch and the filler rod are generally kept inclined at an angle of 70° to 80° with weld plane.
- For welding, a leftward technique is used.

Advantages

- No flux is used; hence there is no danger of flux entrapment.
- The operator can exercise a better control on the welding process as there is clear visibility of the arc and the workpiece.
- This process produces smooth and sound welds with fewer spatters.
- No weld cleaning is required.
- It produces high quality welds in non-ferrous metals.

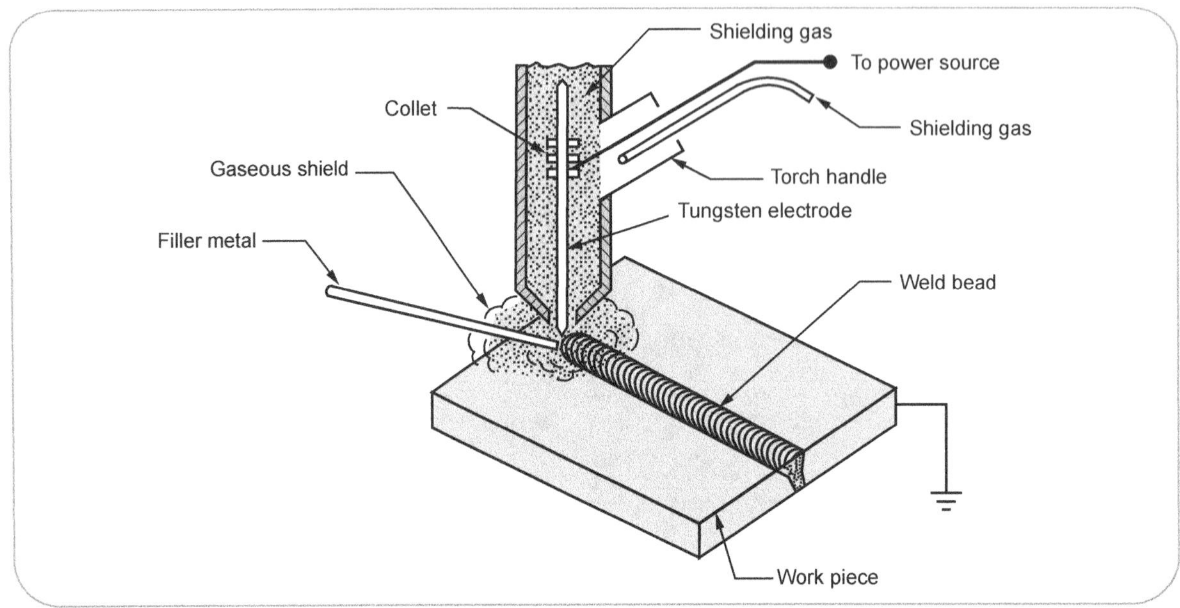

Fig. 7.12.3 Gas tungsten arc welding process (TIG)

Disadvantages

- Cost of equipment is very high.

- By chance, if a filler rod end comes out of the inert gas shield, then it can cause weld metal contamination.

- It is a slow process.

- Separate filler rod is required.

Applications

- Welding of sheet metal and thinner sections.

- Precision welding in aircraft, chemical and instrumental industries.

- Welding of expansion bellows, instrument diaphragms and transistor cases.

- Welding of Al, Mg, Cu, Ni and their alloys, alloy of stainless steel, etc.

7.12.5 Gas Metal Arc Welding (GMAW) Process

- This process is also called as **Metal Inert Gas (MIG) Welding.**

- It is an arc welding process where coalescence is produced by heating the workpiece with an electric arc generated between a continuously fed metal electrode and the workpiece.

- Flux is not used, but the arc and molten metal are shielded by an inert gas.

- The inert gas may be argon, helium, carbon dioxide or a gas mixture.

Principle :

- Fig. 7.12.4 shows the MIG welding operation.

- In this process, the wire is fed continuously from a reel through a gun at constant rate, which also imparts current to the wire.

- The current ranges from 100 to 400 Amp. depending upon the wire diameter and melting point of the wire.

- Depending upon the current being used, the welding gun can be either air or water cooled.

- Bare electrodes are generally used for MIG welding.

- The electrode wire is generally in diameters of a 0.09 to 1.6 mm however sizes upto 3.2 mm are made.

- In MIG welding, the welding area is flooded with an inert gas which will not combine with the metal.

- The flow rate of this gas flow is sufficient to keep oxygen of the air away from the non-metallic surface while welding is being done.

Advantages

- Because of continuously feeding electrode, it is a faster process.

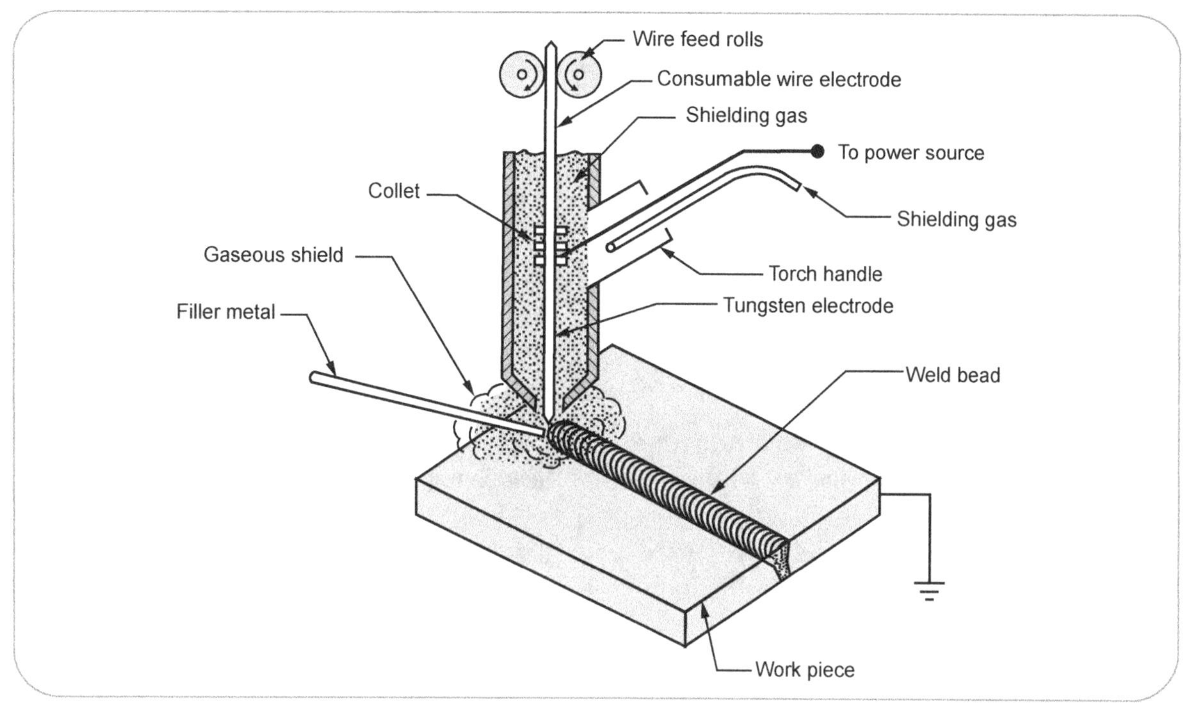

Fig. 7.12.4 Gas metal arc welding (MIG)

- It can produce joint with deep penetration.

- Both thin and thick workpieces can be welded.

- Higher metal deposition rates are achieved by using MIG welding and this process can be easily mechanised.

- As flux is not used, MIG welding produces smooth, neat, clean and spatter free welded surfaces.

Disadvantages

- This process is more complicated.

- Welding equipments are more complex, costly and less portable.

- As air drafts may disperse the shielding gas, MIG welding is not used for outdoor applications.

Applications

- For welding of tool steels and dies.

- For manufacturing of refrigerator parts.

- Also used in industries such as aircraft, automobile, pressure vessel and ship building.

- Welding of carbon, silicon and low alloy steels, stainless steels, Al, Mg, Cu, Ni and their alloys can be done.

7.12.6 Comparison between TIG and MIG Welding Process

SPPU : Dec.-04, May-07

Sr. No.	TIG	MIG
1.	Non consumable electrodes are used.	Consumable electrode wires are used.
2.	Electrodes are made of tungsten or tungsten alloys.	Bare welding wire is made of desired composition.
3.	Electrode only generates an arc and does not melt.	Electrode generates an arc and melt also.
4.	Easier for thin plates and small parts.	Widely used for thick plates (above 4 mm).
5.	Welding torch is water cooled.	Welding torch is air or water cooled.
6.	Used for joining dissimilar metals.	Used for joining similar metals.
7.	It is a slow process.	It is a faster process.
8.	During the process, separate filler material is used.	In this process, metal electrode will act as a filler material.
9.	Cost of equipment is low.	Cost of equipment is high.

7.13 : Soldering Process

SPPU : Dec.-03,04,05,06,08, May-05,10,12,17

- Soldering is a process in which two or more metal items are joined together by melting and flowing a filler metal into the joint.

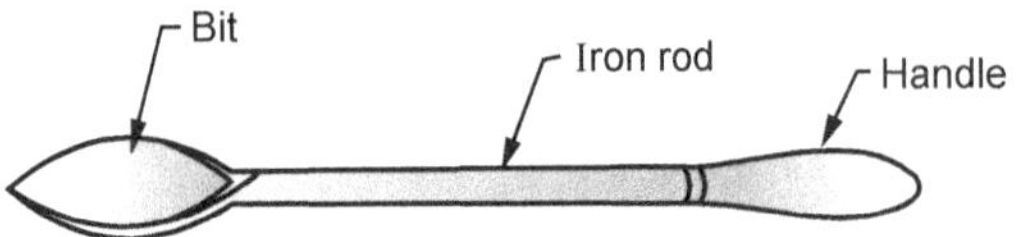

Fig. 7.13.1 Soldering iron

- The filler metal used in the process is called as **solder** which have relatively low melting point.

- Soldering is distinguished from the welding process by the base metals not being melted during the joining process.

- In this process, heat is applied to the parts to be joined, causing the solder to melt and it is drawn into the joint by the capillary action.

- After the metal cools, the resulting joints are not as strong as the base metal, but have adequate strength, electrical conductivity and water tightness for many uses.

- Soldering filler materials are available in many different alloys for various applications.

- Soldering operations can be performed with hand tools. Hand soldering is performed with a soldering iron (Refer Fig. 7.13.1), soldering gun or a torch.

- All soldering techniques requires almost same procedure as discussed below (Refer Fig. 7.13.2) :

    o Clean the metal parts to be joined.

    o Fit the joints and heat the parts.

    o Apply the flux and filler materials.

    o Remove the heat and hold the assembly until the filler metal has completely solidified.

    o Clean the of cooled parts, if required.

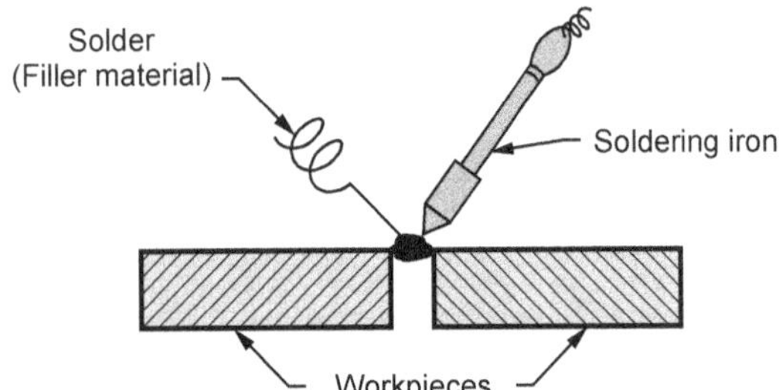

Fig. 7.13.2 Soldering process

7.13.1 Advantages, Disadvantages and Applications of Soldering Process

SPPU : Dec.03, 05, 06, May-08,17

Advantages

- By soldering process variety of dissimilar metals can be joined.

- It is a simple and low cost method.

- Workpieces of different thickness can also be joined.

- The joint formed in the process, do not require any machining.

- Soldering is a low temperature process, hence there is no change in the properties of metals.

Disadvantages

- Soldered joints do not have much strength, so the process should not be used for high load carrying members.

- Since soldering temperatures are low, a soldered joint has limited service at elevated temperatures.

- Corrosion resistance of soldered joint is less.

- Commonly used solder alloys are mixtures of tin and lead but, the used lead is toxic in nature.

Applications

Soldering process is commonly used for joining following components :

- Most frequent application of soldering is assembly of electronic components (chips, IC's, etc.) to printed circuit boards.

- Joints in sheet metal objects like food cans, roof flashing, iron, etc.

- Joints in the wires.

- Now-a-days jewellery components are also assembled and repaired by soldering process.

7.14 : Brazing Process

SPPU : Dec.-03,05,06,08,11,17, May-05,07,08,10,12,18

- Brazing is a metal joining process whereby a filler metal is heated above and distributed between two or more close fitting parts by capillary action.

- In brazing, metallic parts are joined by a non-ferrous filler metal or alloy.

- It is similar to soldering, except the temperatures used to melt the filler metal is above 450° C.

- Similar to soldering, the parts to be joined by brazing are carefully cleaned, the flux is applied and the parts are clamped in position for joining. Refer Fig. 7.14.1.

- The filler metals used in the process are copper based alloys and silver based alloys.

- There are various methods of brazing. Some of them are as follows :

| 1. Torch brazing | 2. Furnace brazing |
| 3. Silver brazing | 4. Immersion brazing |

Torch Brazing

- It is the most versatile method of brazing and finds wide application in industry.

- During brazing, the operator plays the torch flame (by using ordinary gas welding equipment) on the thoroughly cleaned parts.

- To prevent the oxidation of the parts during heating, flux is applied. The filler metal is then hand-fed to the joint area. Refer Fig. 7.14.2.

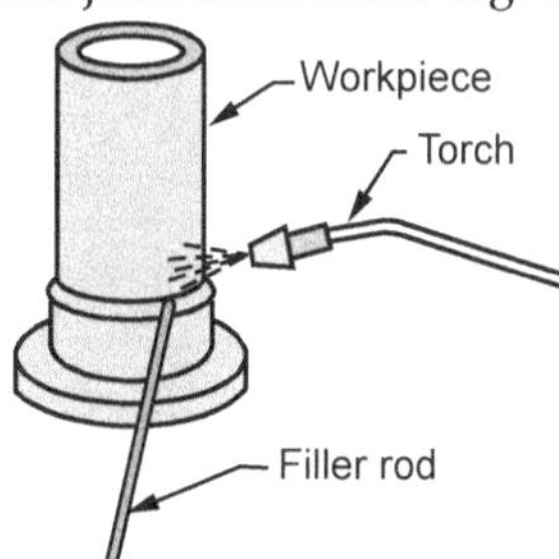

Fig. 7.14.2 Torch brazing

Furnace brazing

- Furnace brazing is similar to ordinary brazing process; only the difference is that, for heating purpose instead of flame, furnace is used.

7.14.1 Advantages, Disadvantages and Applications of Brazing Process

SPPU : Dec.-03, 05, 06, May-08

Advantages

- By brazing process dissimilar metals and non-metals can be brazed.

- Due to uniform heating of parts, it produces less thermal distortion than the welding process.

- Complicated components can also be brazed at low cost.

- It is suitable for mass production.

- Machining of the brazed joint is not required.

- Brazing does not melt the base metal which allows much tighter control over the tolerances.

- Brazing produces a clean joint.

Disadvantages

- Strength of the brazed joints is less as compared to welded joints.

- Brazed joints can be damaged under high service temperatures.

- Brazed joints requires a high degree of base metal cleanliness.

- The joint colour is different than that of base metal which creates an asthetic disadvantage.

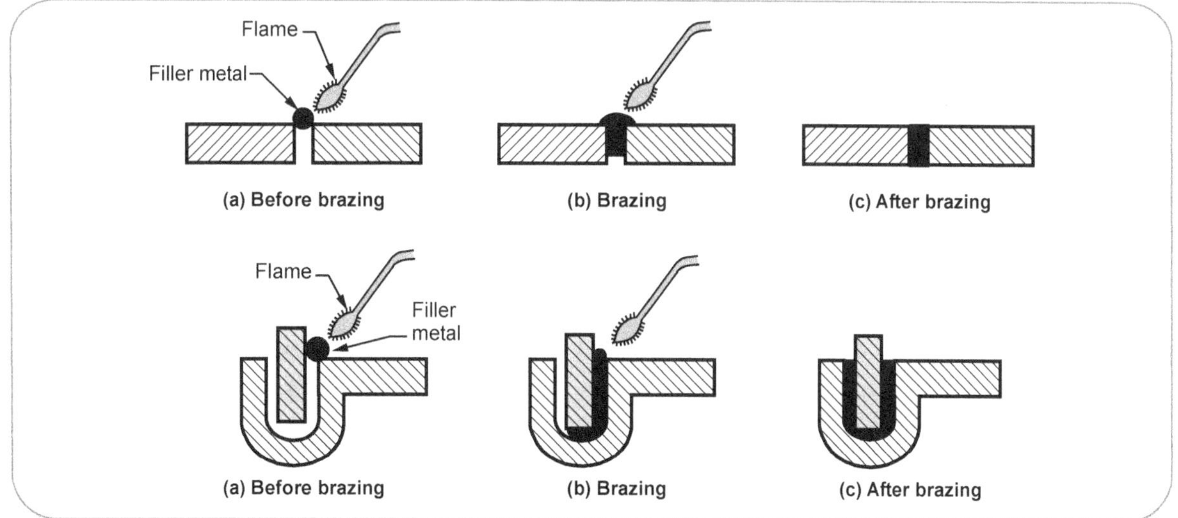

Fig. 7.14.1 Brazing of different components

• Filler metals used in the process are costly.

Applications

Brazing process is used in the joining of following components :

• Brazing can join :

- Non-metals to metals   - Porous metal components   - Dissimilar metals

• Joining of electrical equipments, pipes, heat exchangers, etc.

• Joining of carbide tool tips, steam turbine blades, etc.

7.15 : Comparison between Welding, Soldering and Brazing   SPPU : Dec.-16,17, May-18,19

Sr. No.	Parameter	Welding	Soldering	Brazing
1.	Defination	Welding is a joining process of metals by the application of heat and pressure.	Soldering is a process of joining two or more metal items by melting and flowing a filler metal into the joint below 427 °C.	Brazing is a metal joining process where a filler metal is heated above 427°C and distributed between close fitting parts by capillary action.
2.	Source of heat	By electric energy or by flame or chemical reaction.	Electric soldering iron or flame.	By flame or in furnace.
3.	Strength of joint	Strong	Weak	Lower than welded joints and higher, than soldering.
4.	Surface finish	Poor	Poor	Good
5.	Cost	Expensive	Cheap	Expensive
6.	Application	• Ships, aircrafts. • Automobiles. • Bridges, towers, pressure vessles, • Storage tanks, etc.	• Similar and dissimilar metals. • Electronic component. • Sheet metal object like food can, roof flashing, etc. • Wire joints • Jewellery components, etc.	• Non metals to metals. • Dissimilar metals • Porous metal component. • Joining of carbide tool tips, turbine blades. • Pipes and heat exchangers, etc.

7.16 : Metal Cutting Processes

• **Machine tools** are defined as the machines used for carrying out metal cutting processes (removing the material from the workpiece in the form of chips by using cutting tools) and surface finishing processes (imparting good surface finish to the already machined workpiece, with the negligible removal of the material).

• Machine tool is a power driven machine used for producing the components of required shape and size with desired accuracy and surface finish by removing the material from the workpieces.

• The shapes which are commonly produced through the machine tools used for metal cutting processes are flat, cylindrical, spherical or combination of two or all of them.

• For performing the operations successfully, every machine tool has to perform following functions :

   o Hold, support and guide the workpiece to be machined.

   o Hold, support and guide the cutting tool used for machining.

   o Impart the motion to the cutting tool or workpiece or both of them for producing the desired shape of the workpiece.

○ Regulate the cutting speed and feeding movement between the cutting tool and workpiece.

LATHE MACHINE

7.17 : Lathe Machine

SPPU : Dec.-03,04,05,06,07,08,09,10,11,12,13, May-04,05,06,07,08,09,10,12,13,17

- **Lathe machine** is known as the mother of all the machine tools.

- It is one of the most widely used machine tool.

- Lathe was the first machine tool which came into being as a useful machine for metal cutting.

7.17.1 Working Principle of Lathe Machine

- Lathe is a machine tool which holds the workpiece securely between the two rigid and strong supports, called as centres or in a chuck or face plate, while the workpiece revolves.

- The cutting tool is rigidly held and supported in a tool post and is fed against the revolving workpiece.

- While the workpiece revolves about its own axis, the tool is made to move in parallel or at an inclination with the axis of a material to be cut.

- Hence, the main function of a lathe is to remove metal from a workpiece to give it desired shape and size.

- The material from the workpiece is removed in the form of chips.

- Fig. 7.17.1 shows the working principle of the lathe machine.

- Also, to cut the material properly, the tool material should be harder than the workpiece material.

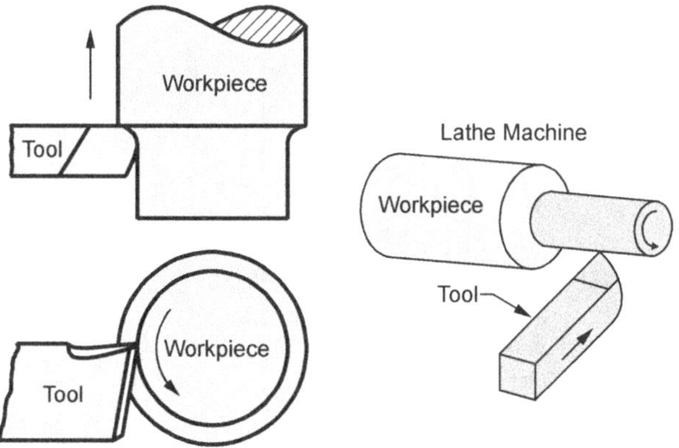

Fig. 7.17.1 Working principle of lathe machine

7.17.2 Basic Elements of Lathe Machine

SPPU : Dec.-03,04,06,08,10,12,13,18, May-05,06,08,10,13,17,19

Fig. 7.17.2 shows the actual lathe machine which consists of following principal parts :

a) Bed   b) Headstock
c) Tailstock   d) Carriage
e) Feed mechanism
f) Thread cutting mechanism (lead screw)

a) Bed :

- It acts as a base of the machine on which, different fixed and operating parts of the lathe are mounted.

- The headstock and tailstock are mounted on either end of the bed and the carriage rests on the lathe bed and slides on it.

- Also, it has to withstand various forces exerted on the cutting tool during the operation.

- Usually, lathe bed is made up of cast iron alloyed with nickel and chromium.

Functions of lathe bed :
- It is used to support all the elements of lathe and to withstand cutting forces during the operation.

b) Headstock :

- The headstock is a box-like casting mounted permanently on the bed to the left hand end of the machine.

- The headstock supports the spindle (hollow rotating shaft used to hold the workpiece) and contains a gear-box by which the spindle may be rotated at various speeds.

Functions of the headstock :
- ○ To support the spindle.
- ○ To carry driving mechanism for the workpiece.
- ○ To give multiple speeds to the spindle.

c) Tailstock :

- Tailstock is also called as **loose headstock** or **puppet head.**

- Tailstock is situated at the right hand end of the bed and can be moved on the innerways of the bed.

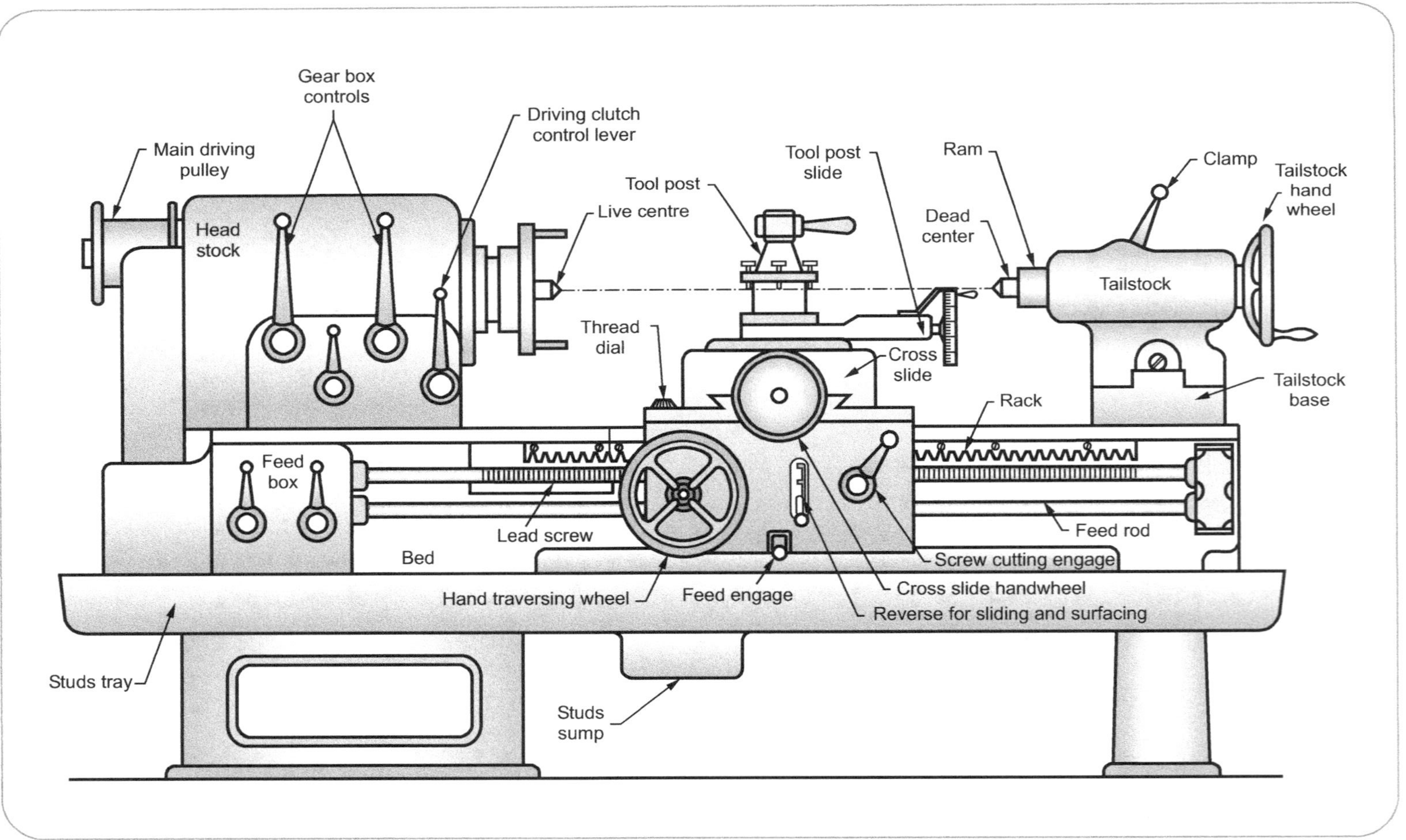

Fig. 7.17.2 Lathe machine

Functions of the tailstock :

- ○ To support the other end of the workpiece when it is being machined between centres.
- ○ To hold a tool for performing operations such as drilling, reaming, tapping, etc.
- ○ To turn a taper on the workpiece by using tailstock set-over method.

d) Carriage :

- The lathe carriage is located between the headstock and tailstock of the lathe.
- It can be moved left or right between the headstock and tailstock either by handwheel or power-feed.
- It carries the cutting tool and precisely controls its movements.

Functions of the carriage : It is used for supporting, guiding and feeding the tool against the workpiece during the operation on the lathe.

- The different parts of carriage are tabulated as follows :

i) Saddle	It slides along the bedways of the lathe and supports the cross-slide, compound rest and tool post.
ii) Cross-slide	It is mounted on the top of the saddle and moves in a direction perpendicular to the axis of the main spindle.
iii) Compound rest	It is supported on the cross-slide and consists of a graduated circular base called as swivel plate which enables cross-slide to swivel at any required angle in a horizontal plane.
iv) Tool post	It is located on the top of the compound rest and used to hold the tool or tool holder in a convenient working position.
v) Apron	It is bolted to the front of the saddle. It houses the gears and controls for the carriage and feed mechanism.

Table 7.17.1 Parts of a carriage

e) Feed mechanism :

- The movement of the tool relative to the workpiece is known as **feed.**
- The lathe tool can have longitudinal, cross and angular feed.
- The feed mechanism has various units through which motion is transmitted from the headstock spindle to the carriage.

Functions of feed mechanism : It is used to transmit the motion from headstock spindle to the carriage by using various units.

f) Thread cutting mechanism (Lead screw) :

- Lead screw is a long threaded shaft which is used as a master screw and it is brought into operation only when threads have to be cut.
- In all other time, it is disengaged from the gear box and remains stationary.

Functions of thread cutting mechanism : It uses the rotation of the lead screw to traverse the tool along the workpiece for producing screw threads. It consists of pair of half-nuts which are capable of moving in or out of mesh with the lead screw.

Fig. 7.17.3 shows the block diagram of a lathe machine.

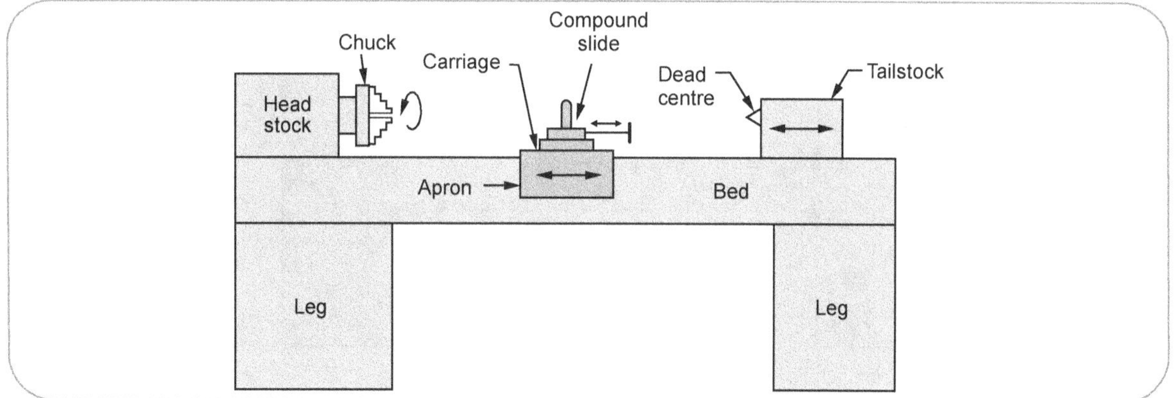

Fig. 7.17.3 Block diagram of a lathe machine

7.17.3 Specifications of Lathe Machine

SPPU : Dec.-05, 09

In order to specify the size of a lathe, following specifications should be included (Refer Fig. 7.17.4) :

(a) Height of the centres : It is measured from the lathe bed.

(b) The length between centres : It is the maximum length of the workpiece that can be mounted between centres.

(c) The length of bed : It includes the length of headstock and tailstock.

(d) The swing diameter over bed : It is the largest diameter of the workpiece that revolves without touching the bed.

(e) Swing diameter over carriage : It is the largest diameter of the workpiece that can revolve over the lathe saddle. It is always less than swing diameter over the bed.

(f) Maximum bar diameter : It is the maximum diameter of bar stock that passes through the hole of the headstock spindle.

(g) Spindle speed.

(h) Spindle nose diameter.

(i) Lead screw pitch.

(j) Motor horse power and R.P.M.

7.17.4 Lathe Machine Operations

SPPU : May-04, 05, 07, 09, 10, 12, Dec.-07, 09, 11, 12

The different operations performed on a lathe are as follows :

1. Straight turning	2. Step turning
3 Eccentric turning	4. Taper turning
5. Facing	6. Drilling
7. Reaming	8. Boring
9 Knurling	10. Thread cutting
11. Parting off	12. Chamfering
13. Grooving	14. Forming

1. Straight turning :

- This operation is performed for producing a cylindrical surface by removing the excess material from the workpiece.

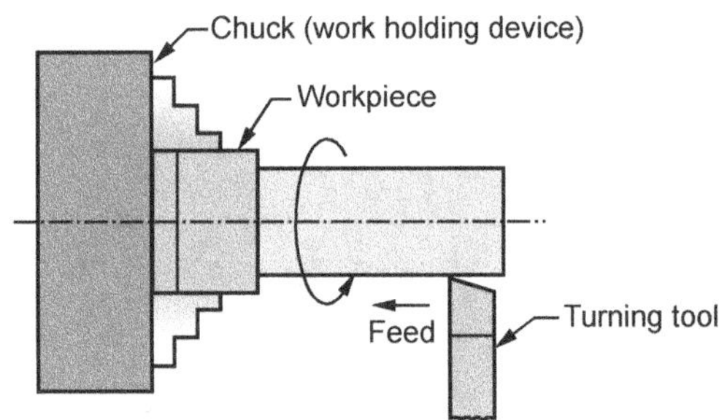

Fig. 7.17.5 Straight turning

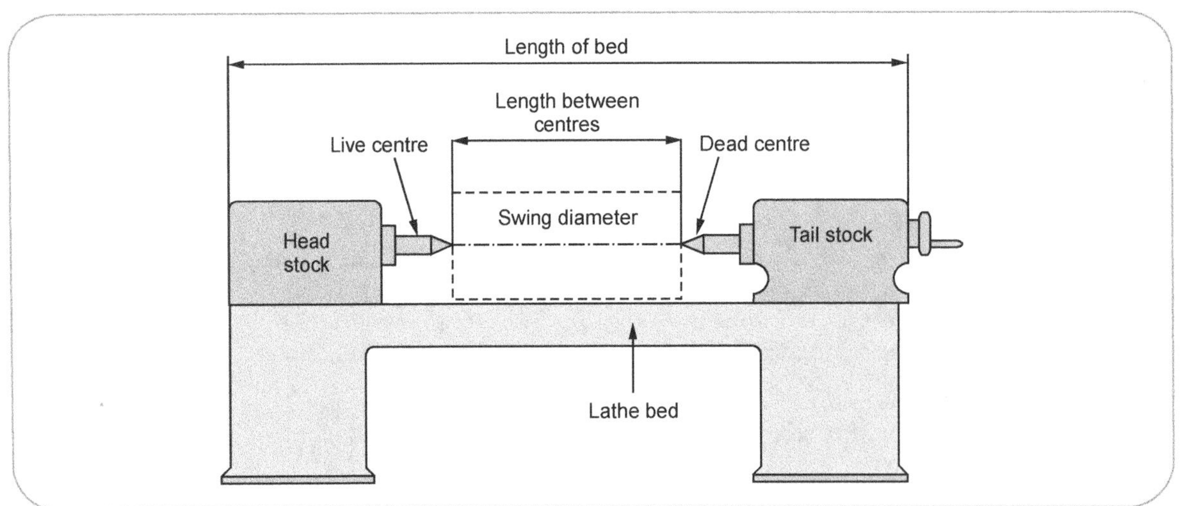

Fig. 7.17.4 Specification of lathe machine

- The cutting tool is held in the tool post and fed into the rotating work parallel to the lathe axis. Refer Fig. 7.17.5.

- It may be rough turning or finish turning.

- The tool used in this operation is called as **turning tool.**

2. Step turning :

- Step turning is also called as **shoulder turning**.

- When workpiece of different diameters are turned, the surface formed from one diameter to the other is called as shoulder. Refer Fig. 7.17.6.

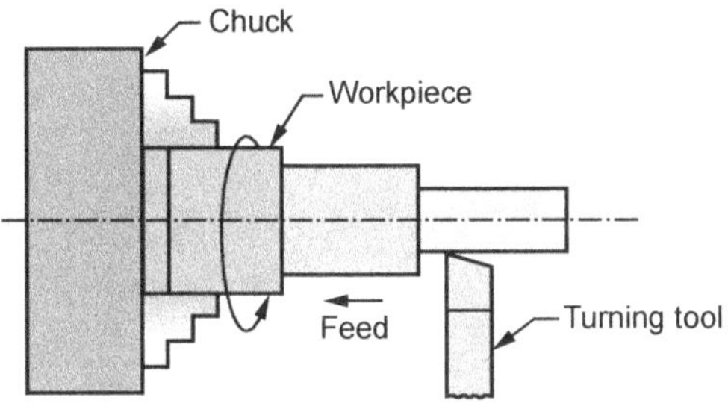

Fig. 7.17.6 Step turning

- The machining of this part of the workpiece is known as shoulder turning.

3. Eccentric turning :

- If a cylindrical workpiece has two separate axis of rotation one being out of centre to the other, then the workpiece is called as **eccentric**.

- The turning of different surfaces of the workpiece is known as eccentric turning. Refer Fig. 7.17.7.

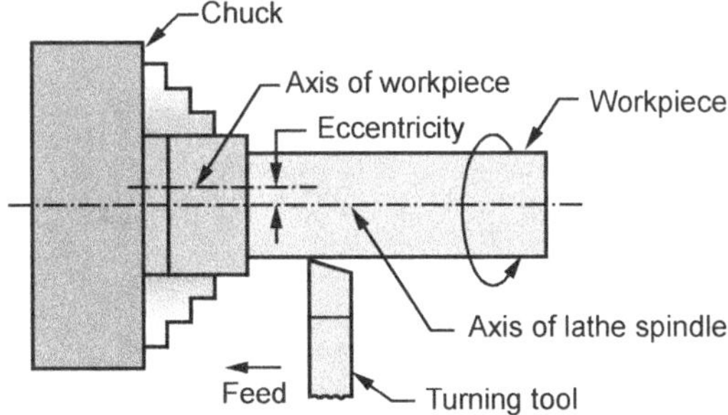

Fig. 7.17.7 Eccentric turning

- For eccentric turning, the workpiece is first mounted on its true centre and turned, then it is remounted on the offset centre and the eccentric surfaces are machined.

4. Taper turning :

- In taper turning, workpiece is rotated on the lathe axis and tool is fed at an angle to the axis of rotation of the workpiece.

- The tool is mounted on the compound rest which is attached to circular base.

- A circular base is graduated in degrees which can be swivelled and clamped at any required angle. Refer Fig. 7.17.8.

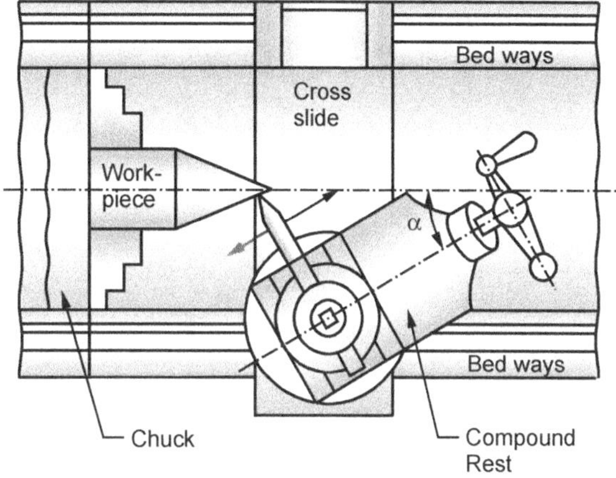

Fig. 7.17.8 Taper turning

- Fig. 7.17.9 shows a tapered workpiece in which,

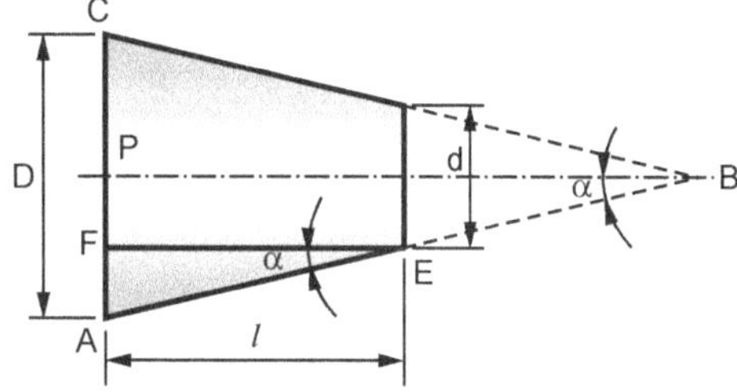

Fig. 7.17.9 Elements of taper

$$D = \text{Large taper diameter in mm,}$$

$$d = \text{Small taper diameter in mm,}$$

$$l = \text{Length of tapered part in mm,}$$

$$\alpha = \text{Taper angle or half taper angle,}$$

$$2\alpha = \text{Full taper angle}$$

$$\tan \alpha = \frac{D - d}{2l}$$

5. Facing :

- It is the operation of machining the ends of a workpiece to produce a flat surface with the axis.

- It involves feeding of the tool perpendicular to the axis of rotation of the workpiece. Refer Fig. 7.17.10.

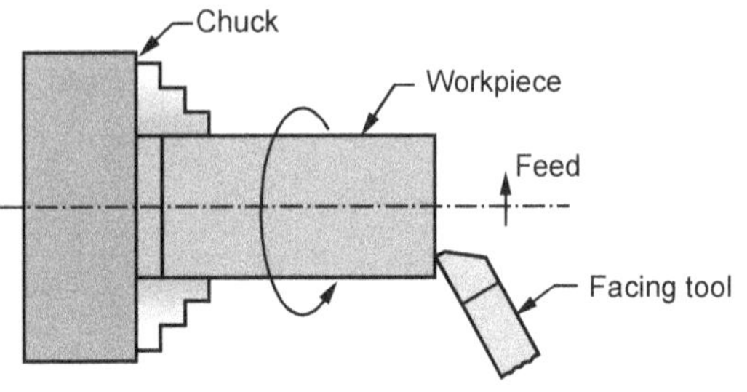

Fig. 7.17.10 Facing

- The tool used for facing is properly ground and mounted in a tool holder of the tool post.

6. Drilling :

- Drilling is the process of producing cylindrical hole in the workpiece.

- In this operation, workpiece is held in a chuck or suitable device like face plate and the drill (tool) is held in the tailstock.

- During operation, the drill is fed by rotating the handwheel of the tailstock in clockwise direction. Refer Fig. 7.17.11.

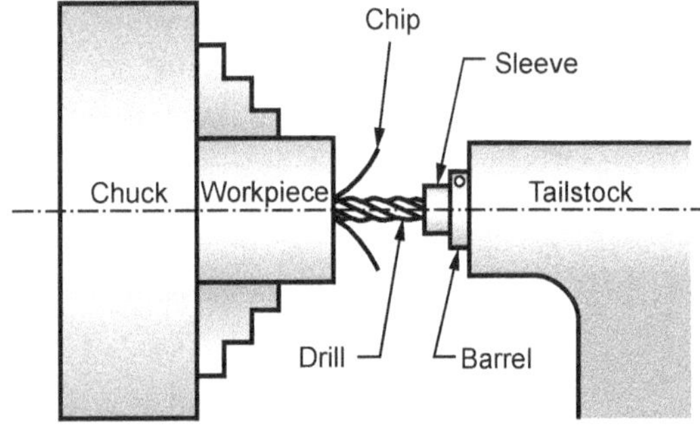

Fig. 7.17.11 Setup for drilling on a lathe

- First a shorter length is drilled by using smaller and shorter drill, followed by producing the required diameter with the help of correct drill size.

- At the end of operation, the drill is taken out by rotating the handwheel of tailstock in anticlockwise direction.

7. Reaming :

- Reaming is a finishing operation because a very small amount of material is removed during the operation.

- For performing reaming a multi-teeth tool is used, which is called as reamer.

- During the operation, the workpiece is held in a chuck or face plate and the reamer shank is fitted in a sleeve or inserted in the tapered hole of the tailstock spindle. Refer Fig. 7.17.12.

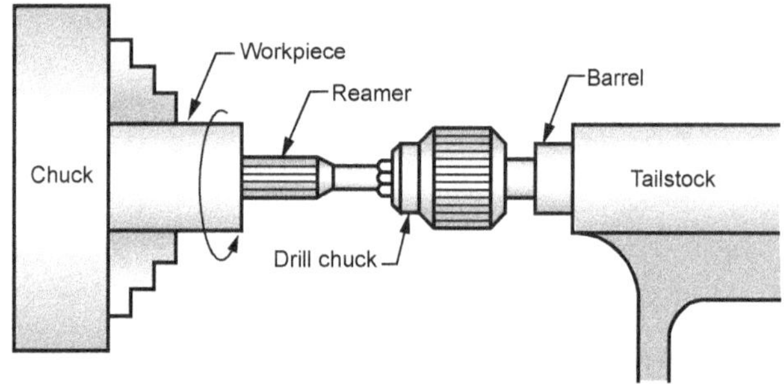

Fig. 7.17.12 Setup for reaming on a lathe

8. Boring :

- It is an operation which is employed for machining internal surfaces, hence also called as **internal turning.**

- Boring is done to enlarge the already drilled hole and bring them to the exact required size.

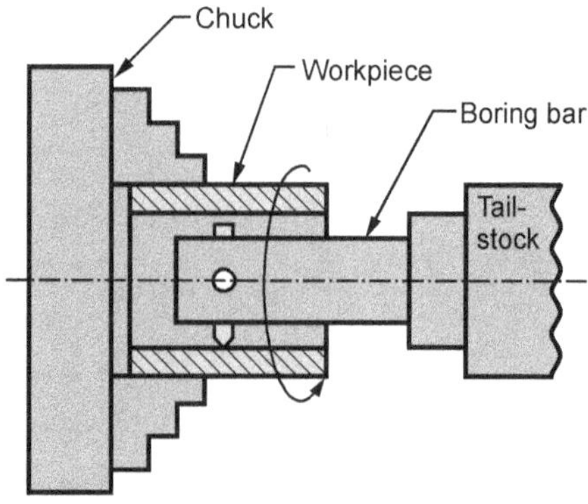

Fig. 7.17.13 Use of boring bars

- Generally, a single point solid boring tool is used for this purpose, which may be either rough or finish.

9. Knurling :

- Outer surfaces of some components such as dumbels, handles, measuring instruments and tools, gauges, etc. are generally provided with rolled impressions on them.

- These depressions are provided for better grip as compared to smooth surface.

- These indentations are known as knurls and the corresponding surface is known as **knurled surface.**

- The operation performed for producing this knurled surface is called as knurling.

- The tool used for knurling is termed as **knurling tool**, which consists of a straight shank fitted with one or two knurling wheels at its front. Refer Fig. 7.17.14 (a) and (b).

10. Thread cutting :

- Cutting screw threads on a centre lathe is called as thread cutting.

- In thread cutting, helical grooves are produced on a cylindrical or conical surface by feeding the tool longitudinally when the workpiece is revolved between the centres (live centre and dead centre).

- For thread cutting, it is necessary that, for every revolution of the workpiece the tool should move parallel to the axis of workpiece by a distance equal to the lead of the screw to be cut.

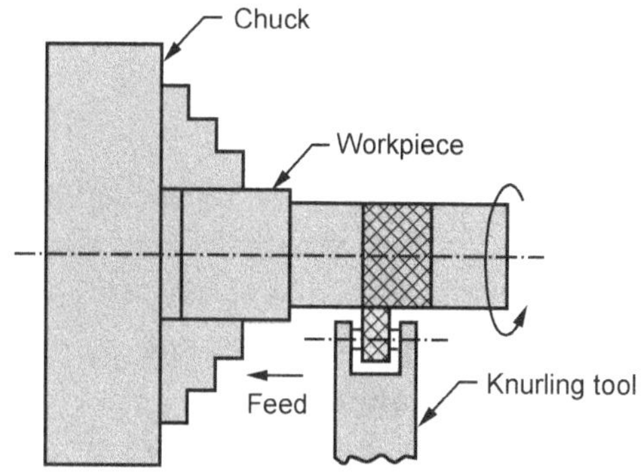

(a) Knurling

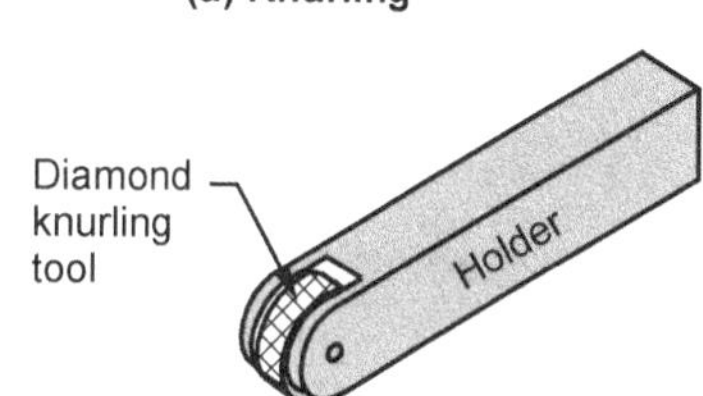

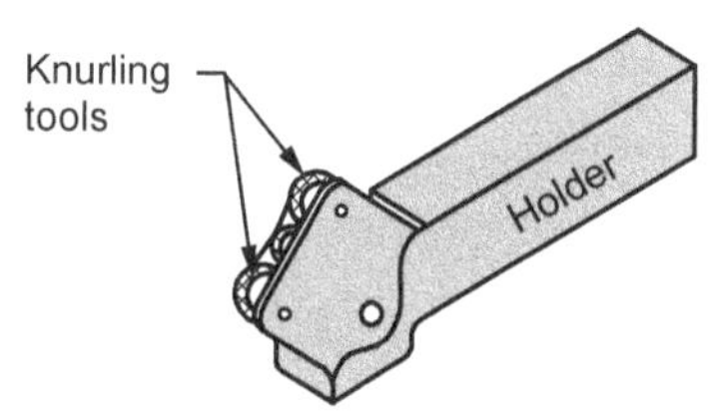

(b) Knurling tool

Fig. 7.17.14

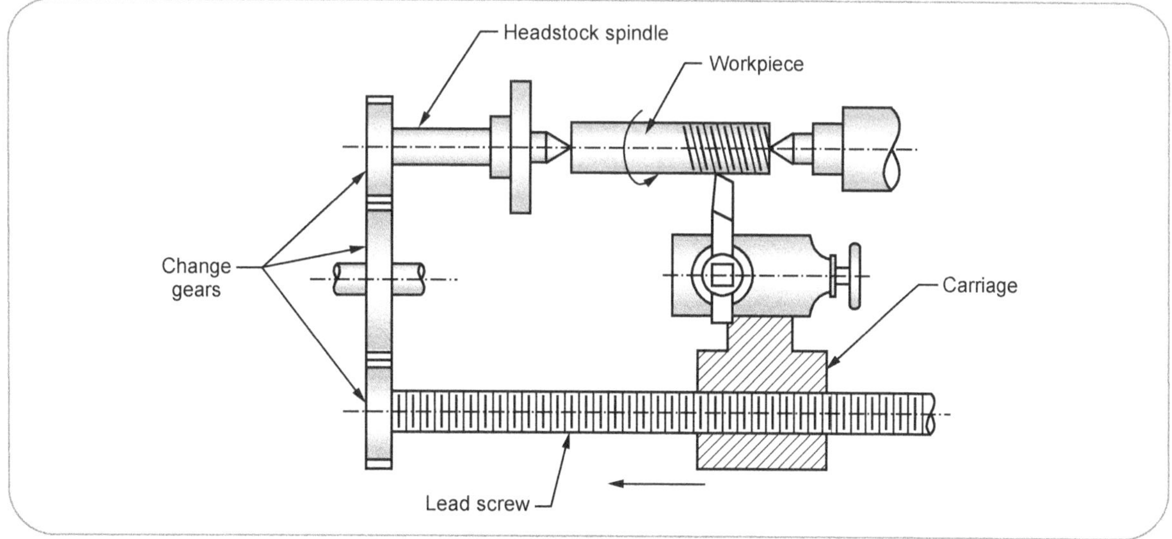

Fig. 7.17.15 Lathe setting for thread cutting

- The motion is transmitted from the lathe spindle to the change gears and finally to the lead screw. Refer Fig. 7.17.15.

11. Parting off :

- It is an operation employed for cutting away a desired length from the bar stock hence, also called as **cutting off**.

- Tools used for this purpose is known as **parting tool** which have a longer point as they are required to cut from the outside surface right upto the centre of the job. Refer Fig. 7.17.16.

- During the operation, the workpiece is held in chuck at one end and supported at the dead centre on the other end.

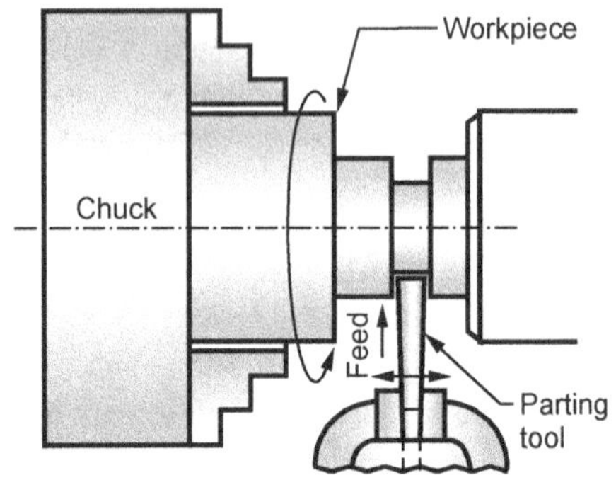

Fig. 7.17.16 Setup for parting off on lathe

12. Chamfering :

- It is an operation of bevelling the extreme end of a workpiece.

- Chamfering is done to remove the burrs, to protect the end of the workpiece from being damaged and to have better appearance.

- Generally, chamfering is performed at the end of all the operations. Refer Fig. 7.17.17.

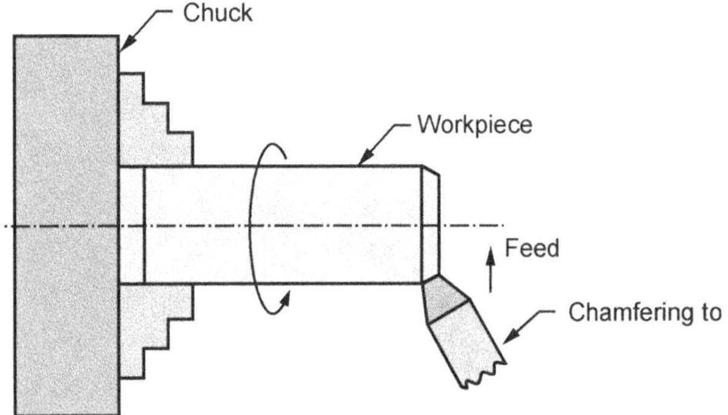

Fig. 7.7.17 Chamfering

13. Grooving :

- It is an operation through which a groove of approximately same width is produced on the job as that of the cutting edge of the tool. Refer Fig. 7.17.18.

- For grooving, the reduction in diameter is effected by cross feed of the tool.

- Longitudinal feed is rarely used because width of the groove is similar to the cutting edge of the tool. Refer Fig. 7.17.18.

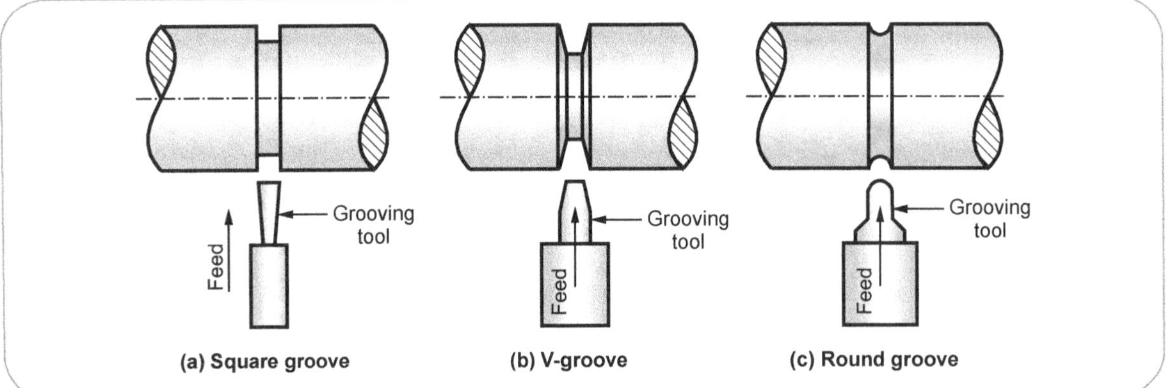

Fig. 7.17.18 Grooving lathe

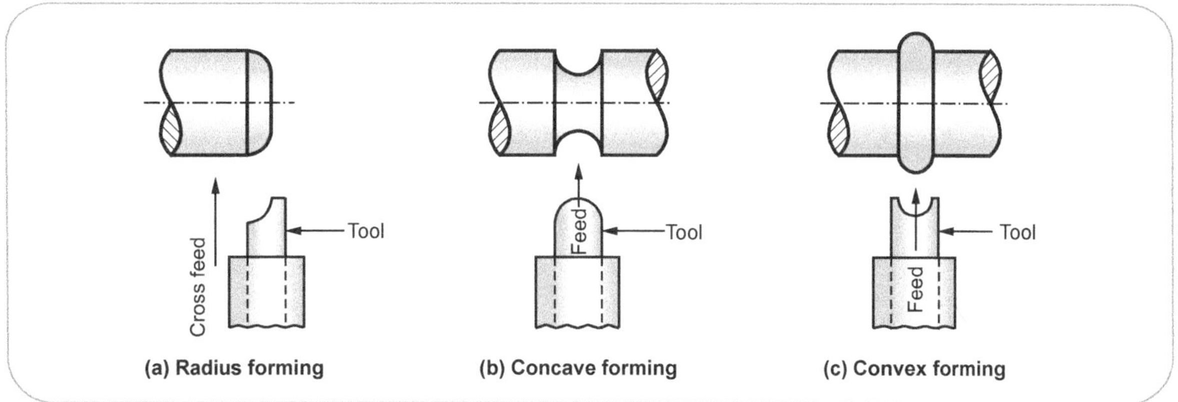

Fig. 7.17.19 Different shapes turned on the lathe using form tools

14. Forming :

- Some components having neither cylindrical nor tapered surfaces are said to have shaped or formed surfaces. Refer Fig. 7.17.19.

- For machining such shapes, special tools are used which are called as **form tools**.

DRILLING MACHINE

7.18 : Drilling Machine

SPPU : Dec.-03,04,06,07,08,09,10,11,12,14,16, May-04,05,06,07,09,10,11,14,17,19

- **Drilling** is an operation through which holes are produced in a solid material by using a revolving tool which is called as **drill.**

- The machine tool used for producing the holes in a solid material is called as **drilling machine.**

- In a drilling machine holes are produced quickly and at low cost, hence drilling machine is a very important machine tool in the industry.

7.18.1 Working Principle of Drilling Machine

SPPU : Dec.-14, May-19

- In a drilling machine, the workpiece is clamped firmly on the worktable by using nuts and bolts. Refer Fig. 7.18.1.

- The tool used to produce holes in solid material called as drill is press fitted in a drill chuck.

- The rotating drill is made of harder material than that of the workpiece and it is fed against the stationary workpiece by hand feed as well as power feed.

- During the operation, the material is removed in the form of chips.

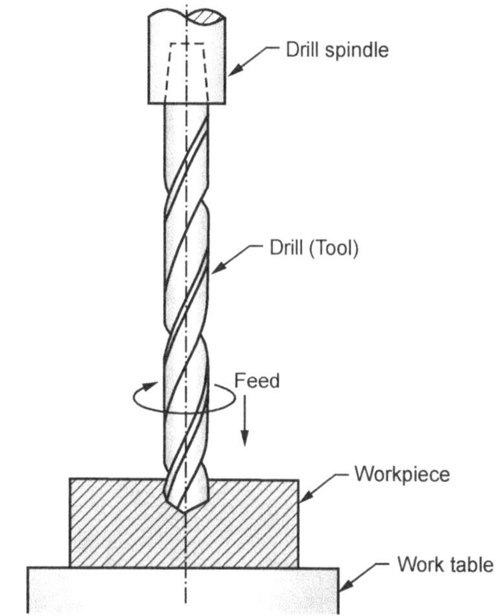

Fig. 7.18.1 Working principle of drilling machine

- While drilling, the feed handle is rotated in anticlockwise direction for providing the feed to the drill and at the end of operation the feed handle is rotated in clockwise direction to remove the drill from the workpiece.

- It is important to note that, during the process high amount of heat is generated, hence continuous supply of coolant is required.

7.18.2 Basic Elements of Drilling Machine

The drilling machine consists of following basic elements. Refer Fig. 7.18.2.

1. Base	2. Column	3. Work
4. Drill head	5. Spindle drive	

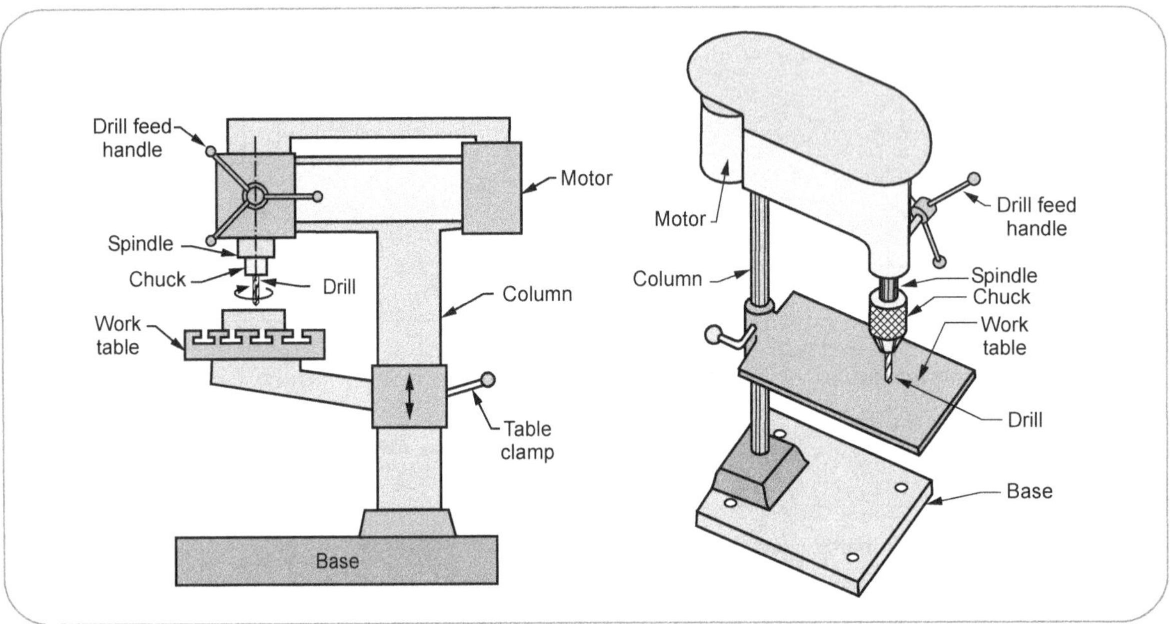

Fig. 7.18.2 Basic elements of drilling machine

1. Base :
- It is the part of the machine which supports the entire structure.
- Generally, the machine base is made from casting.

2. Column :
- It is the vertical member of the machine which supports the table and head containing all the driving mechanisms.
- It may be made of round section or box section.

3. Work table :
- The table is mounted on the column. It may be round or rectangular in shape.
- The worktable is used for clamping the workpiece directly on its face, hence it is provided with T-slots.
- The table has following types of motions :
  - Vertical motion (up and down)
  - Circular motion (swing) about its own axis.

4. Drill head :
- The drill head carries driving and feeding mechanism for the spindle.
- The driving mechanism is used for driving the drill spindle which consists of an electric motor, gear box, etc. whereas, feedig mechanism is used for feeding the drill spindle into the workpiece.
- The drill head is mounted on the top of the column.

5. Spindle drive :
- Spindle is a hollow part, in which drill (tool) is inserted.
- During the operation, spindle rotates as well as move up and down, hence the drill also rotates as well as moves up and down.

7.18.3 Types of Drilling Machines
SPPU : Dec.-09, May-17

- Drilling machines are manufactured in different sizes and varieties to suit different types of workpieces.
- They are classified as follows :

1. Portable drilling machine
2. Sensitive or Bench drilling machine
3. Upright drilling machine
4. Radial drilling machine
5. Gang drilling machine
6. Multi-spindle drilling machine
7. Deep hole drilling machine
8. Automatic drilling machine

1. Portable drilling machine :

- Portable drilling machine is very small, compact and self contained unit carrying a small electric motor inside it. Refer Fig. 7.18.3.

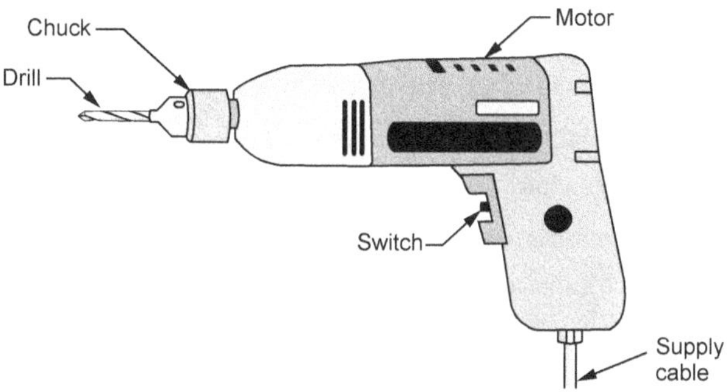

Fig. 7.18.3 Portable drilling machine

- It is generally used for drilling holes in components which cannot be transported to the workshop, because of their size or weight.
- In such cases, the drilling operation is performed on the site by using portable electric drill.
- Portable drills are light in weight and manufactured in different sizes and capacities.
- One of the major advantage of portable drills is that, the holes can be drilled at any desired inclination.
- As small size of drill is used, the machine is operated at high speed.

2. Sensitive or bench drilling machine :

SPPU : May-06

- This type of drilling machine is designed for drilling small holes at high speed in light workpieces.
- It is a small machine having simple construction and operation.
- Fig. 7.18.4 shows sensitive drill machine which consists of a base, which may be mounted on a bench or on the floor.
- The vertical column carries a swivelling table, the height of which can be adjusted along the column.
- Also, the table can be swung to any required position.
- At the top, the column is provided with belt-pulley arrangement.
- One of the pulley is mounted on the motor shaft and other on the machine spindle.
- Different spindle speeds can be obtained by shifting a V-belt to different pairs of driver and

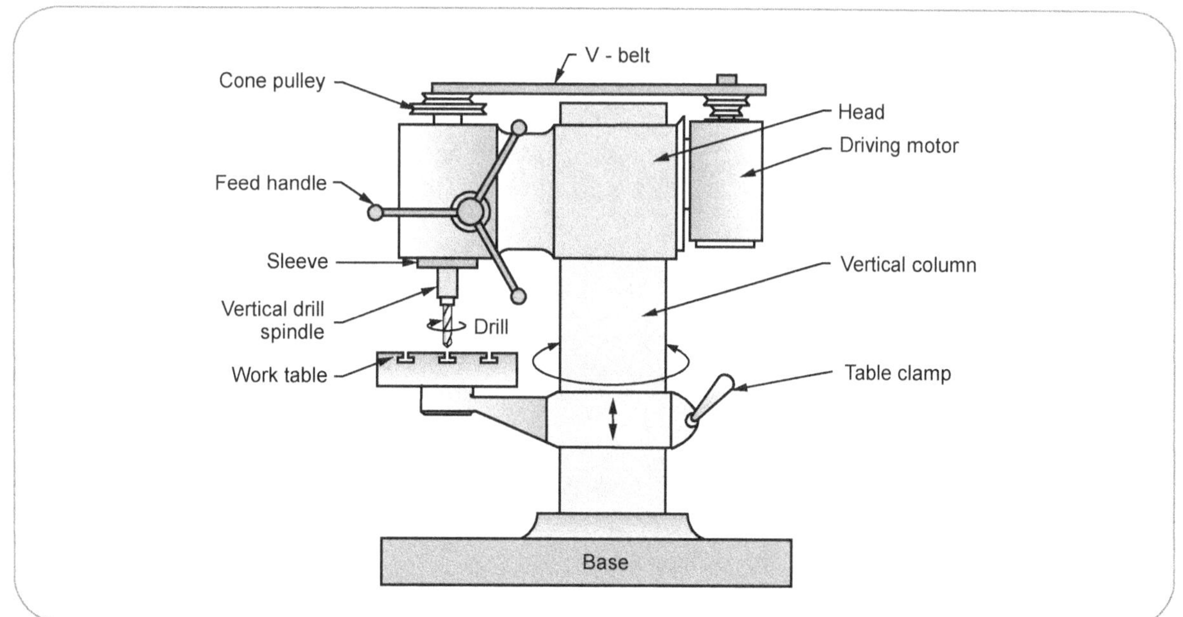

Fig. 7.18.4 Sensitive drilling machine

driven pulleys, while the motor continues to rotate at the same speed.

- There is no arrangement for automatic feed of the drill spindle and hence the drill is fed into the work by hand control.

- For drilling small holes, high speed and hand feed are necessary.

- Hand feed enables the operator to feel the gradual penetration of the drill into the workpiece and sense obstruction, if any.

- As the operator senses the cutting action at any instant, it is known as **sensitive drilling machine.**

3. Upright or pillar drilling machine :

SPPU : May-04,07,09

- Upright drilling machine is similar to sensitive drilling machine but differs in the following manners. Refer Fig. 7.18.5.

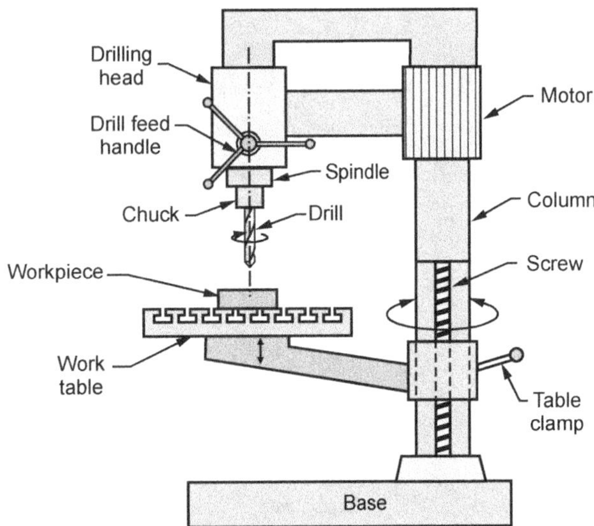

Fig. 7.18.5 Upright drilling machine

- Upright drilling machine is larger and heavier than sensitive drilling machine.

- Upright drilling machine is used for handling medium sized workpieces.

- In this machine, large number of spindle speeds and feeds are available for drilling different types of workpieces.

- The spindle feed is power feed or automatic.

- The table of the machine has different type of adjustments.

4. Radial drilling machine :

SPPU : Dec.-04,07,10, May-14,17

- Radial drilling machine is designed for drilling medium to large and heavy workpieces.

- It consists of a heavy, round and vertical column mounted on a large base.

- To accommodate workpieces of different heights, the column supporting a radial arm can be raised or lowered.

- The arm may be swung around to any required position over the work bed.

- The drill head (contains mechanism for rotating and feeding the drill) is mounted on a radial arm and can be moved horizontally on the guide-ways. Refer Fig. 7.18.6.

- These three movements in a radial drilling machine when combined together, allows the drill to be located at any required point on a large workpiece for drilling the hole.

- When several holes are to be drilled on a large workpiece, the position of the arm and drill head is changed, so that the drill spindle may be moved from one position to another position after drilling the hole without changing the setting of the workpiece.

- Based on the type and number of movements possible, the radial drilling machine can be grouped as :

  - Plain radial drilling machine

  - Semiuniversal radial drilling machine

  - Universal radial drilling machine

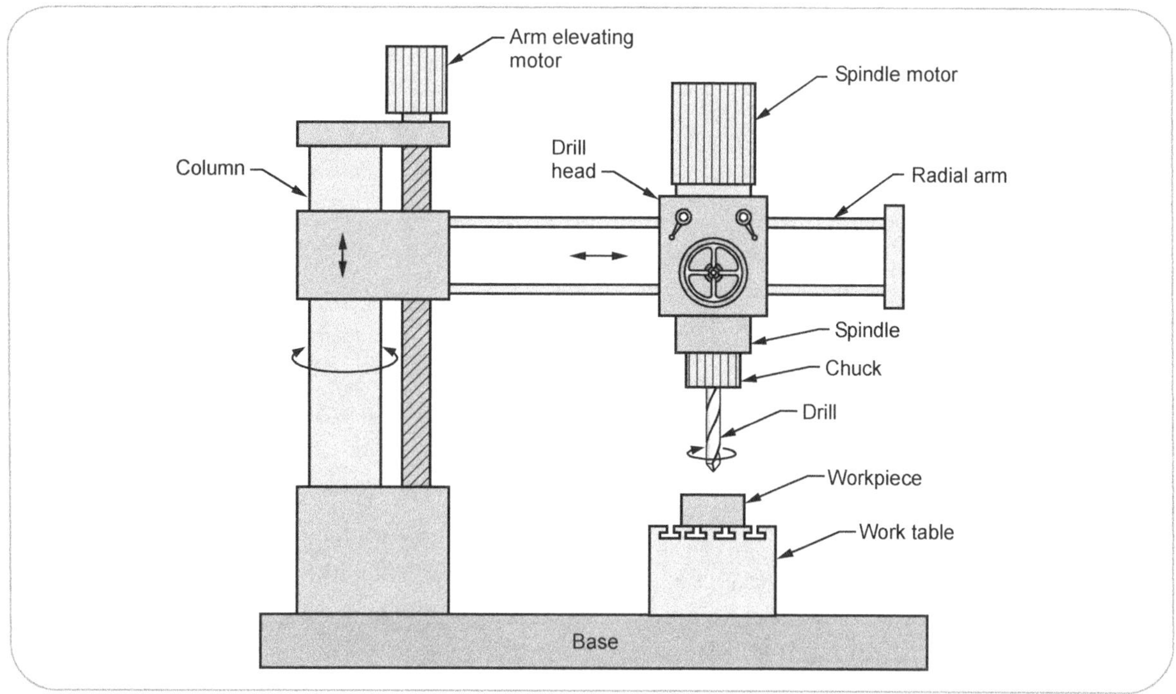

Fig. 7.18.6 Radial drilling machine

7.18.4 Drilling Machine Operations

The different operations that can be performed on a drilling machine are as follows :

1. Drilling	2. Reaming
3. Boring	4. Counter-boring
5. Counter-sinking	6. Spot facing
7. Tapping	8. Trepanning

1. Drilling :

- Drilling is the operation of producing circular hole in a solid metal by using revolving tool which is called as **drill**. Refer Fig. 7.18.7.

- Before drilling, the centre of the hole is located on the workpiece by drawing two lines at right angle to each other and centre punch is used to produce an indentation at the centre.

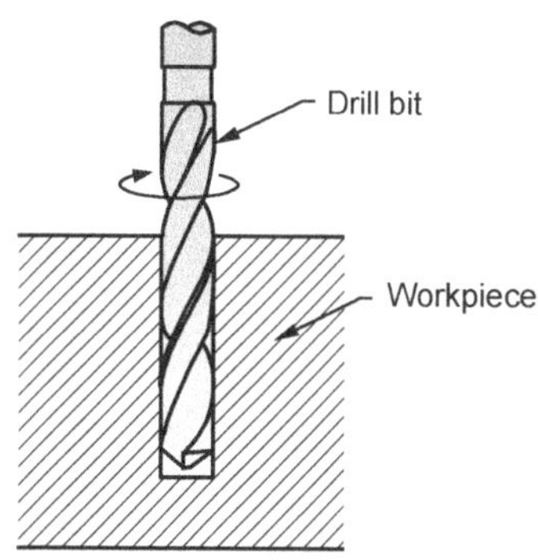

Fig. 7.18.7 Drilling operation

2. Reaming :

- Reaming is the operation of finishing a hole which has been previously drilled and has a coarse surface finish.

- The operation is performed by using a multitooth tool called as **reamer**.

- Reamer cannot produce a hole, it only follows the path which has been previously drilled and removes a very small amount of material. Refer Fig. 7.18.8.

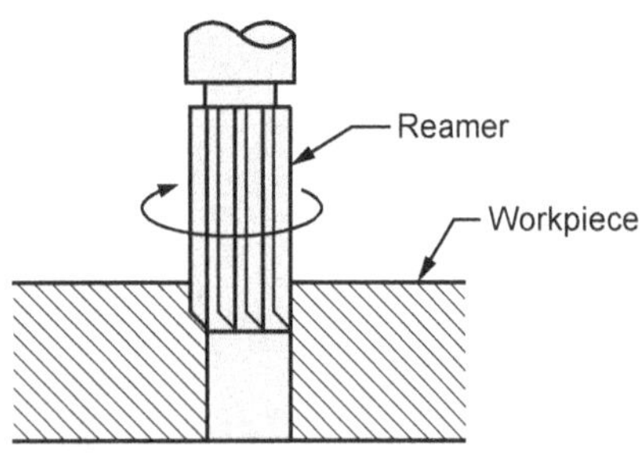

7.18.8 Reaming operation

3. Boring :

- Boring is the process of enlarging a drilled hole by using an adjustable cutting tool which is called as **boring tool.**
- The boring operation is used for the following purposes :
  - To finish a drilled hole accurately and bring it to the required size.
  - To correct the roundness of a hole.
- It is important to note that, the boring tool is having only one cutting edge. Refer Fig. 7.18.9.

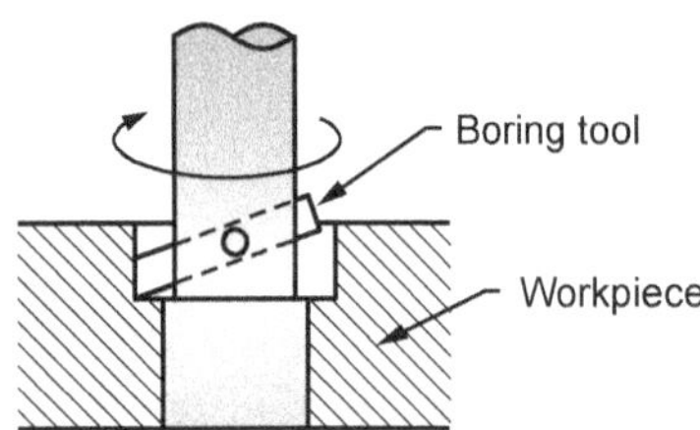

Fig. 7.18.9 Boring operation

4. Counter-boring :

- The operation which is used for enlarging only limited portion of the hole is called as **counterboring** and the tool used for this purpose is called as **counterbore.**
- Fig. 7.18.10 shows counterboring operation in which the enlarged hole forms a square shoulder with the original hole.
- The cutting edges of counterbore may have straight or spiral teeth.

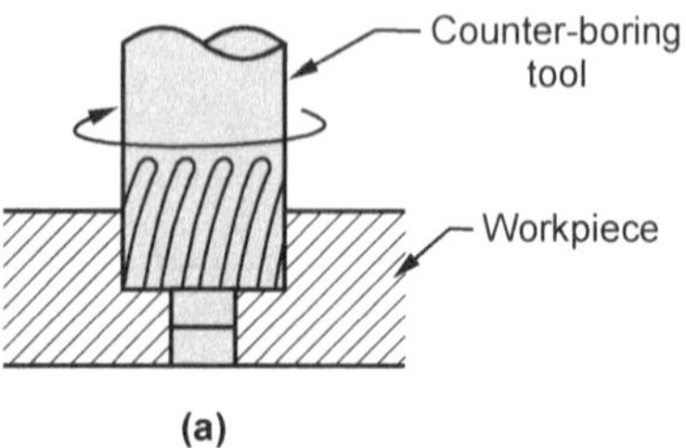

(a)

Fig. 7.18.10 Counter-boring operation

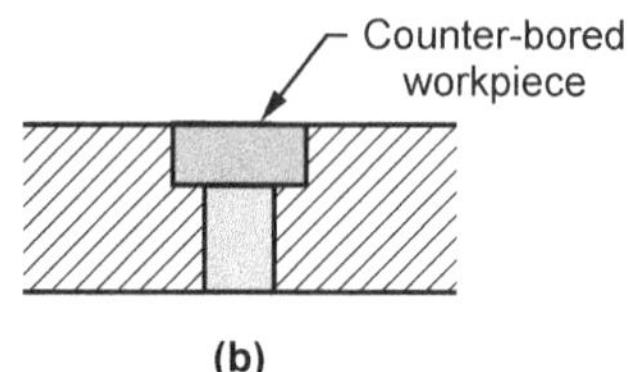

(b)

Fig. 7.18.10 Counter-boring operation

5. Counter-sinking :

- Countersinking operation is used for enlarging the end of a hole and to give it a conical shape for a short distance.
- Fig. 7.18.11 shows countersinking operation in which the tool used is called as countersunk.

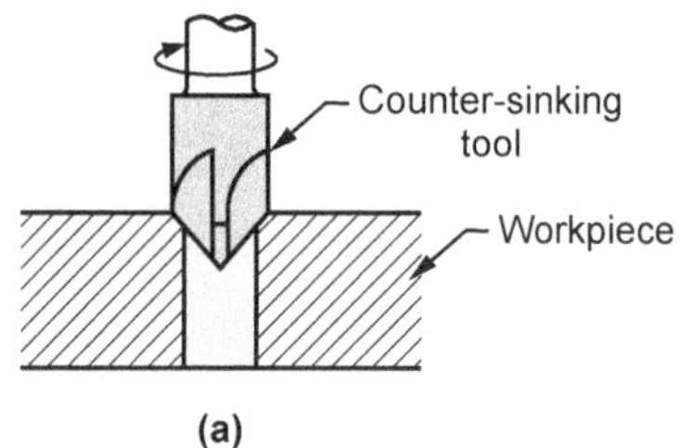

(a)

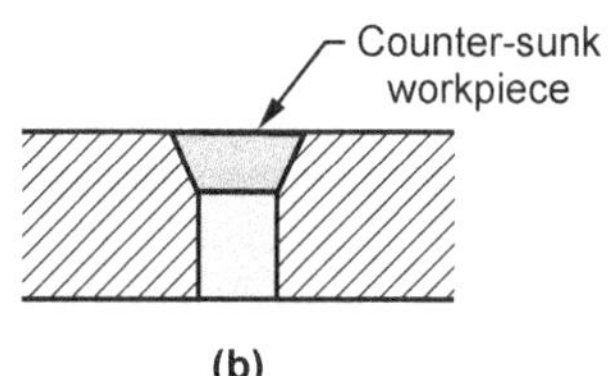

(b)

Fig. 7.18.11 Counter-sinking operation

- The standard countersinks have the included angles of 60°, 82° or 90° and cutting edges of tool are at the conical surface.

6. Spot facing :

- It is the operation of smoothing and squaring the surface around a hole for the seat of a nut or the head of a screw.

- The hole may be spot faced below the rough surface or above it.

- For spot facing counterbore or special spot facing tool is used. Refer Fig. 7.18.12.

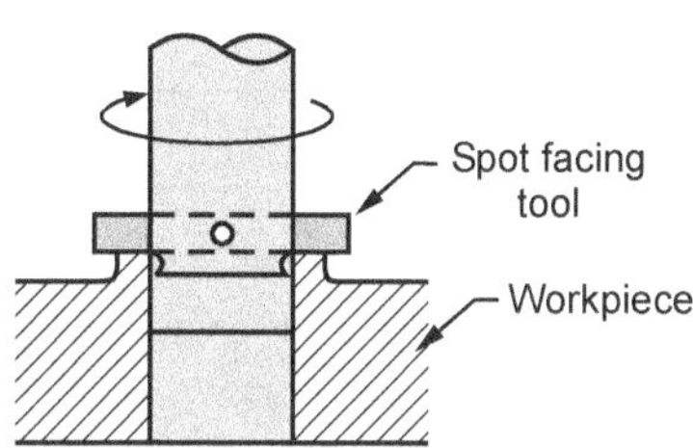

Fig. 7.18.12 Spot facing operation

7. Tapping :

- Tapping is an operation of cutting internal threads by using a cutting tool called as **tap**.

- For tapping purpose, the machine should be equipped with a reversible motor or some other reversing mechanism.

- A tap is considered as a bolt with accurate threads cut on it and the threads act as cutting edges which are hardened and ground.

- Fig. 7.18.13 shows tapping operation in which the tap is screwed into the hole and metal is removed.

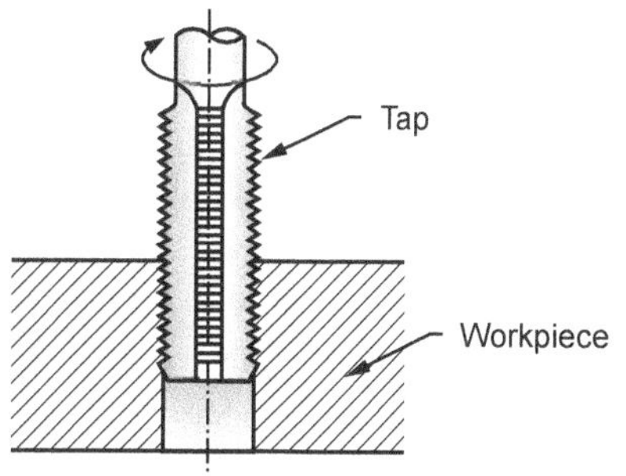

Fig. 7.18.13 Tapping operation

8. Trepanning :

- When large diameter holes are required in a sheet metal, then trepanning operation is carried out.

- It is carried out by using a trepanning tool. Refer Fig. 7.18.14.

- A small hole, to suit the pilot is drilled at the centre of the required position.

- The adjustable arm is extended so that the edge of the high speed-steel cutting tool produces the required hole size.

- The pilot is located in the pilot hole and the tool rotates as it is fed through the workpiece.

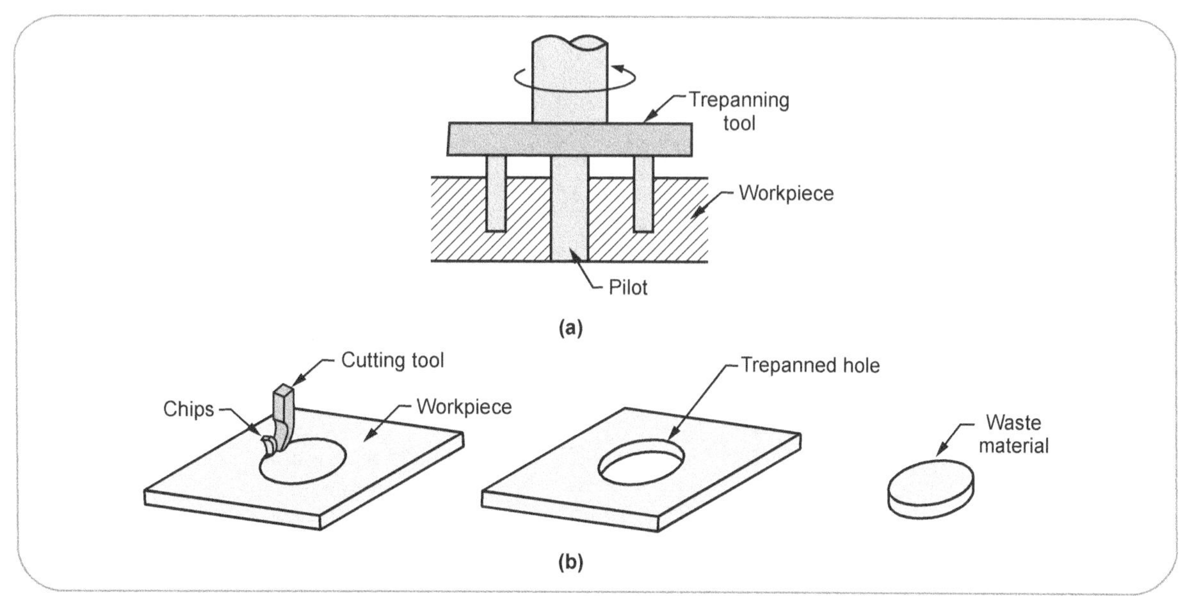

Fig. 7.18.14 Trepanning operation

MILLING MACHINE

7.19 : Milling Machine `SPPU : Dec.-09,10, May-09,10`

- Milling is the machining process in which the metal removal takes place due to the cutting action of a revolving cutter when workpiece is fed past it.

- The revolving cutter is held on a spindle or arbor and the workpiece is clamped on the machine table.

- During the process, to produce the desired shape, the cutter removes the metal in the form of chips from the workpiece surface.

- The machine tool on which the milling operation is performed is called as **Milling machine.**

- Also, the workpiece can be fed in a vertical, longitudinal or cross direction. Fig. 7.19.1 shows the working principle of a milling machine.

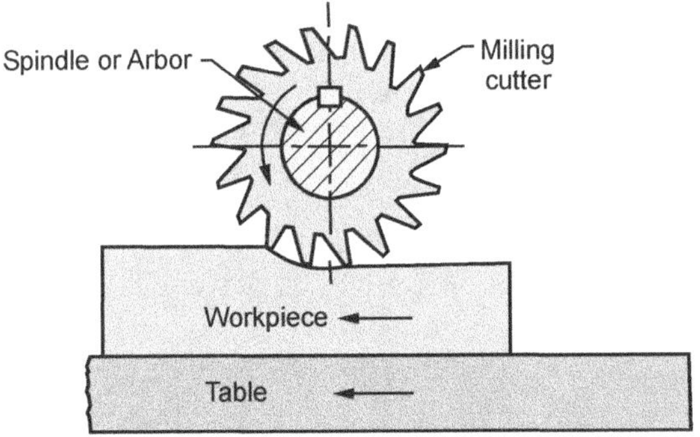

Fig. 7.19.1: Working principle of a milling machine

7.19.1 Basic Elements of Milling Machine
`Dec. 2009`

Column and knee type milling machine is the most commonly used type of milling machine. Fig. 7.19.2 shows block diagram of column and knee type milling machine. It consists of following basic elements.

1. Base	2. Column
3. Knee	4. Table
5. Saddle	6. Over-arm
7. Arbor	

1. Base :

- Base is a heavy casting and provided at the bottom of the machine.

- It acts as a load taking member for all other parts of the machine and a reservoir for the coolant.

- It also carries an elevating screw which supports and moves the knee.

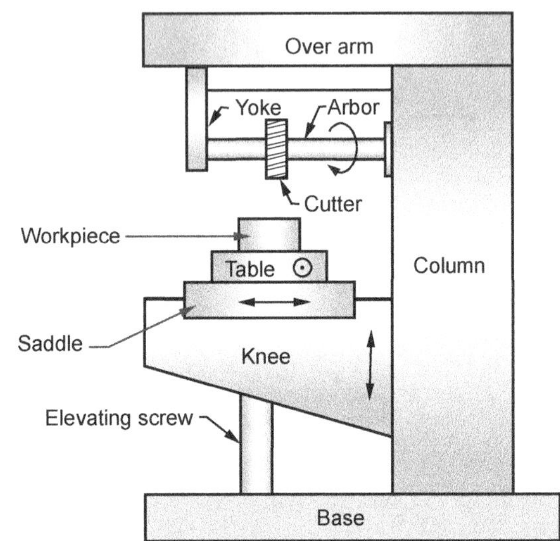

Fig. 7.19.2 : Column and knee type milling machine

2. Column :

- Column is important part of a milling machine because on the column many parts and controls are fitted.

- Front face of the column consists of vertical parallel guideways in which knee slides up and down.

- On the top of the column, over-arm is supported.

3. Knee :

- Knee is a rigid casting which is capable of sliding up and down along the guideways provided on front face of the column.

- With the movement of the knee, the table height or distance between the cutter and workpiece can be adjusted.

4. Table :

- The table acts as a support for the workpiece. It is mounted on the saddle.

- To accommodate the clamping bolts for fixing the workpiece, the top surface of the table carries T-slots.

- Longitudinal (along the length) feed is provided to the table by using hand wheel (↔), vertical feed is provided by raising or lowering the knee (↕) and cross feed by moving the saddle (•).

5. Saddle :

- Saddle is the intermediate part between the table and knee which acts as a support for the table.

- To provide cross feed to the table, the saddle can be adjusted crosswise along the guideways provided on the top of the knee.

6. Over-arm :

- It is a heavy suppart provided on the top of the milling machine.

7. Arbor :

- Arbor is used for holding the milling cutters during the operations.

- One end of the arbor is supported by the over-arm and other end is by the yoke.

7.19.2 Milling Machine Operations   `May 2009`

A large variety of components can be machined on a milling machine which involves various types of operations. These operations can be broadly classified as follows :

1. Plain or slab milling	2. Face milling
3. Angular milling	4. Form milling
5. Straddle milling	6. Gang milling
7. Slot and groove milling	8. Saw milling
9. Side milling	10. End milling

1. Plain or slab milling :

- It is used for machining a flat, plain, horizontal surface which is parallel to the axis of cutter. Refer Fig. 7.19.3.

- During the operation, the workpiece and cutter are secured properly on the machine; and by rotating the vertical feed screw of the table and machine, the depth of cut is adjusted.

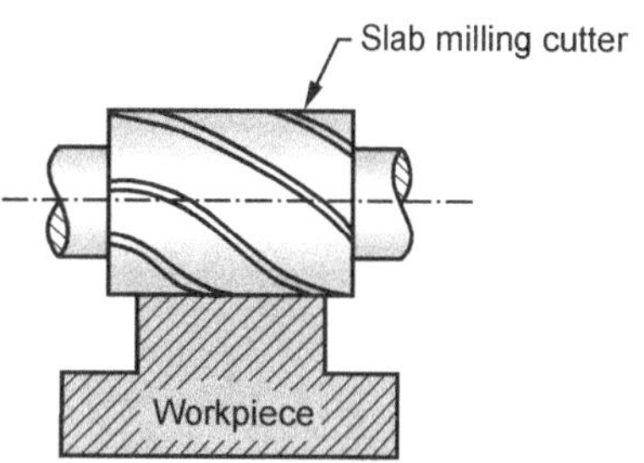

Fig. 7.19.3 : Plain or slab milling

- Following two methods are commonly used for performing this operation :
    a) Up milling
    b) Down milling

a) Up milling :

- Up milling is also called as **conventional milling.**

- In this method, cutter rotates in a direction opposite to that in which the workpiece is fed. Refer Fig. 7.19.4 (a).

b) Down milling :

- Down milling is also called as **climb milling.**

- In this method, the direction of rotation of the cutter and the direction of workpiece fed is same. Refer Fig. 7.19.4 (b).

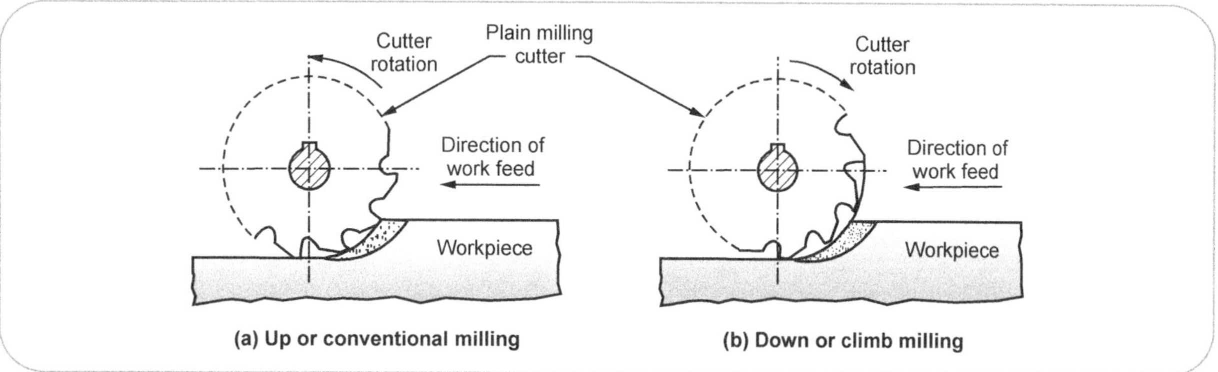

Fig. 7.19.4

2. Face milling :

- Face milling is employed for machining a flat surface which is at right angle, to the axis of revolving cutter. Refer Fig. 7.19.5.

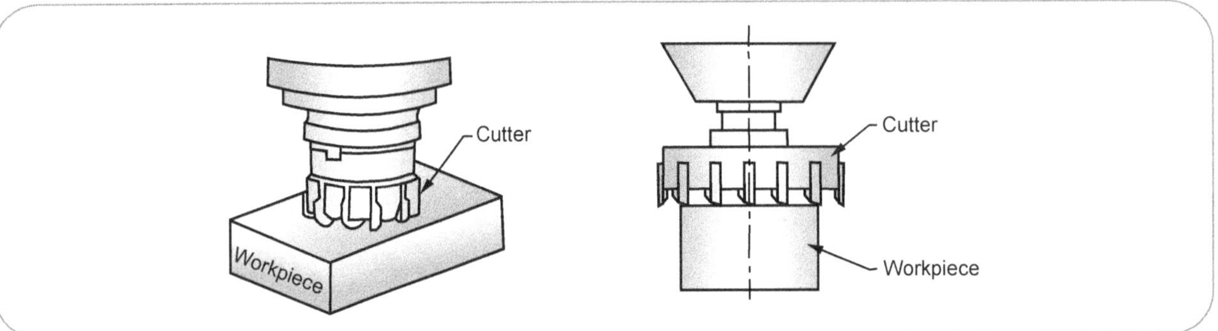

Fig. 7.19.5 : Face miling

- The cutter used in the process is face milling type which is rotated about an axis perpendicular to the workpiece.

- Face mills under 150 mm are called as shell mills and are held on a short arbor.

3. Angular milling :

- Angular milling is used for machining a flat surface at an angle, other than right angle, to the axis of milling machine spindle. For example, V-blocks.

- According to the type of surface (single or two mutually inclined), the single or double angle cutters may be used. Refer Fig. 7.19.6

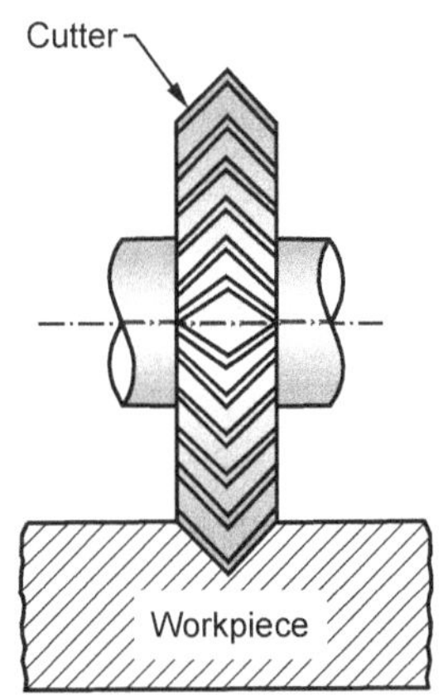

Fig. 7.19.6 : Angular milling

4. Form milliing :

- Form milling is the operation of machining irregular contours by means of form milling cutter. Refer Fig. 7.19.7.

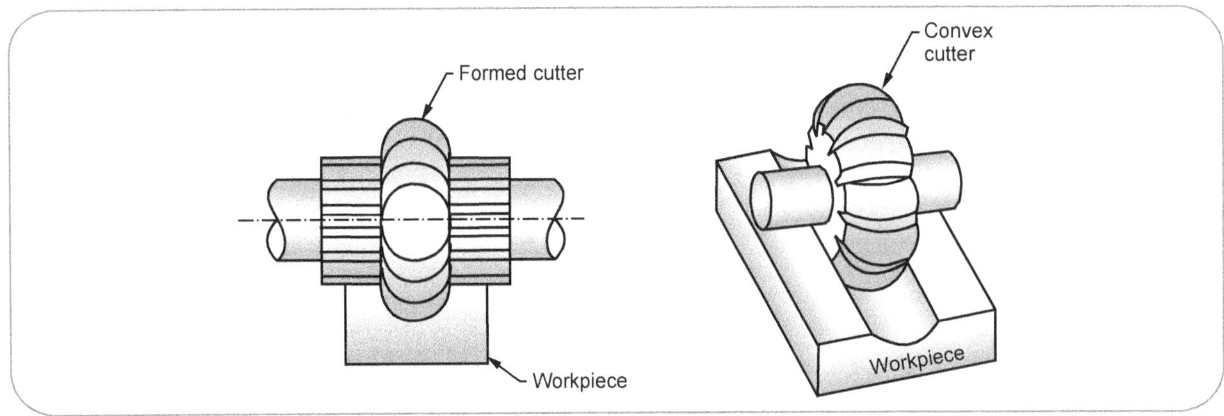

Fig. 7.19.7 : Form milling

- The irregular contour may be concave, convex or of any other shape.

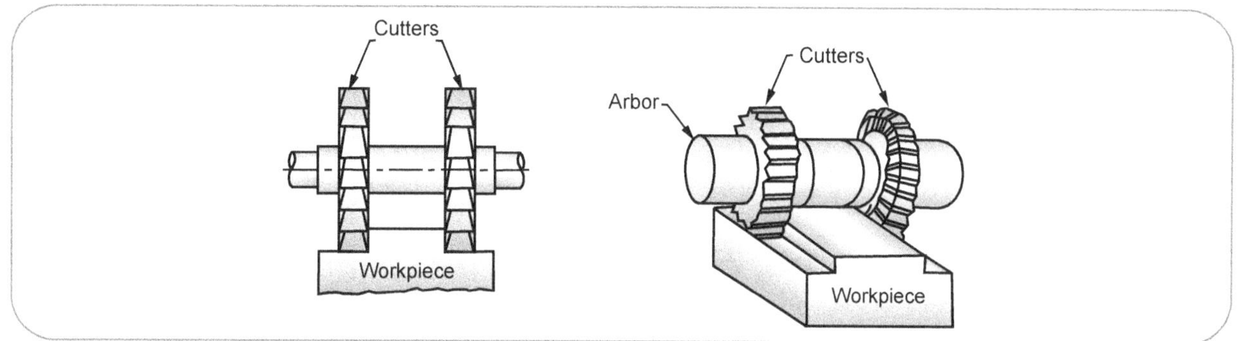

Fig. 7.19.8 : Straddle milling

5. Straddle milling :

- Straddle milling is the operation of flat vertical surfaces on both the sides of a workpiece simultaneously.

- For this purpose two side milling cutters are mounted on the same arbor. Refer Fig. 7.19.8.

6. Slot and groove milling :

- Slot milling is the operation of producing slots in solid workpieces.

- These slots can be of different shapes such as plain slots, T-slots, dovetail slots, etc. Refer Fig. 7.19.9.

- Groove milling is the operation of producing grooves of various shapes like plain grooves, curved grooves, V-grooves, etc.

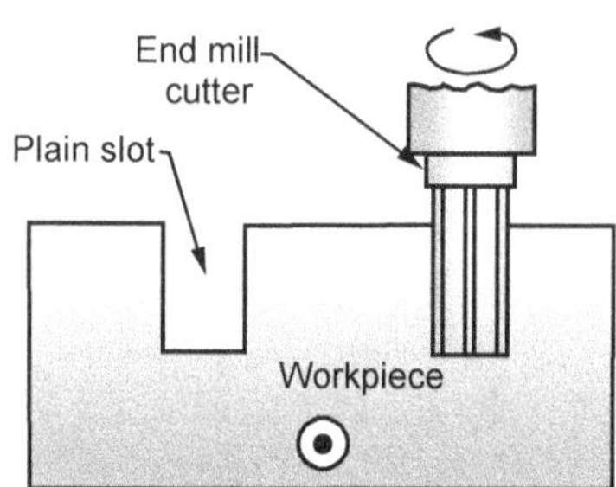

Fig. 7.19.9 : Milling a plain slot by using an end mill cutter

7. Saw milling or slitting :

- Saw milling is the operation of production of narrow slots or grooves on a workpiece by using a saw or slitting milling cutter.

- It is also used for complete parting-off operation.

- The cutter and workpiece are set in such a way that, the cutter is directly placed over one of the T-slots of the table. Refer Fig. 7.19.10

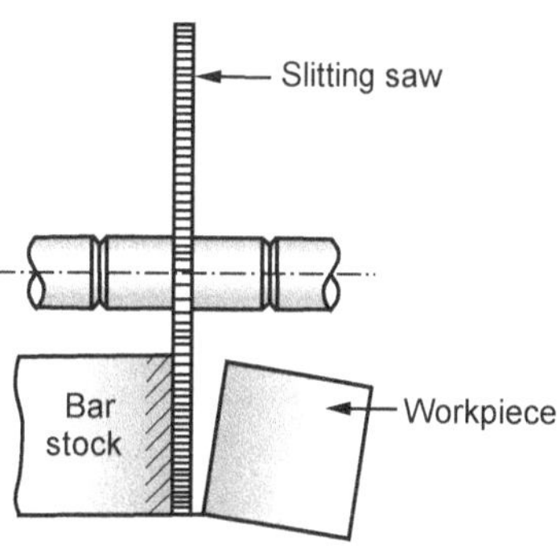

Fig. 7.19.10 : Slitting or saw milling

8. Side milling :

- In this operation a side milling cutter is used for machining a flat vertical surface on one side of the workpiece.

- When two parallel vertical flat surfaces are required to be machined, then two side milling cutters in pairs are used.

- Hence, this operation is similar to **straddle milling**. Refer Fig. 7.19.8.

9. End milling :

- This operation is used to machine and produce a flat surface or a pair of flat surfaces by using an end milling cutter.

- Fig. 7.19.11 shows the end milling operation in which flat surface is produced with the help of an end milling cutter.

- The surfaces produced may be horizontal, vertical or inclined with respect to the top of the machine table.

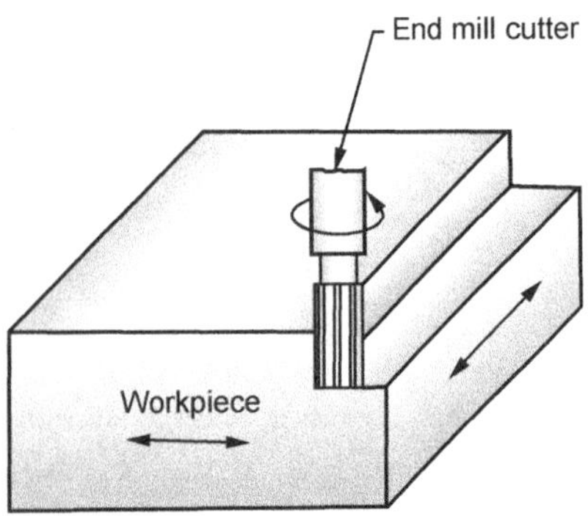

Fig. 7.19.11 : End milling

7.19.3 Types of Milling Machines   SPPU : May-10

A large variety of different types of milling machines are available. Some of the important types are as follows :

1. Horizontal milling machine
2. Vertical milling machine
3. Universal milling machine
4. Hand milling machine

1. Horizontal or Plain milling machine :

- Fig. 7.19.12 shows block diagram of a plain milling machine, in which the vertical column serves as housing for electricals, main drive, bearings, etc.

- For supporting the saddle, worktable and other accessories such as indexing head, knee is provided.

- The arbor carrying the cutter rotates about a horizontal axis.

- Except swivelling, the table can be given motion in three different directions i.e.

    ○ Longitudinal direction ($\leftrightarrow$)

    ○ Vertical direction ($\updownarrow$)

    ○ Cross direction ($\bullet$).

- For vertical movement of the table, the knee alongwith whole unit over it slides up and down in the guideways provided on the column.

- For cross movement of the table, the saddle is moved towards or away from the column alongwith the whole unit over it.

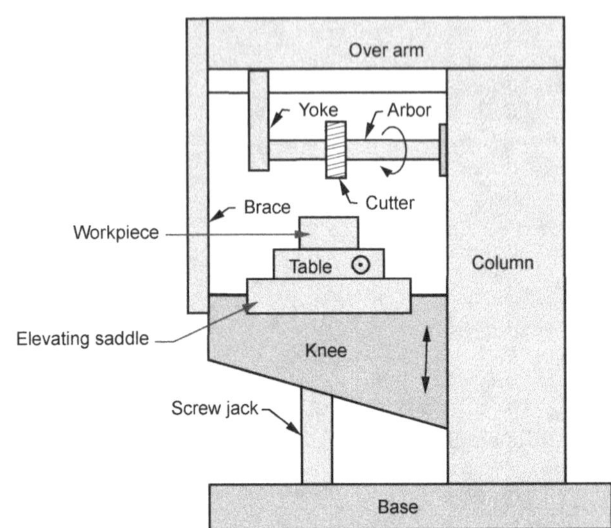

Fig. 7.19.12 : Plain milling machine

2. Vertical milling machine :   Dec. 2010

- In this type of machine spindle position is vertical hence, called as **vertical milling machine.**

- Fig. 7.19.13 shows the block diagram of vertical milling machine with fixed head.

- It consists of a vertical column on a heavy base and overarm is made integral with the column and carrying housing at its front.

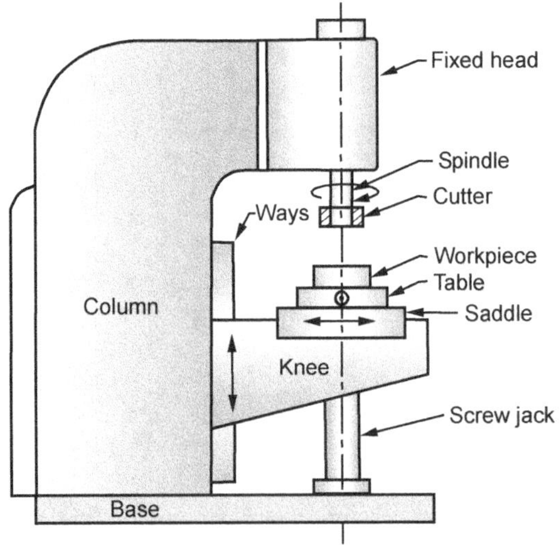

Fig. 7.19.13 : Vertical milling machine with fixed head

- This housing is called as **head** which can be of fixed type or swivelling type.

- In fixed head type, the spindle always remains vertical and can be adjusted up and down.

- Whereas, in swivelling head type, the spindle can be swivelled to any required angle for machining inclined surfaces.

> **Note :** *In vertical milling machine instead of the over-arm, head is present which contains all main drives, bearings, etc. Also there is no arbor in vertical milling machine, hence the milling cutters are directly mounted on the spindle.*

ADVANCED PROCESSES

7.20 : Additive Manufacturing (AM)

- Additive Manufacturing (**AM**) is an appropriate name to describe the technologies that build 3D objects by **adding** layer-upon-layer of material, whether the material is plastic, metal, concrete or one day human tissue.

- Common to **AM** technologies is the use of a computer, 3D modeling software (Computer Aided Design or CAD), machine equipment and layering material.

- Once a CAD sketch is produced, the AM equipment reads in data from the CAD file and lays down or adds successive layers of liquid, powder, sheet material or other, in a layer-upon-layer fashion to fabricate a 3D object.

- The term **AM** encompasses many technologies including subsets like 3D Printing, Rapid Prototyping (RP), Direct Digital Manufacturing (DDM), layered manufacturing and additive fabrication.

- **AM** application is limitless. Early use of AM in the form of Rapid Prototyping focused on preproduction visualization models. More recently, AM is being used to fabricate end-use products in aircraft, dental restorations, medical implants, automobiles, and even fashion products.

7.21 : Generic AM Process

- Additive manufacturing involves number of steps, but different products will involve AM in different ways and to different level.

- Simple products make use of AM for visualization models whereas complex and larger products involve AM during numerous stages.

- AM process generally involve the following eight steps. Refer Fig. 7.21.1.

> i)   CAD model
> ii) Conversion to STL format
> iii) File transfer to AM machine
> iv) Machine setup
> v)   Build
> vi) Removal
> vii) Post-processing
> viii) Application or Use

i) CAD model :

- All kind of AM parts start from a 3D CAD model which describes its geometry.

- For this purpose any CAD solid modeling software can be used, but the output must be 3D solid. Sometimes laser scanning can also be used.

ii) Conversion to STL format :

- Every AM machine excepts only STL (Standard Triangulation Language) file format, so it become a de facto standard and every CAD software can output this file format.

- It acts as a basis for calculation of the slices as it describes external surface of original CAD model.

iii) File transfer to AM machine :

- Converted file to STL format is transferred to the AM machine where general manipulation of the file like correct size, position, etc. are carried out.

iv) Machine setup :

- Prior to build process AM machine must be properly set. These settings are related to thickness of layers, timings, source of energy, etc.

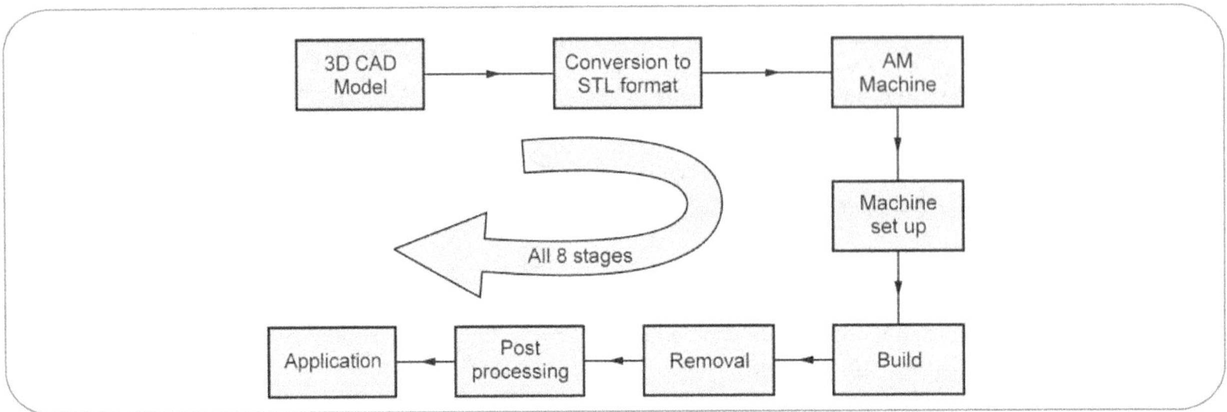

Fig. 7.21.1 : Generic AM process

v) Build :

- In building stage, the layer by layer material is added to form a required product.

- It is an automated process and AM machine carry on without any supervision.

vi) Removal :

- Once the building is completed, the components must be removed which require some interaction with the machine.

- As it deals with high or low temperatures and some moving parts, safety rules must be followed by the operator before removing the component from the machine.

vii) Post-processing :

- After removing the part from the AM machine, it may require additional cleaning before use. Sometimes supporting features may be removed, hence careful and experienced manipulation is required.

viii) Application or use :

- Now the part or component is ready to use. Sometimes it requires additional treatment like painting or priming before actual use.

7.22 : Advantages Disadvantages and Applications of AM

7.22.1 Advantages of AM

- **Complexity free :** It actually costs less to print a complex part instead of a simple cube of the same size. The more complex (or, the less solid the object is), the faster and cheaper it can be made through additive manufacturing.

- **Variety free :** If a part needs to be changed, the change can simply be made on the original CAD file, and the new product can be printed right away.

- **No assembly required :** Moving parts such as hinges and bicycle chains can be printed in metal directly into the product, which can significantly reduce the part numbers.

- **Less lead time :** Engineers can create a prototype with a 3-D printer immediately after finishing the part's STL file. As soon as the part has printed, engineers may then begin testing its properties instead of waiting weeks or months for a prototype or part to come in.

- **Less-skill manufacturing :** While complicated parts with specific parameters and high-tech applications ought to be left to the professionals, even children in elementary school have created their own figures using 3-D printing processes.

- **Few constraints :** Anything you can dream up and design in the CAD software, you can create with additive manufacturing.

- **Infinite shades of materials :** Engineers can program parts to have specific colors in their CAD files, and printers can use materials of any color to print them.

- **Lower energy consumption :** AM saves energy by eliminating production steps, using substantially less material, enabling reuse of by-products, and producing lighter products.

- **Less waste :** Building objects layer by layer, instead of traditional machining processes that cut away material can reduce material needs and costs by upto 90 %. AM can also reduce the "cradle-to-gate" environmental footprints of component manufacturing through avoidance of the tools, dies, and materials scrap associated with CM processes. Additionally, AM reduces waste by lowering human error in production.

- **Reduced time to market :** Items can be fabricated as soon as the 3-D digital description of the part has been created, eliminating the need for expensive and time-consuming part tooling and prototype fabrication.

- **Innovation :** AM enables designs with novel geometries that would be difficult or impossible to achieve using conventional processes, which can improve a component's engineering performance. Novel geometries enabled by AM technologies can also lead to performance and environmental benefits in a component's product application.

- **Part consolidation :** The ability to design products with fewer, more complex parts, rather than a large number of simpler parts is the most important of these benefits. Reducing the number of parts in an assembly immediately cuts the overhead associated with documentation and production planning and control. Also, fewer parts means less time and labor is required for assembling the product, again contributing to a reduction in overall manufacturing costs.

- **Light weighting :** With the elimination of tooling and the ability to create complex shapes, AM enables the design of parts that can often be made to the same functional specifications as conventional parts, but with less material.

- **Agility to manufacturing operations :** Additive techniques enable rapid response to markets and create new production options outside of factories, such as mobile units that can be placed near the source of local materials. Spare parts can be produced on demand, reducing or eliminating the need for stockpiles and complex supply chains.

7.22.2 Disadvantages of AM

- **Slow build rates :** Many printers lay down material at a speed of one to five cubic inches per hour. Depending on the part needed, other manufacturing processes may be significantly faster.

- **High production costs :** Sometimes, parts can be made faster using techniques other than additive manufacturing, so the extra time can lead to higher costs.

- **Considerable effort in application design and setting process parameters :** Extensive knowledge of material design and the additive manufacturing machine itself is required to make quality parts.

- **Requires post-processing :** The surface finish and dimensional accuracy may be lower quality than other manufacturing methods.

- **Discontinuous production process :** Parts can only be printed one at a time, preventing economics of scale.

- **Limited component size/small build volume :** In most cases, polymer products are about 1 cubic yard in size, while metal parts may only be one cubic foot. While larger machines are available, they will come at a cost.

- **Poor mechanical properties :** Layering and multiple interfaces can cause defects in the product.

7.22.3 Applications of AM

1. Aerospace applications

- AM is already being used for a great variety of applications within the aerospace industry. In particular, the design and manufacturing of lighter-weight parts play an important role for the aerospace industry. For instance, the following parts have already been manufactured additively :

a) Structure parts for unmanned aircraft

b) Customized interior of business jets and helicopters

c) Physical 3D mock-ups by Boeing

d) Turbine blades

e) Windshield defrosters.

f) Swirler - fuel injection nozzle for gas turbine applications

- AM-technologies are used for reparation and remanufacture of worn component parts, such as turbine blade tips and engine seal sections.

2. Automotive applications

- AM technology as an important tool in the design and development of automotive components because it can shorten the development cycle and reduce manufacturing and product costs. Some examples for notable applications are as follows :

  a) Testing part design to verify correctness and completeness of parts by BMW, Caterpillar, Mitsubishi.

  b) Parts for race vehicles, e.g. aerodynamic skins, cooling ducts, electrical boxes.

  c) Pre-series components for luxury sport cars, e.g. intake manifolds, cylinder heads by Lamborghini

  d) Replacement of series parts that are defect or cannot be delivered, e.g. cover flaps by Lamborghini.

  e) Assembly assists for series production by BMW, Jaguar etc.

3. Medical field

- Medical devices like hearing aids, prosthetic organs, dental products, etc are nowdays manufactured by AM process.

4. Defence field

- Some defence equipments like rifles, guns and some safety components are also manufactured by using AM process.

5. Manufacturing and tooling field

- AM process is also used to produce plastic mould parts used in vacuum casting process, investment and die casting process.

- Also tool (electrode) used for EDM process can be easily manufactured by AM process.

7.23 : Fundamentals of Rapid Prototyping

- Rapid Prototyping (RP) is a family of fabrication methods to make engineering prototypes in minimum time by using Computer Aided Design (CAD) model of the part.

- Machining is a traditional method of fabricating the prototype part which requires more time (sometimes upto several weeks or more), that depends on the complexity of the part.

Rapid prototyping systems :

- The term 'Rapid Prototyping' (RP) is normally used in the current day to specify a series of processes making use of specialized equipment, software and materials that can use 3D CAD design data to directly fabricate geometrically complex objects.

- Rapid prototyping means developing a prototype as a solid representation of the part from wood, wax or clay models without all of the mechanical properties required for the actual product been used in the manufacturing industry.

- Rapid prototyping systems are generally based upon a Layered Manufacturing (LM) concept in which firstly a 3D model of the object as a CAD file is transferred into the system and then divided into equidistant layers with parallel horizontal lines. The system then generates motion for the material to be added in each layer by the RP machine.

Rapid - prototyping techniques :

- A large number of RP techniques have been developed in the past few years. The major Rapid - Prototyping techniques used are :

  1) Stereolithography (SLA)

  2) Selective Laser Sintering (SLS)

  3) 3 D Printing (3 DP)

  4) Fused Deposition Modeling (FDM)

  5) Laser Engineering Net shaping (LENS)

7.23.1 Three Dimensional Printing (3 DP)

- The term 3D printing originally referred to as a process that deposits a binder material onto a powder bed with inkjet printer heads layer by layer.

- It covers a variety of processes in which material is joined or solidified under computer control to create a 3D object, with material being added together (layer by layer).

- 3D printing was developed at the Massachusetts Institute of Technology (MIT) in 1993.

- In 3 D printing liquid binder is applied to bond the powder particles.

- The printer spreads a layer of powder from the feed box to cover the surface of the build platform and then prints the binder solution onto the loose powder, forming the first cross section of the part.

- When the binder is printed, the powders particles are glued together. The remaining powder is loose and supports the part as it is being printed. When the cross - section is completed the build platform is lowered slightly, and a new layer is spread over its surface. The process is repeated until the whole model is completed. Refer Fig. 7.23.1.

- 3D printing enables small quantities of customized goods to be produced at relatively low costs.

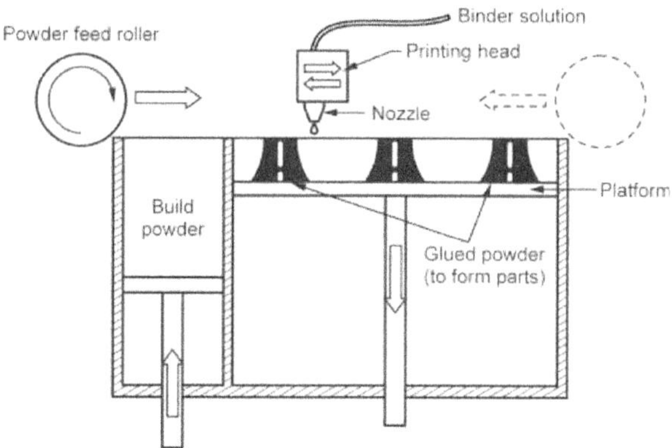

Fig. 7.23.1 : 3D printing

- It is currently used in the production of replacement parts, dental crowns, artificial limbs etc.

7.24 : Micro Machining

- As many of the components used in modern products are becoming smaller and smaller, miniaturization is the central theme in modern fabrication technology.

- Miniaturization is now no longer remaining to micro level accuracy, but deals with nano level accuracy also.

- Most of the improved functionality of advanced mobile phones, data storage devices, holograms for data protection, micro fluidic mixers or medical implant stencils nowadays is based on the development and manufacturing of Micro and Nano Technology (MNT) enabled products.

- Such products are indispensable part of our everyday life - from electronic devices and communication technologies to medicine and airspace.

- The high-throughput mass production of most of the non-electronic micro-components is achieved by the use of serial replication technologies like micro injection moulding, hot embossing, imprinting and their different variations.

- The continuous market demand for improved products integrating more and more functions into a unit product of constantly decreasing size and weight had raised the requirements towards Micro and Nano Technology in terms of process capabilities, production time and cost.

- Thus, innovative manufacturing solutions to meet the constant demand for better and more sophisticated products are the focus of industry

7.24.1 Need for Micro and Nano Machining

- Now a days there is a growing need for industrial product with multi-dimensional, increased and versatile functions and of reduced dimensions.

- Micro machining is the most basic technology for such miniature parts production and components.

- Micro machining is an ability to produce features with the dimensions as small as from 1 μm to 999 μm with the material removal rate at the micro level.

- Lithography based micro machining technology uses Silicon as a material to produce integrated circuitry components and microstructures.

- Also now days in manufacturing industries the demand for micro products and micro components are increased which includes electronics, optics, medical, bio-technology and automotive sectors.

- Example - connectors, switches, micro engines, micro pumps etc.

- These micro-system-based products represent key value-adding elements for many companies and, thus, an important contributor to a sustainable economy.

7.24.2 Lithography

- An IC consists of many microscopic regions on the wafer surface that make up the transistors, other devices, and intra connections in the circuit design.

- In the planar process, the regions are fabricated by a sequence of steps to add layers to selected areas of the surface.

- The form of each layer is determined by a geometric pattern representing circuit design information that is transferred to the wafer surface by a procedure known as lithography.

- Following are the lithographic technologies used in semiconductor processing :

(1) Photolithography   (2) Electron lithography

(3) X-ray lithography   (4) Ion lithography

- As their names indicate, these above process make differences on basis the type of radiation used to transfer the mask pattern to the surface by exposing the photo resist. The traditional technique is photolithography.

Photolithography

- Photolithography, also known as an optical lithography, which uses light radiation for exposing a coating of photo resist on the surface of the silicon wafer; a mask containing the required geometric pattern for each layer separates the light source from the wafer, so that only the portions of the photo resist not blocked by the mask are exposed.

- A photo-resist is an organic polymer that is sensitive to light radiation in a certain wavelength range the sensitivity causes either an increase or decrease in solubility of the polymer to certain chemicals.

- The mask consists of a flat plate of transparent glass on to which a thin film of an opaque substance has been deposited in certain areas to form the desired pattern.

- Thickness of the glass plate is around 2 mm whereas the deposited film is only a few mm thick, for some film materials, less than 1 mm.

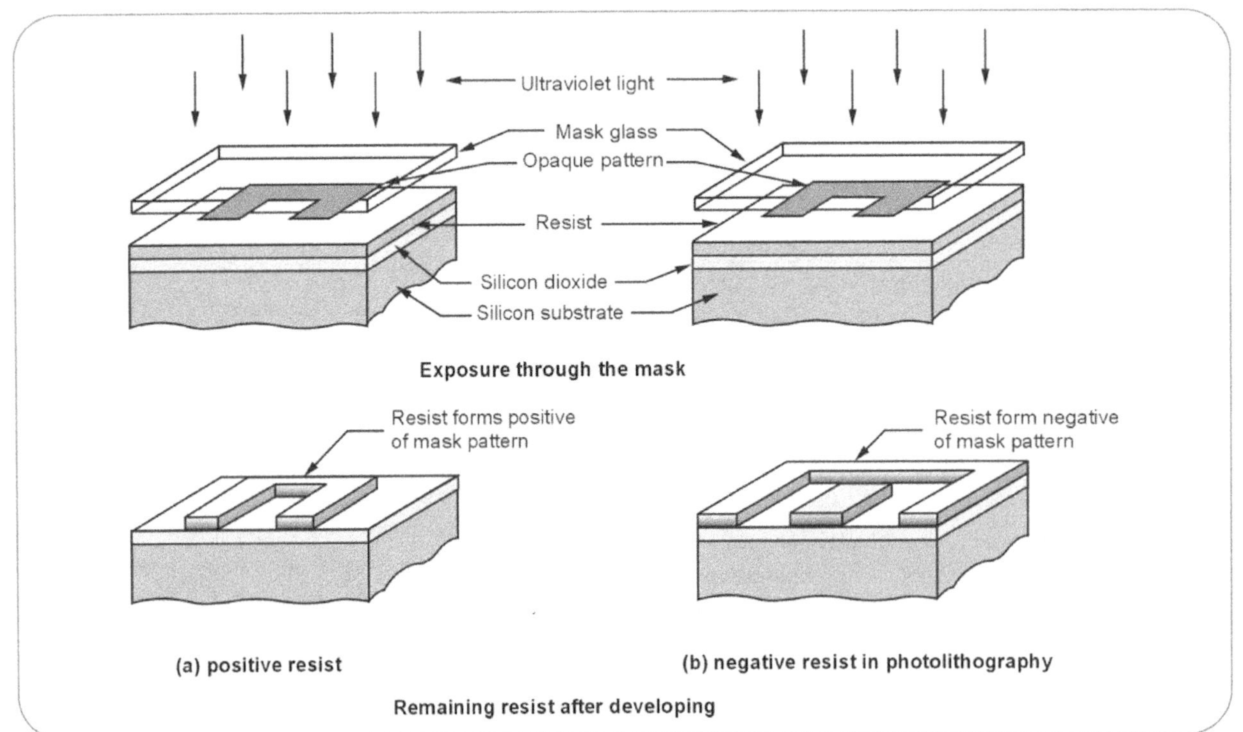

Fig. 7.24.1

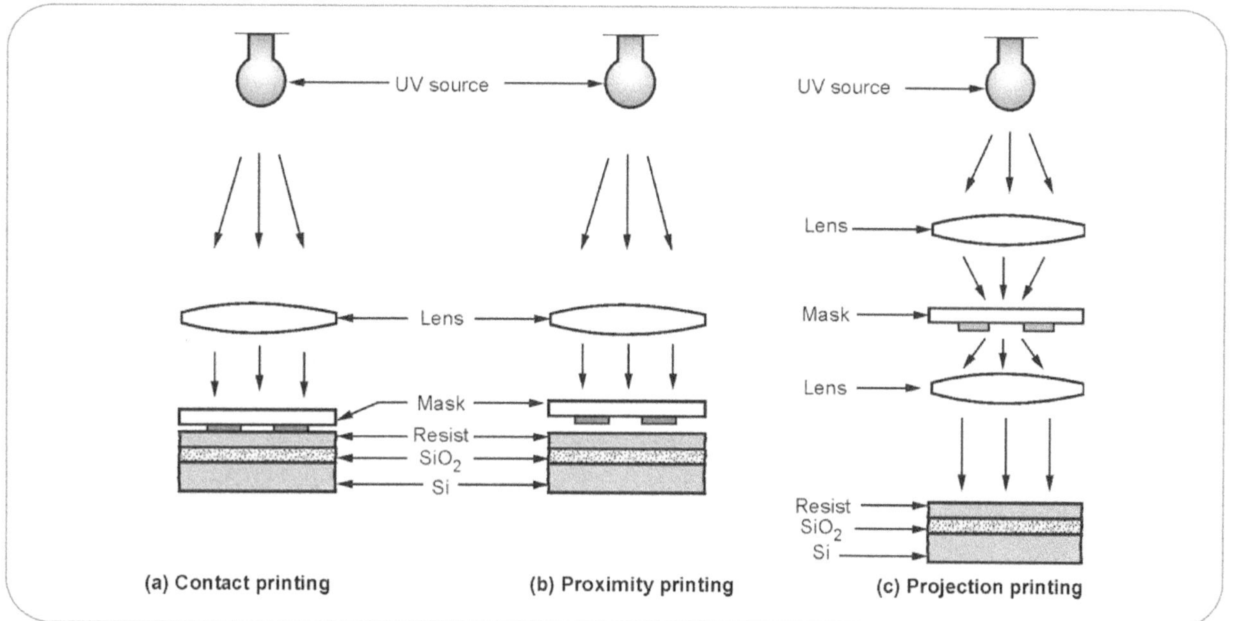

Fig. 7.24.2 Photolithography exposure techniques

- The mask itself is fabricated by lithography, the pattern being based on circuit design data, usually in the form of digital output from the CAD system used by the circuit designer.

- Generally, there are two types of photo-resists available positive and negative.

- A positive resist becomes more soluble in developing solutions after exposure to light. A negative resist becomes less soluble when exposed to light.

- The resists are exposed through the mask by one of the following three exposure techniques :

i) Contact printing

ii) Proximity printing

iii) Projection printing

(i)   In contact printing, the mask is pressed against the resist coating during exposure. This results in high resolution of the pattern onto the wafer surface; an important disadvantage is that physical contact with the wafers gradually wears out the mask.

(ii) In proximity printing, the mask is separated from the resist coating by a distance of 10 to 25 mm. This eliminates mask wear, but resolution of the image is slightly reduced.

(iii) Projection printing involves the use of a high-quality lens (or mirror) system to project the image through the mask onto the wafer. This has become the preferred technique because it is noncontact thus mask does not get wear, and the mask pattern can be reduced through optical projection to obtain high resolution.

- Following steps are carried out as a sequence of photolithography :

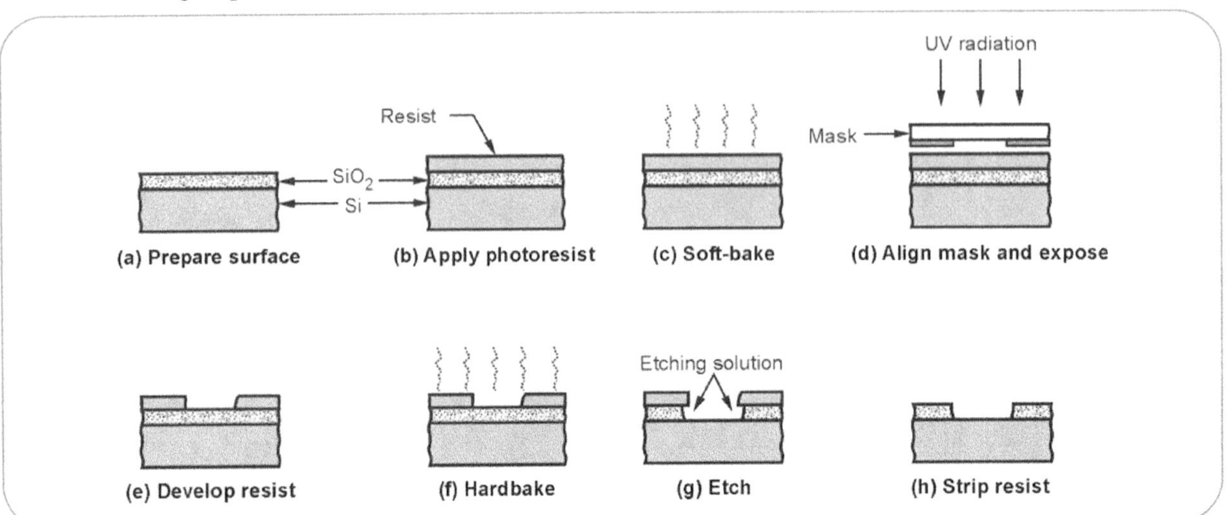

Fig. 7.24.3 : Photolithography process applied to a silicon wafer

i) **Prepare surface :** In this step the wafer is properly cleaned to promote wetting and adhesion of the resist.

ii) **Apply photo-resist :** Photo-resists are applied by feeding a required amount of liquid resist onto the center of the wafer and then spinning the wafer to spread the liquid and due to which a uniform coating thickness is achieved. Desired thickness is around 1 mm which gives good resolution yet minimizes pinhole defects.

iii) **Soft-bake :** Purpose of this pre exposure bake is to remove solvents, promote adhesion, and harden the resist. For this step the temperature is around 90 °C for 10 to 20 minutes.

iv) **Align mask and expose :**
- The pattern mask is aligned relative to the wafer and the resist is exposed through the mask by one of the methods described in the preceding.
- Alignment must be accomplished with very high precision, using optical-mechanical equipment designed specifically for the purpose.
- If the wafer has been previously processed by lithography so that a pattern has already been formed in the wafer, then subsequent masks must be accurately registered relative to the existing pattern.
- Exposure of the resist depends on the same basic rule as in photography; the exposure is a function of light intensity time.
- For the exposure mercury arc lamp or other source of UV light is used.

v) **Develop resist :** The exposed wafer is then immersed in a developing solution, or the solution is sprayed onto the wafer surface.

vi) **Hard-bake :** This baking step expels volatiles remaining from the developing solution and increases adhesion of the resist, especially at the newly created edges of the resist film.

vii) **Etch :** Etching removes the SiO$_2$ layer at selected regions where the resist has been removed.

viii) Strip resist :

- After etching, the resist coating that remains on the surface must be removed.

- Stripping is accomplished using either wet or dry techniques.

- Wet stripping uses liquid chemicals; a mixture of sulfuric acid and hydrogen peroxide (H_2SO_4 - H_2O_2) is common.

- Dry stripping uses plasma etching with oxygen as the reactive gas.

Applications photolithography

- Photolithography produces integrated circuits and other internal computer parts.

- It has been used in recent years to produce nanites as well as microscopic computer systems.

- Photolithography also produces patterns on any small surface and is simply a form of lithography.

- Other forms of lithography include electron beam, X-ray, extreme ultraviolet, ion projection and immersion lithography.

7.25 : Reconfigurable Manufacturing System (RMS)

- RMS is designed at the outset for rapid change in 1st Structure, hardware and software components also to adjust the production capacity.

- It also adjust functionality within a part family with respect to sudden market changes.

- The ideal RMS provides exactly the functionality and production capacity needed and it can be economically adjusted exactly when needed.

- An ideal RMS possesses the following six core characteristics :

  i) Modularity   ii) Integrability

  iii) Scalability   iv) Convertibility

  v) Diagnosability vi) Customized Flexibility

- With these characteristics, RMS increases the speed of responsiveness of manufacturing systems to unpredicted events like sudden change in market or sudden failure of machines.

- The main components of RMS are as follows :

  ■ CNC Machines

  ■ Reconfigurable machine tools

  ■ Reconfigurable inspection machines

  ■ Reconfigurable machine transport systems etc.

- RMS operate according to a set of basic principles. The more of these principles applicable to a given manufacturing system, the more reconfigurable that system.

- The RMS principles are as follows :

  i) RMS capacity is rapidly scalable in small or optimal increments. Also its functionality is rapidly adaptable to the production of new products.

  ii) To enhance the speed of responsiveness of a manufacturing system, RMS characteristics

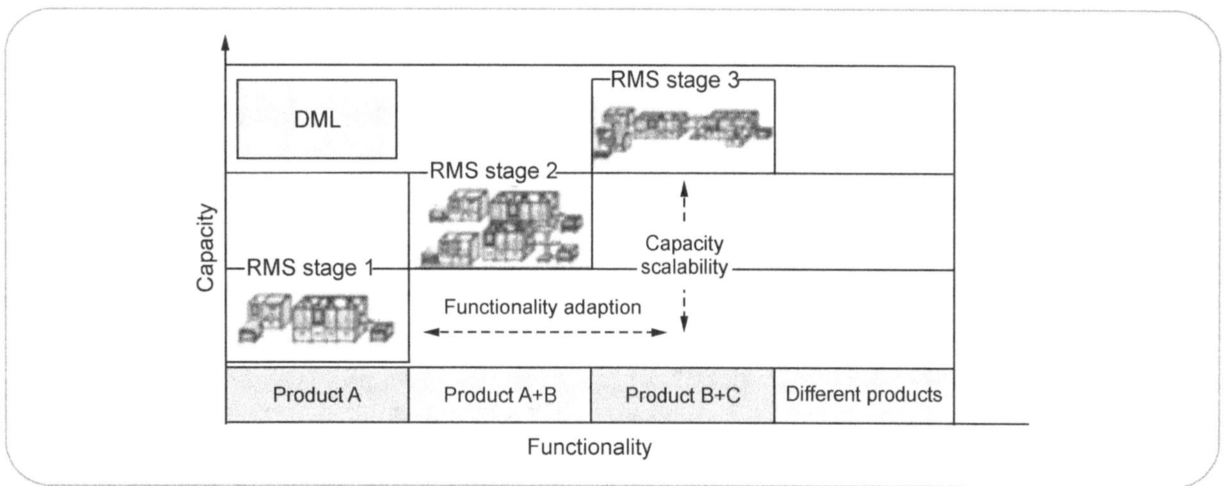

Fig. 7.25.1

should be embedded in complete system and its components.

iii) It is designed around a part family with just enough customized flexibility required to produce all parts in the family.

iv) RMS contains an economic equipment mix of flexible and reconfigurable machines with customized flexibility.

v) RMS possess hardware and software capabilities to cost effectively respond to unpredictable events.

NC and CNC MACHINES

7.26 : Necessity of NC Machine Tool

- When any machine tool is manually operated, the operator controls the relative movements of the workpiece and tool.

- The accuracy of these movements is controlled by reference to some form of measuring device fitted on the machine slide or lead screw.

- The operator has to perform functions like starting and stopping the machine, turning the coolant on and off, etc.

- With manual control accuracy of final workpiece, quality and time required to manufacture depends on the skill, concentration and experience of the operator.

- When many batches of identical parts are required, it is preferable to use jigs, fixtures and templates.

- Automatic machine tools are also used in order to minimize errors and variable quality of manual operation.

- So, to avoid human errors, minimize production cost and due to many other reasons, NC i.e. **Numerical Control** machines comes into the picture.

7.27 : Numerical Control (NC) Machine Tool
SPPU : Dec.-04,05,08, May-06,08,10

- Numerical control is a programmable automation in which actions are controlled by means of coded numbers, letters and other symbols."

- The numerical data which is required for producing a part is maintained on **punched tape.**

- This data is arranged in the form of blocks of information.

- The block contains cutting speed, feed, dimensional information and contour form.

- For preparing this data, part programmer is required which should have the knowledge of tools, cutting fluids, use of machinability data and process engineering.

- Fig. 7.27.1 (a) shows the block diagram for the procedure of production through NC.

- Fig. 7.27.1 (b) shows the block diagram for NC machine tool system.

7.27.1 Basic Elements of NC System
SPPU : Dec.-04,05,08, May-06,08

A numerical control (NC) machine consists of following elements (Refer Fig. 7.27.1 (b)).

> 1. Machine Control Unit (MCU)
> 2. Machine tool and NC tooling
> 3. Part program and drawings.

1. Machine Control Unit (MCU)

- It is the heart of NC machine tool system and consists of many sub-units inside it.

- The first sub-unit is tape reader which receives the coded data from punched tape.

- The tape reader reads this data and passes data to the buffer storage through the decoding circuits.

- The buffer storage stores the received information, till it is required and transfers it to the required area.

- Almost all the operations like tool movements, tool change, speed and feed change and many others can be controlled by MCU.

2. Machine tool and NC tooling

- It is the manufacturing arm of NC machine tool system.

- It receives the raw material and performs different operations like turning, milling, drilling, grinding, etc.

- For performing these operations, it should receive the information from the MCU.

- As per the information, the desired shape and size is modified.

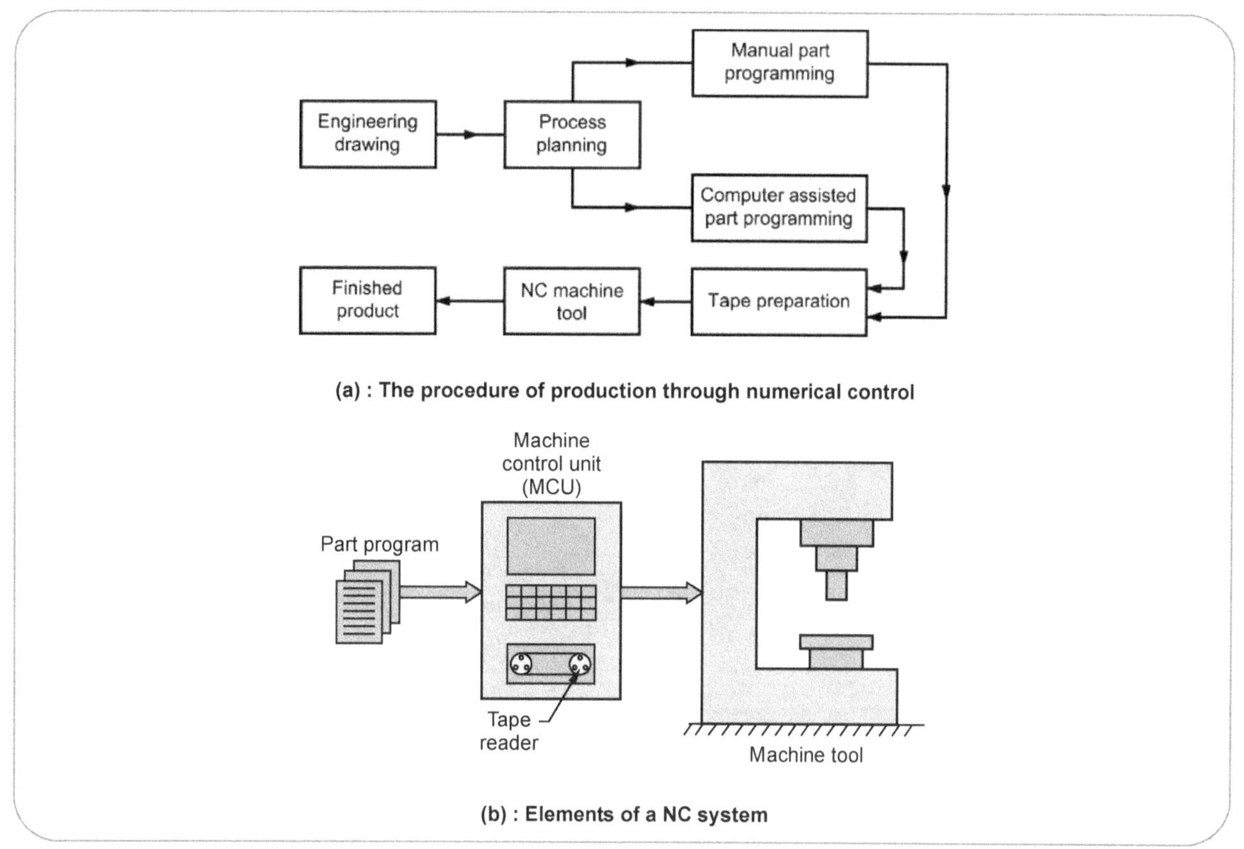

(a) : The procedure of production through numerical control

(b) : Elements of a NC system

Fig. 7.27.1

3. Part program and drawings

- NC machine operates as per coded information, which is Input Data for the machine.

- The feeding of this data may be manually or automatic.

- The manual feeding of input data includes operator and hence chances of error increases.

- Hence, data is fed by automatic means and for this purpose punched tape is mostly used.

- The other input media instead of punched tapes are :

    ○ Punched cards

    ○ Magnetic tapes

    ○ Diskettes, etc.

- Now a days, to enter the program, instead of these media, magnetic cassettes, floppy discs, compact disc (CD) are used.

7.27.2 Advantages, Disadvantages and Applications of NC System

Advantages

- **High productivity :** Due to less set up and lead time, productivity is higher.

- **Less scrap :** As human errors are eliminated, accurate components are machined hence, scrap is reduced.

- **Reduced jigs and fixtures :** Work and tool positioning is done by NC tape, hence less requirement of jigs and fixtures.

- **High quality :** Due to higher accuracy of NC systems, the quality of products is easily controlled.

- **Flexibility in design :** In NC system complicated profiles can be easily produced at faster rate.

- **Utilization of manpower :** In NC system, there is greater utilization of manpower because, after setting a component, operator can perform other operations.

- Reduction in the inventory.

- Safety to the operator and machine tool.
- There is a greater flexibility in the manufacturing.
- Less floor space is required.
- As no jigs and fixtures are required, tooling cost is low.
- Skilled operator is not required.

Disadvantages

- **High initial cost :** Initial investment is high.
- **High maintenance cost :** Maintenance is costly and complicated.
- **Costy control system :** Control systems are also costly.
- **Skilled operator :** For part programming well trained and highly skilled operator is required.
- **Unemployment :** As only one operator is required, there is increase in unemployment.

Applications

- NC system is used where 100 % inspection is required.
- NC system is suitable for machining of parts where frequent changes in design occurs.
- Repetitive production of precise parts in small and medium size production can be done by NC system.
- When accuracy requirement is high, NC system is suitable.
- When high amount of material is to be removed, NC system is preferred.
- For complex machining operations also, NC system is required.

7.28 : Computerized Numerical Control (CNC) Machine Tool

SPPU : May-04,05,07,09,10, Dec.-03,05,06,07,10,

- Computerized Numerical Control (CNC) is a NC system which uses micro-computer as the machine control unit.
- The presence of a microprocessor, RAM, ROM, Input/Ouput (I/O) devices has raised the automation level in NC system.
- In CNC machines, various functions are controlled with the help of part programs which are generally entered through the keyboard.

7.28.1 Basic Elements of CNC System

SPPU : May-04,05,07,10, Dec.-06

A CNC system consists of following basic elements (Refer Fig. 7.28.1) :

> 1. Part program and drawings 2. Input/Output devices
> 3. Memory storage devices   4. Micro-computer (MCU)
> 5. CNC machine tool   6. Feedback device

1. Part Program and Drawings :

- The part program is written by observing the part drawing and the given cutting process parameters like feed, speed, depth of cut, etc.

2. Input/Output Devices :

- The written part program is entered into the micro-computer by using input devices like keyboard, CD, DVD, etc.

3. Memory Storage Devices :

- The entered program is stored in computer memory which can be recalled whenever required.
- Also, the program can be easily edited and modified as per the requirement.

4. Micro-computer (MCU) :

- The entered part program is read by the micro-computer which is the machine control unit of CNC system.
- It controls all the movements of machine tool, actuation of all the drives, coolant supply, etc. of the machine tool.

5. CNC Machine Tool :

- It is the manufacturing arm of CNC machine tool system.
- It receives raw material and performs various operations which are needed.
- For performing these operations it should receive the information from the MCU.
- In CNC system, all the operations like spindle start and stop, tool positioning, tool changing, speed control, etc. are fully automatic.

6. Feedback Device :

- Feedback device receives feedback from the machine tool and gives it to the MCU.

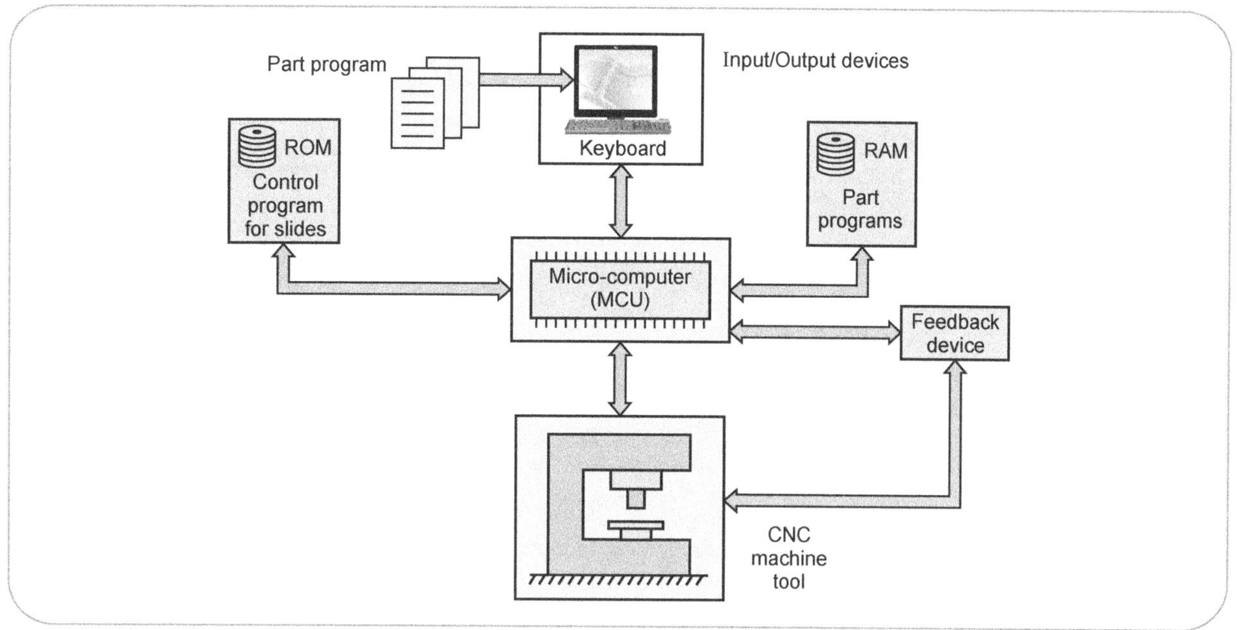

Fig. 7.28.1 : Basic elements of CNC system

- By using feedback device, the cause can be easily detected and rectified.

7.28.2 Advantages, Disadvantages and Applications of CNC System

SPPU : Dec.-03,05,07,10, May-09

Advantages

Advantages of CNC machines are similar to NC machine. Some additional advantages due to additional feature in CNC machine over conventional machines are as follows :

- **Program storage :** As computer is available, multiple programs can be stored in the machine.

- **Reliability of system :** As the data is directly entered with the help of computer, no need to use punched tape. It also improves reliability of the system.

- **Online part programming :** The part program can be done online with editing, if required.

- **Flexibility of system :** The system is too flexible, as new systems can be added at low costs.

- **Metric conversions :** Part program, which is written in inches, can be easily converted into millimeters i.e. metric conversion is easy.

- **Interpolations :** In NC system, there is interpolation for straight and circular path, but in CNC it is available for helical, parabolic and cubic curves also.

- **Expanded tool compensations :** For the purpose of tool offset and tool wear, tool compensation is provided.

Disadvantages

Disadvantages of CNC machines are similar to NC machines.

- **High initial cost :** Initial investment is high.

- **High maintenance cost :** Maintenance is costly and complicated.

- **Costly control system :** Control systems are also costly.

- **Skilled operator :** For part programming well trained and highly skilled operator is required.

- **Unemployment :** As only one operator is required, there is increase in unemployment.

- **Computer problem :** If there is any problem with computer, then the whole machine will get stop.

- **Costly software :** The software required for the operation of CNC machines is also costly.

Applications

- Applications of CNC system are similar to NC system.

7.29 : Comparison of NC and CNC System

Table 7.29.1 gives the comparison between the NC and CNC system.

Table 7.29.1 : Comparison of NC and CNC system

Sr.No.	Parameter	NC system	CNC system
1.	Mode of entering the program	In this system part program is entered by using punch tape, magnetic tapes, punch cards, etc.	In this system, part program is entered by using keyboard, CD, DVD, floppy discs, etc.
2.	Feedback device	No feedback device.	Feedback device is available.
3.	Memory storage ability	Only one program can be stored at a time.	More than one program can be stored.
4.	MCU	Tape reader acts as a MCU.	Micro-computer act as a MCU.
5.	Program editing	The program is entered in the form punch tape, hence editing in program is difficult.	The program is entered through the keyboard, hence editing in program is simple.
6.	Flexibility of system	Less flexible.	More flexible.
7.	Productivity of system	Less	High
8.	Initial cost	Moderate	High

7.30 : Comparison between AM and CNC Machining

Sr. No.	Parameter	AM machining	CNC machining
1.	Material	Developed around polymeric materials, waxes and paper laminates.	CNC machining can be used for soft materials like medium density fiberboard, machinable foams, machinable waxes,etc.
2.	Speed	To make part by AM may take few hours.	CNC machining would take weeks to develop same part.
3.	Complexity	AM machines are generally complex and due to complexity they are more efficient.	CNC machines aren't complex hence they have less efficiency.
4.	Accuracy	Due to its variable resolution along 5 different orthogonal axes AM is more accurate as compared to CNC.	Due to limited resolution along only 3 orthogonal axes CNC machines aren't that accurate.
5.	Programming	In AM machines the part won't be built properly if incorrect programming is done.	In CNC machines incorrect programming can badly damage the machine and may even be a safety risk.
6.	Geometry	AM machines simply breakup a complex 3D problem into series of simple 2D cross-sections with suitable thickness.	This cannot be done in CNC, machining of surfaces must normally be generated in 3D space.
7.	Process	It is an additive process (material added layer by layer).	It is subtractive process (material is removed by machining).

7.31 : IoT (Internet of Things)

- The Internet of Things (IoT) is the network of physical objects i.e. devices, vehicles, buildings and other items embedded with electronics, software, sensors, and network connectivity that enables these objects to collect and exchange data.

- Wikipedia definition : The Internet of Things, also called The Internet of Objects, refers to a wireless network between objects, usually the network will be wireless and self-configuring, such as household appliances.

- WSIS 2005 Definition : By embedding short-range mobile transceivers into a wide array of additional gadgets and everyday items, enabling new forms of communication between people and things, and between things

- A phenomenon which connects a variety of things. Everything that has the ability to communicate.

- The Internet of Things is the intelligent connectivity of physical devices driving massive gains in efficiency, business growth, and quality of life. Fig. 7.13.1 shows IoT ecosystem.

- The Internet of Things refers to the capability of everyday devices to connect to other devices and people through the existing Internet infrastructure. Devices connect and communicate in many ways. Examples of this are smart phones that interact with other smart phones, vehicle-to-vehicle communication, connected video cameras, and connected medical devices. They are able to communicate with consumers, collect and transmit data to companies, and compile large amounts of data for third parties.

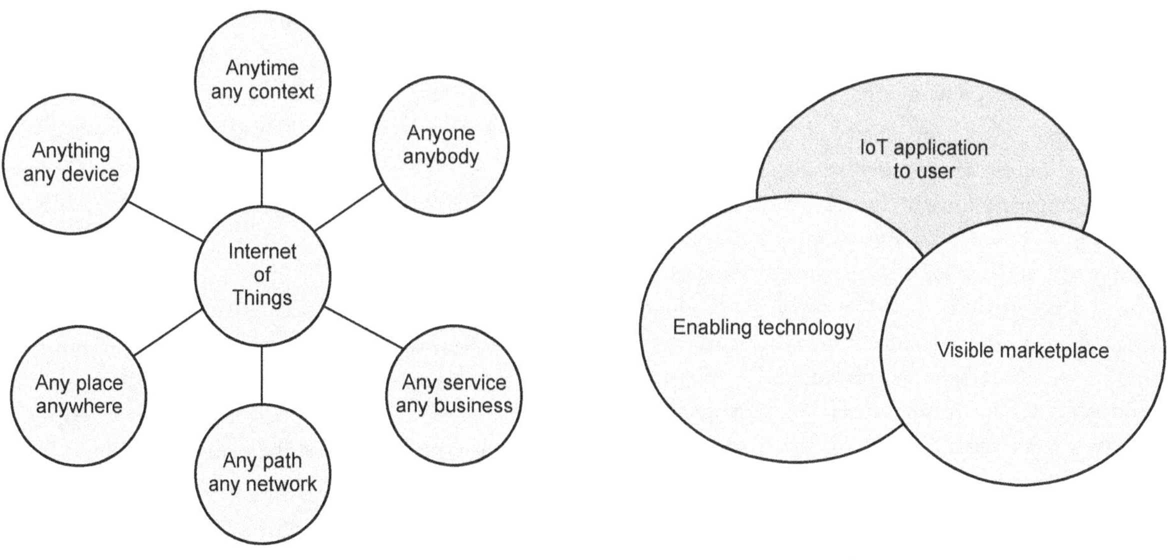

Fig. 7.13.1

- IoT data differs from traditional computing. The data can be small in size and frequent in transmission. The number of devices, or nodes, that are connecting to the network are also greater in IoT than in traditional PC computing.

- Machine-to-Machine communications and intelligence drawn from the devices and the network will allow businesses to automate certain basic tasks without depending on central or cloud based applications and services.

- IoT impacts every business. Mobile and the Internet of Things will change the types of devices that connect into a company's systems. These newly connected devices will produce new types of data. The Internet of Things will help a business gain efficiencies, harness intelligence from a wide range of equipment, improve operations and increase customer satisfaction.

- Ubiquitous computing, pervasive computing, Internet Protocol, sensing technologies, communication technologies, and embedded devices are merged together in order to form a system where the real and digital worlds meet and are continuously in symbiotic interaction.

- The smart object is the building block of the IoT vision. By putting intelligence into everyday objects, they are turned into smart objects able not only to collect information from the environment and interact /control the physical world, but also to be interconnected, to each other, through Internet to exchange data and information.

- The expected huge number of interconnected devices and the significant amount of available data open new opportunities to create services that will bring tangible benefits to the society, environment, economy and individual citizens. In this paper we present the key features and the driver technologies of IoT. In addition to identifying the application scenarios and the correspondent potential applications, we focus on research challenges and open issues to be faced for the IoT realization in the real world.

- However, the IoT is still maturing, in particular due to a number of factors, which limit the full exploitation of the IoT. Some of the factors are listed below :

1. There is no unique identification number system for object in the world.

2. IoT uses Architecture Reference Model (ARM) but there is no further development in ARM.

3. Missing large-scale testing and learning environments

4. Difficulties in exchanging of sensor information in heterogeneous environments.

5. Difficulties in developing business which embraces the full support of the Internet of Things.

Characteristics of the Internet of Things

1. Interconnectivity : Everything can be connected to the global information and communication infrastructure.

2. Heterogeneity : Devices within IoT have different hardware and use different networks but they can still interact with other devices through different networks.

3. Things-related services : Provides things-related services within the constraints of things, such as privacy and semantic consistency between physical and virtual thing.

4. Dynamic changes : The state of a device can change dynamically.

7.13.1 Component of IoT

- The hardware utilized in IoT systems includes devices for a remote dashboard, devices for control, servers, a routing or bridge device, and sensors. These devices manage key tasks and functions such as system activation, action specifications, security, communication, and detection to support-specific goals and actions.

- Major components of IoT devices are as follows:

    1. **Control units :** A small computer on a single integrated circuit containing processor core, memory and a programmable I/O peripheral. It is responsible for the main operation.

    2. **Sensor :** Devices that can measure a physical quantity and convert it into a signal, which can be read and interpreted by the microcontroller unit. These devices consist of energy modules, power management modules, RF modules, and

sensing modules. Most sensors fall into 2 categories : Digital or analog. An analog data is converted to digital value that can be transmitted to the Internet.

a. Temperature sensors : accelerometers

b. Image sensors: gyroscopes

c. Light sensors : acoustic sensors

d. Micro flow sensors : humidity sensors

e. Gas RFID sensors : pressure sensors

3. **Communication modules :** These are the part of devices and responsible for communication with rest of IoT platform. They provide connectivity according to wireless or wired communication protocol they are designed. The communication between IoT devices and the Internet is performed in two ways:

A) There is an Internet-enable intermediate node acting as a gateway;

B) The IoT Device has direct communication with the Internet.

• The communication between the main control unit and the communication module uses serial protocol in most cases.

4. **Power sources :** In small devices the current is usually produced by sources like batteries, thermocouples and solar cells. Mobile devices are mostly powered by lightweight batteries that can be recharged for longer life duration.

• **Communication Technology and Protocol :** IoT primarily exploits standard protocols and networking technologies. However, the major enabling technologies and protocols of IoT are RFID, NFC, low-energy Bluetooth, low-energy wireless, low-energy radio protocols, LTE-A, and WiFi-Direct. These technologies support the specific networking functionality needed in an IoT system in contrast to a standard uniform network of common systems.

Working :

1. **Collect and transmit data :** The device can sense the environment and collect information related to it and transmit it to a different device or to the Internet.

2. **Actuate device based on triggers :** It can be programmed to actuate other devices based on conditions set by user.

3. **Receive information :** Device can also receive information from the network.

4. **Communication assistance :** It provides communication between two devices of same network or different network.

• Fig. 7.31.2 shows working of IoT.

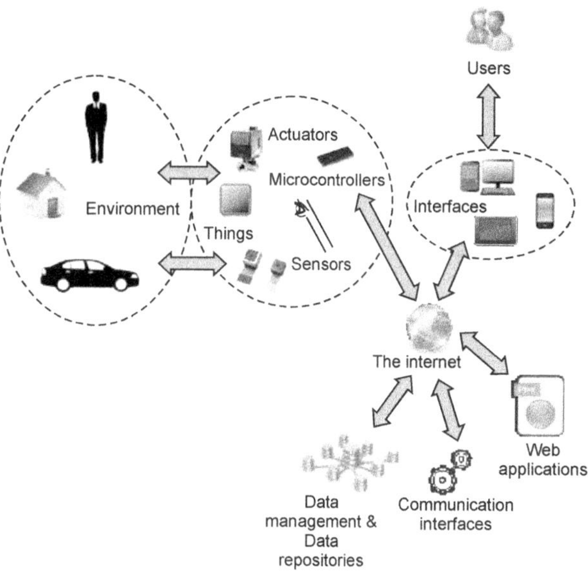

Fig. 7.31.2 Working of IoT

• Sensors for various applications are used in different IoT devices as per different applications such as temperature, power, humidity, proximity, force etc.

• Gateway takes care of various wireless standard interfaces and hence one gateway can handle multiple techologies and multiple sensors. The typical wireless technologies used widely are 6LoWPAN, Zigbee, Zwave, RFID, NFC etc. Gateway interfaces with cloud using backbone wireless or wired technologies such as WiFi, Mobile , DSL or Fibre.

7.31.2 Advantages and Disadvantages of IoT

Advantages of IoT

1. Improved customer engagement and communication.

2. Support for technology optimization

3. Support wide range of data collection

4. Reduced waste

Disadvantages of IoT

1. **Loss of privacy and security :** As all the household appliances, industrial machinery, public sector services like water supply and transport, and many other devices all are connected to the Internet, a lot of information is available on it. This information is prone to attack by hackers.

2. **Flexibility :** Many are concerned about the flexibility of an IoT system to integrate easily with another

3. **Complexity :** The IoT is a diverse and complex network. Any failure or bugs in the software or hardware will have serious consequences. Even power failure can cause a lot of inconvenience.

4. **Compatibility :** Currently, there is no international standard of compatibility for the tagging and monitoring equipment

5. Save time and money

7.31.3 Application of IoT

1. **Home :** Buildings where people live. It controls home and security systems.

2. **Offices :** Energy management and security in office buildings; improved productivity, including for mobile employees.

3. **Factories :** Places with repetitive work routines, including hospitals and farms; operating efficiencies, optimizing equipment use and inventory.

4. **Vehicles :** Vehicles including cars, trucks, ships, aircraft, and trains; condition-based maintenance, usage-based design, pre-sales analytics

5. **Cities :** Public spaces and infrastructure in urban settings; adaptive traffic control, smart meters, environmental monitoring, resource management.

6. **Worksites :** It is custom production environments like mining, oil and gas, construction; operating efficiencies, predictive maintenance, health and safety.

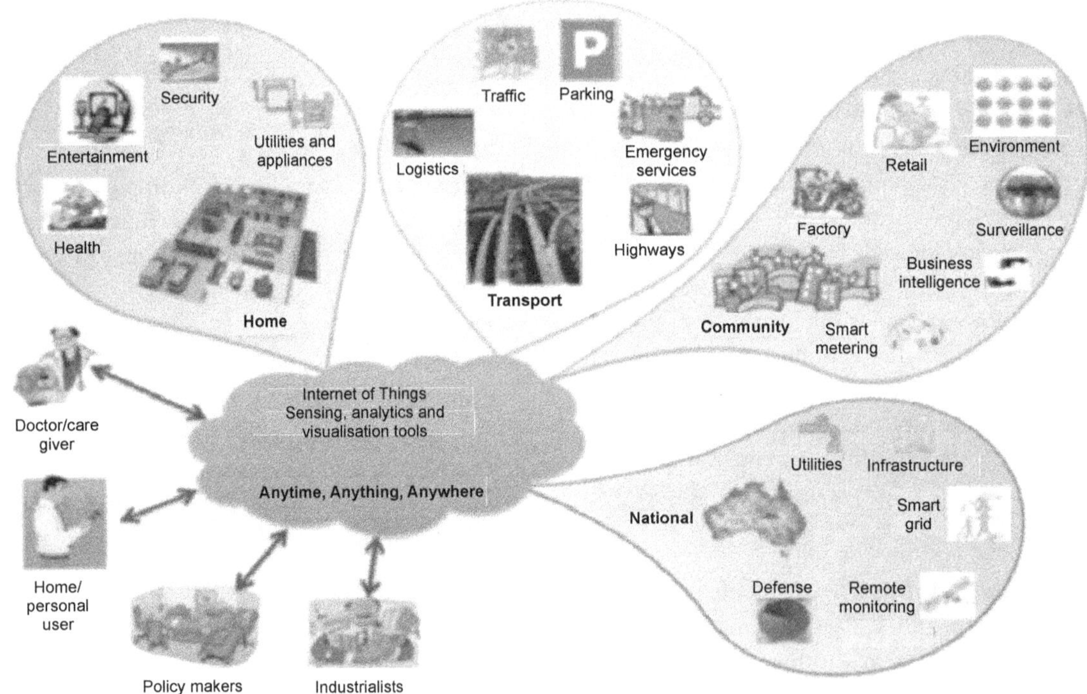

Fig. 7.31.3

Review Questions

1. What is manufacturing process ? Give its classification?

2. What is casting process ? State its advantages, disadvantages and applications.

3. Explain the basic steps in casting process.

4. What is forging process ? State its advantages, disadvantages and applications.

5. Explain drop forging process with neat sketch.

6. Give comparison between casting and forging.

7. Give comparison between hot working and cold working.

8. What is meant by metal forming ? What are its types ?

9. Explain wire drawing and tube drawing with neat sketch.

10. Explain extrusion process in detail.

11. Compare direct extrusion and indirect extrusion.

12. What is sheet metal working ? State its applications.

13. Explain the following sheet metal operations :
    a) Blanking   b) Punching   c) Perforating

14. State advantages, disadvantages and applications of welding process.

15. Give classification of welding process.

16. Explain arc welding process with neat sketch.

17. Compare TIG and MIG welding processes.

18. Explain soldering and brazing process with neat sketch.

19. Write the working principle of lathe machine.

20. Explain the function of following parts of lathe machine.
    a) Bed   b) Headstock   c) Tailstock   d) Carriage

21. Explain the following lathe operations :
    a) Turning   b) Knurling
    c) Thread cutting   d) Facing

22. Explain drilling machine with neat sketch.

23. Explain the following drilling machine operations :
    a) Counter boring   b) Reaming
    c) Tapping   d) Boring

24. Write the working principle of milling machine.

25. Explain the function of different parts of milling machine.

26. Explain the following milling operations :
    a) Plain millimg   b) Face milling
    c) Form milling   d) Straddle milling

27. What is additive manufacturing ? State its applications.

28. Explain Generic of AM process.

29. Write short note on : 3D printing.

30. Explain photolithography micromachining method.

31. Write short note on : RMS.

32. What are the elements of CNC ? Explain.

33. State advantages and disadvantages of CNC.

34. Compare NC and CNC system.

35. Write short note on IoT.

7.32 : University Questions with Answers

May - 2012

Q.1   Explain the following sheet metal processes :
i) Punching   ii) Perforating
iii) Lancing   iv) Blanking
Explain the curling and wire drawing process.
(Refer section 7.9.1)   **[9]**

Dec. - 2012

Q.2   Describe sand casting process.
(Refer section 7.2.3)   **[8]**

Q.3   Describe shearing, bending, squeezing and drawing operations for sheet metal.**(Refer section 7.9)**   **[6]**

Q.4   Explain soldering and brazing processes.
(Refer sections 7.13 and 7.14)   **[8]**

May - 2013

Q.5   Explain with sketches the different stages involved in manufacturing of sand casting.
(Refer section 7.2.3 and Fig. 7.2.2)   **[7]**

Q.6   Draw self explanatory sketches of various operations performed in sheet metal working.
(Refer section 7.9.1)   **[6]**

Q.7   Explain with neat sketch the major parts of a center lathe machine.
(Refer section 7.17.1)   **[7]**

Dec. - 2013

Q.8 *What is sand casting ? Explain its advantages, disadvantages and applications.*
(Refer sections 7.2.1 and 7.2.2) **[7]**

Q.9 *Draw self-explanatory sketches of any three; sheet-metal cutting and any three; sheet-metal forming operations.*
(Refer sections 7.8 and 7.9.2) **[6]**

Q.10 *Draw the block diagram of a lathe machine and explain the functions of its various parts.*
(Refer section 7.17.2) **[7]**

May - 2014

Q.11 *Differentiate between hot working and cold working.* **(Refer section 7.4.3)** **[4]**

Q.12 *Explain the following sheet metal working operations with neat sketch, blanking and drawing.*
(Refer sections 7.9.1 and 7.9.2) **[4]**

Q.13 *Write a note on electric arc welding.*
(Refer section 7.12) **[6]**

Q.14 *Explain with suitable sketch, radial drilling machine. State its advantages.*
(Refer section 7.18.3 (4)) **[4]**

Dec. - 2014

Q.15 *Explain with neat sketch forging process. Give advantages, disadvantages and applications.*
(Refer section 573) **[7]**

Q.16 *Write note on brazing.* **(Refer section 7.14)** **[6]**

Q.17 *Explain working principle and basic elements of drilling machine.*
(Refer sections 7.18.1 and 7.18.2) **[6]**

May - 2015

Q.18 *Identify and explain suitable process to join two copper tubes.* **(Refer section 7.12.4)** **[4]**

Q.19 *Draw self-explanatory sketches of any four sheet metal cutting process.* **(Refer section 7.9.1)** **[4]**

Q.20 *Draw neat sketch of sand casting process setup. State advantages, limitations and engineering applications of the process.* **(Refer section 7.2)** **[6]**

Q.21 *Explain boring operation performed on lathe machine and radial drilling machine.*
(Refer sections 7.17.4 and 7.18.4) **[6]**

Q.22 *Explain taper turning, parting and knurling operation performed on lathe machine.*
(Refer section 7.17.4) **[6]**

Dec. - 2015

Q.23 *Explain hot forging process with neat sketch.*
(Refer section 7.3) **[4]**

Q.24 *Draw self-explanatory sketches of various sheet metal forming processes.* **(Refer section 7.9.1)** **[6]**

Q.25 *Identify and explain suitable manufacturing process to join two mild steel plates.*
(Refer section 7.12) **[4]**

Q.26 *Draw a labelled block diagram of lathe machine.*
(Refer Fig. 7.17.3) **[4]**

Q.27 *Explain drilling operation performed on lathe machine and radial drilling machine.*
(Refer sections 7.17.4 and 7.18.4) **[4]**

May - 2016

Q.28 *Draw neat sketch of sand casting process setup. State applications of the process.*
(Refer sections 7.2.1 and 7.2.3) **[4]**

Q.29 *Identify and explain suitable manufacturing process to join two copper tubes.*
(Refer section 7.12.4) **[6]**

Q.30 *Explain hot forging process with neat sketch.*
(Refer section 7.3) **[4]**

Q.31 *Draw self-explanatory sketches of various sheet metal cutting process.*
(Refer section 7.9.1) **[4]**

Q.32 *Explain drilling operation performed on lathe machine and radial drilling machine.*
(Refer sections 7.17.4 and 7.18.4) **[4]**

Dec. - 2016

Q.33 *Draw self-explanatory sketches of blanking, forming and drawing operation on sheet metal.*
(Refer sections 7.9.1 and 7.9.2) **[6]**

Q.34 *Compare arc-welding, brazing and soldering process.* **(Refer section 7.15)** **[6]**

Q.35 *Explain reaming, counter sinking and tapping operations on drilling machine.*
(Refer section 7.18.4) **[6]**

May - 2017

Q.36 *What is sand casting process ? Draw neat sketch of sand casting process setup and explain steps involved in the process.*
(Refer sections 7.2, 7.2.2 and 7.2.3) **[7]**

Q.37 *Draw neat sketch of soldering process setup. Explain the process in brief. State application of the process.* **(Refer sections 7.13 and 7.13.1)** **[6]**

Q.38 *Draw a block diagram of a radial drilling machine and explain tapping operation with sketch.*
(Refer sections 7.18.3 and 7.18.4) **[6]**

Q.39 *Daw block diagram of lathe machine. Explain function of headstock, tail stock and carriage.*
(Refer section 7.17.2) **[7]**

Dec. - 2017

Q.40 *What is welding ? Draw neat sketch of arc welding and brazing process setup.*
(Refer sections 7.11, 7.12 and 7.14) **[7]**

Q.41 *Differentiate between arc welding, brazing and soldering process (6 points).* **(Refer section 7.15)**
[6]

Q.42 *Explain drilling, reaming, boring, tapping operation performed on radial drilling machine.*
(Refer section 7.18.4) **[7]**

May - 2018

Q.43 *What is welding ? Draw neat sketch of arc welding and brazing process setup.*
(Refer sections 7.11, 7.12 and 7.14) **[7]**

Q.44 *Explain reaming, counter sinking, tapping operation performed on radial drilling machine.*
(Refer section 7.18.4) **[6]**

Q.45 *Differentiate between arc welding, brazing and soldering process (6 points).* **(Refer section 7.15)**
[6]

Q.46 *Explain drilling, reaming, boring, tapping operation performed on radial drilling machine.*
(Refer section 7.18.4) **[7]**

Dec. - 2018

Q.47 *Differentiate between hot and cold working process. (Four points). Draw neat sketch of hot forging process setup.*
(Refer sections 7.4.3 and 7.3) **[7]**

Q.48 *Explain any three sheet metal working process with neat sketches.* **(Refer section 7.9.1)** **[6]**

Q.49 *Draw block diagram of a lathe machine. Explain function of headstock, tailstock and carriage of lathe machine.* **(Refer section 7.17.2)** **[7]**

May - 2019

Q.50 *Draw block diagram of lathe machine. Explain function of headstock, tailstock and carriage.*
(Refer section 7.17.2) **[7]**

Q.51 *Explain punching, piercing perforating, notching operations in sheet metal working.*
(Refer section 7.9.1) **[6]**

Q.52 *Compare welding, soldering and brazing process.*
(Refer section 7.15) **[6]**

Q.53 *Explain working principle of drilling machine with block diagram and explain any three operations performed on it.*
(Refer sections 7.18.1 and 7.18.4) **[7]**

□□□

Notes

Unit - VI
Engineering Mechanisms and their Application in Domestic Appliances

Chapter - 8   Engineering Mechanisms & their Applications in Domestic Appliances   (8 - 1) to (8 - 36)

UNIT - VI

8   Engineering Mechanisms & their Applications in Domestic Appliances

Syllabus

Introduction to Basic mechanisms and equipment : *Pumps, blowers, compressors, springs, gears, Belt-Pulley, Chain-Sprocket, valves, levers, etc. Introduction to terms : Specifications, Input, output, efficiency, etc.*

Applications of : *Compressors - Refrigerator, Water cooler, Split AC unit; Pumps - Water pump for overhead tanks, Water filter/Purifier units; Blower - Vacuum cleaner, Kitchen Chimney; Motor - Fans, Exhaust fans, Washing machines; Springs - Door closure, door locks, etc.; Gears - Wall clocks, watches, Printers, etc.; Application of Belt-Pulley/Chain-Sprocket - Photocopier, bicycle, etc.; Valves - Water tap, etc.; Application of levers - Door latch, Brake pedals, etc.; Electric/Solar energy - Geyser, Water heater, Electric iron, etc. (simple numerical on efficiency calculation)*

Contents

Mind Map - Engineering Mechanisms & their Applications in Domestic Appliances

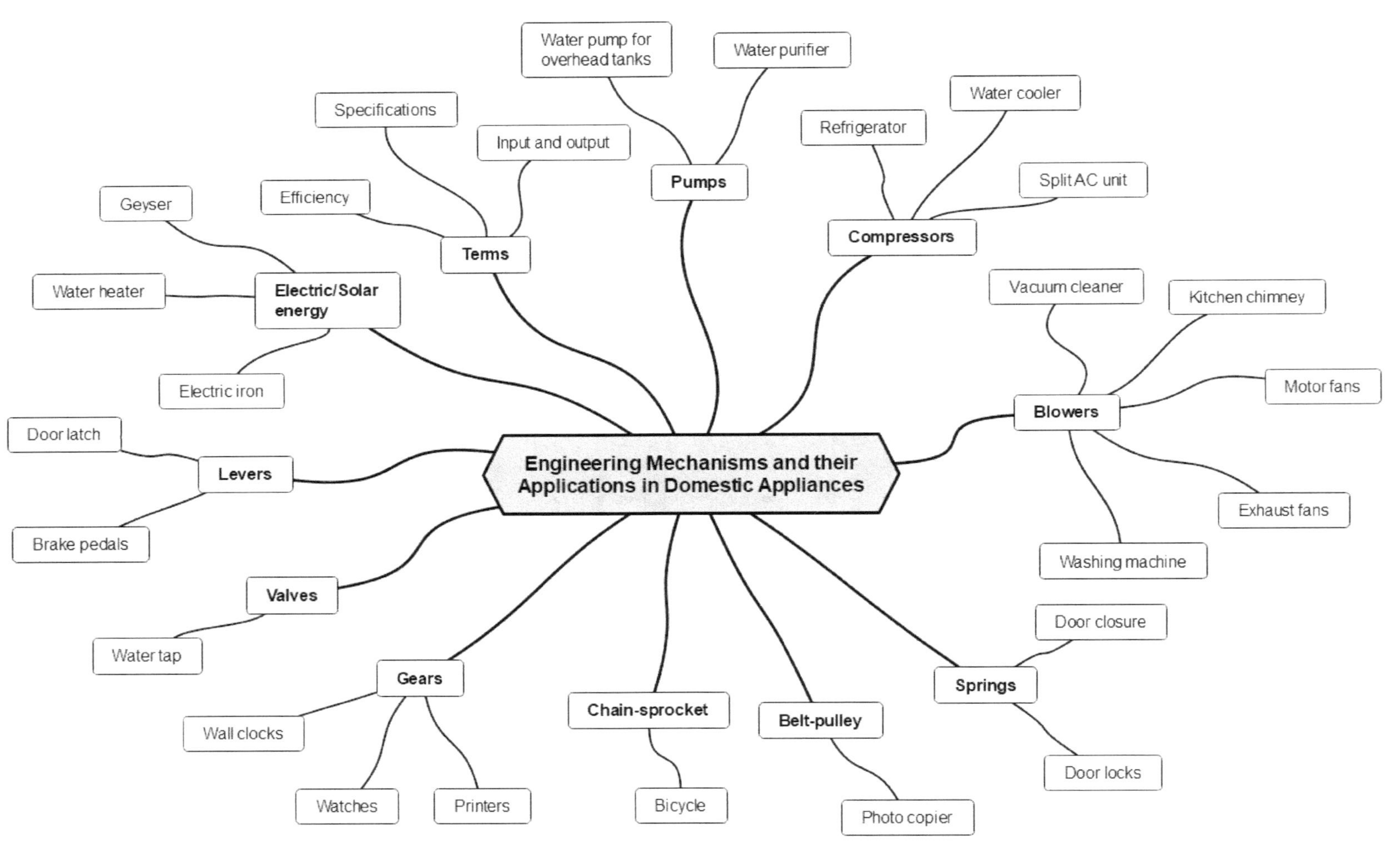

8.1 : Introduction

- In engineering, a mechanism is a device that transforms input forces and movement into a desired set of output forces and movement.

- Machines are devices used to accomplish work. It includes various mechanisms.

- Basically, mechanism is the heart of a machine. It is the mechanical portion of a machine that has the function of transferring motion and forces from a power source to an output.

- Mechanisms are assembly of rigid or flexible members called as links connected together by joints.

- Mechanisms generally consist of moving elements that can include :
  - Gears
  - Belt and pulley
  - Springs
  - Chain and sprocket
  - Levers etc.

- In this chapter we are going to study various mechanisms and their respective applications in domestic appliances.

PUMPS

8.2 : Pumps

- Pump is a mechanical device which when connected in a pipeline converts the mechanical energy supplied to it into the hydraulic energy and transfers it to the liquid.

- Hence, pump increases the energy of flowing fluid. Almost all the pumps increases the pressure energy of the liquid.

- The main difference between turbine and pump is that, in case of turbine flow of fluid takes place from high pressure to low pressure whereas in case of pump flow of fluid takes place from low pressure to high pressure.

8.2.1 Classification of Pumps

Generally pumps are divided into two major groups :

> 1. Positive-displacement pumps
> 2. Rotodynamic pumps (Dynamic pressure pumps)

1. **Positive-displacement pumps :** In these pumps, the fluid is sucked and actually displaced or pushed due to the thrust exerted on it by a moving member (piston) which results in lifting the liquid to the desired height. The commonly used positive-displacement pump is a *reciprocating pump.*

2. **Rotodynamic or dynamic pressure pumps :** These pumps have a rotating element (impeller) through which as the liquid passes its angular momentum changes hence the pressure energy of liquid is increased. These pumps do not push the liquid as in case of positive-displacement pump. The commonly used rotodynamic pump is a *centrifugal pump.*

8.3 : Reciprocating Pumps

- Reciprocating pump is a positive-displacement type pump in which liquid is displaced by a piston-cylinder arrangement which is driven by crank and connecting rod mechanism.

- Reciprocating pump is suitable for small capacities and high heads.

- Reciprocating pumps can be classified as follows :

> 1. Single acting pump
> 2. Double acting pump

8.3.1 Construction of Reciprocating Pump

Fig. 8.3.1 shows the arrangement of single acting reciprocating pump. The various components of reciprocating pump are as follows :

> 1. Suction pipe and delivery pipe
> 2. Piston and cylinder
> 3. Crank and connecting rod mechanism
> 4. Sump (Reservoir)

1. **Suction pipe and delivery pipe :** Suction pipe is connected to the sump of the pump and through

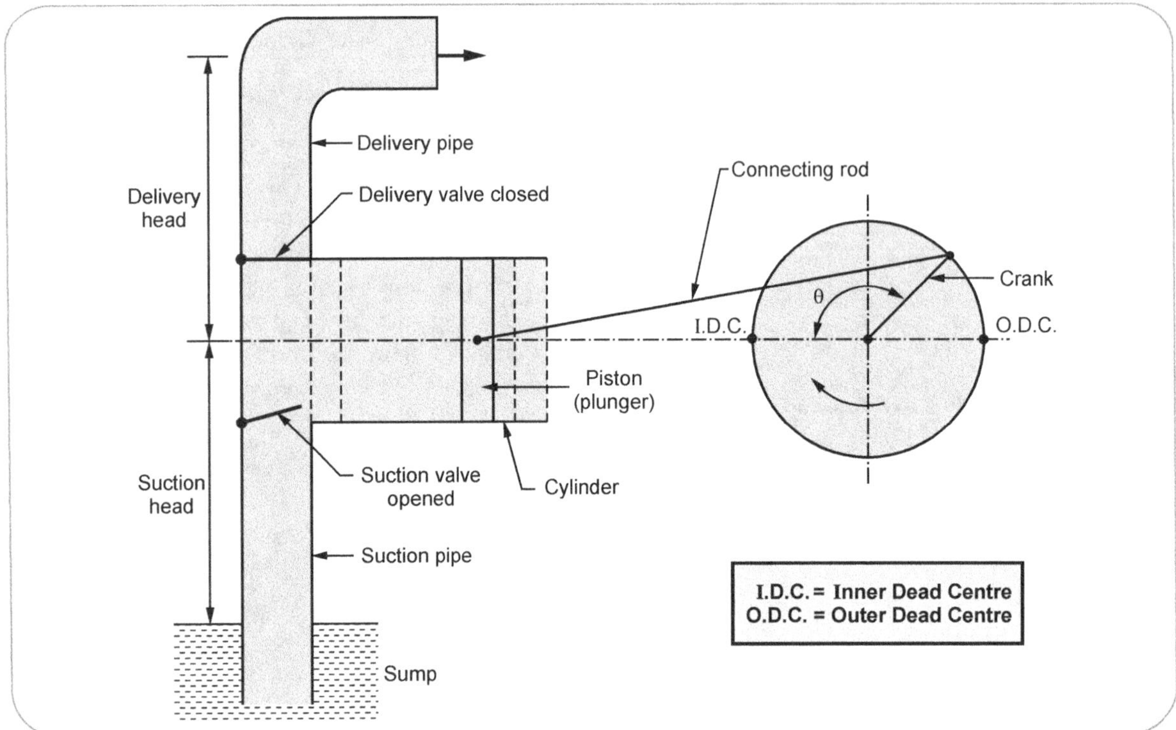

Fig. 8.3.1 Single acting reciprocating pump

this pipe liquid is sucked into the pump. Delivery pipe is connected to the discharge end and it carries liquid at high pressure through some height. It also consists of a non-return valve.

2. **Piston and cylinder :** The piston reciprocates inside the cylinder. The connecting rod transfers rotary motion of the crank into reciprocating motion of the crank into reciprocating motion of piston.

3. **Crank and connecting rod mechanism :** The crank is mounted on a crankshaft and it is driven by either I.C. engine or electric motor. The crank is connected to the piston by connecting rod hence the rotary motion of crank is converted into reciprocating motion of piston.

4. **Sump :** It is the reservoir through which the liquid is pumped into the system.

8.3.2 Single Acting Reciprocating Pump

SPPU : Dec.-07, 11, 12, 16, May-11

• Fig. 8.3.2 shows the arrangement of single acting reciprocating pump which has the same construction as discussed above.

• Initially the crank is at IDC and starts rotating in clockwise direction. When the crank rotates, the piston moves towards right side and on the left side of the piston vacuum is created. This vacuum opens the suction valve and the liquid will be forced from sump to the left side of the pistion. When the crank reaches to ODC, the piston is on extreme right side the suction stroke is completed.

• At the end of suction stroke, the cylinder is full of liquid. When the crank rotates from ODC to IDC in clockwise direction, the liquid will be compressed and high pressure will be built in the cylinder. Because of high pressure delivery valve opens and liquid is delivered through the delivery pipe.

• At the end of delivery stroke, the crank is at IDC and piston is on extreme left position.

8.3.3 Applications of Reciprocating Pumps

SPPU : Dec.-11, 16

Reciprocating pumps are used in following applications :

- It is used as a feed water pump in boilers, hydraulic jacks, kerosene pumps, hand operated pumps, etc.

- It is used in industries and agriculture field.

- It is also used in service stations for pressure washing.

8.4 : Centrifugal Pumps   **SPPU : Dec.-16, May-17**

- The hydraulic machines which convert mechanical energy into hydraulic energy are called as pumps. The hydraulic energy is generally in the form of pressure energy.

- If the mechanical energy is converted into pressure by using centrifugal force acting on the fluid, the hydraulic machine is known as **centrifugal pump**.

- It works on the principle of forced vortex flow which means that when a certain mass of liquid is rotated, the rise in pressure of the rotating liquid takes place.

- The main parts of a centrifugal pump are as follows (Refer Fig. 8.4.1) :

1. Impeller	2. Casing
3. Suction pipe	4. Delivery pipe

1. **Impeller :** It is a rotating part of a centrifugal pump. It consists of backward curved vanes. It is mounted on a shaft which is connected to the shaft of an electric motor.

2. **Casing :** It is an air-tight passage surrounding the impeller. It is designed in such a way that the kinetic energy of the water discharged at the outlet of impeller is converted into pressure energy before the water leaves the casing and enters the delivery section. The casing is of spiral type in which the area of flow increases gradually. This increase in area reduces the velocity of flow and increases the pressure of the liquid flowing through the casing.

3. **Suction pipe :** It is a pipe whose one end is connected to the inlet of the pump and other end

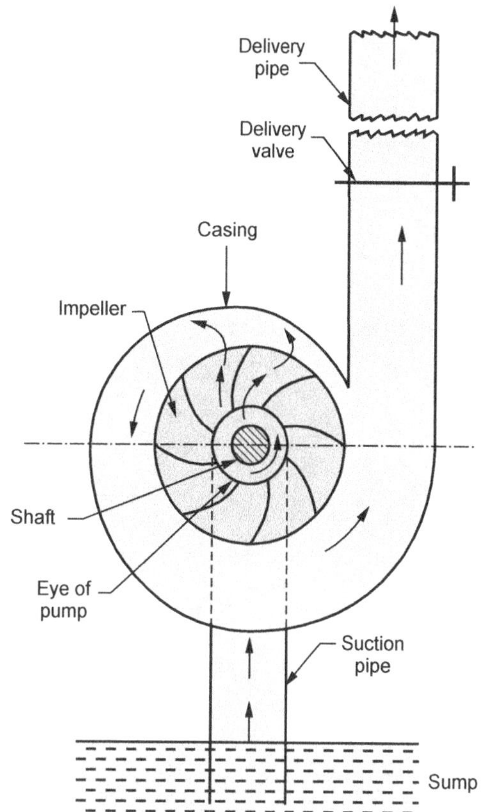

Fig. 8.4.1 Centrifugal pump

dips into the water sump. A non-return valve is fitted at the lower end of the suction pipe. It opens only in upward direction. To remove dust, dirt, etc. a strainer (filter) is also fitted at the lower end of the suction pipe.

4. **Delivery pipe :** It is a pipe whose one end is connected to the outlet of the pump and other end delivers the water at a desired height. To control the flow of liquid a delivery valve is connected to this pipe.

8.4.1 Applications of Centrifugal Pump

SPPU : Dec.- 16

- It is used in domestic water supply.

- It is used in energy and oil refineries, power plants.

- It is used in building services for pressure boosting, fire protection sprinkler systems, air conditioners, etc.

- It is used in boiler feed applications, waste water management, irrigation, etc.

- It is also used in chemical and process industries.

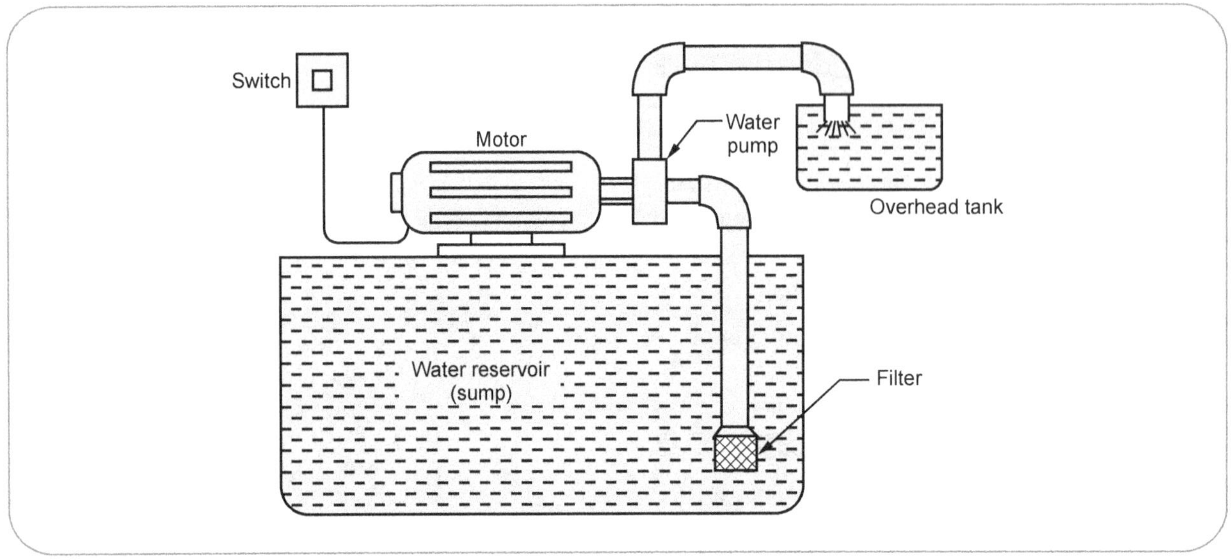

Fig. 8.5.1 Water pump for overhead tanks

8.4.2 Comparison between Centrifugal Pump and Reciprocating Pump

Sr. No.	Centrifugal pump	Reciprocating pump
1.	The flow rate is continuous and smooth.	The flow rate is fluctuating.
2.	It can supply large quantity of liquid.	It can supply small quantity of liquid only.
3.	These pumps run at high speed.	These pumps run at low speed.
4.	The working of these pumps is smooth and without much noise.	The working these pumps is complicated and with much noise.
5.	Cost of centrifugal pump is low.	Cost of reciprocating pump is high.
6.	The cost of maintenance and installation is also low.	The cost of maintenance and installation is also high.
7.	Efficiency of these pumps is high.	Efficiency of these pumps is low.
8.	It requires smaller floor area.	It requires larger floor area.

8.5 : Applications of Pump

8.5.1 Water Pump for Overhead Tanks

- When overhead tank gets empty then water pump will turn ON, conditionally there is minimum level of water in underground water (pump).

- When overhead tank gets full or underground tank gets empty, then pump will turn OFF.

- When the water levels inside the ground tank or the overhead tank reaches a certain level, the auto control system comes into play and the water pump connected starts or stops.

- When municipal water supply starts and the ground level tank fills, the sensor activates the auto system and the pump gets started. It sucks the liquid from the sump and delivers to the overhead tank. Refer Fig. 8.5.1.

- When the overhead tank is full, the sensor will switch off the system and pump stops working.

8.5.2 Water Filter/Purifier Unit

- Water purification is the process of removing undesirable chemicals, biological contaminants, suspended solids and gases from water.

- Water purification may reduce the concentration of particulate matter including suspended particles, bacteria, viruses and fungi as well as reduce the concentration of a range of dissolved and particulate matter.

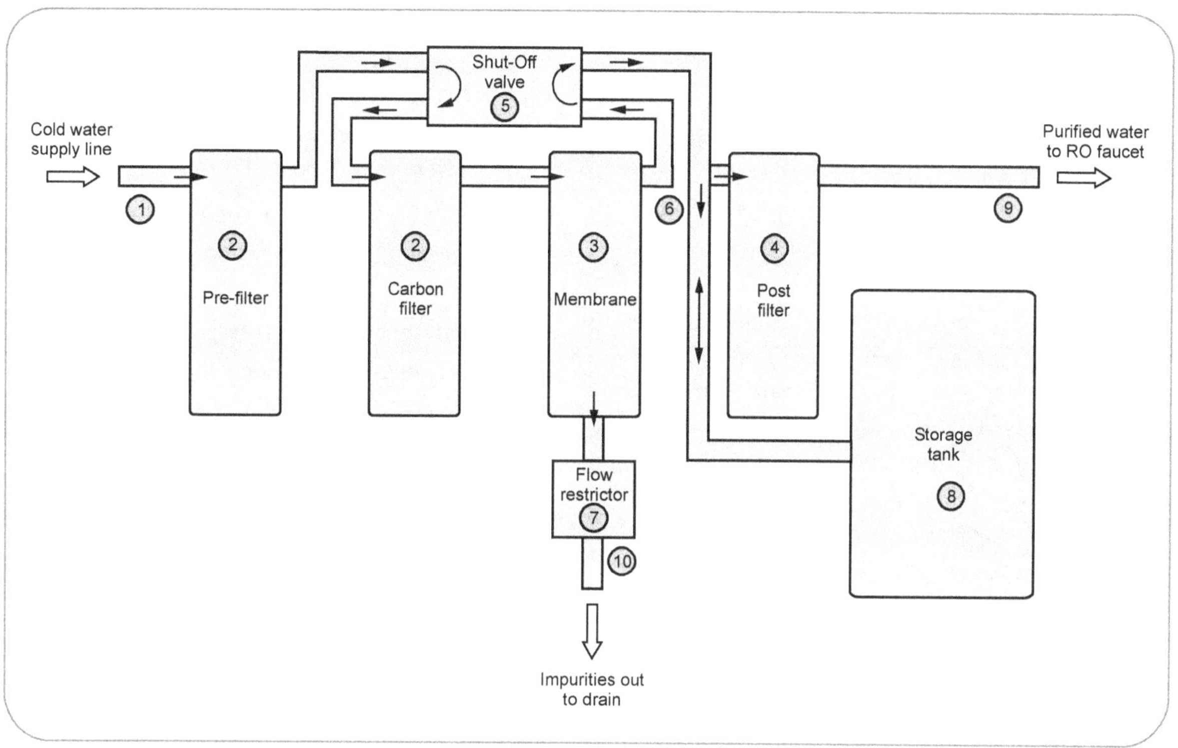

Fig. 8.5.2 Water purifier unit (RO)

- The commonly used water filter or purifier unit in India is RO (Reverse Osmosis).

- Reverse osmosis water purification process is a simple and straight forward water filtration process. It is accomplished by water pressure pushing tap water through a semi-permeable membrane to remove impurities from water. This is a process in which dissolved inorganic solids (such as salts) are removed from a solution (such as water). Refer Fig. 8.5.2.

Basic components of reverse osmosis system :

1. **Cold water line valve :** It is a valve that fits onto the cold water supply line.

   The valve has a tube that attaches to the inlet side of the RO pre-filter. This is the water source for the RO system.

2. **Pre-filters :** Water from the cold water supply line enters the reverse osmosis pre-filter first. There may be more than one pre-filter used in a reverse osmosis system, the most common being sediment and carbon filters.

   These pre-filters are used to protect the RO membranes by removing sand silt, dirt, and other sediment that could clog the system. Additionally, carbon filters may be used to remove chlorine, which can damage the RO membranes.

3. **Reverse osmosis membrane :** The reverse osmosis membrane is the heart of the system. The semi-permeable RO membrane is designed to remove a wide variety of both aesthetic and health-related contaminants.

   After passing through the membrane, the water goes into a pressurized storage tank where treated water is stored.

4. **Post filters :** After the water leaves the RO storage tank, but before going to the RO faucet, the treated water goes through a final post filter.

   The post filter is usually a carbon filter. Any remaining tastes or odors are removed from the product water by post filtration polishing filter.

5. **Automatic Shut Off Valve (SOV) :** To conserve water, the RO system has an automatic shut off valve. When the storage tank is full, the

automatic shut off valve closes to stop any more water from entering the membrane and blocks flow to the drain.

Once water is drawn from the RO faucet, the pressure in the tank drops; the shut off valve then opens to send the drinking water through the membrane while the contaminated waste water is diverted down the drain.

6. **Check valve :** A check valve is located in the outlet end of the RO membrane housing. The check valve prevents the backward flow of treated water from the RO storage tank. A backward flow could rupture the RO membrane.

7. **Flow restrictor :** Water flowing through the RO membrane is regulated by a flow restrictor. There are many different styles of flow controls, but their common purpose is to maintain the flow rate required to obtain the highest quality drinking water (based on the gallon capacity of the membrane).

The flow restrictor also helps maintain pressure on the inlet side of the membrane.

8. **Storage tank :** The standard RO storage tank holds from 2 - 4 gallons of water. A bladder inside the tank keeps water pressurized in the tank when it is full.

9. **Drain line :** This line runs from the outlet end of the reverse osmosis membrane housing to the drain. The drain line is used to dispose of the wastewater containing the impurities and contaminants that have been filtered out by the reverse osmosis membrane.

8.6 : Compressors

- In various industrial applications compressed air is required. A device or machine providing air at high pressure is called as **air compressor.**

- An air compressor takes in air at atmospheric pressure, compresses it and delivers the high pressure air to a storage vessel called as **receiver.**

- From receiver compressed air may be conveyed by the pipe-line to a place where the supply of compressed air is required.

Applications of compressed air　`SPPU : Dec.-08, 10`

The compressed air is used in various industrial applications some of them are as follows :

- It is used to operate pneumatic drills, air motors, hammers, riveting and nut tightening, etc.

- Cleaning of workshops and automobiles.

- Supercharging of I.C. engines and in gas turbine power plants.

- In paint industries for spray painting.

- In refrigeration and air-conditioning industry.

- In construction roads dams, tunnels, bridges, etc.

- For spraying fuel in high speed diesel engines.

- For driving mining machinery.

- For conveying sand materials, concrete, etc.

- For operating air brakes.

- In paper industries and printing machinery.

8.6.1 Classification of Air Compressors
`SPPU : May-04, 07, 09`

Air compressors may be classifed as follows :

1. According to the type of motion

a) **Reciprocating air compressors :** In these compressors air is compressed by the reciprocating action of piston in a cylinder. It provides high pressure air with intermittent discharge.

b) **Rotary air compressors :** These compressors have rotating element to compress the air. It provides the air at low pressure but in large and continuous quantity. These compressors are further classified as follows :

　i) **Positive displacement compressors :** These compressors have two sets of mutually engaging surfaces. Air is trapped between these lobes and squeezing action takes place. For example : Roots blower, vane blower, etc.

　ii) **Non-positive displacement compressors :** In these compressors kinetic energy is converted into pressure energy by the arrangement of impeller and fixed rings. For example : Centrifugal compressor, axial flow compressor, etc.

2. According to the number of stages

a) **Single stage compressor :** In these compressors, air is sucked and compressed in a single cylinder.

b) **Multi-stage compressor :** In these compressors, air is sucked and compressed in two or more cylinders. Generally, number of stages is equal to the number of cylinders.

3. According to the working position (side) of piston

a) **Single acting compressors :** In these compressors, air is sucked and compressed on one side of piston.

b) **Double acting compressors :** When suction and delivery of air takes place on both sides of piston then it is called as double acting compressor. It handles double the air than single acting air compressor.

4. According to the discharge pressure

a) **Low pressure compressors :** Delivery pressure is less than 10 bar.

b) **Medium pressure compressors :** Delivery pressure is between 10 to 80 bar.

c) **High pressure compressors :** Delivery pressure is greater than 80 bar.

5. According to the capacity of compressor

a) **Low capacity compressors :** Discharge capacity is less than 0.15 m^3/s.

b) **Medium capacity compressors :** Discharge capacity is between 0.15 to 5 m^3/s.

c) **High capacity compressors :** Discharge capacity is greater than 5 m^3/s.

8.6.2 Reciprocating Compressors

SPPU : May-05, 10, 11, 12, 14, 17, Dec.-05, 07, 08, 10, 11

- Fig. 8.6.1 shows the principal parts of reciprocating air compressor, which is a single stage and single acting type.

- In these compressors, crank is connected to prime mover or electric motor.

- The inlet and outlet valves are opened and closed by the pressure difference on both sides of valves.

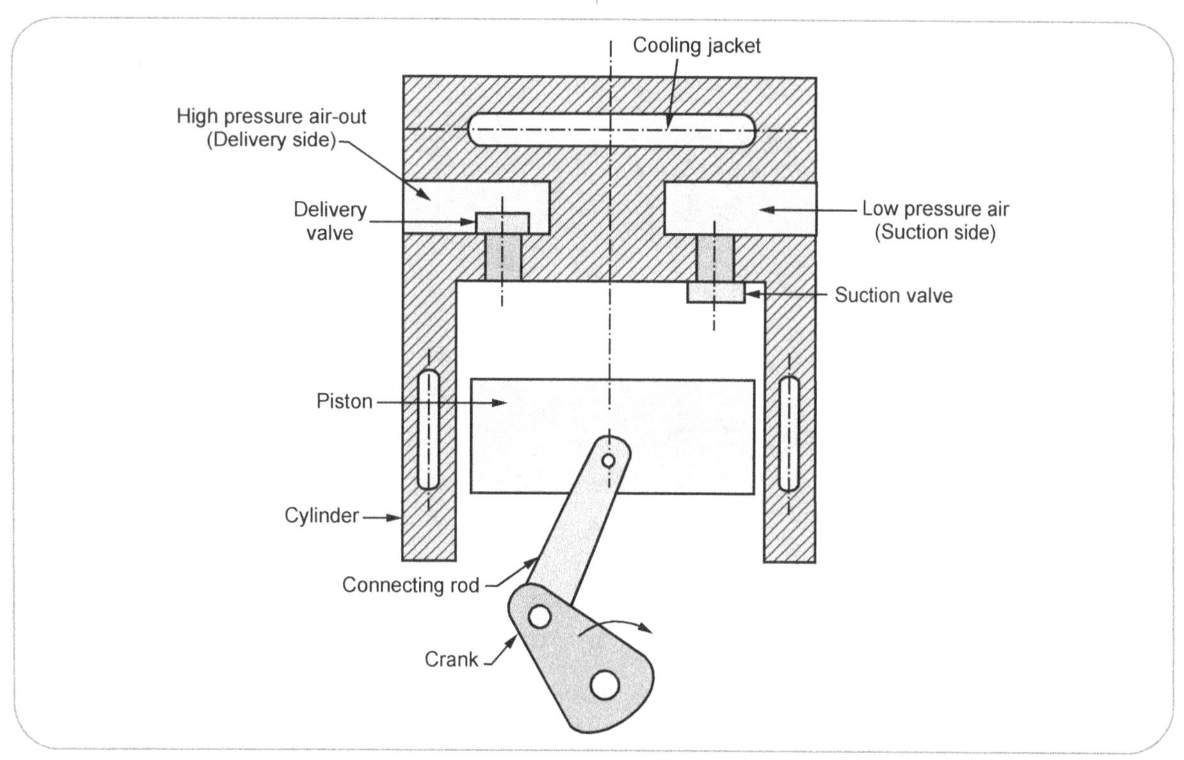

Fig. 8.6.1 Reciprocating air compressor

- The operation of reciprocating compressor is divided in two strokes i.e. suction stroke and delivery stroke.

- During suction stroke, piston moves downwards due to which pressure in the cylinder falls below atmospheric and inlet valve opens and air is sucked.

- During delivery stroke, piston moves upwards with compression of air in the cylinder. At that time, both inlet and delivery valves are closed.

- At the end of this stroke pressure increases above the receiver pressure, hence delivery valve opens and air is discharged to the receiver. (Receiver is a vessel which acts as a storage tank).

- These pumps are widely used in industries, agriculture. They are suitable for high pressure heads with low flow rates.

8.7 : Applications of Compressors

8.7.1 Refrigeration

- **Refrigeration** is defined as the branch of science that deals with the process of reducing and maintaining the temperature of that space or material below the temperature of surroundings.

- The system maintained at lower temperature is called as **refrigerated system** and equipment used to produce this is called as **refrigerator.**

- The cooling effect produced by refrigerator is termed as **refrigerating effect** and the working substance used to produce this effect is known as **refrigerant.**

- The heat withdrawn from the refrigerated space is rejected to atmosphere which acts as a natural heat reservoir.

8.7.1.1 Vapour Compression Refrigeration System
SPPU : May-06, 14, Dec.-10

- Vapour compression refrigeration system is most commonly used method of refrigeration for refrigerators, air-conditioners, etc.

- In this system, the liquid refrigerant boils in evaporator which at low pressure, by absorbing latent heat.

- The vapours formed are condensed in condenser which is at high pressure by rejecting latent heat.

- Thus, in this cycle heat is transferred in the form of sensible heat as well as latent heat which gives higher COP.

- Fig. 8.7.1 shows the simple vapour compression refrigeration system.

- In this system commonly used refrigerants are NH_3, R-11, R-12, R-22. In modern refrigerators Freon-22 and 134-a refrigerants are used.

- Simple vapour compression refrigeration system consists of four different processes.

1. Compression	2. Condensation
3. Expansion	4. Evaporation (Vaporization)

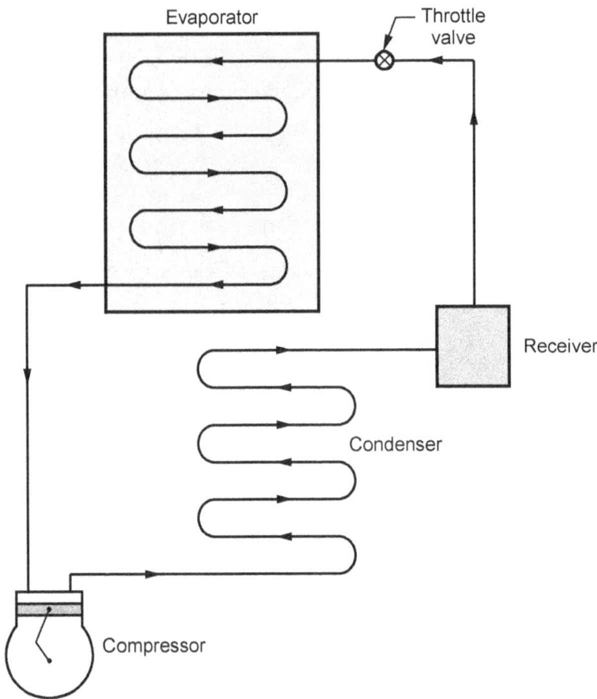

Fig. 8.7.1 Vapour compression refrigeration system

1. Compression : During the suction stroke of compressor, low-pressure vapour in dry state is drawn from the evaporator. Here, the temperature and pressure of vapour increases until the vapour temperature is greater than the condenser temperature.

2. Condensation : During condensation, high pressure refrigerant vapour enters the condenser where the cooling medium absorbs the heat and converts the vapour into liquid.

3. Expansion : After condensation, the liquid refrigerant is stored in the receiver and from receiver it is passed to evaporator through expansion or throttle valve. This valve reduces the pressure by keeping the enthalpy constant (Throtting process).

4. Evaporation (Vapourization) : After expansion, low pressure liquid refrigerant enters in evaporator where considerable amount of heat is absorbed by it and converted into vapour.

This low pressure vapour is sucked by the compressor and the cycle repeats.

8.7.1.2 Household Refrigerator

SPPU : Dec.-03, 04, 13, May-04, 08, 10, 12

- A household refrigerator is the application of vapour compression refrigeration system

- Fig. 8.7.2 shows the arrangement of household refrigerator.

- A houshold refrigerator consists of following main elements :

> 1. Compressor   2. Condenser
> 3. Receiver   4. Capillary tube
> 5. Evaporator

1. Compressor : In household refrigerators generally hermetically sealed type of reciprocating compressor is used. It is mounted on the rear bottom side of the refrigerator. The compressor is driven by electric motor.

2. Condenser : In household refrigerators condenser is placed on the rear side. The heat transfer takes place by circulation of air naturally. It is made of tubing of copper.

3. Receiver : In the receiver the refrigerant is stored under high pressure.

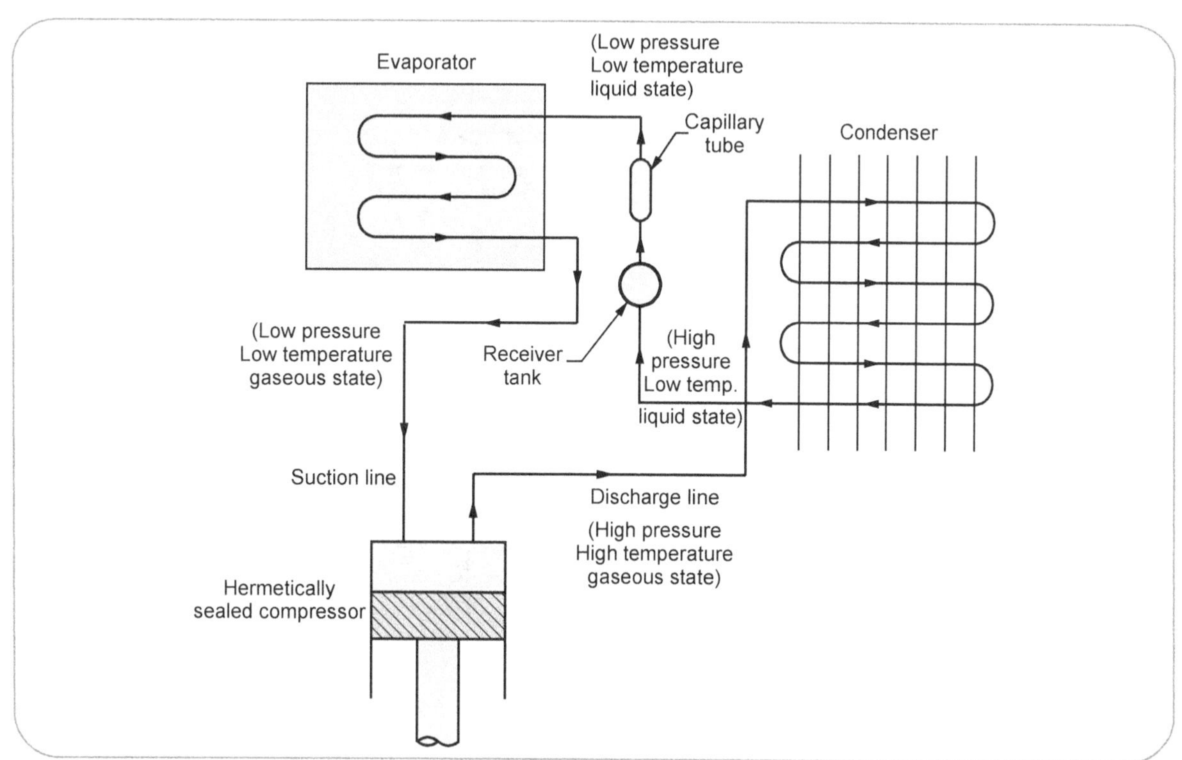

Fig. 8.7.2 Household refrigerator

4. Capillary tube : In household refrigerators, capillary tube will act as a throttle valve where expansion of refrigerant takes place.

5. Evaporator : The household refrigerators are commonly equipped with two evaporators. Out of them one is used for frozen food and other for regular refrigerating temperatures. But now-a-days new models of refrigerators provides much more storage capacity using the same outside dimensions than the older type refrigerators. Modern refrigerators also provide more effective and thinner insulation. Evaporator coils are generally made of aluminium or stainless steel.

- The working of household refrigerator is similar to vapour compression refrigeration system as discussed in previous section.

8.7.2 Split AC Unit

- Air conditioning is defined as the simultaneous control of temperature, humidity (moisture content in air), motion (circulation and air movement) and purity (air filtering and cleaning) of air within an enclosed space.

- Air conditioning is widely used in industries, offices, hospitals, commercial places, etc. for human comfort.

- The name of split air conditioners indicates that the conditioning system is splitted in two parts i.e. fan coil unit (fitted inside the room) and condensing unit (fitted outside the room).

- Basically air conditioner works by transforming a refrigerant compound from a gas to a liquid and back again in continuous cycle. Refer Fig. 8.7.3.

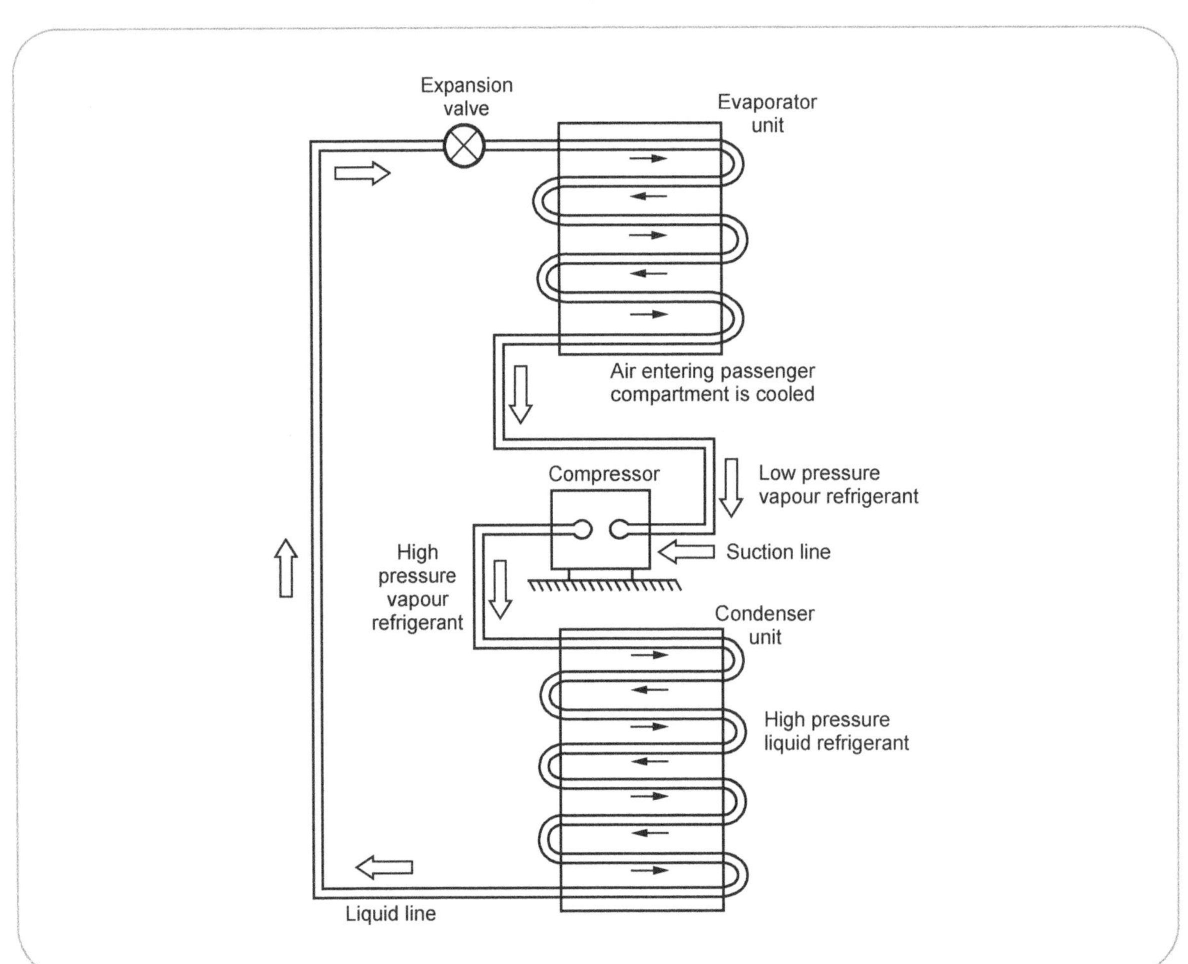

Fig. 8.7.3 AC refrigeration system

- Any AC system consists of four important mechanical parts : A compressor, condenser expansion valve and an evaporator.

Functions of compressor and other elements

- The compressor is the heart of the cooling cycle.

- The cycle begins when the compressor draws in cool and low pressure refrigerant gas from the indoors.

- The compressor which is motor driven, squeeze the refrigerant, raise the temperature of gas and pressure as well so that it exists the compressor as a hot and high pressure gas.

- The compressor pushes the hot gas to the finned condenser coil in the outdoor side of AC where fans blow cool outside air over the coil and through the fins, extracting the heat from the refrigerant and transferring it to the outside air.

- When sufficient heat is extracted from the refrigerant, it condenses into a warm liquid that passes under high pressure to an expansion valve which turns the refrigerant into the cool and low pressure liquid.

- The refrigerant goes from expansion valve to evaporator coil located in the indoor or room side of the AC unit.

- When the refrigerant enter the evaporator coil where the pressure is low, it is chemically compelled to evaporate into a gas.

- This process requires heat which comes from the room's warm air being blown over the evaporator coil by another fan.

- Now, the refrigerant back to a cool and low pressure gas and it is drawn back into the compressor to continue the cycle.

8.7.3 Water Cooler

- Water cooler also called as **water dispenser** is a machine that cools and dispenses water with a refrigeration unit.

- It is generally located near the restroom due to closer access to plumbing.

- A drain line is also provided from the water cooler into the sewer system.

- The most popular water cooler is the **bottle fed water dispenser**. Refer Fig. 8.7.4.

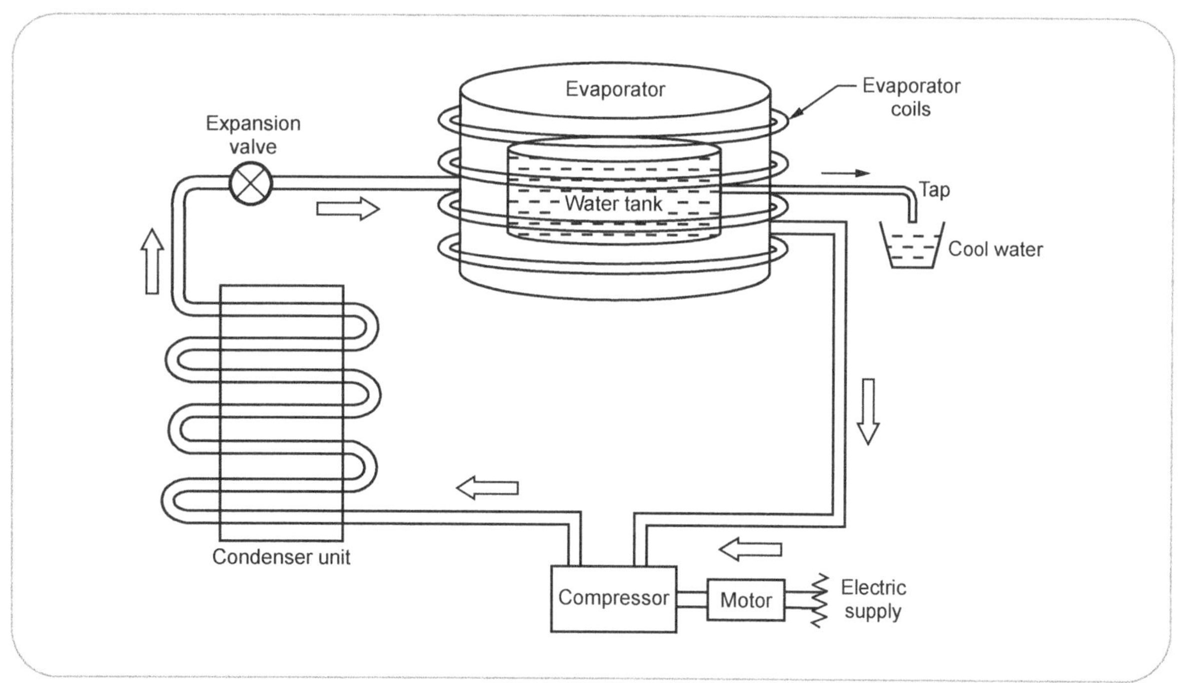

Fig. 8.7.4 Water cooler

- This type of machine gets its water supply from an inverted bottle of water placed on top of water cooler.

- Inside the water cooler is a valve that prevents the water from the bottle flooding the water cooler.

- The water inside the water cooler is fed into reservoir where it is cooled using refrigerant.

- The refrigerant (cooling medium circulated in pipes) changes from a liquid to gas as it moves in the pipe towards the reservoir because of the pressure in the pipes created by compressor inside the water cooler.

- The function of elements like compressor, evaporator, condenser and expansion valve is same as in case of refrigerator or AC unit.

- Water coolers are commonly used in schools, collages, hospitals, malls, theatres, industries, etc.

8.8 : Blowers and Fans

Fan

- By definition, a fan is machine that is used to create flow within a fluid, such as air.

Fig. 8.8.1 Fan

- It consists of vanes or blades that rotate and act on air. This rotating assembly of blades and hub is known as an impeller, a rotor, or a runner. The impellers help in directing the air flow, and producing air at low pressure.

- Most fans are powered by electric motors, but other sources such as hydraulic motors and internal combustion engines can also be used.

Blower

- A blower is defined as a machine which is used to produce large volumes of gas with a moderate increase in pressure.

- Similar to fans, blowers are also used to create air, but they only provide air at a specified position.

Fig. 8.8.2 Blower

- It consists of a wheel with small blades on its circumference, and a casing to direct the flow of air out toward the edge. The casing in the center of the wheel uses centrifugal force to propel the air forward into the open.

- Both, fans and blowers, are mechanical devices used for circulation of air. Based on this, they are differentiated from each other, wherein a fan circulates air around an entire room, or space, and a blower only focuses on the specific or given area.

- Generally, a fan is an electrical device that moves air, whereas a blower is a mechanical device that consists of a fan, and which channels the air from the fan and directs it to a specific location or point. Also, a fan circulates the air around an entire room or a large area, while a blower is only positioned to a specific direction or point.

8.9 : Applications of Blower

8.9.1 Vacuum Cleaner

Principle : It works on the principle of pressure diference between the two locations.

Construction : It consists of mainly three parts :
i) Motor   ii) Cetrifugal fan   iii) Filter

Working :

- An ideal vacuum cleaner has a centrifugal fan which is connected to a motor as shown in the Fig. 8.9.1.

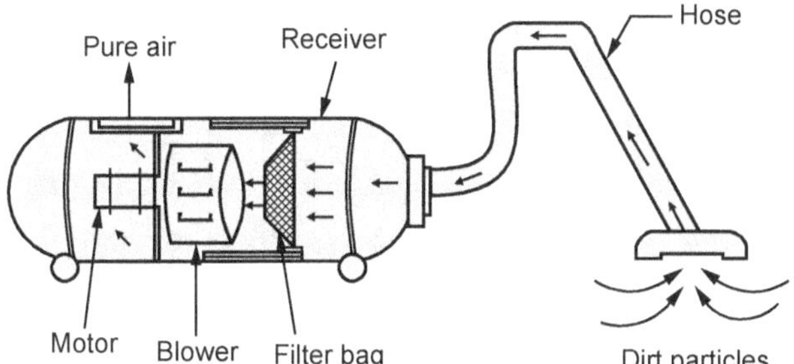

Fig. 8.9.1 Vacuum cleaner

- It has suction and discharge connection, on suction side, fliter bag is fitted to collect the dirt and the discharge have air purifier filter opened to the atmosphere.

- When electric power is given to the motor which rotates the blower and air is sucked from the suction side which carries mist, dirt and small solid particles to the suction filter. All these particles get trapped in the filter and the dirt free air is pushed out from the discharge.

Applications :

1. **Used in industrial area :** Automotive, Agriculture, Chemical, Textiles etc.

2. **House floor space :** Carpet, Rugs, Bare floors

3. **Small areas :** Stairs, corners, car interiors

4. **Multi purpose :** Homes, offices, hotels, hospitals

8.9.2 Kitchen Chimney

Principle : It is used to remove organic and inorganic substances like smoke, soot, water vapour, oil fumes and bits of food.

Construction : It consist of motor, filters, hose pipes, hood etc.

Working : • The mechanism used in the kitchen chimney is very simple.

- Oil droplets and smoke is been emitted during the cooking process which are drawn out from the chimney.

- These particles get trapped in the filters while the smoke and moisture are channeled out through the ducts and released into the environment as shown in the Fig. 8.9.2.

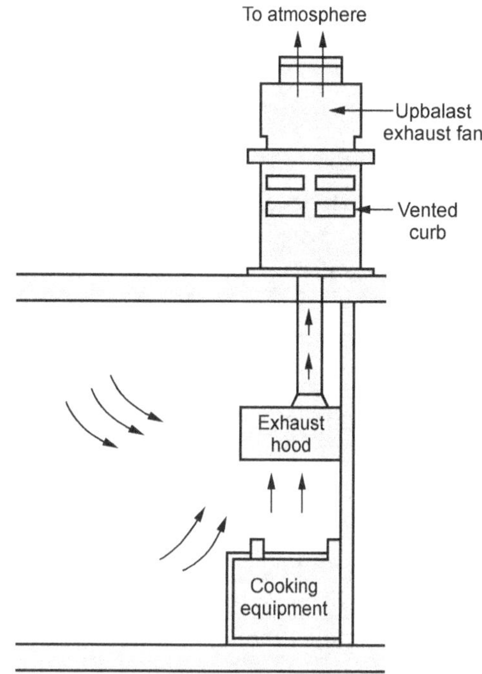

Fig. 8.9.2 Kitchen chimney

- The chimney can be ducted or ductless.

- In ducted chimney it requires proper plumbing for the ducts to be laid out. It is generally placed directly above the stone to ensure maximum ventilation.

- Ductless chimney is similar to ducted chimney except it have carbon filter which eliminates bad odor, smoke and humid air. The recycled air is then released back into the house which is not emmited out has done in ducted chimney.

- There are 3 types of filter
   i) Mesh or cassetle filter ii) Baffle filter
   iii) Charcoal filter

8.9.3 Motor Fans

Principle : If a current carrying conductor is placed in a magnetic field it experiences a force and start to rotate.

Construction :

- It consists of electric motor, capacitor, blades or paddles, metal arms, flywheel, rotor, motor housing etc.

- Capacitor is needed to get up enough torque to run the motor. The blades are usually made from plastic, plywood or aluminium. The metal arm holds the blade and connect it to the motors.

- The flywheel is attached to the shaft of the motor.

Working :

- The capacitor of the fan torques up the electric motor which causes it to start. Refer Fig. 8.9.3.

- As soon as the electrical current reaches the motor, it enters coils of wire which are wrapped around the metal base, a magnetic field is produced that expands the force in a clockwise motion which converts the electrical energy into mechanical energy and causes the motor coil to spin.

- Thus, the blades attached to the motor also start gaining motion with the spinning of the coil.

- The rotation mechanism of the fan is such that it attract the warm air upwards and as it rises up the blades of the fan slice this air and pushes it down.

- This being a continuous process causes the air to circulate in the entire room.

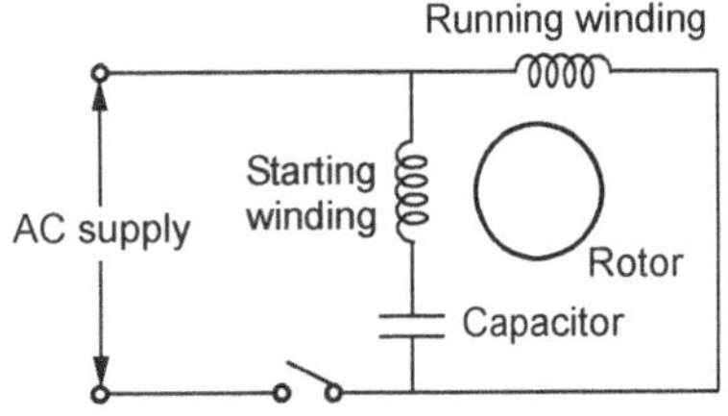

Fig. 8.9.3 Fan wiring diagram

Applications :

1. Water treatment aeration 2. Air ventilation

3. Air drying 4. Food processing

5. Construction industries

8.9.4 Exhaust Fans

Principle : It is a ventilation device which draws out the polluted air from the room and replaces it with fresh air.

Construction : It consists of blades, electric motor, capacitor, rotor etc.

Working :

- An exhaust fan has a rotating arrangement of blades which are driven by a motor. Refer Fig. 8.9.4.

- The blades take the humid air and pushes out of the room and entering the fresh air from elsewhere in the room.

- The effectiveness of the fan depends on the size of the room and occupants in the room. The number of blades of the fan vary from 3 to 5.

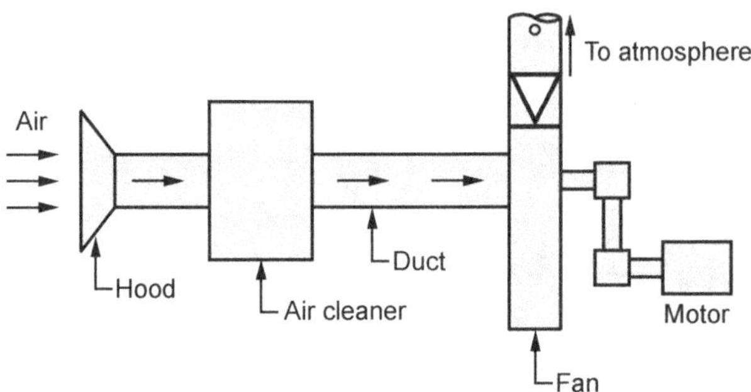

Fig. 8.9.4 Working of exhaust fan

Applications :

1. Kitchens 2. Bathroom

3. Industries 4. Hospitals

8.9.5 Washing Machine

Principle : It works on the principle of configuration which is the fictiotius force that pulls out from the center on the body while moving in the circular path.

Construction : It consists of

i) Water pump : It circulates the water through machine

ii) Water inlet control valve

iii) Drum iv) Motor

v) Agitator vi) Drain pipe

vii) Printed circuit board

viii) Timer

Working : • Now a days the washing machine uses the universal motor because it can be operated on AC or DC Power.

- These types of motor have permanent magnets field poles that are controlled by the microprocessor which makes it easy to control speed and direction.

- The poles winding around the iron core passes the current through each winding which attracts and repels the permanent magnet on the rotor that is

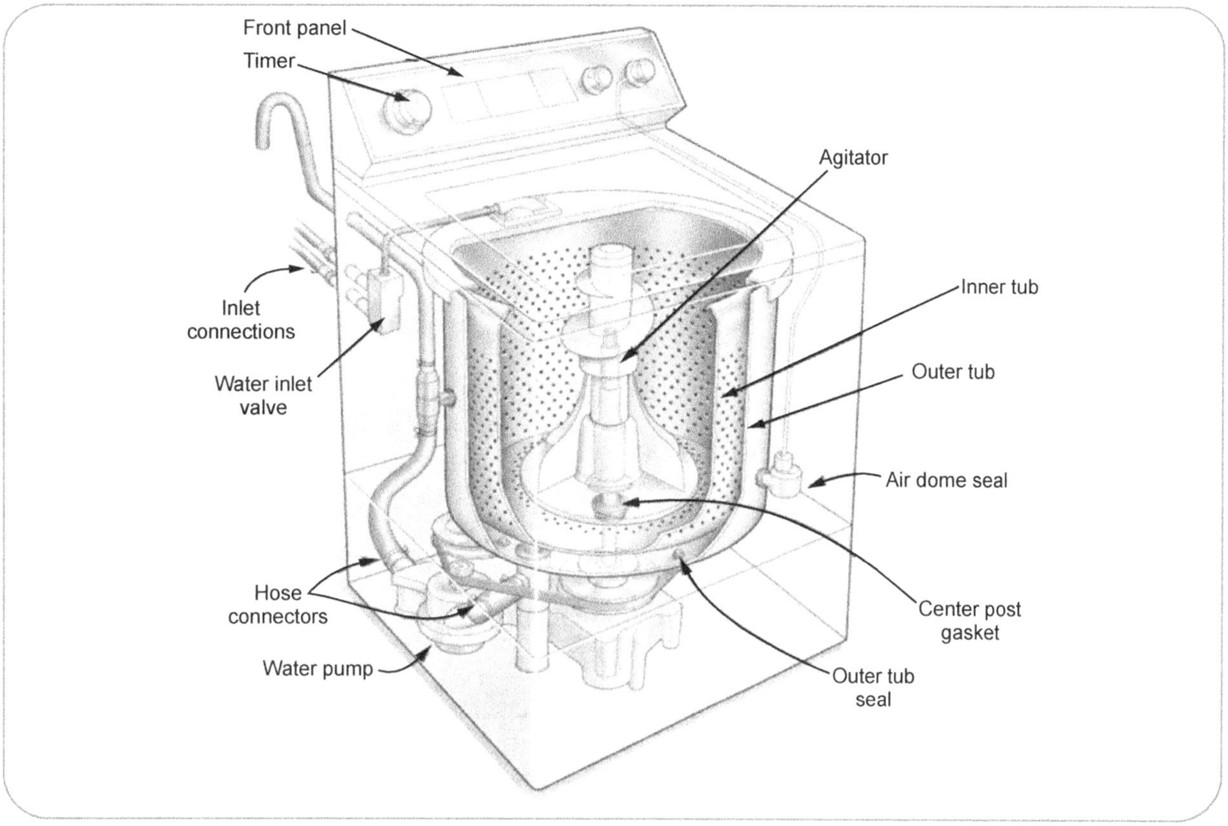

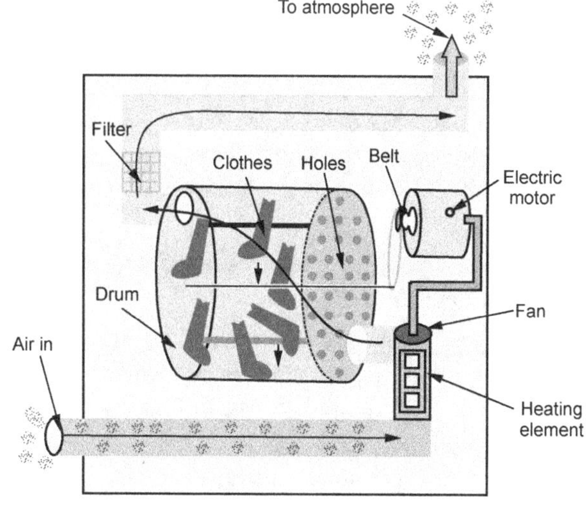

Fig. 8.9.5 Washing machine

been attached to the drum and causes it to rotate at a desired speed. Refer Fig. 8.9.5.

8.9.5.1 Working of Dryer

- The best drying conditions involve heat, air movement, and low humidity (or the constant removal of moisture).

- Clothes drying machines (also called **tumble dryers**) combine these things to dry clothes quickly and efficiently inside a large, rotating metal drum.

- Unfortunately, dryers like this cost a lot of money to run and they can shrink, damage, or slowly degrade the clothes (check washing labels on your clothes very carefully before putting them in a dryer like this).

- The basic idea is to blow hot dry air into one side of the drum as it tumbles the clothes around and extract moist wet steam from another part of the drum at the same time. Step-by-step, here's how it works. (In this picture, we're looking from the side and the front of the machine is on the left.)

1. The heart of the machine is a large metal **drum** with paddles around its inside rim. In large machines, such as those in launderettes, the drum always rotates in the same direction.

Fig. 8.9.6 Working of dryer

2. Cold air is drawn into the machine through an **air intake.** Often it's at the front of the machine to stop it getting dirty and dusty (as it would around the back).

3. A **fan** sucks the air in and pulls it toward a heating element.

4. The fan is powered by an **electric motor.**

5. As cool air passes over the **heating element,** it's warmed and turned to hot dry air.

6. Warm air from the heating element enters the drum, typically through large **holes** at the back. In launderette machines, the entire drum is full of small holes and hot air rises up from below.

7. The drum is rotated slowly by a **belt** connected to the electric motor, made from something like rubber.

8. As the drum rotates, the paddles lift and tumble wet **clothes** until they reach the top of the drum. Then gravity makes them fall back down through the hot, dry air. Dryers work most efficiently when the washing tumbles through the hot air this way

9. The air that leaves the dryer passes through a **lint filter** that catches dust and bits of fluff. Some dryers have a second (exhaust) fan to help extract the moist air.

10. Exhausted air passes up through a **vent hose** either mounted permanently in the ceiling (as here) or temporarily poked through an open window.

8.10 : Springs

- Spring is defined as elastic body or elastic machine element which deflects under the action of the load and recovers its original shape when load is removed.

- Depending upon the application, spring can take any shape and form.

- Springs are widely used in engineering field. Some of the important applications are as follows :

  ○ Springs are used to absorb shocks and vibrations (vehicle suspension springs, buffer springs in elevators, railway and machinery springs).

  ○ They are used to store energy (springs in toys, pens, clocks, cameras).

  ○ They are used to maintain contact between two machine elements (jigs and fixtures, cam and follower mechanism in I.C. engines).

  ○ They are used to measure the magnitude of force (springs in weighing balances and engine indicators).

  ○ They are used to provide operating force (spring in clutch and brakes, control valves).

8.9.1 Types of Springs

There are many types of springs which are commonly used in engineering applications. Some of the important types are as follows :

> 1. Helical springs
>
> 2. Conical and volute springs
>
> 3. Torsion springs
>
> 4. Leaf or laminated springs

1. Helical springs :

- Helical spring is made from a wire, usually of circular cross-section, bent in the form of a helix.

- According to the type of load and deflection requirement, the cross-section of wire may be square or rectangular.

- Helical springs are classified as :

According to the type of axial force :

     i) **Helical compression springs :** In this type of helical springs, the external force tends to

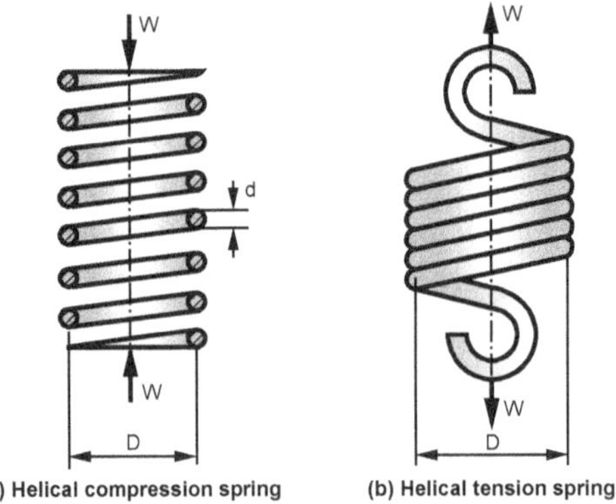

(a) Helical compression spring   (b) Helical tension spring

Fig. 8.10.1

shorten the spring or the spring is compressed. Refer Fig. 8.10.1 (a).

ii) Helical tension springs : In this type of helical springs, the external force tends to lengthen the spring or the spring is elongated. Refer Fig. 8.10.1 (b).

2. Conical and Volute springs :

- Conical and volute springs are used in special applications where a telescopic spring or a spring with stiffness that increases with load is required.

- The conical spring is wound with a uniform pitch as shown in Fig. 8.10.2 (a).

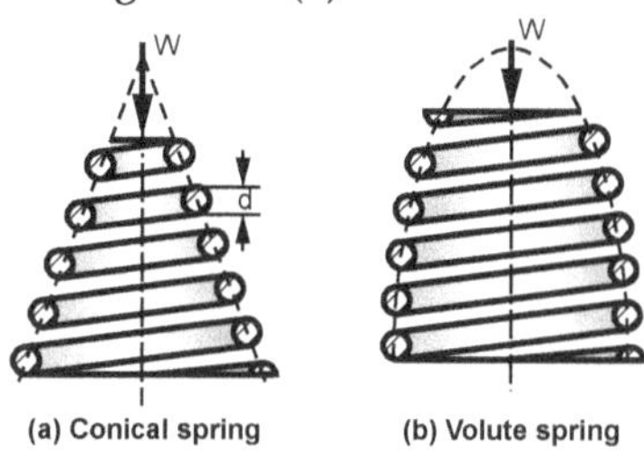

Fig. 8.10.2 : Conical and volute springs

- Volute springs are wound in the form of paraboloid with constant pitch and lead angles. Refer Fig. 8.10.2 (b).

3. Torsion springs :

- Torsion springs may be of **helical type** or **spiral type.**

- Helical type torsion springs are used in applications where the load tends to wind up the spring, for example in electrical machines. Refer Fig. 8.10.3 (a).

- Spiral type torsion springs are used where the load tends to increase the number of coils, for example in watches and clocks. Refer Fig. 8.10.3 (b).

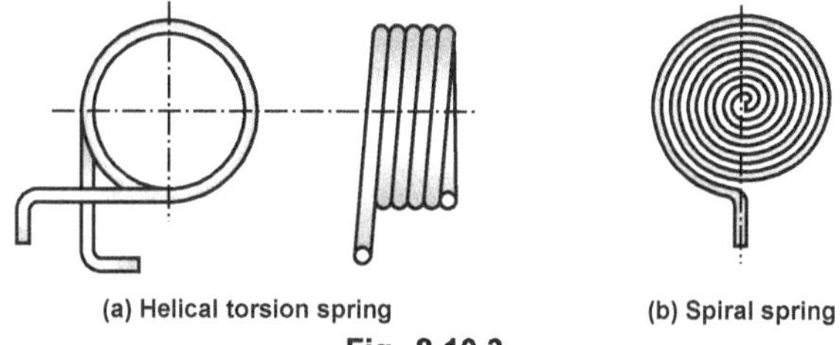

Fig. 8.10.3

4. Leaf or Laminated springs :

- Leaf spring consists of number of flat plates (leaves) of semi-elliptical shape and different lengths held together by using clamps and bolts. Refer Fig. 8.10.4.

- These type of springs are widely used in automobiles and railroad suspensions.

- In these springs, tensile and compressive stresses are produced.

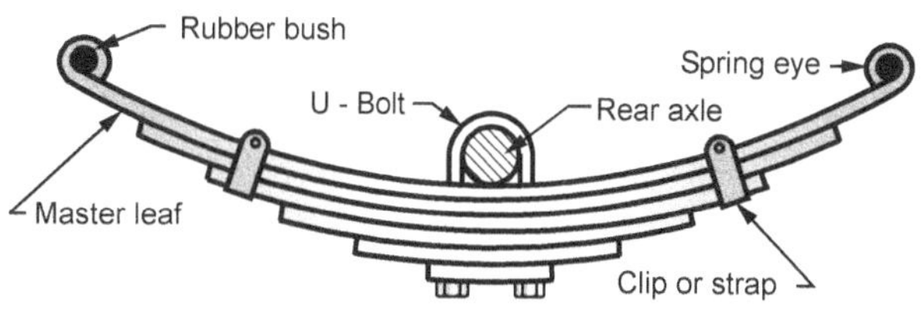

Fig. 8.10.4

8.11 : Applications of Spring

8.11.1 Door Closure

Principle : The basic principle of a door closure is to close the door automatically.

Construction : The basic components of the door closure are transmission gear, supporting guide parts, reset spring, one-way valve, rack plunger, throttle valve, end cover, sealing ring and connecting rod.

Working : • The door closure is simple and easy to work with as one end of the hydraulic arm is attached to the door and the other end is attached to the door frame.

- When the door is opened, the door closure pulls the door and closes it because of the spring which is there in the sealed tube.

- It includes a fluid-filled chamber which releases the pressure to close the door in a slow manner rather than banging it.

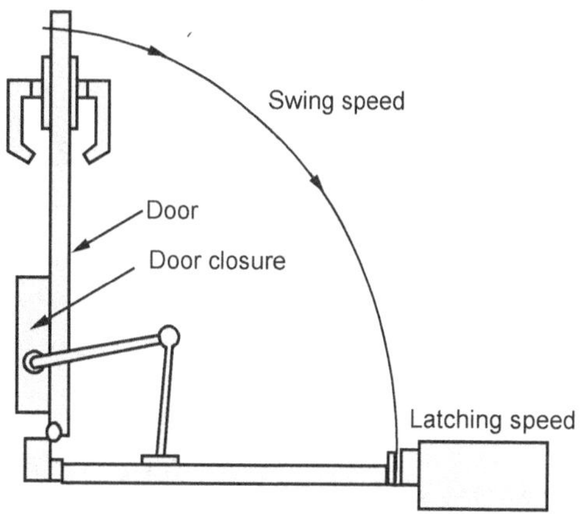

Fig. 8.11.1 Door closure

Applications :

1. Hospitals   2. Retail establishment   3. Schools and collages and many more.

8.11.2 Door Locks

Principle : The principle of the spring in the door locks is to create a tension.

Construction : The basic components of the door locks are hollow cylinder, spring, key, upper pin, lower pin, etc.

Working :

- Modern lock and keys are made of steel.

- It consists of a row of pins, ususally five in number.

- Each pin has its own cylinder arrangement as shown in the Fig. 8.11.2.

- When the lock is locked it holds the two pieces of metal together with the help of spring force.

- The pins are of varying length i.e. to open the lock we need to align all the pins at equal level.

- Thus there are notches provided on the key which are made to push the pin upward to the correct amount in there respective cylinder to separate the two blocks from one another.

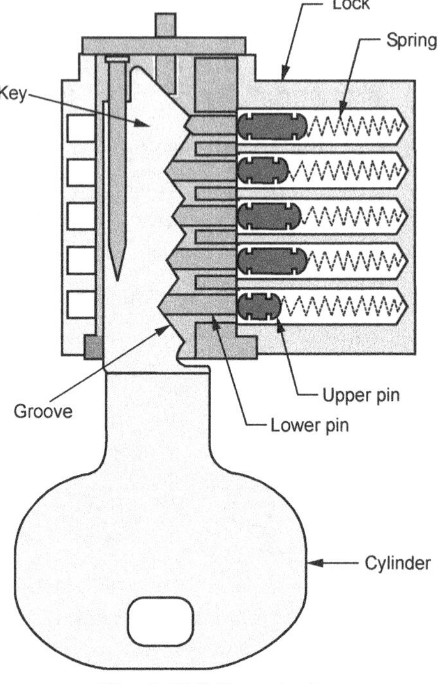

Fig. 8.11.2 Door locks

8.12 : Gears

- In case of belt drive there is a slipping phenomenon which reduces the speed ratio of a system.

- In precision machines or mechanisms like watch gear mechanism, definite speed ratio is required which is obtained by **gear drive** or **gear wheels.**

- Gear drive is a positive drive and it is provided when the distance between the driver and follower is very small.

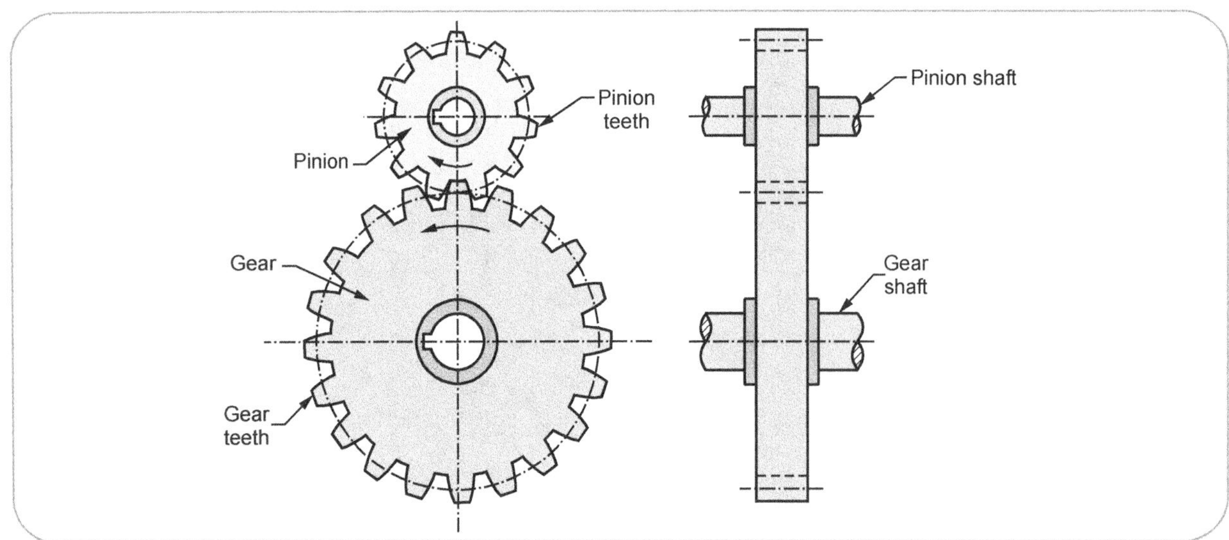

Fig. 8.12.1 Gear drive

- **Gears** are defined as toothed wheels which can transmit power and motion from one shaft to another shaft by means of successive engagement of teeth.
- It is important to note that, both the gears which are engaged, always rotate in opposite direction.
- The gear drive consists of two wheels. The smaller wheel is called as **pinion** and larger wheel is called as **gear**. Refer Fig. 8.12.1.
- Both pinion and gear are mounted on separate shafts.

> **Note :** Types of Gear are already discussed in Chapter 6.

8.12.1 Gear Train

- A gear train is a combination of two or more gears which is used to transmit motion from one shaft to another shaft.
- Gear train is necessary when it is required to obtain large speed reduction within a small space.
- The nature of the gear train used depends on the required velocity ratio and the relative position of the axes of shafts.
- Gear trains are commonly used in various machines (lathe, milling, drilling, etc.), automobiles, clocks, ships, watches, etc.

Simple Gear Train

- If only one gear is mounted on each shaft and there is no relative motion between the axis of shafts, then it is called as simple gear train.

- Fig. 8.12.2 shows arrangement of simple gear train by using spur gears. The gears are represented by their pitch circles.
- In Fig. 8.12.2 (a) gear 1 drives gear 2 hence gear 1 is called as **driver** and gear 2 is called as **driven** or **follower**. It is important to note that, the direction of rotation of driving and driven gear is opposite to each other.
- If the distance between the two gears is large, then the **intermediate gears** are used. Refer Fig. 8.12.2 (b) and (c).
- If the number of intermediate gears are odd then the motion of driver and driven gear is same. Refer Fig. 8.12.2 (b).
- Similarly, if the number of intermediate gears are even then the motion of driver and driven gear is opposite. Refer Fig. 8.12.2 (c).

8.13 : Applications of Gear

8.13.1 Wall Clocks

Principle : It is a device that performs regular movements in equal intervals of time.

Construction : Basic components of the wall clock are electric stepping motor, battery, microchip, quartz crystal oscillator, gears used to turn hour, minute and second hands at different speeds, shaft to hold the hands in place etc.

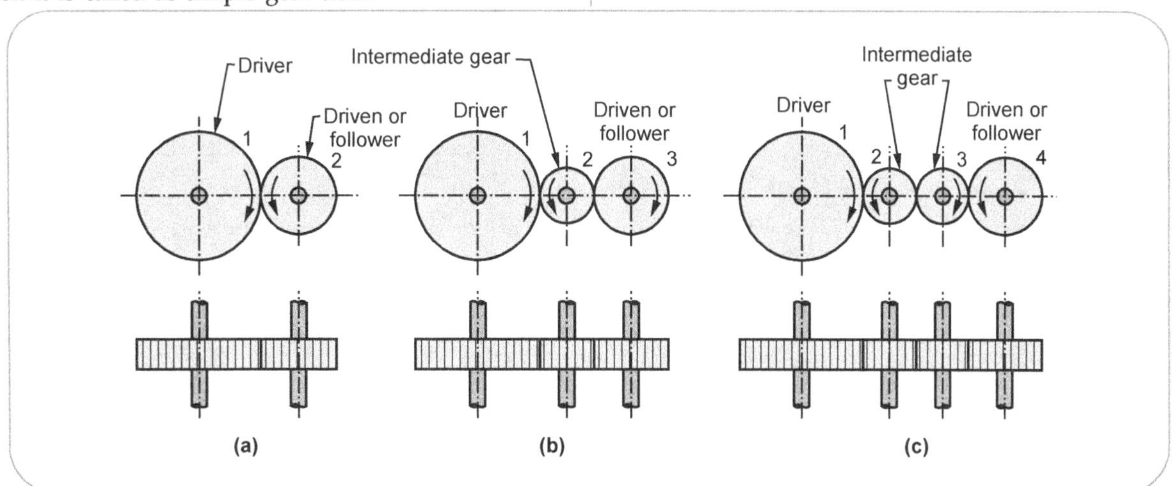

Fig. 8.12.2 Simple gear train by using spur gears

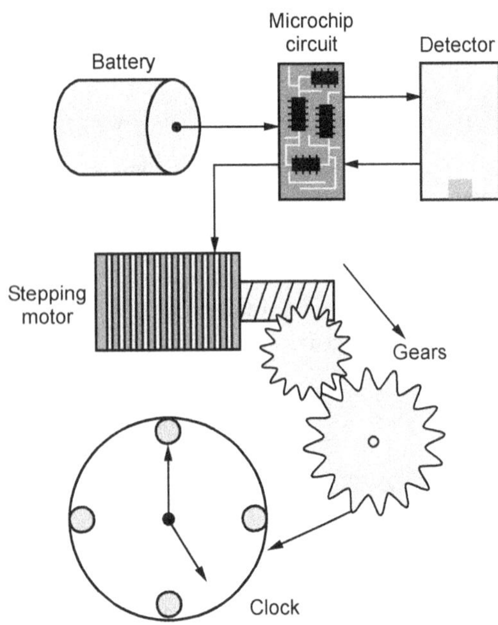

Fig. 8.13.1 Block digram of clock

Working :

- In wall clocks we have a particular arrangement of the gears which are used to move hour, minute and second hand in various ratio by using the battery source.

- As shown in the Fig. 8.13.1 the battery sends an eletrical signal to the quartz crystal when oscillates at a precise frequency.

- The circuit counts the numbers of vibrations and used them to generate eletric pulses.

- These pulses are given to the stepping motor which drives the gear wheel that spins the clock's second, minute and hour hand.

8.13.2 Printers

- Laser printers are a lot like photocopiers and use the same basic technology. The first laser printers were actually built from modified photocopiers.

- In a photocopier, a bright light is used to make an exact copy of a printed page. The light reflects off the page onto a light-sensitive drum; static electricity makes ink particles stick to the drum; and the ink is then transferred to paper and "fused" to its surface by hot rollers.

- A laser printer works in almost exactly the same way, with one important difference : because there is no original page to copy, the laser has to write it out from scratch.

- The information stored in computer is in electronic format : each piece of data is stored electronically by a microscopically small switching device called a transistor.

- The printer's job is to convert this electronic data back into words and pictures: in effect, to turn electricity into ink.

- The electronic data from computer is used to control a laser beam and it's the laser that gets the ink on the page, using static electricity in a similar way to a photocopier.

- When we print something, computer sends a vast stream of electronic data (typically a few megabytes or million characters) to laser printer.

- An electronic circuit in the printer figures out what all this data means and what it needs to look like on the page. It makes a laser beam scan back and forth across a drum inside the printer, building up a pattern of static electricity. The static electricity attracts onto the page a kind of powdered ink called toner. Finally, as in a photocopier, a fuser unit bonds the toner to the paper.

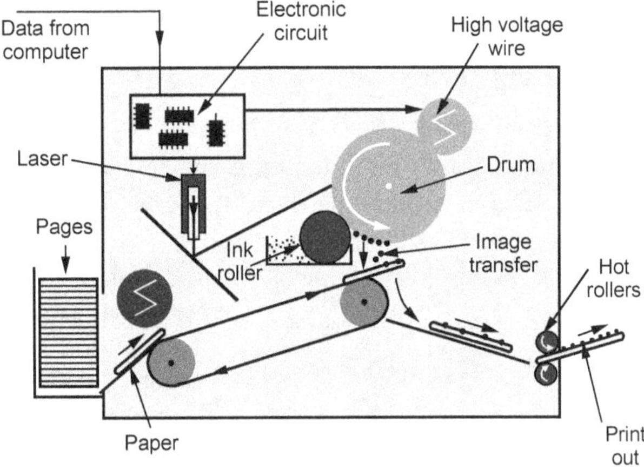

Fig. 8.13.2 Printer

1. Millions of bytes (characters) of **data** stream into the printer from your computer.

2. An **electronic circuit** in the printer (effectively, a small computer in its own right) figures out how to print this data so it looks correct on the page.

3. The electronic circuit activates the **corona wire.** This is a high-voltage wire that gives a static electric charge to anything nearby.

4. The corona wire charges up the **photoreceptor drum** so the drum gains a positive charge spread uniformly across its surface.

5. At the same time, the circuit activates the **laser** to make it draw the image of the page onto the drum. The laser beam doesn't actually move: it bounces off a moving mirror that scans it over the drum. Where the laser beam hits the drum, it erases the positive charge that was there and creates an area of negative charge instead. Gradually, an image of the entire page builds up on the drum: where the page should be white, there are areas with a positive charge; where the page should be black, there are areas of negative charge.

6. An **ink roller** touching the photoreceptor drum coats it with tiny particles of powdered ink (toner). An inked image of the page builds up on the drum.

7. A sheet of **paper** from a hopper on the other side of the printer feeds up toward the drum. As it moves along, the paper is given a strong positive electrical charge by another corona wire.

8. When the paper moves near the drum, its positive charge attracts the negatively charged toner particles away from the drum. The image is transferred from the drum onto the paper but, for the moment, the toner particles are just resting lightly on the paper's surface.

9. The inked paper passes through two hot rollers (the **fuser unit**). The heat and pressure from the rollers fuse the toner particles permanently into the fibers of the paper.

10. The **printout** emerges from the side of the copier.

8.13.3 Watches

- A **mechanical watch** is a watch that uses a clockwork mechanism to measure the passage of time, as opposed to quartz watches which function electronically via a small battery.

- A mechanical watch is driven by a mainspring which must be hand-wound periodically.

- Its force is transmitted through a series of gears to power the balance wheel, a weighted wheel which oscillates back and forth at a constant rate.

- A device called an escapement releases the watch's wheels to move forward a small amount with each swing of the balance wheel, moving the watch's hands forward at a constant rate.

- The escapement is what makes the 'ticking' sound which is heard in an operating mechanical watch.

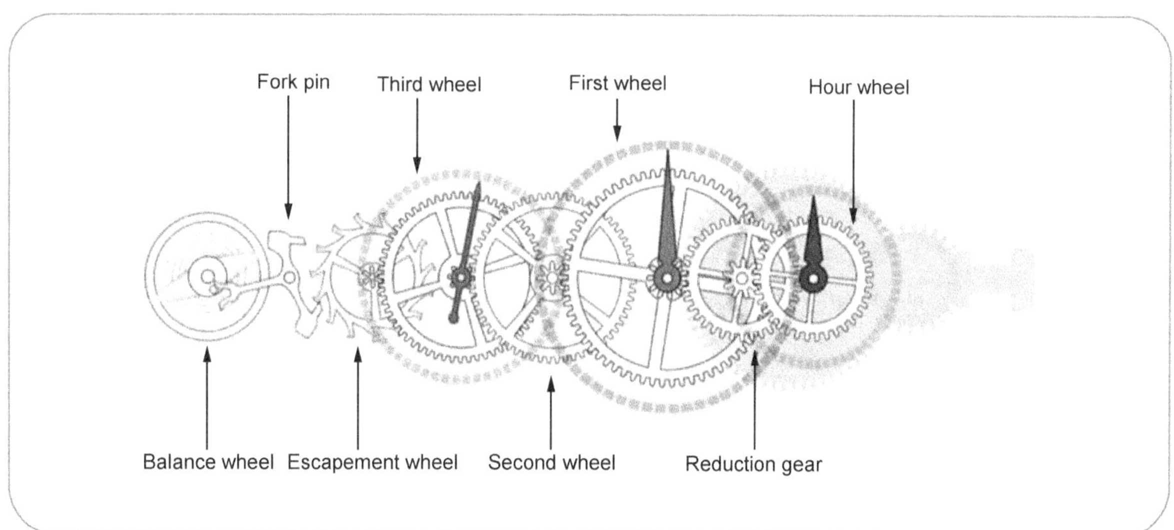

Fig. 8.13.3

- Mechanical watches are typically not as accurate as modern electronic quartz watches, and they require periodic cleaning by a skilled watchmaker. Refer Fig. 8.13.3.

- The internal mechanism of a watch, excluding the face and hands, is called the movement.

- All mechanical watches have these five parts :

i)   A mainspring, which stores mechanical energy to power the watch.

ii) A gear train, called the wheel train, which has the dual function of transmitting the force of the mainspring to the balance wheel and adding up the swings of the balance wheel to get units of seconds, minutes, and hours. A separate part of the gear train, called the keyless work, allows the user to wind the mainspring and enables the hands to be moved to set the time.

iii) A balance wheel, which oscillates back and forth. Each swing of the balance wheel takes precisely the same amount of time. This is the timekeeping element in the watch.

iv) An escapement mechanism, which has the dual function of keeping the balance wheel vibrating by giving it a push with each swing, and allowing the watch's gears to advance or 'escape' by a set amount with each swing. The periodic stopping of the gear train by the escapement makes the 'ticking' sound of the mechanical watch.

v)   An indicating dial, usually a traditional clock face with rotating hands, to display the time in human-readable form.

8.14 : Belt Drive

- The belt drives are used to transmit power from one shaft to another shaft by using pulleys which rotate at the same speed or at different speed.

- A belt drive consists of following three elements. (Refer Fig. 8.14.1) :

> 1. Driving pulley or head pulley
>
> 2. Driven pulley or tail pulley
>
> 3. Endless belt

- The power is transmitted from driving pulley to driven pulley because of frictional grip between the belt and pulley surface.

8.14.1 Speed Ratio of Belt Drive   SPPU : Dec.-09

- Speed ratio or **velocity ratio** is defined as the ratio between the speed of the driver and the driven shaft.

- Generally, speed ratio is given by

$$\text{Speed ratio} = \frac{\text{Speed of driving pulley}}{\text{Speed of driven pulley}}$$

Let,   d and D = Diameters of driving (small) and driven (big) pulley in mm,

   n and N = Speed of driving (small) and driven (big) pulley in r.p.m.,

   t = Thickness of belt in mm

$$\text{Speed ratio} = \frac{n}{N} = \frac{D}{d}$$

If thickness of belt is considered then,

$$\text{Speed ratio} = \frac{n}{N} = \frac{D + t}{d + t}$$

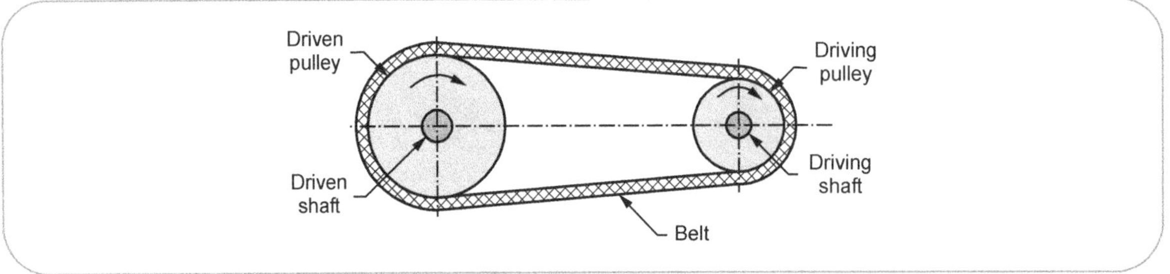

Fig. 8.14.1 Belt drive

8.14.2 Types of Belts

SPPU : Dec.-04, 08, May-05

- There are many types of belts used in variety of applications in industrial and automotive sector.

- Amongst the wide variety of types of belts, the following four types are commonly used (Refer Table 8.14.1) :

Sr. No.	Type	Characteristics
1.	**Flat belt** Fig. 8.14.2 (a)	• The flat belts have rectangular cross section. • The width of the belt is substantially higher compared to its thickness. • They are used where moderate amount of power is to be transmitted from one pulley to another, when the two pulleys are not more than 8 meters apart. • To avoid running out of the belt from the pulleys the pulleys are provided with proper crowning. • Such pulleys are called "**crown pulleys**".
2.	**V-Belt** Fig. 8.14.2 (b)	• The V-belts have trapezoidal cross section. • They are used where great amount of power is to be transmitted from one pulley to another, when the two pulleys are very near to each other. • To have higher surface contact between the belt and pulley, the pulleys are provided with a groove. • Such pulleys are called "**grooved pulleys**".
3.	**Circular Belt** Fig. 8.14.2 (c)	• The circular belts have circular cross section. • They are used where great amount of power is to be transmitted from one pulley to another, when the two pulleys are more than 8 meters apart. • To avoid running out of the circular belt from the pulley, the pulleys are provided with a groove.

Table 8.14.1 Types of belts

8.14.3 Types of Flat Belt Drives

SPPU : Dec.-09, 10, 12, 16, May-10, 16

• The commonly used types of flat belt drives to transmit power are discussed in Table 8.14.2.

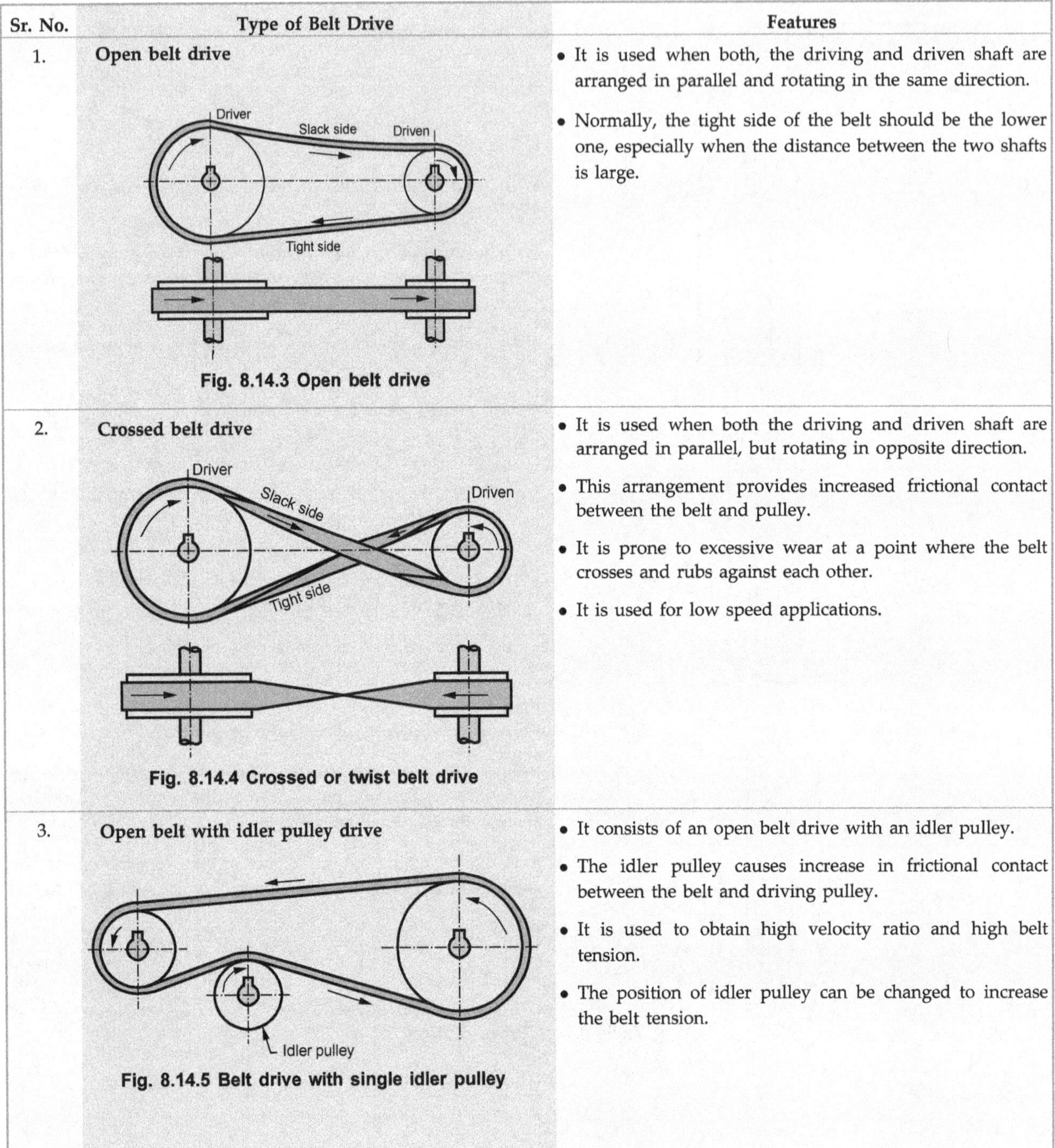

Sr. No.	Type of Belt Drive	Features
1.	**Open belt drive**	• It is used when both, the driving and driven shaft are arranged in parallel and rotating in the same direction. • Normally, the tight side of the belt should be the lower one, especially when the distance between the two shafts is large.
2.	**Crossed belt drive**	• It is used when both the driving and driven shaft are arranged in parallel, but rotating in opposite direction. • This arrangement provides increased frictional contact between the belt and pulley. • It is prone to excessive wear at a point where the belt crosses and rubs against each other. • It is used for low speed applications.
3.	**Open belt with idler pulley drive**	• It consists of an open belt drive with an idler pulley. • The idler pulley causes increase in frictional contact between the belt and driving pulley. • It is used to obtain high velocity ratio and high belt tension. • The position of idler pulley can be changed to increase the belt tension.

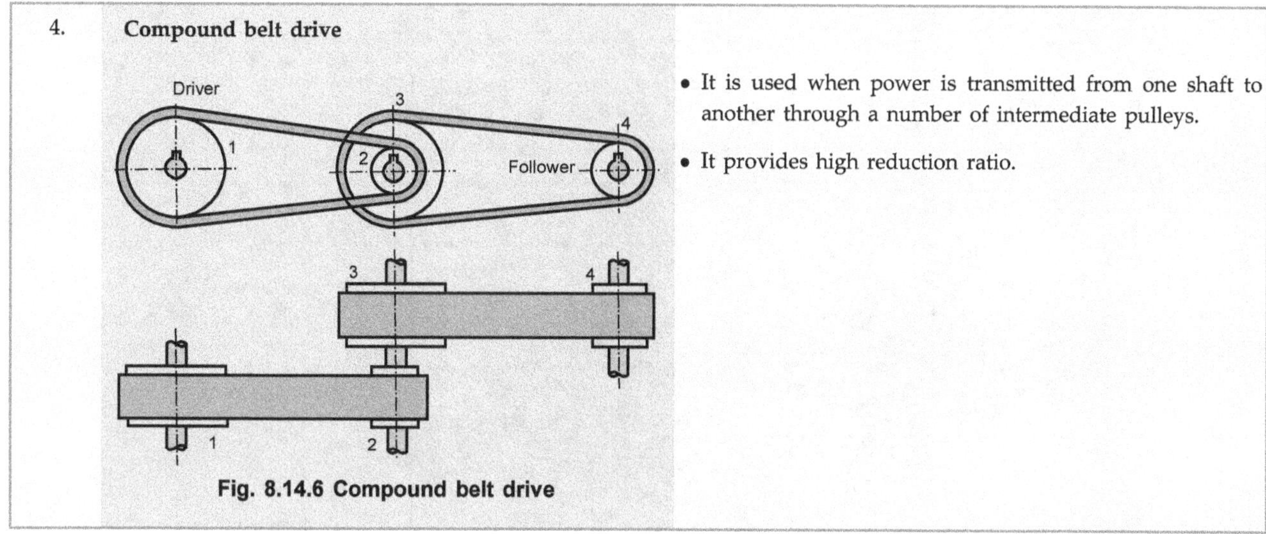

Fig. 8.14.6 Compound belt drive

Table 8.14.2 Types of Flat belt drive

8.14.4 Advantages and Disadvantages of Belt Drives

Advantages

- Belt drives can be used for long centre distances.
- As belts are flexible, they possess a capacity to absorb shocks and vibrations.
- For belt drive, lubrication is not required.
- Belt drives are not affected to dust and dirt.
- Belt drives do not require precise alignment of shaft and pulley.
- Belt drives are relatively cheap and easy to maintain.

Disadvantages / Limitations

- Some of the belts have low power transmitting capacity.
- Belt drives can not be used at extremely high speeds.
- As compared to gear drives, belt drives have shorter life.
- As compared to gear drives, belt drives occupy more space.
- Belt drive is not a positive drive.

8.15 : Applications of Belt-Pulley

8.15.1 Photo Copier

Principle : It works on the principle of "opposite charges attract" and the tendency to become more electrically conductive.

Construction : A photo copier consist of the following components.
1) Belt/Photocopier drum   2) Toner   3) Corona wires

4) A light source   5) Fuser

Working :

- The master copy is placed on the glass surface of the photocopier which is scanned by the bright light beam and the eletrical shadow of the master copy is formed on the photo conductor.

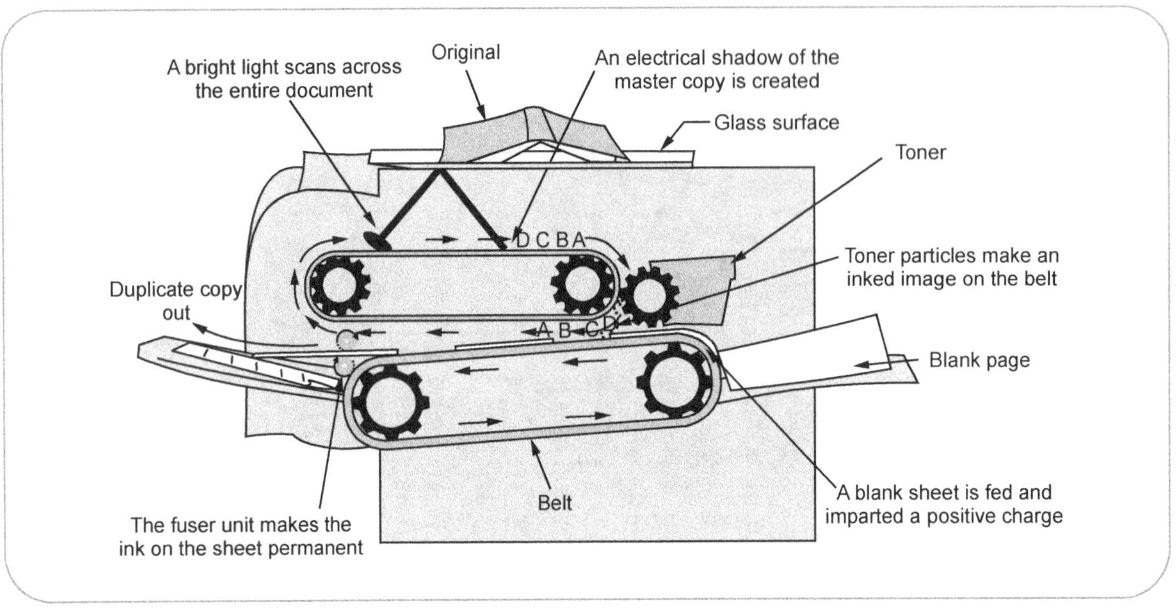

Fig. 8.15.1 Photo copier

- As the conveyor belt moves it takes the eletrical shadow with it and the negatively charged toner product strick to the eletrical shadow and inked impression of the master copy is made on the conveyor belt.

- A blank piece of paper is fed from the other side of the photocopier which moves towards the photoconductor belt and strong positive charge is imparted which pulls the negatively charged toner particle towards it.

- Thus the duplicate image of the master copy is formed on the blank paper.

8.15.2 Solved Examples

Ex. 8.15.1 : *A flat belt is used to transmit 15 kW power from a pulley running at 1440 r.p.m. to another pulley running at 480 r.p.m. The belt velocity is 20.35 m/s. Calculate : (i) The diameter of both the pulleys (ii) Torque required to transmit the power*

Sol. : Given data : $P = 15 \, kW = 15 \times 10^3 \, W$,

$n = 1440$ r.p.m., $N = 480$ r.p.m., $V_B = 20.35$ m/s

To find : (i) d and D   (ii) T

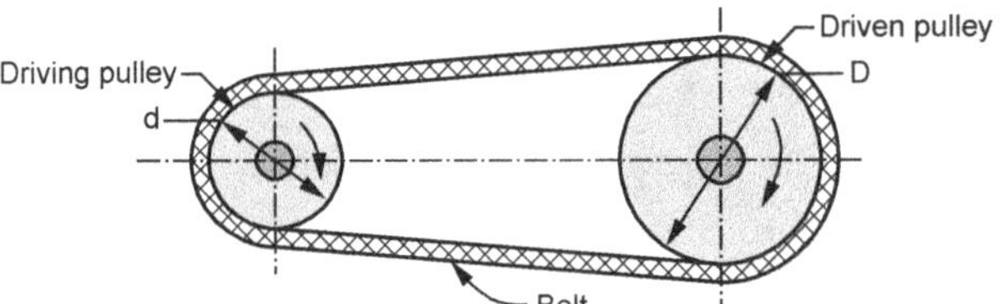

Fig. 8.15.2 Belt drive

Step - 1 : Calculate the diameter of both the pulleys

(1) Velocity of belt (V_B)

$$V_B = \frac{\pi \cdot d \cdot n}{60 \times 1000}$$

$$\therefore \quad 20.35 = \frac{\pi \times d \times 1440}{60 \times 1000}$$

$$\therefore \quad d = 269.9 \, mm \approx 270 \, mm \qquad \text{... Ans.}$$

(2) Speed ratio (S.R.)

$$S.R. = \frac{n}{N} = \frac{D}{d} \qquad \therefore \quad \frac{1440}{480} = \frac{D}{270}$$

$$\therefore \quad D = 810 \, mm \qquad \text{... Ans.}$$

Step - 2 : Calculate the torque required to transmit the power

Power transmited is given as,

$$P = \frac{2\pi N T}{60}$$

$$\therefore \quad 15 \times 10^3 = \frac{2\pi \times 480 \times T}{60}$$

$$\therefore \qquad T = 298.4155 \text{ N-m} \qquad \text{... Ans.}$$

Ex. 8.15.2 : *An electric motor running at 1000 r.p.m. is used to drive a motor rotating at 500 r.p.m. The torque required to transmit the power is 124.1408 Nm. If the diameter of driving pulley is 230 mm, Calculate : (i) The diameter of driven pulley (ii) Velocity of belt (iii) Power to be transmitted by belt.*

Sol. : Given data : n = 1000 r.p.m., N = 500 r.p.m., d = 230 mm, T = 124.1408 N-m.

To find : (i) D   (ii) V_B (iii) P

Step - 1 : Calculate the diameter of driven pulley

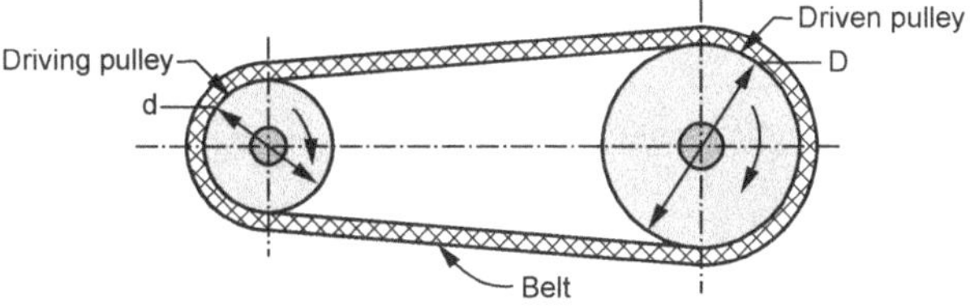

Fig. 8.15.3 Belt drive

The speed ratio is,

$$\text{S.R.} = \frac{n}{N} = \frac{D}{d} \qquad \therefore \frac{1000}{500} = \frac{D}{230}$$

$$\therefore \qquad D = 460 \text{ mm} \qquad \text{... Ans.}$$

Step - 2 : Calculate the velocity of the belt

The velocity of the belt,

$$V_B = \frac{\pi \cdot d \cdot n}{60 \times 1000}$$

$$= \frac{\pi \times 230 \times 1000}{60 \times 1000}$$

$$= 12.0427 \text{ m/s} \qquad \text{... Ans.}$$

Step - 3 : Calculate the power to be transmitted by the belt

Power transmitted is given as,

$$P = \frac{2\pi N T}{60}$$

$$= \frac{2\pi \times 500 \times 124.1408}{60}$$

$$= 6499.9970 \text{ W}$$

$$\therefore \qquad P \approx 6500 \text{ W} \qquad \text{... Ans.}$$

Ex. 8.15.3 : *A belt drive is used to transmit 20 kW power from an electric motor to an exhaust fan. The diameter of motor and fan pulley are 250 mm and 1000 mm respectively. The speed of motor shaft is 750 r.p.m and thickness of belt is 6 mm. Determine : (i) Speed of the exhaust fan pulley (ii) Velocity of the belt (iii) Torque required to transmit the power.*

Sol. : Given data : P = 20 kW = 20×10^3 W, d = 250 mm, D = 1000 mm, n = 750 r.p.m., t = 6 mm.

To find : (i) N   (ii) V_B (iii) T

Step - 1 : Calculate the speed of the exhaust fan pulley

Speed ratio is,

$$\text{S.R.} = \frac{n}{N} = \frac{D + t}{d + t}$$

$$\therefore \qquad \frac{750}{N} = \frac{1000 + 6}{250 + 6}$$

$$\therefore \qquad N = 190.8548 \text{ r.p.m} \qquad \text{... Ans.}$$

Step - 2 Calculate the velocity of the belt

Velocity of the belt is,

$$V_B = \frac{\pi \cdot d \cdot n}{60 \times 1000}$$

$$= \frac{\pi \times 250 \times 750}{60 \times 1000} = 9.8174 \text{ m/s ... Ans.}$$

Step - 3 : Calculate the torque required to transmit the power

We know that,

$$P = \frac{2\pi \cdot N \cdot T}{60}$$

$$\therefore \quad 20 \times 10^3 = \frac{2\pi \times 190.8548 \times T}{60}$$

$$\therefore \quad T = 1000.6870 \text{ N.m} \qquad \text{... Ans.}$$

8.16 : Chain Drive   `SPPU : Dec.-05,06,08, May-17`

- In case of belt drive, while transmitting motion, slipping may occur (non-positive drive). Hence, to avoid slipping steel chains are used.

- The chains are made up of number of rigid links which are hinged together by using pin joints, to provide the necessary flexibility of wraping around the driving and driven wheels.

- The driving and driven wheels have projecting teeth of special profile and fit into the corresponding recesses in the chain links as shown in Fig. 8.16.1.

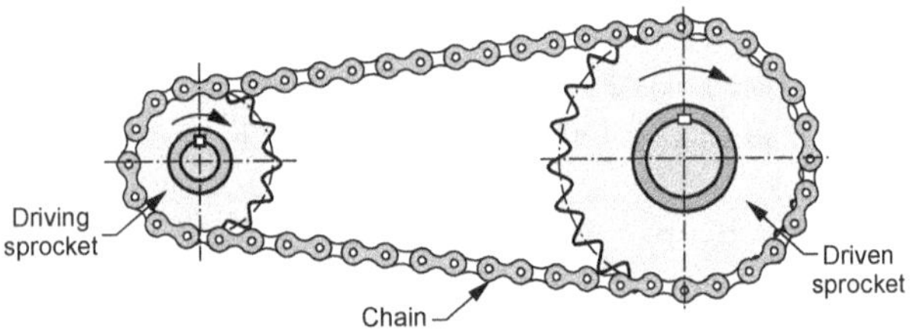

Fig. 8.16.1 Chain drive

- The toothed wheels are called as **sprockets.**

- The chain drive is commonly used to transmit motion from one shaft to other shaft for a short centre distance.

- Chain drives are commonly used in bicycles, motor cycles, conveyors, road rollers, rolling mill, agricultural machinery, etc.

- For effective functioning of chain drive lubrications are required.

8.17 : Applications of Chain Drive

8.17.1 Bicycle

Principle :   Transfering of power from the pedals to drive - wheel of a bicycle.

Construction :   The basic parts of the bicycles are :
i) Frame   ii) Steering

iii) Seating   iv) Brakes

v) Chain and sprocket   vi) Wheels and tires

vii) Suspension

Working :
- Chain and sprockets are used for transmitting force from the output drive shaft to another shaft.

- A sprocket is a toothed wheel that is used to transmit the motion from one shaft to another and chains are used to transmit force from one sprocket to another.

- In case of bicycle the human effort will rotate the crank through which power is transmitted to the chain and sprocket arrangement which drive the wheels of the bicycle.

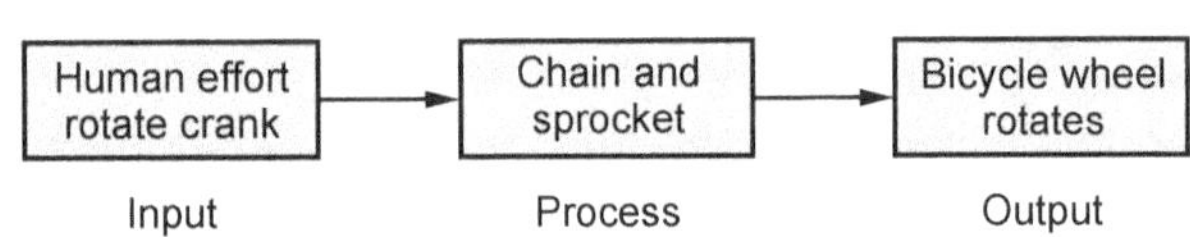

Fig. 8.17.1

8.18 : Valves

- Valve is device for controlling the flow of fluids (liquids, gases, slurries, etc.) in a pipe or other enclosure.

- Control is by means of a movable element that opens, shuts or partially obstructs an opening in a passageway.

- Valves have many applications :

  - Controlling water for irrigation.

  - Residential uses like ON/OFF and pressure control to dish and clothes washers and taps in the home.

  - Industrial uses for controlling the process.

  - Military and transport sectors.

  - In veins for controlling blood circulation etc.

8.19 : Applications of Valves

8.19.1 Water Tap

Principle : The principle of the valve is to regulate, direct and control the flow of a fluid by opening and closing the water tap.

Construction : The main components of the water tap are cartridges, flow restrictors, washers, flanges, O rings, caps, handle etc.

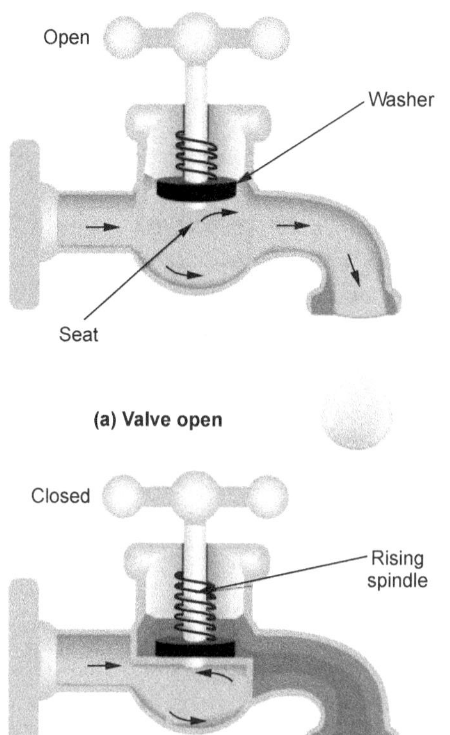

Fig. 8.19.1 Water tap

Working :

- Valves are used to regulate the flow of the water. It uses a soft rubber or a washer which is screwed down a valve seat in order to stop the flow.

- As shown in the Fig. 8.19.1 when the plunger of the tap is rotated in upwarad direction the spring get compressed and valve attacted with it get lifted up and water flows out of the tap and vice versa.

- There are different types of valves. The common ones are butterfly valve, cock or plug valve, gate valve, globe valve, poppet valve, spool valve etc.

Applications :

1. Bathroom   2. Kitchen

3. Flushing of toilets   4. Village pumps, etc.

8.20 : Levers

- A lever is a mechanical device in the form of a rigid rod or bar pivoted about a point known as **fulcrum** and used to overcome a load by the application of a small effort or to transfer the force in a desired direction.

- Fig. 8.20.1 shows the construction of a simple lever in which 'W' is the load or force produced by the lever and 'P' is the effort required to produce that load.

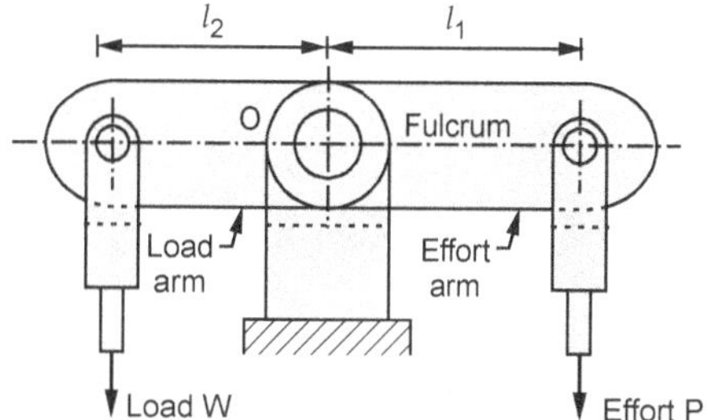

Fig. 8.20.1 : Construction of lever

- A perpendicular distance of effort P from the fulcrum is called as **effort arm** and a perpendicular distance of load W from the fulcrum is called as **load arm**.

- Let, W = Load to be produced in N

  P = Required effort in N

  l_1 = Length of effort arm in mm

  l_2 = Length of load arm in mm

- Considering moment about the fulcrum 'O',

$$\sum M_O = 0 = P \times l_1 - W \times l_2$$

or $P \times l_1 = W \times l_2$

$\therefore$ $\dfrac{W}{P} = \dfrac{l_1}{l_2}$ ...(i)

- The ratio of load to effort (W/P) is called as **Mechanical Advantage** (M.A.) of the lever and the ratio of effort arm to load arm $\left(\dfrac{l_1}{l_2}\right)$ is called as **leverage**.

- It means, mechanical advantage is equal to the leverage.

- The equation (i) is used to exert a large load by the application of small effort by increasing length l_1 and reducing length l_2.

- In some cases, it is not possible to increase length l_1 due to space restriction hence compound levers are used.

8.21 : Applications of Lever

8.21.1 Brake Pedals

Principle : The principle of the brake pedal is to move a piston in the master cylinder, with the application of small force.

Construction : The braking system consist of
i)   Brake pedal
ii) Brake master cylinder
iii) Brake liner
iv) Rotor/Drums
v)   Wheel cylinders
vi) Brake pads

Working :
- When we press the brake pedal by the foot which is connected by the lever that pushes a piston into a cylinder fitted with the hydraulic brake fluid.

- When the fluid reaches another cylinder at wheel which is much wider hence there will be increase in the force which tends to push the brake pad towards the brake disc. Refer Fig. 8.21.1.

- When the brake pad touches the brake disc, the friction generated which slows down the wheel.

8.21.2 Door Latch

- The traditional door knob has a bolt or spindle running through it that sits just above a cylinder, to which the spindle is connected.

- Turning the knob pulls the cylinder in the direction of the turn. The end of the cylinder is the latch which protrudes into a space carved out of the door frame, and which prevents the door from being opened if the knob is not turned.

- A spring or similar mechanism causes the latch to return to its protruding state whenever the knob is not being turned.

- The keyhole covers, usually circular, through which keys pass to enter the lock body.

- If the door handles have a square or rectangular plate on which the handle is mounted this is called the backplate.

- The backplate can be plain (for use with latches), pierced for keyholes (for use with locks), or pierced and fitted with turn knobs and releases (for use with bathroom locks).

- The plate on the front edge of the lock where the latch bolt protrudes is called the faceplate.

- The most common and basic type of doors handle is the Lever latch door handle on a backplate found in residential houses and commercial and public buildings.

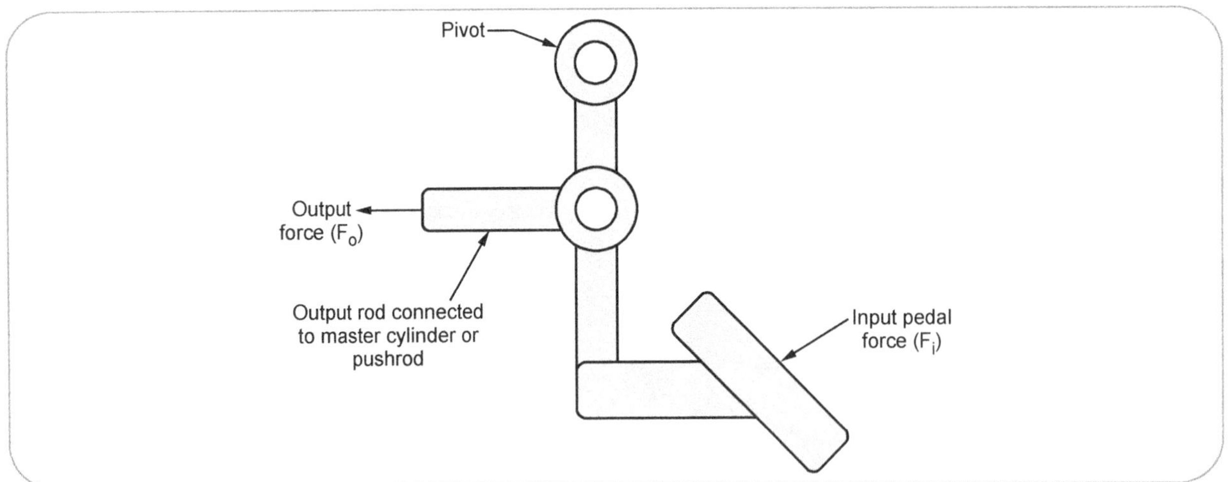

Fig. 8.21.1 Brake pedal

- Doors fitted with this handle have a latch that keeps the door shut. The door handle has only a lever handle which operates this latch.

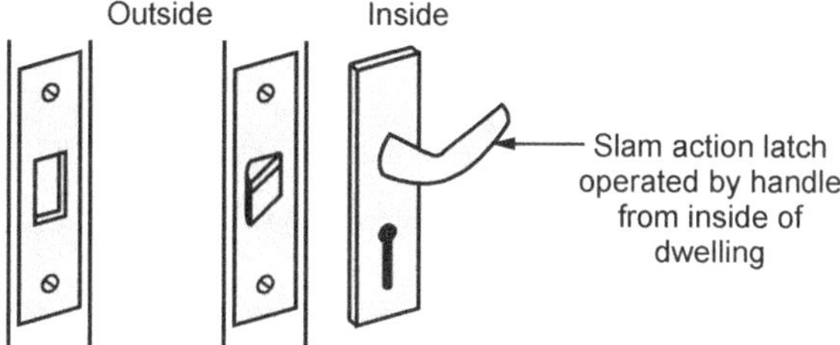

Fig. 8.21.2 Door latch

- Pushing the handle down rotates the spindle, operating the tubular latch mechanism inside the door, allowing it to be opened.

- This type of door handle is used on interior doors that do not require to be locked. The lever latch handle is easy to install and use, and is available in a variety of styles and finishes.

8.22 : Basic Terms

i) Specifications :

- Specification basically is an act of specifying.

- It is a detailed description of requirements, dimensions, materials, etc as of a proposed building, machine, bridge, etc.

- Also it is an act of making specific.

- For example : Specification of vehicle or specifications of engine.

ii) Input and output :

- Input is nothing but anything which is taken in to operate any system or process.

- Example : For a motor input is electrical energy (current) which is converted into mechanical energy (rotation of shaft), for a vehicle input is a fuel, etc.

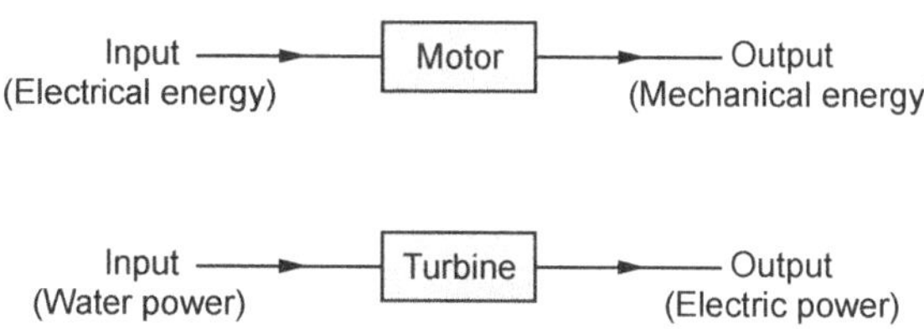

Fig. 8.22.1 Input and output

- Output is defined as the amount of energy, work, goods or services produced by a machine, industry or an individual in a period.

- Example : For a turbine input is water power (hydraulic energy) and output is electric energy.

iii) Efficiency :

- Efficiency is defined as the comparison of what is actually produced with what can be achieved with the consumption of resources.

- It signifies a level of performance that describes using the least amount of input to achieves the highest amount of output.

- Basically it is the state or quality of being efficient.

- Mathematically, efficiency is the ratio of output power of system to the input power of system. It is denoted by 'η'.

- It is always expressed in percentage (%) and it is always less than 1 (100 %)

$$\eta = \frac{\text{Output power}}{\text{Input power}} \times 100$$

- If for a system η = 80 % means system is 80 % effective or there is loss of 20 % in the system.

The relation between input, output and losses can be given as,

$$\text{Input = Output + Losses}$$

or $$\text{Losses = Input – Output}$$

8.23 : Applications of Electric/Solar Energy

8.23.1 Electric Geyser

Principle : The principle of the geyser is to convert the electrical energy into the heat energy.

Construction : The main component of the geyser are water tank, inlet and outlet pipe, heater, insulating material and the heating is been controlled by the thermostats as shown in the Fig. 8.23.1.

Working :

- It works on the principle of converting electrical energy into heat by using heating elements to raise temperature of water.

- As shown in the figure it consist of two pipes one for the inlet of the cold water and another for the outlet of the hot water.

- As the water get stored into the water tank the heater get started up and the temperature of the water rises to the set temperature by the thermostat.

- Once the temperature of the water is obtained, the heater switches off automatically and the hot water can be taken out by the outlet channel.

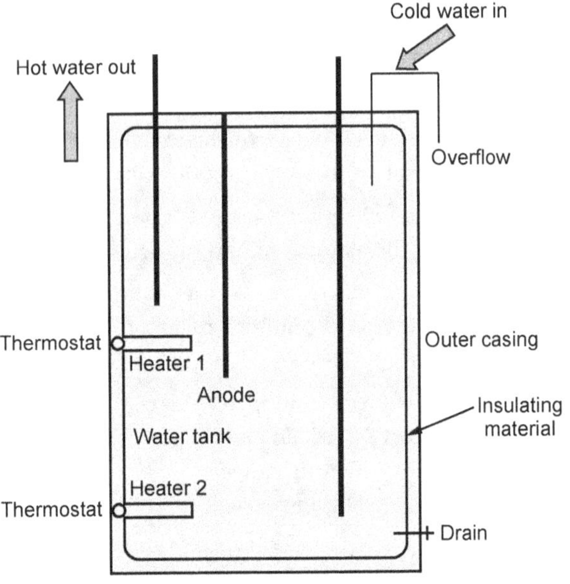

Fig. 8.23.1 Electric geyser

Applications

1. Bathroom

2. Kitchen

3. Industries, etc.

8.23.2 Solar Water Heater

Principle : It works on the conversion of sunlight into heat for heating the water by using a solar thermal collector.

Construction : The main components for the solar water heater are solar panels, electric meter, storage tank, connecting pipe, circulating pump, etc.

Working :

- In a solar water heater, the water gets heated up by the solar thermal energy which is the absorbed by the solar panels or the collectors.

- As the heat energy is absorbed by the panel which is transferred to the water in the tank and thus the hot water with lower density moves upward and the cold water with higher density will move down in the tank due to gravity.

- The collectors are arranged in a series-parallel combination to get higher quantity of hot water.

- Hence this hot water can be stored in the tank and supplied to the various parts of the building.

Applications

1. Residential houses

2. Textile industries

3. Swimming pools

4. Hospitals

5. Hotels and restaurants, etc.

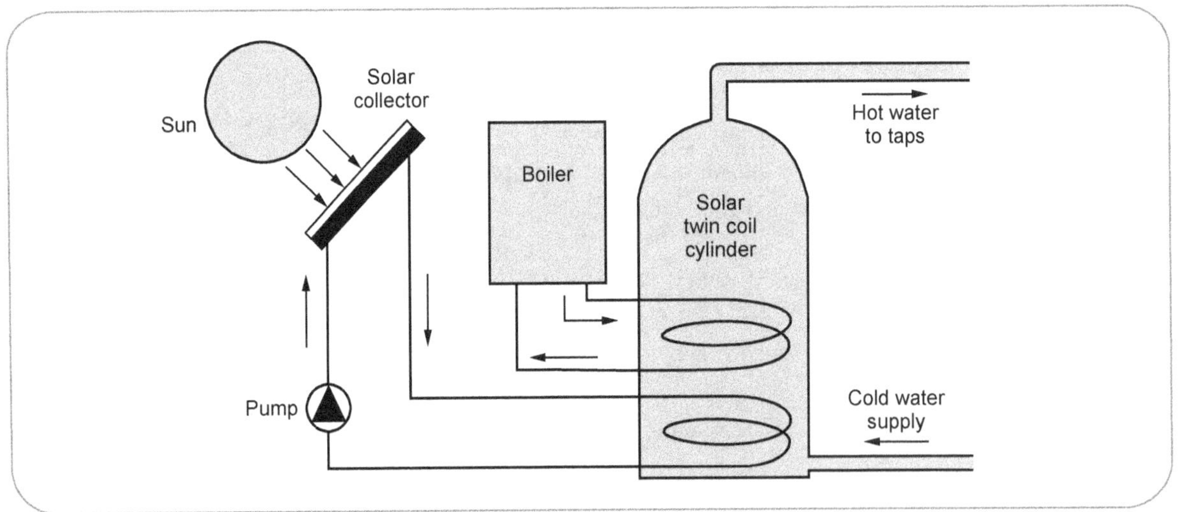

Fig. 8.24.1 Solar water heater

8.23.3 Electric Iron

Principle :   It works on the simple phenomenon of the conversion of electric energy to heat which is transferred to the base plate to remove the crease from the clothes.

Construction :   The components of the electric iron are

1. Base plate

2. Pressure plate

3. Heating element

4. Cover plate

5. Thermostat

6. Capacitor

Working :

• It takes the electric circuit from the main and uses it to heat up a coil inside it and this heat is transferred to the base plate which is used to remove crease from the clothes.

• The main component of the electric iron is the thermostat which uses the bimetallic strip that is made up of two different types of metals with different coefficient of expansion.

• The bimetallic strip is connected to a contact spring through pins. At moderate temperature, the contact point remain in contact with the bimetallic strip as shown in Fig. 8.25.1 and as the temperature exceed to a certain limit the strip begins to bend towards the metal thus it disconnects with the contact point.

• As soon as the temperature drops the strip will again acquire the original shape and the current flow.

• This cycle is repeated till we switch off the power.

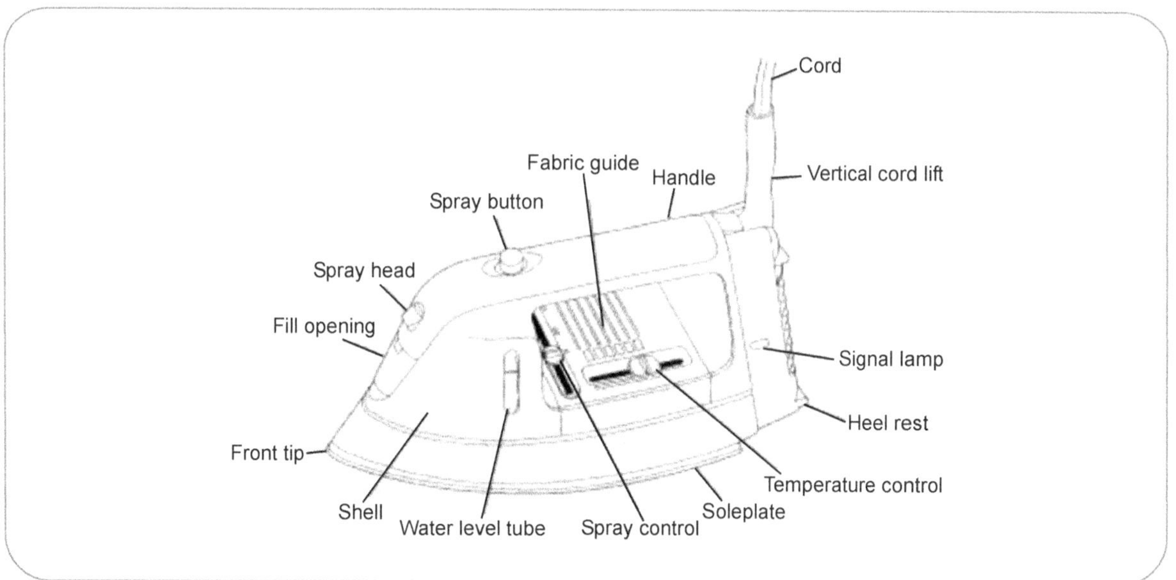

Fig. 8.25.1 Electric iron

Note : For numericals on efficiency of pump and compressor please refer chapter 2.

Review Questions

1. What is pump ? What are its types ? Explain the centrifugal pump.

2. Explain the water pump for overhead tank.

3. With neat sketch explain water purifier unit.

4. What is air compressor ? What are its types ?

5. Explain the application of compressor to refrigerator and split AC unit.

6. With neat sketch explain water cooler.

7. What is blower ? State its function .

8. Explain the following applications of blower

   a) Vocuum cleaner

   b) Kitchen chimney

   c) Exhaust fan

9. Explain the function and applications of spring.

10. What are the types of spring ?

11. Explain door lock as an application of spring.

12. Write short note on : Door closure.

13. What is gear ? Explain gear ratio.

14. Explain the following application of gears

   a) Wall clock   b) Watches   c) Printers

15. Explain with neat sketch working of photocopier

16. With neat sketch explain water tap.

17. What is lever ? Explain door latch with neat sketch.

18. Write short note on :

   a) Geyser   b) Water heater

   c) Electric iron.

SOLVED SAMPLE QUESTION PAPER - As PER 2019 PATTERN

Systems in Mechanical Engineering
F.E. (Common to All Branches) Sem - I (In Sem)

Time : 1 Hour]   **[Maximum Marks : 30**

Instructions to the candidates :

1) Answer Q.1 or Q.2, Q.3 or Q.4.

2) Neat diagrams must be drawn wherever necessary.

3) Figures to the right indicate full marks.

4) Make suitable assumptions if necessary.

Q.1 a) *Explain hydroelectric power plant with neat sketch.* **(Refer section 1.3)**   **[6]**

b) *What is hybrid power plant ? State its advantages. Explain use of solar energy with block diagram.*
(Refer sections 1.5.3 and 1.5.1)   **[5]**

c) *In an Impulse water turbine water power is 150 kW. Out of this 85 % power is used to drive the runner. If mechanical losses (in bearing and all) are 20 % , find overall efficiency and shaft power of turbine.*
(Refer example 2.11.3)   **[4]**

OR

Q.2 a) *Write short note on : i) Reciprocating compressor* **(Refer section 2.8.2)**
ii) Pelton wheel **(Refer section 2.2.2)**   **[6]**

b) *State the advantages and disadvantages of nuclear power plants.* **(Refer section 1.4)**   **[4]**

c) *What is wind mill ? What are its types ?* **(Refer sections 2.10 and 2.10.1)**   **[5]**

Q.3 a) *Explain following terms : (i) System sourrounding and Boundary (ii) Kelvin Plank Statement of second low of thermodynamics.* **(Refer sections 3.2 and 3.11)**   **[4]**

b) *Calculate rate of heat transfer per m^2 through wall of 200 mm thick inner layer of 'A', a central layer of 'B' 100 mm thick and a outer layer of 'C' 100 mm thick. Temperature of gas in the furnace is 1670 °C with $h_{in} = 74 \ W/m^2 \ °C$ and outside surface temperature of 'C' is 70 °C.*
Given : $K_A = 1.25 \ W/m°C$, $K_B = 0.074 \ W/m°C$, $K_C = 0.55 \ W/m°C$
Assume steady state, 1-D flow of heat. **(Refer example 4.7.5)**   **[6]**

c) *What is thermodynamic system ? Explain various types of thermodynamic systems with example.*
(Refer sections 3.2 and 3.2.1)   **[5]**

OR

Q.4 a) *Explain four stroke petrol engine with neat sketch.* **(Refer section 4.8.4)**   **[5]**

b) *The efficiency of a Carnot engine rejecting heat to a cooling pond at 28 °C is 30 %. If the cooling pond receives 1050 kJ/min. What is the power developed by the cycle in kW. Also find the temperature of the source.*
(Refer example 3.13.2)   **[6]**

c) *Give the classification of steam generators.* **(Refer section 4.10.1)**   **[4]**

❑❑❑

SOLVED SAMPLE QUESTION PAPER - As PER 2019 PATTERN

Systems in Mechanical Engineering
F.E. (Common to All Branches) Sem - I (End Sem)

Time : $2\frac{1}{2}$ Hours]   [Maximum Marks : 70

Instructions to the candidates :

1) Answer Q.1 or Q.2, Q.3 or Q.4, Q.5 or Q.6, Q.7 or Q.8.

2) Neat diagrams must be drawn wherever necessary.

3) Figures to the right indicate full marks.

4) Make suitable assumptions if necessary.

Q.1 **a)** *What is meant by articulated vehicle ? Explain the neat sketch.* **(Refer section 5.7.2)**   **[5]**

    b) *With neat sketch explain the construction and working of electric vehicles.* **(Refer section 5.11)**   **[7]**

    c) *What are the basic components / system of an automobile ?* **(Refer section 5.3)**   **[5]**

OR

Q.2 **a)** *Explain spcifications of any two wheeler with suitable example.* **(Refer section 5.5.1)**   **[7]**

    b) *How are automobiles classified ?* **(Refer section 5.4)**   **[5]**

    c) *Explain the cost analysis of vehicle.* **(Refer section 5.13)**   **[5]**

Q.3 **a)** *What are the different layouts of vehicle ?* **(Refer section 6.4)**   **[9]**

    b) *State the functions of braking system.* **(Refer section 6.7)**   **[4]**

    c) *What is the necessity of cooling system ? What are the different methods of it ?*

    (Refer sections 6.8 and 6.9)   **[6]**

OR

Q.4 **a)** *With neat sketch explain suspension system (leaf spring arrangement).* **(Refer section 6.6)**   **[4]**

    b) *State the functions of clutch. Explain single plate clutch with neat sketch.* **(Refer section 6.14)**   **[5]**

    c) *Write short note on a) Air bag* **(Refer section 6.26)** *b) ABS* **(Refer section 6.7.4)**

    c) Seat belt **(Refer section 6.25)**   **[9]**

Q.5 **a)** *Compare direct extrusion and indirect extrusion.* **(Refer section 7.6.3)**   **[5]**

    b) *Explain the function of following parts of lathe machine.*

    a) Bed   b) Headstock   c) Tailstock   d) Carriage **(Refer section 7.17.2)**   **[6]**

    c) *What is additive manufacturing ? State its applications.* **(Refer sections 7.20 and 7.22)**   **[6]**

OR

Q.6 **a)** *Write short note on : IOT* **(Refer section 7.31)**   **[6]**

    b) *What is casting process ? State its advantages, disadvantages and applications.* **(Refer section 7.2)**   **[5]**

(S - 2)

c) *Explain arc welding process with neat sketch.* **(Refer section 7.12)** **[6]**

Q.7 **a)** *Write short note on : a) Geyser b) Water heater   c) Electric iron.* **(Refer section 8.23)** **[9]**

b) *Explain the water pump for overhead tank.* **(Refer section 8.5)** **[5]**

c) *Explain the function and applications of spring.* **(Refer section 8.10)** **[4]**

OR

Q.8 **a)** *Explain the following application of gears a) Wall clock b) Watches c) Printers* **(Refer section 8.13)** **[9]**

b) *With neat sketch explain water tap.* **(Refer section 8.19)** **[5]**

c) *With neat sketch explain water cooler.* **(Refer section 8.7.3)** **[4]**

❑❑❑

Notes